AF829112

Disaster and Risk Research: GADRI Book Series

Series Editors

Stefan Hochrainer-Stigler, Risk and Resilience Program, International Institute for Applied Systems Analysis (IIASA), Laxenburg, Niederösterreich, Austria

Hirokazu Tatano, Disaster Prevention Research Institute, Kyoto University, Uji, Kyoto, Japan

Wei-Sen Li, National Science and Technology Center for Disaster Reduction, New Taipei City, China

Andrew Collins, Department of Geography and Environmental Sciences, Northumbria University, Newcastle upon Tyne, UK

Khalid Mosalam, Pacific Earthquake Engineering Research Center, University of California, Berkeley, Berkeley, CA, USA

Charles Scawthorn, SPA Risk LLC, Berkeley, CA, USA

Lori Peek, Natural Hazards Center, University of Colorado Boulder, Boulder, CO, USA

Disaster and Risk Research: GADRI Book Series is published under the auspices of the Global Alliance of Disaster Research Institutes (GADRI). The global status of disaster research reflects the major strides made in the disaster sciences. These volumes present the forefront of disaster research, including key scientific findings, methodologies, policy recommendations and case studies. Whilst disaster risk needs to be managed in an integrated manner, persistently isolated applications of knowledge, practice and policy are falling short of ensuring disaster-resilient societies. Responding to this deficit calls for measurement, tools, techniques and institutional structures that can realistically support comprehensive risk assessment and management across multiple hazard landscape. As such, disaster research is now faced with a multi-disciplinary, multi-stakeholder challenge. Contributions to this series therefore address many varied and critical opportunities to advance the subject area. A cross-cutting vision shared across the Disaster and Risk Research volumes is to improve the future of scientific and technological guidance with clearly identifiable roadmaps to ensure human safety and security. The Global Alliance of Disaster Research Institutes was established in March 2015, directly after the United Nations World Conference on Disaster Risk Reduction (WCDRR 2015) in Sendai, Japan, based on the belief that a multi-institutional alliance can strengthen disaster research and its influences around the world. GADRI has a mandate to support the implementation of the Sendai Framework for Disaster Risk Reduction 2015-2030 and is a member of the Scientific and Technical Advisory Group (STAG) of the United Nations Office for Disaster Risk Reduction (UNDRR). In addition, GADRI provides a platform for scientific communities from different disciplines, backgrounds and countries, helping them share their knowledge, findings and views. This approach yields more holistic and farther-reaching insights, which can contribute to further steps in effective disaster risk management.

More information about this series at http://www.springer.com/series/16177

Mahua Mukherjee · Rajib Shaw
Editors

Ecosystem-Based Disaster and Climate Resilience

Integration of Blue-Green Infrastructure in Sustainable Development

Springer

Editors
Mahua Mukherjee
Department of Architecture and Planning
Indian Institute of Technology Roorkee
Roorkee, Uttarakhand, India

Rajib Shaw
Graduate School of Media and Governance
Keio University
Shonan Fujisawa Campus
Fujisawa, Kanagawa, Japan

ISSN 2524-5961 ISSN 2524-597X (electronic)
Disaster and Risk Research: GADRI Book Series
ISBN 978-981-16-4814-4 ISBN 978-981-16-4815-1 (eBook)
https://doi.org/10.1007/978-981-16-4815-1

© The Editor(s) (if applicable) and The Author(s), under exclusive license to Springer Nature Singapore Pte Ltd. 2021
This work is subject to copyright. All rights are solely and exclusively licensed by the Publisher, whether the whole or part of the material is concerned, specifically the rights of translation, reprinting, reuse of illustrations, recitation, broadcasting, reproduction on microfilms or in any other physical way, and transmission or information storage and retrieval, electronic adaptation, computer software, or by similar or dissimilar methodology now known or hereafter developed.
The use of general descriptive names, registered names, trademarks, service marks, etc. in this publication does not imply, even in the absence of a specific statement, that such names are exempt from the relevant protective laws and regulations and therefore free for general use.
The publisher, the authors and the editors are safe to assume that the advice and information in this book are believed to be true and accurate at the date of publication. Neither the publisher nor the authors or the editors give a warranty, expressed or implied, with respect to the material contained herein or for any errors or omissions that may have been made. The publisher remains neutral with regard to jurisdictional claims in published maps and institutional affiliations.

This Springer imprint is published by the registered company Springer Nature Singapore Pte Ltd.
The registered company address is: 152 Beach Road, #21-01/04 Gateway East, Singapore 189721, Singapore

Preface

The *Global Alliance of Disaster Research Institutes* (GADRI) is a global research network and forum of disaster research institutes and organizations of disaster research for promoting capacity development through sharing knowledge. Research collaboration for disaster risk reduction and resilience to disasters, data and information sharing and exchange for scientific research across the globe are few of the primary objectives of GADRI.

The GADRI activities are guided by its Board of Directors and implemented efficiently by the GADRI Secretariat. Publication of GADRI Book Series on *Disaster and Risk Research* is one such remarkable activity. The present book on *Ecosystem-Based Disaster and Climate Resilience: Integration of Blue-Green Infrastructure in Sustainable Development* is part of the *Disaster and Risk Research: GADRI Book Series.*

The purpose of this book is to introduce readers to the challenging context of development, urbanization, disaster risk and climate change and opportunity creation for natural ecosystem-based risk reduction (Eco-DRR) and climate change adaptation (EbA). An attractive proposition of the book is bringing appropriate focus on mainstreaming blue-green infrastructure (BGI). Blue-green infrastructure (BGI) has a root in nature-based solution which is strongly gaining recognition as a multifunctional solution and can respond as one common strategy to targets set by the Sustainable Development Goals (SDGs, UNDP), Climate Agreements (IPCC-AR6, UNEP), the Sendai Framework for Disaster Risk Reduction (SFDRR, UNISDR) and the New Urban Agenda proposed at Habitat III (UNCHS).

This book attempts to study various concepts of nature-based solutions, dimensions of urbanization, implications of disaster risk and climate change and recent advances in the policy perspectives and policy implementation mechanisms in a systematic way.

Critical role of Eco-DRR and EbA for community resilience to multiple hazard-induced and environmental risks such as rising heat, water stress and pollution is gaining attention. Science investigation, technology and planning intervention help to better understand the complex multi-hazard risk scenario, and trends and convergence between challenges and beneficial impacts of BGI on Eco-DRR and EbA.

Evidences of implementation of BGIs are presented as analysed case studies from different parts of the world where participatory integration of BGI is experimented with at various levels of success. It envisaged that shared tacit experiences from the realm of practice will further strengthen explicit knowledge. The emerging paradigm of looking back at nature and to integrate the ecosystem services and resilience services can be seen as a major attitudinal shift of adaptive governance for sustainable development at regional to local scale. The road map for the future redevelopment/development is expected to be an effective communication on policy, governance, science and technology, role of private sectors, civil society and communities.

We will be happy if the readers find this book useful and relevant. Fifty-eight experts from 11 countries with multidisciplinary backgrounds have shared their experiences to enrich the content of the book. We thank all the authors for their contributions.

Professor Hirokazu Tatano, Secretary General of Global Alliance for Disaster Research Institutes (GADRI), and Professor, DPRI, Kyoto University, is instrumental for this book. Suggestion from Professor Andrew Collins and other colleagues from GADRI, especially GADRI Board of Director Members, was immensely helpful. Regular and timely inputs from Professor Subhojyoti Samaddar of DPRI, Kyoto University, contributed significantly to shape the book. Coordination efforts of Ms. Wilma James and others from GADRI Secretariat are appreciated. We acknowledge help from the Springer Japan editorial services for publication of this book.

The book is a sincere contribution towards nature-centric urban and regional resilience; its wide coverage on different aspects of ecosystem-based disaster and climate risk resilience will be a valuable resource material for student, researcher, academia, policy-maker, development practitioner and community. The book is dedicated to the UN Decade on Ecosystem Restoration 2021–2030 to "prevent, halt and reverse the degradation of ecosystems worldwide".

Roorkee, India Mahua Mukherjee
Fujisawa, Japan Rajib Shaw

About This Book

The book on *Ecosystem-Based Disaster and Climate Resilience: Integration of Blue-Green Infrastructure in Sustainable Development* is part of *Disaster and Risk Research: GADRI Book Series.*

This book has 23 chapters dealing with various aspects: nature-based risk resilience, concepts and relevance with special reference to blue-green infrastructure for ecosystem-based disaster risk resilience (Eco-DRR) and climate change adaptation (EbA). The chapters are organized in three parts in addition to an overview on uncertainties in urbanizing world and nature-based resilience building in the beginning and a concluding discussion on forward looking lens to mainstream blue-green infrastructure (BGI). The thematic grouping of chapters is presented as

Part I: *Policy Analysis, Policy Framing and Recognition of Nature-Based Solution*
Part II: *Science Investigation, Technology and Planning Intervention*
Part III: *Case Studies.*

Importance of policy, science (investigation, analysis and design) and implementation planning as brought out in several chapters highlights the importance of co-design and co-delivery in the participatory mode. Each part of the book is designed to be self-contained, yet linked with its predecessor, successor or both, as the case may be.

Contents

Editors and Contributors

About the Editors

Prof. Mahua Mukherjee is Head, Centre of Excellence in Disaster Mitigation and Management (CoEDMM), IIT Roorkee, and Faculty in the Department of Architecture and Planning, IIT Roorkee, since 2003. She is currently Member of Board of Directors of the UNDRR-APSTAAG and the GADRI (hosted by DPRI at Kyoto University). She is Secretary General for the South Asia Alliance for Disaster Resilience Institutes (SAADRI). She has pursued B.Arch. and Ph.D. from Jadavpur University and M.Tech. in Building Science and Technology from IIT Roorkee followed by a career in architecture profession, NGO and academic research in the field of sustainable urban development. Her graduated research interest includes risk resilience to urban climate and climate responsive campus and housing. Her exposure to international academia in Lund University, Sweden, through SIDA Fellowship and Penn State University as a Fulbright Fellow influenced her intellectual pursuit. She as Visiting Associate Professor was in DPRI, Kyoto University, in 2016. She has organized courses, competitions, research seminars, conferences and workshops regularly in the field of risk resilience and sustainable development. Her publications include peer-reviewed international journals, conference proceedings and chapters. She delivered talks on different platforms including UN organizations on ranges of issues. Sponsored and consultancy researches conducted by her entail non-structural elements' seismic safety, climate responsive and earthquake resilient housing, blue-green infrastructure network and building regulation for resilience.

Rajib Shaw is Professor in Graduate School of Media and Governance in Keio University, Japan. Earlier, he was Executive Director of the Integrated Research on Disaster Risk (IRDR), a decade-long research programme co-sponsored by the International Science Council (ICSU) and the United Nations Office for Disaster Risk Reduction (UNDRR). He is also Senior Fellow of Institute of Global Environmental Strategies (IGES), Japan, and Chairperson of SEEDS Asia and CWS Japan, two Japanese NGOs. He is also Co-founder of a Delhi (India)-based social entrepreneur start-up Resilience Innovation Knowledge Academy (RIKA). Previously, he served as Professor in the Graduate School of Global Environmental Studies of Kyoto University. His expertise includes community-based disaster risk management, climate change adaptation, urban risk management, and disaster and environmental education. He was Chair of the United Nations Global Science Technology Advisory Group (STAG); current Co-chair of the Asia Pacific Science Technology Academic Advisory Group (APSTAAG); and Coordinating Lead Author (CLA) for the IPCC 6th Assessment Report (Asia Chapter). He is Editor of a book series on disaster risk reduction, published by Springer. He has published more than 48 books and over 400 academic papers and chapters. He is the recipient of prestigious "Pravasi Bharatiya Samman Award (PBSA)" in 2021 for his contribution in education sector. PBSA is the highest honour conferred on overseas Indians and persons of Indian origin from the President of India. More of his work can be seen in: www.rajibshaw.org.

Contributors

Basant Raj Adhikari Institute of Engineering, Tribhuvan University, Kathmandu, Nepal

Neelima Alam Technology Mission Division (Energy, Water and Others), Department of Science and Technology (DST), New Delhi, India

Sanjay Bajpai Technology Mission Division (Energy, Water and Others), Department of Science and Technology (DST), New Delhi, India

Jayanta Bandyopadhyay ORF Kolkata, Kolkata, India

Shweta Bhardwaj National Institute of Disaster Management, Ministry of Home Affairs, Government of India, New Delhi, India

Piyalee Biswas Technology Mission Division (Energy, Water and Others), Department of Science and Technology (DST), New Delhi, India

Miranda Booth Humanitarian, Emergency and Disaster Management Studies Program, Charles Darwin University, Darwin, NT, Australia

Talbot Brooks Delta State University, Cleveland, USA

Pashupati Chaudhary Agriculture and Forestry University, Rampur, Chitwan, Nepal

Sunita Chaudhary International Centre for Integrated Mountain Development (ICIMOD), Kathmandu, Nepal

Abhinav Daalia Management Consultant Mckinsey & Company, Mumbai, India

Anamitra Anurag Danda ORF Kolkata, Kolkata, India

Sameer Deshkar Department of Architecture and Planning, Visvesvaraya National Institute of Technology, Nagpur, Maharashtra, India

Shalini Dhyani CSIR-National Environmental Engineering Research Institute, Nagpur, Maharashtra, India

Tashi Dorji International Centre for Integrated Mountain Development (ICIMOD), Kathmandu, Nepal

Siva Ram Edupuganti IIT Roorkee, Roorkee, India

Vidushi Gautam Department of Civil Engineering, Delhi Technological University, New Delhi, India

Nilanjan Ghosh ORF Kolkata, Kolkata, India

Anil Kumar Gupta National Institute of Disaster Management, Ministry of Home Affairs, Government of India, New Delhi, India

Sugata Hazra School of Oceanographic Studies, Jadavpur University, Kolkata, India

Tomohiro Ichinose Faculty of Environment and Information Studies, Keio University, Fujisawa, Kanagawa Prefecture, Japan

Ikuko Imoto Keio Research Institute at SFC, Keio University, Fujisawa, Kanagawa Prefecture, Japan

Jun Ishii Fukui Prefectural Satoyama-Satoumi Research Institute, Mikatakaminaka-gun, Fukui Prefecture, Japan

Muhammad Barkat Ali Khan Department of Geography, Higher Education Department, Peshawar, KP, Pakistan

Atul Kumar Indian Institute of Technology Roorkee, Roorkee, India

Jeevan Madapala Indian Institute of Technology Roorkee, Roorkee, India

Rudrodip Majumdar National Institute of Advanced Studies, Bangalore, Karnataka, India

Norio Maki Kyoto University, Kyoto, Japan

Peter J. Marcotullio Department of Geography, Hunter College, City University of New York, New York, NY, USA;
Institute for Sustainable Cities At Hunter College, New York, NY, USA

Bhogendra Mishra Science Hub, Balaju, Nepal;
Policy Research Institute, Kathmandu, Nepal

Bijon Kumer Mitra Institute for Global Environmental Strategies (IGES), Hayama, Japan

Mahua Mukherjee Indian Institute of Technology Roorkee, Roorkee, Uttarakhand, India

Vickyson Naorem IIT Roorkee, Roorkee, India

Karan Narang Project Controls, Bechtel Oil, Gas&Chemicals, Houston, TX, USA

Ujjawal Nathani Department of Civil Engineering, Delhi Technological University, New Delhi, India

Ngoc Huy Nguyen Oxfam International, Hanoi, Vietnam

Celeste Norman Farco Architecture, Brisbane, QLD, Australia

Renuka Poudel International Centre for Integrated Mountain Development (ICIMOD), Kathmandu, Nepal

Shobha Poudel Science Hub, Balaju, Nepal;
Policy Research Institute, Kathmandu, Nepal

Atta-ur Rahman Department of Geography, University of Peshawar, Peshawar, Pakistan

Aditya Rahul IIT Roorkee, Roorkee, India

Bindiya Rashni South Pacific Regional Herbarium and Biodiversity Centre, Institute of Applied Science,, University of the South Pacific, Laucala Campus, Suva, Fiji

Shirish Ravana UN-SPIDER, Vienna, Austria

Harini Santhanam National Institute of Advanced Studies, Bangalore, Karnataka, India

Raju Sarkar Department of Civil Engineering, Delhi Technological University, New Delhi, India

Michael T. Schmeltz Department of Public Health, California State University, East Bay, CA, Hayward, USA

Sandeep Sharma Indian Institute of Technology Roorkee, Roorkee, India

Rajib Shaw Graduate School of Media and Governance, Keio University, Fujisawa, Kanagawa, Japan

Chetna Singh Department of Regional Planning, School of Planning & Architecture, New Delhi, India

Vibhas Sukhwani Graduate School of Media and Governance, Keio University, Fujisawa, Japan

Akhilesh Surjan Humanitarian, Emergency and Disaster Management Studies Program, Charles Darwin University, Darwin, NT, Australia

Chris Tapnio Yale School of the Environment, Yale University, New Haven, CT, USA

Ai Tashiro Graduate School of Environmental Studies, Tohoku University, Sendai, Japan

Kamakshi Thapa Department of Architecture and Planning, Visvesvaraya National Institute of Technology, Nagpur, Maharashtra, India

Thi My Thi Tong Vietnam Institute of Economics, Vietnam Academy of Social Sciences, Hanoi, Vietnam

Noralene Uy Department of Environmental Science, Ateneo de Manila University, Quezon City, Philippines

Deepthi Wickramasinghe Department of Zoology and Environmental Science, University of Colombo, Colombo, Sri Lanka

Wanglin Yan Graduate School of Media and Governance, Keio University, Fujisawa, Japan

Chapter 1
Uncertainties in Urbanizing World and Nature-Based Resilience Building

Mahua Mukherjee and Rajib Shaw

Abstract Disaster risks and climate change, induced extreme events are spiralling. In urban areas, loss of lives, livelihoods, properties and services are increasing. In this chapter, the authors discuss global urbanization, its process, imperatives and growing impact on nurture and people and uncertain urban living. The history of anthropogenic progress is also an account of inequality and imbalance. Context and need for the concept of sustainable development and dilemma attached with it are corroborated to understand the differential impact of risk on human communities and nature from different societies. Degradation of natural ecosystem is one of the most significant consequences of the development; Climate change scenario agreeably considers the same. The authors introduce the subsequent chapters in the book on "Ecosystem-Based Disaster and Climate Resilience'. The most common thread of these chapters are the nature-based solutions, their relevance in multi-hazard ecosystems, challenges and mainstreaming policy-level strategies. Contribution of science and technology for integrated application of ecosystem-based solutions and experiences from case studies on nature-based risk resilience are reported. The chapter concludes with an optimistic note that this can reach various segments of society and can be of use towards sustainable and resilient society re/development.

Keywords Urbanization · Sustainable development dilemma · Ecosystem · Disaster and climate resilience

1.1 Introduction

The recent global phenomenon for urbanization is not new; this can be traced back to the days of early human civilization. Diverse tribes came together in search of better lives and non-agricultural activities like trade became priorities for livelihood.

M. Mukherjee (✉)
Indian Institute of Technology Roorkee, Roorkee, Uttarakhand 247667, India
e-mail: mahuafap@iitr.ac.in

R. Shaw
Graduate School of Media and Governance, Keio University, Fujisawa, Japan

© The Author(s), under exclusive license to Springer Nature Singapore Pte Ltd. 2021

M. Mukherjee and R. Shaw (eds.), *Ecosystem-Based Disaster and Climate Resilience*, Disaster and Risk Research: GADRI Book Series,
https://doi.org/10.1007/978-981-16-4815-1_1

Interactive exchanges took place through collaboration and competition, politics and governance, rituals and games, infrastructure and amenities; urban expression started to get distinction from rural one. Livelihood opportunities and quality of living are major forces to attract people to cities; this inference has been drawn by many including Aristotle to modern days' urbanists (Cunningham 2011; Polo 2017; Jacobs 1992).

Several ancient world cities are existing till date. In the beginning of recorded human history, rise of towns and cities started in strategic locations; interconnectedness, cooperation, trade and cultural pursuits are notable common strings of characters. These were governed by a group of people with power from religious and/or monarchial seats. Only few old cities could survive whereas the rest are forgotten in time and newer cities came to existence and fame. Scale and number of cities grew; cities next to trade routes flourished. With advanced technology, fortification were built to bring security and prosperity to these cities. Grandeur of built edifices and transformation of surroundings were more accounted for in cities compared to harmonious co-existence with nature in rural. In several societies, linkage with nature and natural elements were protected by rituals, habits, practices and laws. Elite citizens held privileges whereas common people had limited access to amenities like sanctuaries, groves, royal parks or experiencing the same. Survivor cities show site selection had been crucial for their existence. The thriving spirit of human beings is best established through the urban image kaleidoscope.

Brilliant minds and laborious practitioners engaged their inquisitive minds to solve real-life problems since civilization started; the stepping stones for science and technology started developing in the medieval period which has matured today. Arabic knowledge gradually entered Europe during the Moorish period. The confidence and new knowledge helped the society to transform gradually and extending the concept of urban living quality improvement for wider population crept in. Means to control uncertainties of ambient environment and nature were derived and applied. Physical comfort received high attention. Urbanization spree travelled from Asia to Europe.

Courageous individuals and leadership or power seats could see opportunities within and beyond tangible physical and intellectual world. Coupled with commercial interests, greed and sheer curiosity, numerous adventurous voyages towards other parts of the globe not only satisfied the hunger for knowing the unknown, also helped the European society to be the global leaders at the end of medieval period. New trade routes opened; religion of Christianity and other exchanges spread. And in short, বণিকের মানদণ্ড দেখা দিল পোহালে শর্বরী রাজদণ্ডরূপে ['*Boniker mandondo pohale sarbari dekhs dilo rajdondo rupey*'] which meant 'the trader's weighing scales transformed after the dark night as the ruler's scepter' (Tagore 1904). This ensured development of Europe and certain other parts of the world at the cost of subjugation of other countries who by now were turned into European colonies.

Modern world history experienced a rising wave of urban development which is heavily influenced by renaissance and industrial revolution in eighteenth century. Logic- and science-driven inventions, planning and discoveries supported dynamic demand and supply relation, as well as, fueled the imbalance which exists till date. Paradigm shift in thinking process and activities vital to every sphere of life including art, architecture and culture started taking place. Shift from Mechanical energy to Electrical energy led industries to grow multifold and this increased production supported a larger population.

In the nineteenth century, people embraced industrialization as a way of liberation. Acceptance of mass as significant stakeholders in urban ambience came through challenges; class struggle as Marxism proclaim. Yet, that was the first time in human history, when 'common man' became important. Newer version of urbanization came into being where housing, public spaces, transportation, water and waste management infrastructure, were extended for commoners' market gradually. Aided by industrial setup and better living standard demands by public saw gradual introduction and implementation of standards, code of practices, laws, city planning and building byelaws for the sake of better living. Radical change of Land use land cover to accommodate larger number of population, their livelihoods, infrastructure and amenities got priority over natural ecosystem function perspective. These brought drastic changes in the perception about the need and want in Europe and North America and consumerism entered in the commoners' lives. Society started to believe they can control nature and associated uncertainties. Industry 1.0 helped the rich people to grow richer in European societies; since 1870s, when Industry 2.0 allowed mass production with electricity-fed assembly lines, developed countries could share the benefit with much larger populations.

As many countries of Asia, Africa and Latin America came under control of the new imperial rulers, prioritized political positioning and natural resource extraction for import to Europe's mainland became the new normal. Biodiversity-loss, deforestation, unprecedented rate of mining, i.e. in a word merciless disturbances on natural resources created long-term imbalanced impact. En masse migration and slave trade eased affordable or cheap labour; and collectively inventions and innovations in science and technology, continuous resource and labour supply thus helped industrial economy of Europe and North America to grow. Democracy was hailed against the authoritarian rules in Europe by this period, yet reflection of the same was not far-reaching.

Urbanism has seen unprecedented growth in the twentieth century, especially in the latter half. Impact of the urbanization created a wider imbalance among the word population never heard before (Foy and Rogers 2008). So critical review of the progress probably drove the concept of sustainable development in global platform. The need for alternative development paths towards better world were unanimously accepted yet the dilemma persisted on choosing the right path for whom and when and at what cost.

1.2 Impact of Growing Urban

In the seminal work "The World Economy: A Millennial Perspective", Maddison (2001) brought significant insight on population statistics which influenced growth account of the world. Growth means quantitative change, whereas development is a qualitative notion for change. In 0 CE, global population is estimated at 0.23 billion whereas the 2001 global population is 6.15 billion, i.e. 26.64 times growth in population the world has observed over two millennia (Worldometer 2020). In 1700, Maddison estimated population is 0.6 billion (Maddison 2001) and in next 300 years, the global population has increased by another 5.55 billion, unbelievably 9.25 times. Industrialization has fueled and supported both population growth and urbanization. Since the middle of eighteenth century, industrial production of modern man-made materials like cast iron, steel, reinforced concrete and glass started in Europe and North America. Late nineteenth century saw the advancement in electricity and mechanical ventilation which freed the human society from the bounds of natural light and ventilation. New industrial and utilitarian buildings like factory, railway stations and housing required mass production and standardized components. Pollution, appalling living conditions of the working class and uncontrolled urban growth created pressure on governance.

First half of Twentieth century was marked by two world wars creating short and long-term impacts on global development. Social demographic imbalance was created as male members were sent to war; even colonized countries had to send their men un/willingly. Environmental upheaval was immense as the need for timber for trench, wharves, warehouses and war-time communication services burdened forest reserves around the world. Similarly, mines and agricultural fields were plundered thus affecting livelihoods in countries that even were not part of the war directly. New technology and materials propelled demand and supply. Deforestation, monoculture, alien and modified species, over-exploitation of soil and reduced biodiversity, thus widened ecological imbalances.

Crisis kept on surfacing as a warning, like Great Depression of 1929–1930 and Dust Bowl episode in the decade of 1930s leading to Great Migration in 1935s in the United States of America. But developed and richer societies were too confident to take note of them or to be deterred by them. The second world war brought decisive changes in the world of development. Rationality, science and technology and productivity brought sharp improvement in the life of work and leisure. New materials, vertical transportation and other services helped the privileged societies to scale new heights in built environment, i.e. man-made urban ecosystem. Communication, transportation and Material Science were emerging areas: telecommunications, broadcasting, travel by airplane, surface transportation like underground trains, petroleum-based clothes and products, host of alloys, sky-scrapers and many other things rang a new era. Role of Multi-national companies that facilitate Defense supply, consumption-driven "growth" economy took root, changed lifestyle of few societies. 1960s saw starting of very important Industry 3.0, when information technology and computational system entered the civil society of North America and Europe. Warfare strategic computational tools to control the world war was modified and improved to be made useful for development of society during peacetime. Population growth, increase in consumption, wealth creation by few and unprecedented resource depletion defined this era. Population leapt from 1.6 billion in 1900 to 2.5 billion in 1950, 3.7 billion in 1970 and 7.8 billion in 2020. 70 years from 1900 saw a rise in population of 2.1 billion (176%) and 5.3 billion (312%) between 1950 and 2020.

After consecutive two world wars, end of colonialism started in middle of twentieth century and this brought changes in global political alliances. Wide-scale war destruction provided affected countries opportunities to redevelop and unprecedented rate of development started. Rejigging of power balance, feminism, energy, security etc. was happening in different intensities in different corners of the planet. Experts warned about the destructive nature of this development. Limited global resource was believed to impose serious constraints on growth path. *Limits to Growth* (Meadows et al. 1972) and *Mankind at the Turning Point* (Mesarovic and Pestel 1974) from the Club of Rome, a think-tank, are examples of similar such philosophies which were based on state of the art data computations and simulations. Yet growth continued (Meadows et al. 1992). Urban Population in 1960 was 1.02 billion (33.61% of total population), in 1990 2.28 billion (43.03% of global population) and in 2019 climbed up to 4.2 billion, i.e. 55.714% of global population; and estimated urban areas across the world through to 2050 is going to be two-third of global population (Our World in Data 2020; World Bank 2020; UN World Urbanization Prospects 2020).

Advancement in anthropocentric world has brought destruction to natural world. The amount of Human-made materials like concrete, asphalt, metal and plastic used in the built environment is surpassing the weight of all living things on the planet fast (Elhacham et al. 2020). Large, tall and rapid construction brought the engineer, the machine's symbolic master in prominence as compared to intellectual capacity of the nineteenth-century architects; and this continues and manifests more as advanced building technology demands integration and automation of services, efficiency, security and data. Imbalanced and unequal distribution of resources, scarcity of natural materials for shelters, reduction in small-scale farming and home food production to give place to widespread cultivation ensuring food security from fertilizer-fed fields and processed food came to stay.

Fueled by aspirational sentiment, the developing countries gradually caught up with the fancy of Europe and North America's vision of urbanizing world; yet access to infrastructure or resources (both material and finance) and preparedness for the same have wide disparity. With large growing population without access to basic food, healthcare, shelter or education, the developing countries are lacking in skill and means to develop. Traditional historical wisdom for development is not of any immediate help as the present problems are far more complex than evidence of past urbanization. Urban planning, today, is linked with environmental performances. In order to bring differences, transportation, energy, water, health and other infrastructure will necessarily require resources and to be decoupled from emission and pollution. So urbanization today symbolize Frankenstein for many countries, as they themselves have the seeds of destruction sewn into with diminishing quality of living as returns.

1.3 Sustainable Development Dilemma

Developments are propelled by urbanizations and their impacts are present across lives. On a general note, it can be inferred that urbanization influences agricultural employment reduction and create stronger economy and industry- and service-based livelihoods. Inequalities and differential urbanizing patterns exist in the developed and developing world (Our World in Data 2020; Mukherjee 2002; Bloom et al. 2008; Michaels et al. 2012). Developed countries' urbanization is characterized by sparsely built density except for the central business district zones, controlled population rise, duly supportive infrastructure and amenities, strong economy and finance, cleaner and less polluted ambient environment and ample livelihood options. These are in stark contrast with urban centres of developing countries which are largely unplanned, high population density, meagre available options for shelter, food and livelihood, polluted environment, highly stressed infrastructure and no or very little access to urban amenities by poorer citizens. Anthropogenic development is purely anthropocentric at core, yet glaring inequality and imbalance in the development do exist even within a city.

Industrialization boosted confidence for which people started to believe that to govern nature's order and exploit natural resources for production is within legitimate right of human beings. People, for maximum economic benefit, accepted that transformation of landscape is both necessary and acceptable; society attached value to industrial products marketed by multi-national companies. Colonialism became a strategic tool to exploit labour and natural resources from far away locations from developed countries' geographic boundaries. Spread of colonialism facilitated continuous supply of natural resources and cheap labour to feed European industries. It continued even as late as middle of the twentieth century, evidences of which is lying strewn every part of once-colonized areas. 'Encroachment and exploitation' of the Himalayan Forest (Pant 1922) is one such example where British rulers had to recognize the community forestry after century-long struggle (Guha 2001); but not before creating irreparable destruction to the Himalayan mountain forest ecosystem. Before and after the First world war, the Himalayan forest supplied timber for railroad expansion in Middle East Asia and North Africa under British rule for decades. In the present-day Uttarakhand State of India, the local community fought for their traditional rights of forest products (Uttarakhand Forest Department 2007). Though this struggle was anthropocentric in nature, later on, the first National Park of India came up in the same state in 1936 as the Hailey Park. The Park is now known as Jim Corbett National Park and comes under IUCN Category II Protected Areas.[1]

---

[1] 1. Large natural or near natural areas set aside to protect large-scale ecological processes, along with the complement of species and ecosystems characteristic of the area, which also provide a foundation for environmentally and culturally compatible spiritual, scientific, educational, recreational and visitor opportunities.

https://www.iucn.org/theme/protected-areas/about/protected-areas-categories/category-ii-national-park

The differential pattern of growth gave rise to many development theories. The Modernization theory (Peet and Hartwick 1999) is rooted in rationalization; culture of worshipping commodity, supported by production and market institutions aiming continual growth of mass consumption is in the core of the Modernization theory. This favours free enterprise and hails the market economy as positive forces of progress. Human beings are demigods as they keep inventing ways of producing new merchandise and marketing the same. Developed countries of today believe, practice and market modernization theory as 'Progress', which provided justification for free market, exploitation of developing societies and ravaging the biosphere. Definition of sustainable development provided in Bruntland Commission Report can be criticized as to propagating the Modernization Theory in guise. Socialism-centric Dependency theory, on other hand, is based on Marxist analysis of Core (the developed world)—Periphery linkages (the developing world) and finds the same incompatible as they are externally imposed and maintains an imbalanced exploitative economic relationship (So 1990). Exponents of Dependency theory believe the 'core' maintains its control over the 'periphery' even in the post-colonial age; and to charter an autonomous, independent path of development based on socialism, developing countries shall break their linkages with the core (So 1990; Peet and Hartwick 1999). In practice, implementation of dependency theories is marred by the greed for power and wealth by individuals and society. Another relevant development philosophy to be discussed here is Gandhian school of thought on 'need and want'; this has global agreement in principle, but seems too difficult to follow or to implement. Gandhiji advocated optimal consumerism on several occasions; he said "We have Sufficient for Everybody's Needs, not for Greed" (Gandhi 1960) or "Distinguish between real needs and artificial wants and control the latter" (Gandhi 1953). Probably this underlines both the problem and solution on implementability of the philosophy; it asks assertion both at individual as well as community level.

Difference between developed and developing countries widened in numbers including population and per capita economic growth, infrastructure and services. Overcrowding and uncontrolled urbanization came to stay at the fringe; slum conditions establish inability of governments to supply proper municipal, medical and educational services. Even before second world war, ecological views of developmental degradation and resource limits started doing rounds. After the war, United Nations was formed with Peacekeeping Mission in 1945. The second world war impacted colonized countries to a great extent and gradually their struggle against subjugation can find fruition and they could achieve their freedom and self-governance right.

Next few decades were fueled by growth, both planned and unplanned. Awareness about pollution grew, many nations began to take actions to protect the environment within their borders with the realization that pollution did not stop at their borders. Energy came at the centre of development. Increasingly evident ecological consequences of human activities brought dilemmas and arguments on how the society shall grow in future. Majority facilitated the development modelled after 'Manhattan', barring few minority voices (Schumacher 1973; Meadows et al. 1972; Mebratu 1998; Hirsch 2005). Industry 3.0, third industrial revolution phase, started in developed countries around 1960s. IT and computer technology, which ruled next few decades till date, were deployed to automate the production and in science and technology. Now it has pervaded every sphere of human lives.

The United Nations Conference on Human Environment, UNCHE 1972 held at Stockholm linked environmental protection with sustainable development; for the first time in human history, to enjoy a clean and healthy environment and responsibility to preserve the same was recognized as human right. The concepts discussed and plans developed by the Stockholm Conference are respected widely and they have influenced every international conference and treaty on the environment till date. Many nations and non-governmental organizations (NGOs) attended the Stockholm Conference and agreed upon a joint Declaration containing Principles, Action Plans and Resolutions concerning environment and development. One of the ground-breaking issues that emerged from the conference is the recognition of poverty alleviation for protecting the environment. Mrs. Indira Gandhi, then Indian Prime Minister, brought forward the connection between ecological management and poverty alleviation emphatically. In addition, the Declaration noted (a) wildlife ecosystem management to ensure their continued survival, (b) develop economies of developing nations in an environmentally responsible manner and provisioning for financial and technological assistance to enable them for the same, (c) to end to the discharge of pollution into the environment as there is an appreciable risk of effects on the climate and (d) to institute international standards for pollutants based on scientific research.

The recommendation of the Action Plan drawn in the UNCHE 1972 advocated for creation of a network of national and international pollution monitoring agencies, which guided the United Nations to create the United Nations Environment Programme (UNEP) in 1972. UNEP became responsible for coordination of environmental initiatives and providing support to developing nations on environmental issues. Signs of irreparable damaging impact of development were gradually visible and admissible. The European Union (of today) created the Environmental and Consumer Protection Directorate and placed the first Environmental Action Program in 1973.

'Sustainable development is development that meets the needs of the present without compromising the ability of future generations to meet their own needs' (WCED 1987) is an often-repeated definition till date for Sustainable Development. The Brundtland Commission Report (WCED 1987) brought poverty, population growth and development on a single platform for discussion; yet the report left the global community divided as the critics felt the Commission placed environmental issues firmly under the political and economic agenda. It didn't challenge effectively the consumerism-oriented development of the developed countries. For developing countries, sustainable development seemed like an ideology advocated by developed countries to sustain the gap between developed and underdeveloped countries through restriction on developing countries' growth. The Brundtland Report, named after the chairperson of the Committee, focused primarily on the needs and interests of humans and advocated for redistribution of resources towards poorer nations to create means to secure their basic needs to achieve global equity for future generations.

Essential contradiction in the term "sustainable development" is between opposing imperatives of environmental sustainability and economic growth and development. While government and private sector organizations adopted the term sustainable development, academic and NGOs found it desirable to use "sustainability" in similar contexts, as the term focuses on the ability of humans to continue to live within the concept of carrying capacity of ambient environment. Inherent contradiction grew wide as sustainability advocated conservation and protection of resources, whereas, development essentially requires exploitation of resources. Less developed countries have aspiration to improve their living standards through economic development and that will deteriorate quality of the global environment for future generations. Sustainability can be realized only if the developed countries can restrict their growth, transfer technology and finance to developing countries so that they need not recreate the same mistakes which are endangering the environment (Robinson 2004).

Valuation of Ecosystem services and pricing the same as a commodity may help the market-oriented modernization theory of development without forfeiting the opportunity of the future generations' growth (Mukherjee and Takara 2018; Ely and Pitman 2012; Sukhdev et al. 2014). The issues of growth and development continued to challenge societies. The first UN Conference on Environment and Development or the Earth Summit, held at Rio de Janeiro in 1992, was an opportunity to look back at the declaration and the action plan adopted in UNCHE 1972. The UNCED 1992 led to the formulation of.

- the Agenda 21—a comprehensive blueprint for global action required for countries to draw up a national strategy to affect the transition to sustainable development
- the Rio Declaration on Environment and Development—27 principles defining the rights and responsibilities of states in this area
- the United Nations Framework Convention on Climate Change (UNFCCC)—legally binding convention aimed at preventing global climate change
- the Convention on Biological Diversity (CBD)—another legally binding convention aimed at prevention of biologically diverse species' eradication and
- Principles of Forest Management—a set of principles to support the sustainable management of forests worldwide.

The Earth Summit concluded with agreements on internationally important issues like climate change, biodiversity and forests. These conventions were signed by the representatives of more than 150 countries. Later on, the Summit led to establishment of the UN Commission on Sustainable Development. Once the UNFCCC created foundations of pro-environment development paradigm, increased interest and research collaboration arguably paved the way for further understanding of global warming. This has led to such agreements as the Kyoto Protocol, a result of the third Conference of the Parties, COP 3, in the year 1997 and the Paris Agreement in the 21st Conference of the Parties, COP21, in 2015. But significant members are not taking their roleplaying as seriously as they committed. The UN systems are under severe criticism as the global development is haywire. As the utilitarian approach of conservation strategies provides scope to bank upon Ecosystem services, an array of equitable new approaches for next generation received due attention like preservation and conservation of natural areas, renewable resource extraction, clean development mechanism facilitating non-polluting activities, urban green infrastructure etc. Developed countries are increasing the ecological footprint and sadly the people of developing countries are embracing the urban lifestyles which have large ecological footprints too. The IPCC brought back the world's attention to undertake environmental planning and management in a way that facilitates human aspirations for economic and social improvement without bringing any permanent harm to ecological processes. Based on the National Footprint Accounts of 2011 edition (NFA 2012), Boruckea et al. (2013) presented a study for the ecological footprint of people or activities and the bio-capacity of the earth or its regions; and it pegs the global hectare (gha) per capita (measurement unit for ecological footprint), for Afghanistan 0.54, Nepal and Congo, DR 0.76, India 0.87, Japan 4.17, Qatar 11.68, United States 7.19 and Viet Nam 1.39.

Elhacham et al. (2020) in their new study find that this year may mark the point when total weight of everything made by humans outweighs natural things. Millennium Development Goals (MDGs) had achieved mixed success; despite MDGs were promoted as the expression of solidarity with the world's poorest and most vulnerable, cross-sectoral investment in terms of meaningful and effective implementation of action plans were not too many. After a decade and half, the Sustainable Development Goals (SDGs) brought in place of the MDGs in 2015 and the 17 targets are delineated to be achieved by the year 2030. Anthropocentric-development goals are now broadened enough to include other species and ecosystems and every strata of people are included for participatory development. The SDGs recognize urbanization and climate change issues strongly. The then Secretary General, Ban-Ki-Moon opined "It (2030 Agenda for Sustainable Development) is a universal, integrated and human rights-based agenda for sustainable development. It balances economic growth, social justice and environmental stewardship and underlines the links between peace, development and human rights" (UNDP and WB 2016). The Agenda advocated for 5 Ps- People, Prosperity, Planet, Peace and Partnership. The Global Indicator Framework for Sustainable Development Goals was developed and adopted in 2017 (Inter-Agency and Expert Group on SDG Indicators (IAEG-SDGs) 2017). SDG Monitoring and Reporting Toolkit for UN Country Teams (UN 2021) was developed to monitor the progress towards reaching the targets and these helped to monitor the comparative performances of different countries and regions in terms of SDGs' implementation. The Sustainable Development Goals Report 2020 (UN 2020) comments on distinct two phases before and after Pandemic Covid 19. Based on data set, the Report analyzed that before the COVID-19 pandemic, scale and speed of progress was not sufficient to meet expected target set for the Goals by 2030. Some positive impacts like decrease in students drop out of school, reduction of communicable diseases, improved access to safe drinking water and increase in women's representation in leadership roles are reported. Simultaneously, decline of food security, deterioration of natural environment at an alarming rate and dramatic levels of inequality persisted in all regions. The COVID-19 pandemic brought unprecedented negative impact on health, economic and social crisis threatening lives and livelihoods. This really creates doubts about achieving the Goals by 2030. The Pandemic though brought several global improvements in the natural environment.

Increasing disaster risk, climate change induced extreme events and the Pandemic Covid 19 brought the focus of development to resilience-integrated sustainable development. The following subsequent sections deliberate on sustainable and resilient development and most importantly elaborates on how the connect with nature can benefit development.

1.4 Risk and Resilience

Sustainability, i.e. "quality of being able to continue over a period of time" (Cambridge University Press 2020), is seriously disrupted when any disaster and/ or extreme event strikes. The UNDRR (2020a) reports 7348 major recorded disaster events which claimed 1.23 million lives, created impact on 4.2 billion people (many on more than one occasion) and resulted in approximately US$2.97 trillion in global economic losses in the period 2000 to 2019. The vicennial loss during 2000 to 2019 is much more than the previous vicennium. Between 1980 and 1999, 4212 disaster events were linked to natural hazards worldwide resulting far less losses (UNDRR 2020a).

Disaster Risk is defined as "potential loss of life, injury, or destroyed or damaged assets which could occur to a system, society or a community in a specific period of time, determined probabilistically as a function of hazard, exposure, vulnerability and capacity" (UNDRR 2020b). Risk is a compound impact factoring four components, i.e. hazard, vulnerability, coping capacity and exposure. Existing socio-political, economic and environmental (both built and natural) conditions defines vulnerability of any society. Structural and non-structural measures can reduce loss of lives, property, services, infrastructure and business. Perception about risk and its assessment varies within and across societies. Mapping of the prevailing hazards, traditional and S&T-based advanced knowledge, resilience culture and technical advancement help to mitigate underlying risk factors partially. Improvement in understanding complex function and inter-relation of the risk components augmented accuracy in scientific tools to carry out assessment and increased socio-economic interventions for risk reduction is continuing; yet only part of the potential losses can be avoided. Guidelines, codes and planning based on known probabilities of hazards, vulnerability and exposure can help us in this management. After any disastrous event, the gap becomes evident which needs our attention for further improvement. This is called residual risk which demands our attention for continuous improvement at every stage of disaster management, i.e. relief, response, recovery, management and mitigation; and different aspects of knowledge in multi-hazards, participatory implementation, among others become useful tools to further reduce the residual risk.

Adaptation offers strategies enabling a system to resolve problems brought by anthropogenic activities, disaster risk and climate change. It accepts that economic forces are unstoppable and would continue to grow. Many developed countries adopted many adaptations to improve risk resilience of their societies; Hafen City–Hamburg in Germany is one such comprehensive example which for the last four decades are gradually implementing the engineered strategies to take care of the climate change induced seawater level rise (Eleftheriou and Knieling 2017). Developing countries find mitigation more affordable in comparison to adaptation, as this can be more structured and non-structured and less expensive and resource-consuming. Mitigation explores avenues to attenuate climate changes' harmful impact through reduction of greenhouse gas (GHG) emission. It advocates shift in development philosophy which is primarily linked with growth.

With increased industrialization and urbanization, climate change (CC) induced extreme events are increasing, which is influenced by higher concentration of greenhouse gases (GHGs). The GHGs contribute to global warming, imbalance in earth's energy budget and influence extreme weather events significantly. The Intergovernmental Panel on Climate Change (IPCC) was established in 1988 for understanding the CC-induced risk and its assessment and to provide policymakers with regular scientific assessments and state of the art knowledge about CC. In order to address CC issues, the IPCC published first set of CC scenarios in 1992, called IS92. Afterwards, the IPCC published the Special Report on Emissions Scenarios (SRES) since 2000; they are generated using Global Climate Model projecting the future climate change scenarios based on assumptions for differential development patterns. The SRES scenarios were an improvement upon the IS92 scenarios and the Third Assessment Report (TAR) (IPCC 2001) and the Fourth Assessment Report (AR4) (IPCC 2007) used the SRESs. The scenarios of SRES did not consider any current or future measures to limit GHG emissions; they could provide a common baseline for reference as many groups are researching climate science in the last two decades.

Ecosystem-based adaptation (EbA) is the use of biodiversity and ecosystem services as part of an overall adaptation strategy to help people adapt to the adverse effects of climate change. EbA has been practised, promoted and advocated for some years now (Uy and Shaw 2012). It is an overall strategy to help people adapt to the adverse impacts of climate change and emphasizes adaptive capacity and resilience of natural and human systems to prepare for even abrupt and unanticipated changes brought about by climate change.

Anthropocentric activities are one of the primary causes of rapid increase in temperature. In 2007, the IPCC responded to calls for improvements to SRES by catalyzing the process that produced the Representative Concentration Pathways (RCPs). The RCPs are used in the IPCC Assessment Report Five (AR5) (IPCC 2014) in preference to SRES. A climate projection is based on several assumptions including development pattern, influential climate policy and provides probable likelihood scenario of the future climate in various temporal scales (few decades to centuries). The Projection scenario is iteration of the process to better understand the uncertainties and alternative futures emphasizing on collective decisions on growth patterns.

In its 5th Assessment Report (AR5) (IPCC 2014), the IPCC says that "each of the past three decades has been successively warmer at the Earth's surface than any of the previous decades in the instrumental record and the decade of the 2000s has been the warmest". The AR5 introduced four RCPs based on radiative forcing or climate forcing, which means the residual insolation (sunlight) absorbed by the Earth and energy radiated back to space quantified at the tropopause and expressed in units of watts per square meter of the Earth's surface. Total radiative forcing is determined by both positive forcing (additional insolation absorbed) from GHGs and negative forcing from aerosols through reradiating the solar energy. The RCPs described in the AR5 are (IPCC 2014):

(a) RCP 8.5—High emissions; this RCP is consistent with a future with no policy changes to reduce emissions.
(b) RCP 6—Intermediate emissions; Radiative forcing is stabilized shortly after year 2100, which is consistent with the application of range of technologies and strategies for reducing GHG emissions.
(c) RCP 4.5—Intermediate emissions; here radiative forcing is stabilized shortly after year 2100, consistent with a future with relatively ambitious emissions reductions. And,
(d) RCP 2.6—Low emissions; here radiative forcing reaches 3.1 W/m^2 before it returns to 2.6 W/m^2 by 2100. In order to reach such forcing levels, ambitious GHG emission reductions would be required over time.

The 2015 Paris Agreement (UNFCCC 2015) generated a stimulating and encouraging milieu on global environment among the signatory countries initially; yet this positive environment was short-lived as signatory countries are struggling to implement the required and committed policies to meet the targets. The Shared Socio-economic Pathways (SSPs) were developed later on based on the futuristic scenario and they tried to capture (Hausfather 2018) socio-political trends of countries globally and presented five scenarios, SSP1 to SSP5, without any reference to climate policies or their implementation. So from that point, both the SSPs and the RCPs brought no specific emphasis on the dynamics of climate policy implementation. The RCPs were developed by both climate models and integrated assessment models (IAMs) (Van Vuuren et al. 2011). The SSPs were developed *ex-post* to be consistent with the RCPs by '[…] integrating the descriptions of socio-economic development with the climate change projections and with assumptions about climate mitigation and adaptation policies' (Ebi et al. 2014). Climate Scientists (Hewitt et al. 2020; Rogelj et al. 2016) agree that the Scenario planning approaches have wide acceptance to understand and communicate climate change outcomes globally; yet neither the RCPs nor the SSPs provide any information about the dynamics of climate change actors and policies in their real-world context. The IPCC Assessment Report 6 (AR6) which is due in 2020–2021, has based their work on integration of SSPs and RCPs.

In this context of global uncertainty over sustainable and resilient (disaster and climate change-induced risk) policy discussion, adoption and implementations by different countries will be utterly significant for the global warming and extreme events (Revi 2008). Escalating Disaster risk is bringing complexity in the real-world context. Nature-based solutions are now being hailed as mitigation solutions packaged for both climate change scenario calming and responding for disaster risk reduction (UNDRR-APSTAAG 2020). The UN has decided to facilitate more for nature protection and facilitating nature-based resource development in the decade from 2021 to 2030. The United Nations General Assembly (UNGA) proclaimed 2021–2030 as the UN Decade on Ecosystem Restoration on March 1, 2019. The UNEP and FAO Factsheet (UNEP and FAO 2020), talks about the UN Decade on Ecosystem Restoration 2021–2030 to "prevent, halt and reverse the degradation of ecosystems worldwide" while addressing sustainable development targets, Sendai Framework, climate change, food and water security and livelihood options. In the following sections discourse on future built environment with nature-based mitigation and adaptation paradigm is presented.

Ecosystem-based disaster risk reduction (Eco-DRR) (Marisol et al. 2016) advocates for translating the ecosystem services into resilience services towards disaster risk. Mitigation and Ecosystem-based Adaptation (EbA) (UNEP 2019) promotes nature-based solutions and supports preservation and conservation of blue and green natural elements through enhancement and nurturing. Blue-green elements can be categorized under four verticals depending on their spatial scale: Bits of nature, Patch, Corridor and Matrix (Kato and Ahern 2008). The scale becomes relevant for Micro (Cluster of few buildings), Local (Neighbourhood Area), Meso (part of city or whole town) and Macro (region or urban agglomeration) level of intervention for preservation and conservation of nature. State of the art knowledge and creative ideation are helpful for implementation. Confidence on the potential of the BGIs as a resilience tool in addition to triple bottom line benefits and its adaptation as alternative or complementary to existing grey infrastructure are increasing. Decision support systems like GUS-3CC (Mukherjee and Takara 2018) on the strategy adoption depending on status, potential and eminence of blue, green natural elements, their types and health can strengthen the confidence further.

Mandatory networked blue and green elements in Smart City Master Plan development with target for disaster risk and CC-induced extreme events resilience, development regulation for protection, conservation, restoration, new introduction of nature will help both the developed and developing countries' urbanization alike. Green resilient infrastructure like green road, coastal mangrove forest, vegetative slope protection, ecologically efficient public spaces etc. are gradually chartering the development agenda of global cities. National highways to rural roads, sustainable water and waste management in cities, regional lake and forest health preservation are some of the focus areas of application. Many a time, eco-restorations initiatives are experiencing lesser success as an appropriate method in tune with biophysical conditions of the site selected, science and technology interventions, participatory management, capacity building and socio-economic conditions of the community have substantial influence on successful restoration. Lessons from these failures can help the Eco-DRR and EbA to mature further. Deforestation can be controlled with the efficacy of the governance.

Role of science and technology to promote nature-based resilience is enormous. Increase in instrumented studies in last two decades provide scientific information on the performance of natural and man-made ecosystem and their components' interventions. Surface transformation is cited as a major source and reason for long-term biophysical changes in the urban world (Oke et al. 2017). Change of albedo influences ambient and surface temperature, thus decreasing albedo due to urbanization is one of the major reasons for urban heat island (UHI). Reflective coating, cool green roof, selected natural surfaces are used in urban areas to combat decreasing albedo (Akbari et al. 2008). Greening leads to reduction in near ground temp resulting in energy usage reduction by community.

Presence of Water body is an essential condition for lives under water and on earth; SDG 14 and 15 reaffirm the same (UN 2015). Hot water disturbs aquatic habitat ecosystem as with rising temperature dissolved oxygen decreases. So maintaining the temperature of water is vital for them. As water temperature is crucial for life under water, this may be noted with interest, that runoff temperatures from permeable pavements depends on surface material characteristics of pavement, (Roseen et al. 2012). In bio-retention pond, outlet position is important as the outlet at bottom releases the coolest water (Janke et al. 2021). Problem with water retention ponds and wetlands are far more spatially distributed; parameters like thermal pollution due to large surface area, shallow ponding depth and type of media, drain configuration and shading influence wetland ecosystems' functions. Contribution from vegetation comes in terms of shade, cooling, moisture balance etc. One full-grown tree uses 20–30 kw of solar energy for evapotranspiration as latent heat and thus having cooling equivalent to 10–15 AC units (Stone 2006). Vegetated area and green forest create temperature difference of significance. Parks can moderate temperature through greater evaporation (300%) than surrounding areas (Spronken-Smith et al. 2000). Small, distributed vegetated area spread throughout an area extend the thermal benefit of vegetation beyond the physical boundary more effectively than a large, continuous vegetated area of the same size in a central location (Honjo and Takakura 1990; Shashua-Bar and Hoffman 2000). Seasonal rainfall interception by mature open-grown trees are studied closely in Davis, California by Xiao et al. (2000). Gross precipitation (Pg) measured above the vegetation canopy is more than the net precipitation (Pn) that actually reaches the ground. It is the sum of throughfall and stemflow.[2]

1.5 Nature-Centric Resilience

The present-day development restricts and controls the role of non-human components, both organic and inorganic, i.e. biotic-abiotic. Anthropocentric development is

[2] Free throughfall (Th) is the raindrops (a fraction of precipitation) that reaches the ground surface through the gaps in the canopy leaves and branches without hitting the canopy surfaces. Throughfall (TH) is the portion of the precipitation that reaches the ground directly through gaps in the vegetation canopy and drips from leaves, twigs and stems. It is the sum of free throughfall (Th) and canopy drip (D). Stemflow (ST) is the portion of precipitation intercepted by the canopy and reaches the ground by flowing down the stems or tree bole Subramanyam K (2013).

now facing stiff challenges from anthropogenic malice. Disaster and climate change induced risks are gradually taking centre stage. To address this critical context, the book on "Ecosystem-Based Disaster and Climate Resilience' with emphasisis on Integration of Blue-Green Infrastructure in Sustainable Development is a sincere contribution towards nature-centric resilience.

Introductory chapter provides specific overview of the key issues and development at global and regional level related to natural ecosystem, disaster risk reduction and Climate change adaptation. Uncertainties in urbanizing world and nature-based resilience building is being critically reviewed by Mukherjee and Shaw in order to create right contexts and relevance for the subsequent discussions.

Policy influence on nature-based solutions as well as the impact of nature-based solution's integration on policy framing is a major focus area in Part 2. As Policy Analysis and Policy Framing are significant for ensuring implementation of Ecosystem-based Disaster Risk Reduction (Eco-DRR) and Ecosystem-based Adaptation (EbA), Part 2 of the book discusses a basket approaches for Policy Analysis and Policy Framing for disaster and climate risk resilience, integration of blue green as infrastructure, peri-urban ecosystem services, role of innovations to reduce disaster risks for region and country level. Regions of the Hindu Kush Himalayas, countries like Australia, India, Pakistan, Philippines, Nepal and Vietnam are being presented in the book (Chaps. 2 to 9). Status, policy perspectives, opportunities, implementability, evaluation and assessment process, progress and challenges are meticulously discussed in these chapters to make resilience using BGI a reality.

In Chap. 2, Chaudhury et al. discuss the presence of EbA in the Hindu Kush Himalayas. Limited and scattered documentation of EbA implementation is a challenge in the region. In reality, the mountain communities adapt to the harsh realities of climate change and EbA shape their wellbeing a lot. The chapter explores nature and progress of EbA in the region, assess its status and progress using thematic analysis approach using the Aichi targets as Benchmarks. EbA mainstreaming challenges and barriers in the region and policy and practice at national and international scales from a mountain perspective is shared in the chapter.

Evaluation of Ecosystem-based approaches for disaster and climate risk resilience and policy perspectives in Pakistan is presented by Khan et al. in Chap. 3. This chapter discusses a range of disasters and climate risks the country is exposed to. Performance of diverse agro-ecological zones of Pakistan in a changing climate scenario, degraded local ecosystems like mountain forest and coastal mangrove forest, increased risk of soil erosion, flood and landslides are recorded in the chapter and explained how they pose serious threat to blue-green infrastructure. Based on analysis of the present status of the mountain and forest ecosystems with reference to DRR and Climate Risk, the authors presented suggested Policy measures and approaches for blue-green infrastructure utilization.

Poudel et al. discussed the Ecosystem-based approaches and policy perspectives in Nepal in Chap. 4. In this discussion, attention to vulnerable Mountain ecosystems in climate change perspective is provided. Impact of increasing temperatures and disruptive precipitation to floods, droughts and landslide hazards are increasing; the chapter records how these events are devastating the local communities' lives and livelihoods. The authors explained how Eco-DRR and EbA can be implemented through Policy to reduce the vulnerability of the affected communities. This study reviewed the integration of Eco-DRR and EbA approaches into spatial and economic planning and associated challenges in Nepal.

In Chap. 5, a systematic discussion on evolution in India's policy frameworks on Nature-based solutions across various strategically important thematic areas and sectors is presented by Bharadwaj and Gupta. Need to integrate cost-effective and flexible opportunities created by BGI is supported to be integrated into developmental planning and actions. The authors present a diagonal review of strategies and interventions for Eco-DRR and EbA through the existing policy mechanisms. Pros and cons to mainstream the ecosystem-based approaches through developmental planning and practice is argued in this chapter. Supporting detailed case analysis of climate change, disaster management and sectoral developmental policy contexts in India is reviewed and concisely presented.

Thi and Hui presented a discourse in Chap. 6 on Ecosystem-based approaches and policy perspectives toward integrated blue-green solutions in Vietnam. The country's present and future economic status and target is derived keeping in view the principles and standards of the global economy and market. Green and ecosystem-based development format is reflected in the National Vision for Socio-Economic Development, 2020 for Vietnam which accommodates multiple scenarios of low to high roadmap. Promoting nature-based solutions for climate and disaster risk resilience is creating an essential basis for policies and practices on environmental protection. This chapter provides an analysis of the existing policy and practices of ecosystem-based approaches and their implementation challenges with various case studies from the field and infers significant roleplaying of policy for further integration of blue-green infrastructure at both national and local levels.

Nature-based solutions, a critical and urgently significant concept, harnesses potential advantage of nature's innate ability to build high quality, resilient and lower-cost infrastructure. Uy and Tapino brought the status report on nature-based solutions implemented in the Philippines, the supporting policy framework for the same and lessons learnt in Chap. 7. The authors prescribe a proactive approach to introduce Blue-green infrastructure as a substitute to strengthen existing infrastructure systems by preserving, enhancing, or restoring natural elements. Through the case study in Polillo, Quezon, Philippines, the authors establish that political commitment at multiple levels, securing funding and private sector engagement can strengthen blue-green infrastructure mainstreaming; The chapter also emphasizes on evidence sharing to help science and technology and stakeholders for informed decision making to advance its implementation.

From a policy perspective, peri-urban areas (PUA) held a very special place in urban development. In the Chap. 8, Noman et al. deliberated contribution of Peri-urban ecosystem services (BGI) to Urban resilience. Current anthropogenic urban development influences spatial planning priorities within urban areas and neglect natural ecosystems for satisfying anthropogenic needs. The authors advocate to take offload from urban areas and to promote PUA for Eco-DRR integration and they expect urban resilience can be achieved better by this. Spatial planning practices in

PUA can incorporate blue and green infrastructure (BGI) as development pressure is less and allow growth of new typologies and frameworks in urban centres. Challenge associated with relaxed zoning and legal frameworks of law in the PUA and fragmented and ad hoc development and developer-led approaches to growth may create inflexible responses to changing risk if concerns are not reflected. Strengthening links between urban and rural areas through PUA to be considered as an effective strategy for resilience planning; this will improve scope for the preservation of natural environments. The chapter supports stronger and more effective BGI to reduce negative effects of human development and improve resilience against risk significantly.

The Chap. 9 discussed water-related risks, one of the imperatives of lives and roles of innovation. In this chapter, Biswas et al. discussed importance and status of water quality and critical water scarcity scenarios in rural and urban India. A pertinent discussion on innovations as policy instruments to reduce challenges and risks related to water is followed by a discourse on innovative approaches to address water unavailability and water quality. Challenges with waters are many and in a systematic way, the authors raised the emerging problem of inland and coastal salinity, seawater intrusion, health-related disasters due to poor water quality and arsenic menace. Holistic and integrated intervention plans for reduction of risks are broached up with examples of innovative interventions and implemented case studies. Reducing disaster risk in wetlands and other water-based ecosystems are discussed further. This chapter emphasizes on the evidence-based facts on application of science and technology policy which may facilitate innovation as an important tool for Eco-DRR and EbA.

Science Investigation, Technology and Planning Intervention play major role for implementation of Nature-based solutions; scientific community contributes through scientific research in aspects like hazard and risk, advanced tools and techniques for assessment and promoting nature-based solutions. Urban risk assessment tools and techniques, future heat risk mapping, scaling-up process for nature-based solutions, earned value analysis using Building Information Model, smart urban areas management are discussed through Chaps. 10–14. The comprehensive approach for Ecosystem-based solutions, green building infrastructure development, hot-spot identification for stressed areas as well as potential intervention for mitigation and adaptation solutions and mainstreaming nature-based resilience received profound attention in these chapters.

Marcotullio and Schmeltz, in Chap. 10, reflect on heat risk in South Asia and role of ecosystem mitigation. Increasing heatwave events of South Asian cities are increasing in intensity and are projected to increase in frequency, as the climate continues to change. The authors explored interrelations between large and growing urban populations in the region and state and trends of growing heat risk in terms of heat waves and Urban Heat Island (UHI). The authors made use of the Modified UN levels and the SSPs for urban population projections and simulation platforms like the HadGEM2, a coupled Earth System Model and the Inter-Sectoral Impact Model Inter-comparison Project (ISIMIP) ensemble to derive insights. As human well-being is at risk, the chapter attempts to inform urban planners about the same; also explores potential for ecosystem mitigation in South Asia, especially integration of blue-green infrastructure to address these climate change related hazards and the future heat shock events.

Implementation of Ecosystem-based Solutions would require continuous support from S&T side as the complex urbanization process is influencing an integral change in natural fabric influencing climate. The global and the local level urban risk assessment tools and techniques are discussed in Chap. 11 by Rahul et al. The authors concentrate their focus on two major focus areas of urban risks- increasing heat risk and instances of flooding, their causes and impacts. Deliberation on parametric quantitative estimation of these multiple risks for assessment and investigation, sources and variety of different datasets and uncertainties associated with data help the readers to understand challenges the scientist are attempting to resolve. The authors introduce methodologies to identify susceptible hot spots, to investigate the intensity and possible reason, potential of ecosystem-based solutions for risk attenuation and specific problems associated with them to estimate the suitability of intervention strategy.

In Chap. 12, Dhyani et al. broach into sustainability aspect of nature-based solutions and present the necessity of scaling up and mainstreaming of Nature-based Solutions (NbS) for resilience. To create the appropriate context for discussion, methodology for case study reporting and comparative analysis are adopted for selected Indian cities. Population and urban growth brought irretrievable impacts on natural ecosystems and pressure will increase further in future. The study is conducted in Tier II and Tier III cities where potential for BGI integration is argued to be more than Tier I cities, yet such explorations are quite limited in numbers. Fragmentation, degradation, loss of ecosystems, biodiversity and ecosystem services are few verticals against which the authors explored mainstreaming opportunities in this chapter. Scientifically planned NbS for implementation can ensure sustainable interventions in Indian megacities and growing small cities.

Building information modelling (BIM) is information rich and object-based digital tool which represent physical and functional characteristics of a construction project and is used to document building designs, construction and operation. Buildings and infrastructure are moving towards sustainability through green rating. Sarkar et al. presented BIM-based Modelling in sustainable development of green building from stakeholder's perspective to satisfy the interest of built environment in Chap. 13. It is expected sustainability of built environment will release pressure off from nature-based ecosystem and thus indirectly help the Eco-DRR and EbA.

Road to Eco-DRR and EbA is always experimental. In Chap. 14, Maki presents comprehensive planning approach for smart urban area management in Japan. Despite the fact that Land use/Land cover (LULC) change with development and gray infrastructure are inevitable, investment in NbS can benefit. Eco-DRR policy in Japan is traditionally incorporated; the urban planning act regulates green area demarcation and maintenance. Compact city planning in Japan is promoted to address depopulation and land use regulation. Policy of comprehensive urban flood management with urban green farm, detention basin, household rainwater tank urban park and open green area in the city are discussed in this chapter. Explorations of issues like obstacle to Eco-DRR policy implementation, understanding future impact of disaster on society, pre-disaster recovery planning techniques establish further the importance of planning for Eco-DRR and EbA in disaster-prone societies like Japan.

Evidence-based science and technology discussion has better acceptability with the society and its different stakeholders; keeping this important methodological approach in focus, illustrated examples of BGI integration and mainstreaming are explored in the Chaps. 15–22. The key issues and challenges for nature-based solutions, i.e. BGI integration, role of governance and community in its implementation, challenges in decision making for building climate risk resilience from present and historical perspectives are presented in these chapters. Variety of aspects through the Case study lens are brought forward; Ecosystem-based mitigation and adaptation for cyclone and tsunami resilience, sustainable water management in water-stressed urban areas, application of tools and techniques like Remote Sensing Image Analysis, biomonitoring of freshwater quality in remote places, review of decision support system to maintain matrix scale BGI like mangrove forest and retarding basin and post-disaster assessment of efficacy for Eco-DRR and EbA. Examples are selected and presented from different geographical regions and cities of Bangladesh, Fiji, India, Japan and Sri Lanka.

Severe water scarcity experienced in many cities of the world, again and again, prove importance of sustainable management of water supply as a system. Thapa et al. bring water management case study from Shimla, capital city of Himachal Pradesh of India in Chap. 15. Analysis of the water scarcity problem in the city leads to greater understanding of the present status; and the solution leads to ecosystem-based sustainable, integrated and comprehensive management strategies of water resources. Rapid LULC change in the hill city of Shimla due to urban growth, pressure from tourism are analyzed. The authors suggest nature-based solution packages which include water and spring shed development, water harvesting, appropriate crop and irrigation and revival of traditional systems wherever possible in Shimla city.

Another capital city Dehradun from hill state of Uttarakhand in India is presented in Chap. 16 by Kumar et al. in which application of remote sensing images are made both for problem analysis and Eco-DRR solutions. Selected city, Dehradun, is prone to multi-hazards; rapid and unplanned urbanization is adding to the concern. The chapter reports space-based applications to map changing development patterns of the city over decades. Using indices for vegetation, water bodies and build-up and overlaid on land surface temperature, analysis for hotspots for risk concentration and potential application of BGI are presented. Calibrated geospatial information led to the integration of different maps which helps to suggest a holistic approach of Eco-DRR in the urban area for resilient sustainable development.

Third Case Study from India, discussed in Chap. 17, concentrates on water stress management for Nagpur Metropolitan Area, India. Sukhwani et al. share historical research trends on Ecosystem-based approaches. Environmental profiling of the study area includes mapping LULC and Hydrogeological changing pattern, water reservoirs location and status and population within the Nagpur Metropolitan

Area. Based on the analysis, the chapter discusses needs for water conservation, urban–rural linkages and nexus considerations, enhancement of wastewater reuse and resource recovery and enabling transboundary cooperation through multi-stakeholder engagement.

A pertinent query on challenges in decision making for building resilience to climate risks is raised by Danda et al. in the Chap. 18. The context of Sundarbans mangrove forest acted aptly to unfold the crucial discussion related to complexity of decision making. It is argued with evidence in the chapter that even if several parameters and background is similar, the risk perception may vary depending on perspective of stakeholders. So the same event may offer opportunities for multiple valid decisions and this makes the decision making quite challenging. This paper explores three criteria (manage or accommodate, resist or protect and strategic and managed retreat) framework and analyzes decision-making mechanisms to enhance resilience to climate risks. The framework-based analysis for the Indian Sundarbans derives that retreat options are unlikely to be implemented in absence of due consideration of political risk.

In the next chapter, Wickramasinghe presents the Sri Lankan experience of Eco-DRR. The Chap. 19 address hazard-proneness of the tropical island nation of Sri Lanka and vulnerability of its communities and nature. Evidences of communities' nature-based resilience are presented with historical and cultural perspectives. Complexity with development progress and disaster events have impacted heavily on natural habitats including coastal vegetation and wetlands. Role of community to adapt to a greener alternative in order to embrace stronger resilience is established in the chapter. Issues and challenges in mainstreaming Eco-DRR through national policy and actionable agendas are presented.

Ichinose et al. in Chap. 20 shared an analytical and historical perspective of the Watarase retarding basin from Japan. It is the largest retarding basin in Japan. Initially built to contain the copper mine pollution and with the advent of time, the basin area was enlarged which involved lengthy socio-political deliberations and governance issues. The change in the character and area of the basin with time is recorded in the chapter. Now as wetland it provides many different ecosystem services including biodiversity. It continues to have an important flood control function; even in recent Typhoon Hagibis in 2019, the basin prevented downstream flooding. People are visiting the place as it has generated a reputation as traditional example of ecosystem-based disaster risk reduction in Japan.

Role of nature-based resilience have many evidences where traditional communities have tended these practices. In the Chap. 21, Tashiro presents post-disaster recovery analysis of a disaster-affected community from the 2011 Great East Japan Earthquake and Tsunami in Japan, where she finds the important role played by the community. The issues of self-efficacy of EbA strategies and residents' health damaged by natural disasters are inquired apart from post-disaster environmental management in this study. Questionnaire survey on residents' health conditions and behaviours established positive influence of residents' self-efficacy for greenspace management during the post-disaster recovery phase with health and behaviour. The independent variables analyzed in the study are health conditions (self-reported stress, physical activity and self-rated health (SRH)), green management behaviour (weeding experience) and awareness of Ecosystem-based Disaster Risk Reduction (Eco-DRR). This study highlights public health approach to post-disaster environmental management and important role of EbA strategies in a post-disaster rural context.

Freshwater Biomonitoring is an EbA proposed for building climate resilience communities in Fiji by Rashni in Chap. 22. This case represents the risk scenario of small islands and community capacity building towards the same. An innovative, user-friendly, community-based tool- 'Traffic Light Bioindicator Guide' is introduced among the people to assess impacts of climate change coupled with ongoing catchment transformations posing threats to the integrity of Fijian rivers and indigenous communities in remote villages of Polynesia. The chapter has reported the process for the pilot freshwater biomonitoring project, in which training was provided barring age, education and gender of the community people with the aim of developing local community practitioners. Successful trial of the 'Traffic Light Bioindicator Guide' is now included in the Fiji RiverCare Toolkit. Considering the success of this as a model process for the communities in Fiji and Oceania, it will further empower communities to mitigate climatic and non-climatic pressures on aquatic ecosystems and livelihood.

In the concluding chapter, Chap. 23, Mukherjee and Shaw discuss promotion of the nature-based next-generation built environment. Integration of BGI for resilience and implication of Pandemic Covid 19 on society, livelihoods and ecosystems are brought into perspective. How climate change scenarios with the context of nature-based solutions provide impetus in policy planning and favour BGI inclusion are argued. Due to proactive preparedness, loss of lives from disastrous events are reducing; yet resilience for infrastructure is increasingly becoming relevant and troublesome. Loss of property and services are now tremendously affecting countries' growth. Pro-nature adaptive governance, mainstreaming, spatial and economic planning and funding mechanism are important verticals for resilient future. The chapter ends with a forward-looking optimist note on mainstreaming BGI Blue-Green Infrastructure.

Expected Readership

The book has brought several aspects of nature-based resilience and adaptation within its covers. Eco-DRR, EbA and blue-green infrastructure are emerging as very relevant keywords for resilient and sustainable society. Policy, planning, science and technology investigation, innovations, implementations case studies are different aspects touched in three parts of the book. The major focus in the book is on Asia and Pacific, which has the highest number of disaster events, climate change induced extreme events. So any discussion from this region will bring useful learning to other parts of the world. The primary target of the book is any stakeholder who is part of the mission on nature-based resilience, be it policymakers, academia, development practitioners. Urban Planners, Practitioners, Funding Agencies and Policymakers will expectedly take keen interest in Part One of the Book for the collective knowledge into policy perspectives and insights in decision making on NbS. For the academicians, researchers and students from the domain of sustainable development, environment, disaster risk reduction and climate change studies, the Book offers dedicated Part Two to have a better idea on the current trend of research in the field and will provide advanced knowledge on tools and techniques. Local governing bodies, development managers and community leaders will get inspired from the Case Studies, discussed in Part Three, on nature-based Eco-DRR and EBA and specifically BGI. The book is expected to excite one and all people with a nature-centric attitude to advocate for the ecosystem-prioritized development for our future generations and other species. This will ensure better sustainable and resilient future for the Earth.

References

Akbari H, Menon S, Rosenfeld A (2008) Global cooling: increasing world-wide urban albedos to offset CO_2. Clim Change 94:275–286

Bloom DE, Canning D, Fink G (2008) Urbanization and the wealth of nations. Science 319(5864):772–775

Boruckea M, Moore D, Cranstonb G et al (2013) Accounting for demand and supply of the biosphere's regenerative capacity: the National Footprint Accounts' underlying methodology and framework Ecol Ind 24(2013):518–533

Cambridge University Press (2020) Cambridge Advanced Learner's Dictionary & Thesaurus. Cambridge University Press. Accessed on 22 Dec 2020

Cunningham F (2011) The virtues of urban citizenship. City Cult Soc 2(2011):35–44

Ebi KL, Hallegatte S, Kram T, Arnell NW, Carter TR, Edmonds J, Kriegler E, Mathur R, O'Neill BC, Riahi K, Van Vuuren DP, Zwickel T, Winkler H (2014) A new scenario framework for climate change research: background, process, and future directions. Clim Change 122(3):363–372

Eleftheriou V, Knieling J (2017) The urban project of HafenCity. Today's Urban and Traffic profile of the area. Executive summary of methodology and traffic research conducted in the region. Transport Res Procedia. 24:73–80

Elhacham E, Ben-Uri L, Grozovski J et al (2020) Global human-made mass exceeds all living biomass. Nature 588:442–444

Ely M, Pitman S (2012) Green infrastructure life support for human habitats. A review of research and literature: prepared for the Green Infrastructure Project, Botanic Gardens of Adelaide, Department of Environment, Water and Natural Resources

Foy D, Rogers J (2008) Efficiency in cities: a preliminary assessment of Potential. Living Cities Report. USDAE

Gandhi MK (1953) The students. Navajivan Trust Publication, Ahmedabad

Gandhi MK (1960) Trusteeship. Ahmedabad: Navajivan Trust Publication, p 3

Guha R (2001) The Prehistory of Community Forestry in India. Environ History 6(2):213–238 (Special Issue: Forest History in Asia. Oxford University Press on behalf of Forest History Society and American Society for Environmental History)

Hausfather Z (2018) Explainer: how 'shared socioeconomic pathways' explore future climate change. https://www.carbonbrief.org/explainer-how-shared-socioeconomic-pathways-explore-future-climate-change. Accessed on 30 Dec 2020

Hewitt RJ, Roger C, Dmitry VK, Klaus H (2020) Beyond shared socioeconomic pathways (SSPs) and representative concentration pathways (RCPs): climate policy implementation scenarios for Europe, the US and China. Informa UK Limited, trading as Taylor & Francis Group

Hirsch F (2005) Social limits to growth. Taylor & Francis, London

Honjo T, Takakura T (1990) Simulation of thermal effects of urban green areas on their surrounding areas. Energy Build 15(3–4):443–446

IAEG-SDGs (2017) Inter-agency and expert group on SDG indicators

IPCC (2001) IPCC third assessment report (TAR). Climate Change 2001

IPCC (2007) Towards new scenarios for analysis of emissions, climate change, impacts, and response strategies. IPCC Technical Summary, 2007. The Fourth Assessment Report (AR4). Climate Change 2007

IPCC (2014) The IPCC's Fifth Assessment Report (AR5). Climate Change 2014

Jacobs J (1992) The death and life of great American cities. Vintage Books, New York

Janke B, Herb W, Mohseni O, Stefan H (2021) Application of a runoff temperature model (MINUHET) to a residential development in plymouth, MN. Report prepared for Minnesota Pollution Control Agency. St. Paul, Minnesota

Kato S, Ahern J (2008) Learning by doing: adaptive planning as a strategy to address uncertainty in planning. J Environ Planning Manage 51(4):543–559

Maddison A (2001) The world economy: a millennial perspective. OECD Development Centre Studies

Marisol E, Udo N, Renaud FG, Sudmeier-Rieux K (2016) Ecosystem-based disaster risk reduction and adaptation in practice. Springer

Meadows DH, Meadows DL, Randers J, Behrens I, William W (1972) the limits to growth; a report for the club of Rome's project on the predicament of mankind. Universe Books, New York

Meadows D, Meadows D, Randers J (1992) Beyond the lmits. Earthscan Publications, London

Mebratu D (1998) Sustainability and sustainable development: historical and conceptual review. Environ Impact Assess Rep 18(6):493–520

Mesarovic M, Pestel E (1974) Mankind at the turning point. E. P. Dutton & Co., New York

Michaels G, Rauch F, Redding SJ (2012) Urbanization and structural transformation. Q J Econ 127(2):535–586

Mukherjee M (2002) From dawn to dusk—transportation of Rural women to and from the metropolis—Calcutta, India. In: Fernando P, Porter G (eds) Balancing the load—women, gender and transportation. Zed Books, London, pp 225–234

Mukherjee M, Takara K (2018) Urban green space as a countermeasure to increasing urban risk and the UGS-3CC resilience framework. Int. J. Disaster Risk Reduction 28:854–861

NFA (2012). Global footprint network 2011—The National Footprint Accounts. Global Footprint Network, San Francisco, California

Pant GB (1922) The forest problem in Kumaun. Allahabad: Kumaun Parishad, 1922; Reprint NainiTal: Gyanodaya Prakashan, 1983

Peet R, Hartwick E (1999) Theories of development. Guilford Press, New York, pp 85–86

Oke TR, Mills G, Christen A, Voogt JA (2017) Urban climate. Cambridge University Press

Our World in Data (2020) https://ourworldindata.org/urbanization#urban-populations-tend-to-have-higher-living-standards. Accessed on 20 Dec 2020

Polo M (2017) The travels of Marco Polo. Peacock Books, New Delhi

Robinson J (2004) Squaring the circle? Some thoughts on the idea of sustainable development. Ecol Econ 48:369–384

Schumacher EF (1973) Small is beautiful; economics as if people mattered. Harper & Row, New York

Revi A (2008) Climate change risk: An adaptation and mitigation agenda for Indian cities. Environ Urban 20(1):207–229

Rogelj J, Den Elzen M, Höhne N, Fransen T, Fekete H, Winkler H, Schaeffer R, Sha F, Riahi K, Meinshausen M (2016) Paris agreement climate proposals need a boost to keep warming well below 2 °C. Nature 534(7609):631–639

Roseen R, Ballestero T, Houle J, Briggs J, Houle K (2012) Water quality and hydrologic performance of a porous asphalt pavement as a storm-water treatment strategy in a cold climate. J Environ Eng 138(1):81–89

Shashua-Bar L, Hoffman ME (2000) Vegetation as a climatic component in the design of an urban street: An empirical model for predicting the cooling effect of urban green areas with trees. Energy Build 31(3):221–235

Spronken-Smith RA, Oke TR, Lowry WP (2000) Advection and the surface energy balance across an irrigated urban park. Int J Climatol 20:1033–1047
So AY (1990) Social change and development. Sage, Newbury Park, p 262
Stone B (2006) Developing design-oriented strategies to combat regional scale climate change. WIT Trans Ecol Environ 89
Subramanyam K (2013) Engineering hydrlogy, 4th edn. McGraw Hill Education, New Delhi
Sukhdev P, Wittmer H, Miller D (2014) The economics of ecosystems and biodiversity (TEEB): challenges and responses. In: Helm D, Hepburn C (eds) Nature in the balance: the economics of biodiversity. Oxford University Press, Oxford
Tagore R (1904) Shivaji Utsav. VisvaBharati Publication, Sanchayita
UN (2015) Transforming our world: the 2030 Agenda for sustainable development. A/RES/70/1. sustainabledevelopment.un.org.
UN (2020) Sustainable development goals report 2020
UN (2021) SDG monitoring and reporting toolkit for UN Country Teams. https://unstats.un.org/sdgs/unct-toolkit/. Accessed on 2 Jan 2021
UNDRR. (2020a) The human cost of disasters 2000–2019. Accessed on 15 Dec 2020
UNDRR (2020b) https://www.undrr.org/terminology
UNDRR-APSTAAG (2020) Status of science and technology in disaster risk reduction in Asia-Pacific, United Nations Office for disaster risk reduction—Asia-Pacific Science, Technology and Academia Advisory Group. https://reliefweb.int/report/world/status-science-and-technology-disaster-risk-reduction-asia-pacific-2020
United Nations Development Programme (UNDP) and the World Bank Group (WBG) (2016) Ban-Ki-Moon. From Transitioning from the MDGs to the SDGs
UNEP (2019) Intergovernmental science-policy platform report on biodiversity and ecosystem services (IPBES)
UNEP and FAO (2020) UNEP and FAO factsheet. https://wedocs.unep.org/bitstream/handle/20.500.11822/30919/UNDecade.pdf. The UN Decade on Ecosystem Restoration 2021–2030
UNFCCC.(2015). Paris Agreement. https //unfccc.int/process-and-meetings/the-paris-agreement/the-paris-agreement. Accessed on 11 Jan 2021
UN World Urbanization Prospects (2020) https://population.un.org/wpp/Publications/Files/WPP2019_Highlights.pdf. Accessed on 19 Dec 2020
Uttarakhand Forest Department (2007) The magnificent forests of Uttarakhand. ISBN 81-903988-0-6
Uy N, Shaw R. (2012) Ecosystem based adaptation, community environment and disaster risk management, vol 12. Emerald Publisher, London, p 263
Van Vuuren DP, Edmonds J, Kainuma M, Riahi K, Thomson A, Hibbard K, Hurtt GC, Kram T, Krey V, Lamarque J-F, Masui T, Meinshausen M, Nakicenovic N, Smith SJ, Rose SK (2011) The representative concentration pathways: an overview. Clim Change 109(1–2):5–31
World Bank (2020) https://databank.worldbank.org. Accessed on 21 Dec 2020
Worldometer (2020) https://www.worldometers.info/world-population/#ref-1. Accessed on 19 Dec 2020
World Commission on Environment and Development (1987) Our common future. Oxford University Press, Oxford
Xiao Q, McPherson EG, Ustin SL, Grismer ME, Simpson JR (2000) Winter rainfall interception by two mature open-grown trees in Davis, California. Hydrol Processes 14(4)

Part I
Policy Analysis, Policy Framing and Recognition of Nature-Based Solution

Chapter 2
Ecosystem-Based Adaptation (EbA) in the Hindu Kush Himalaya: Status, Progress and Challenges

Sunita Chaudhary, Basant Raj Adhikari, Pashupati Chaudhary, Tashi Dorji, and Renuka Poudel

Abstract Ecosystem-based Adaptation (EbA) has been gaining attention in science, policy and practice as an effective way to address climate change and contribute to sustainable development. In Hindu Kush Himalaya (HKH), EbAs are implemented to enhance resilience of mountain communities to the harsh realities of climate change. However, very little documentation exists on nature and progress of EbA in the region, which are often fragmented and scattered. We analyzed the status, progress, benefits and challenges in EbA implementation. EbAs are focused on restoration (17%), mainstreaming in policy and plans (17%), ecosystem conservation (14%), flood risk management (12%), livelihoods (10%), capacity building (10%) and ecological risks assessment (7%). Though EbA varies across the countries, ecosystem conservation and livelihoods diversification is the focus. Major drivers of changes considered are climate change, floods, drought and landslides. Improved resilience through restoration, capacity building, better networking and better wellbeing are some of the notable benefits. However, awareness and mainstreaming of EbA in policies and plans are limited. Limited cooperation among the countries and stakeholders and short-lived donor-driven agendas are also the challenges. An effective and impactful EbA requires an integrated approach encompassing different sectors with vertical and horizontal cooperation and collaboration at the regional scale.

Keywords Ecosystem services · Adaptation · Aichi targets · Challenges · Barriers · Hindu Kush Himalayas

---

S. Chaudhary (✉) · T. Dorji · R. Poudel
International Centre for Integrated Mountain Development (ICIMOD), Kathmandu, Nepal
e-mail: sunita.chaudhary@icimod.org

B. R. Adhikari
Institute of Engineering, Tribhuvan University, Pulchowk Campus, Kathmandu, Nepal

P. Chaudhary
Agriculture and Forestry University, Rampur, Chitwan, Nepal

© The Author(s), under exclusive license to Springer Nature Singapore Pte Ltd. 2021
M. Mukherjee and R. Shaw (eds.), *Ecosystem-Based Disaster and Climate Resilience*, Disaster and Risk Research: GADRI Book Series,
https://doi.org/10.1007/978-981-16-4815-1_2

2.1 Introduction

The global climate change is an established phenomenon today (IPCC 2014). With the rapidly changing climate, the number and frequency of natural hazards and disasters have been increasing across the globe destroying life and properties (Deo and Garner 2014; Timalsina and Songwathana 2020). Climate change is regarded as the major driver of biodiversity loss and degradation of ecosystem services with significant impacts on millions of people dependent on nature (MEA 2005). To deal with the increased disasters induced by climate change, the engineering structures such as dikes, flood gates, dams are constructed (Jones et al. 2012). This is especially the case in developed countries that can afford financial and technical resources. In developing countries, the physical structures requiring huge investments are not always feasible, as priorities are more on dealing with poverty and development (Reid and Adhikari 2018). There has been, therefore, a growing realization that the Ecosystem-based Adaptation (EbA) could be an ideal adaptation solution, especially for developing countries. EbA is a cost-efficient and effective approach to deal with climate change and a pathway towards sustainable development (Swiderska et al. 2017). EbA, as defined by Convention on Biological Diversity (CBD), is the *use of biodiversity and ecosystem services to help people adapt to the adverse effects of climate change as part of an overall adaptation strategy* (CBD 2009).

EbA has gained increased attention in science, policy and practice since the mid-2000s. In 2001, CBD recognized the role of biodiversity and ecosystems for adaptation and acknowledged that ecosystem approaches could be the foundation of mitigation and adaptation. CBD defined the term EbA in 2008 and further elaborated to focus on 'sustainable management, conservation and restoration of ecosystems, as part of an overall adaptation strategy that takes into account the multiple social, economic and cultural co-benefits for local communities (CBD 2010). These are well reflected in the Strategic Plan for Biodiversity 2011–2020 which was adopted in 2010 with 20 Aichi targets by the Conference of the Parties to the Convention on Biodiversity (CBD) (CBD 2010). Since then, EbA has been increasingly used in science and policy documents (Chong 2014) and projects are implemented across the globe (Monty et al. 2017). For example, UN Environment World Conservation Monitoring Centre (UNEP-WCMC) in collaboration with International Union for Conservation of Nature (IUCN) and International Institute for Environment and Development (IIED) implemented projects in 13 sites in 12 countries around the world between 2015 and 2018 to develop policy guidance for EbA implementation (Reid and Adhikari 2018). Similarly, countries are mainstreaming and implementing EbA adaptation to climate change (Seddon et al. 2016a, b). Some countries such as Germany have already integrated EbA in municipal climate change strategies (Zolch et al. 2018).

In the Hindu Kush Himalaya (HKH), EbAs are implemented to help the mountain communities adapt to the harsh realities of climate change and shape their wellbeing. The HKH covers 4.2 million square kilometres across eight countries: Afghanistan, Bangladesh, Bhutan, China, India, Nepal, Myanmar and Pakistan (see

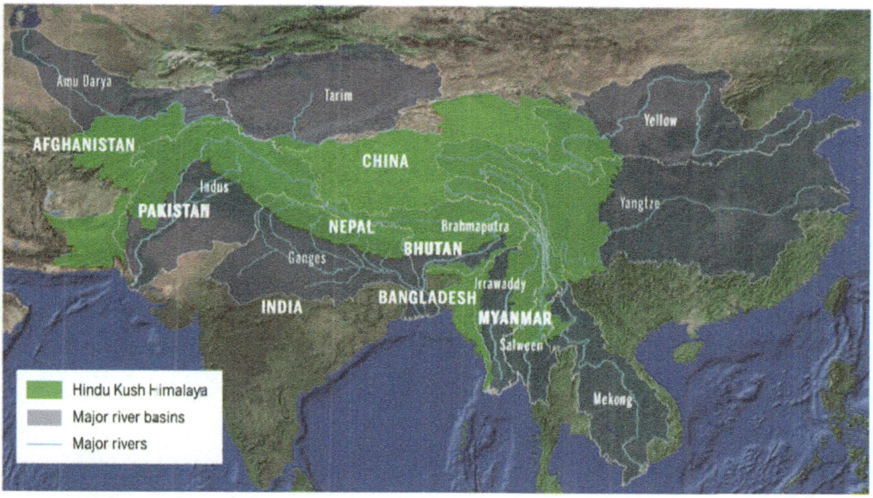

Fig. 2.1 The Hindu Kush Himalaya Region with major rivers and river basins. *Source* Sharma and Molden 2019). Map used with permission

Fig. 2.1) (Wester et al. 2019). The region, with hundred mountain peaks over 6000 m, is the source of ten major rivers of Asia. It hosts four global biodiversity hotspots (Mittermeier et al. 2004) and has diverse cultures, languages and traditional knowledge systems. This diversity provides ecosystem services that directly support the livelihoods of 240 million people in the region (Wester et al. 2019). However, the region is facing rapid changes such as climate change, land use, globalization, unplanned development and urbanization with implications on the environment and people living in and beyond the region (Wester et al. 2019). Actions are in place at local, national and regional scales to sustain the Himalayan ecosystems and improve livelihoods. EbA is one of those actions gradually emerging in the region.

A growing number of EbA are in operation to increase the resilience of the socio-ecological system in HKH region. For example, EbA in Bhutan and Myanmar focus on building resilience to the growing urban system, while the projects in Nepal are focused on enhancing capacity, knowledge and technology for resilient mountain ecosystems and their people. There are also a handful of EbA related projects in HKH countries such as China, India and Pakistan (IUCN 2020) with attempts to integrate the approach in their policy and strategies (Reid and Adhikari 2018). Nepal has briefly mentioned the term 'ecosystem services' in its National Adaptation Plan of Action (NAPA) (Chaudhary and McGregor 2018). In general, documentation of such approaches are very fragmented and scattered. The limited knowledge on the progress and effectiveness of EbA has been hindering the integration of the approach into policy and practice (Ojea 2015). In this context, this review aims to analyze the status of EbA, its types, types of disaster targeted and the progress made in HKH countries. In doing so, we aim to discuss the challenges and benefits related to EbA and recommend actions for further socio-ecological resilience in the region. The chapter is guided by the following research questions:

1. What is the nature of EbA in the Hindu Kush Himalaya (HKH)?
    (a) What types of EbA are in practice?
    (b) What are the similarities and differences in EbA across the countries?
    (c) What types of drivers of change and hazards are considered for EbA?
2. How progressive has been EbA in the HKH countries?
3. What are the benefits and challenges of EbA in the region?
4. What are the recommendations for science, policy and practice of EbA in the region?

2.2 Materials and Methods

This review is based on a systematic review of literature based on an approach (Fig. 2.2) proposed by Brink et al. (2016) and Triyanti and Chu (2018) in the Hindu-Kush Himalaya. This review approach has been already applied for a systematic literature review on EbA for urban areas and Eco-DRR.

2.2.1 Data Collection: Literature Search, Screening and Categorization

The data collection consists of three stages: The first stage is collection of literature, followed by screening and cleaning in the second stage and categorisation of the cleaned data for analysis in the third stage.

Literature collection: Published literature of three different types: peer-reviewed articles, Sixth National Reports submitted to CBD and others (reports, book chapter, booklet and issue brief) were collected. Scopus and Google Scholar, both advanced academic search tools (Chaudhary et al. 2015), were used for academic literature. While the non-academic literature were searched through Google search engine. This included reports, websites, books, issue briefs and brochures. The literature search was guided by a set of criteria:

1. Keywords: EbA, HKH, ecosystem services, country name, DRR, Eco-DRR and early warning as follows:
    (a) Ecosystem-based adaptation (EbA), ecosystem services, Himalaya and/or mountain
    (b) EbA, Hindu Kush Himalaya (HKH) and/or country name (like Afghanistan, Bangladesh, Bhutan, China, India, Nepal, Myanmar and Pakistan)
    (c) EbA, DRR, country name

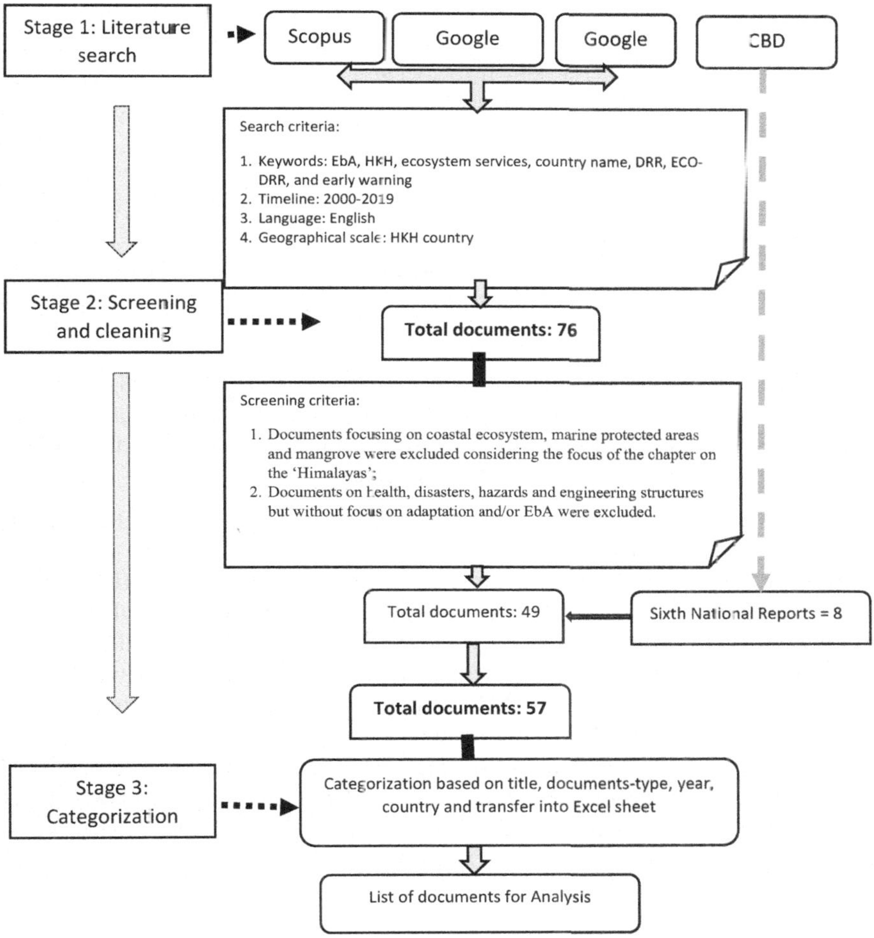

Fig. 2.2 Methodological framework of the review

   (d) EbA, ECO-DRR, country name
   (e) EbA, early warning system, country name.

2. Timeline: As the term EbA was coined in 2000 (Seddon et al. 2016a, b), a timeline from 2000–2019 was used.
3. English language.
4. Geographical scale: Hindu Kush Himalaya.

With this, altogether 76 documents were collected.

In the second stage, screening and cleaning of the literature were done by screening the title and abstract of 76 documents using the following pre-defined criteria:

Table 2.1 Composition of literature

Document	No.
Article	33
Book chapter	3
Booklet	3
Thesis	1
Report	8
Science Brief	1
Sixth National Reports to CBD	8
Total	57

Documents focusing on coastal ecosystem, marine protected areas and mangroves were excluded considering the focus of the chapter on the 'Himalayas';
Documents on health, disasters, hazards and engineering structures that did not focus on adaptation and/or EbA were excluded.

After screening, 27 documents were excluded. This gave us a total of 49 documents for further review and analysis.

Besides, the sixth national reports (6NRs) submitted to CBD by each HKH country were considered for review. The sixth national report provides a final review of progress in the implementation of the Strategic Plan for Biodiversity (2011–2020)[1] including the Aichi Targets to implement the Convention. The plan has five strategic goals with 20 Aichi Targets (CBD 2020). Altogether, including CBD reports, a total of 57 documents were used for a thorough review.

In the third stage, the literature were categorized based on title, country, published year, document type, abstract and project name with donors (if any) in an excel sheet for analysis. The documents included peer-reviewed articles, book chapters, booklets, theses, reports, science briefs and CBD reports. Table 2.1 shows the composition of literature.

2.2.2 Data Analysis

The literature review included a descriptive analysis and a qualitative content analysis (i.e. thematic analysis). Thematic analysis is a qualitative approach that focuses on identifying, analyzing and interpreting patterns of meaning (i.e. theme) in qualitative data (Castleberry and Nolen 2018).

Descriptive analysis: Each document was thoroughly reviewed and information on the country, year of publication, type of EbA and hazards were collected and

[1] The Strategic Plan for Biodiversity including 20 Aichi targets were adopted during the tenth meeting of the Conference of the Parties, held from 18 to 29 October 2010, in Nagoya, Aichi Prefecture, Japan for the 2011–2020 period. The plan is an overarching framework on biodiversity conservation and management (CBD 2010).

saved in an Excel sheet. Then, a detailed analysis guided by research questions was performed in excel. This gave us findings on the status of EbA (country-wise), type and trend of EbA publication over the years.

Thematic analysis: The contents of each document were analyzed through coding of themes. The themes were guided by the research questions: Benefits of EbA, Challenges and Barriers of EbA and recommendations. The coded themes were collected, analyzed and interpreted with reference to the research questions.

Sixth National Report (6NRs) to CBD: The eight 6NRs to CBD submitted by all HKH countries were analyzed to track the progress made by each country towards EbA. Lo (2016) thoroughly reviewed the linkages between the Aichi Targets and EbA and Eco-DRR and identified the Aichi Targets (5, 7, 10, 11, 13, 14 and 15) closely linked to EbA and Eco-DRR. A brief snapshot on the links is given in Annex 1. Following Lo (2016), we selected the Aichi Targets 5, 7, 10, 11, 13, 14 and 15 to analyze the country's progress towards EbA in the region. Countries report their progress on Aichi Targets qualitatively with different levels of progress (see Table 2.2). For each category, we assigned a term and value (very good = 3, good = 2, fair = 1 and low = 0).

Following the table, the progress reported by each country was analyzed descriptively in Excel and interpreted accordingly (see Sect. 3).

Table 2.2 Level of progress category with terms and values assigned

Level of category (as per CBD)	Terms used	Value assigned
 On track to exceed the target	Very good	3
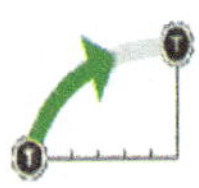 On track to achieve the target	Good	2
 Progress towards target but at an insufficient rate	Fair	1
 Moving away from the target	Low	0

2.3 Findings

2.3.1 Status of EbA in the Hindu Kush Himalaya

The status in the HKH is shown by the type of EbA, type of hazards considered and similarities and differences in EbA activities across the region.

What types of EbA are in the region?

EbAs in the Hindu Kush Himalaya (HKH) are mostly focused on 12 different activities, aiming to maximize the multiple benefits and increase the resilience of both nature and society to climate change and disasters (Fig. 2.3). Most of the EbAs are focused on restoration (17%) and mainstreaming in policy and plans (17%). This is followed by ecosystem conservation (14%), flood risk management (12%) and livelihoods (10%), capacity building (10%) and ecological risks assessment (7%). Sustainable fishery, traditional knowledge and invasive species management were also considered important for increasing resilience in the region. This shows that maintaining healthy ecosystems through conservation and management play an important role in adaptation. Similarly, the assessment and management of different

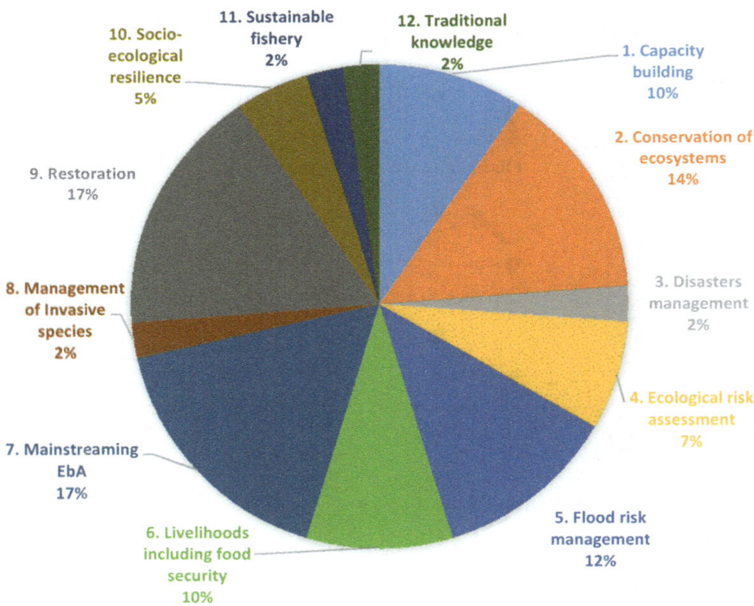

Fig. 2.3 Ecosystem-based Adaption (EbA) activity in the Hindu Kush Himalaya region, as a percentage of total number of activities, as reported in different literatures

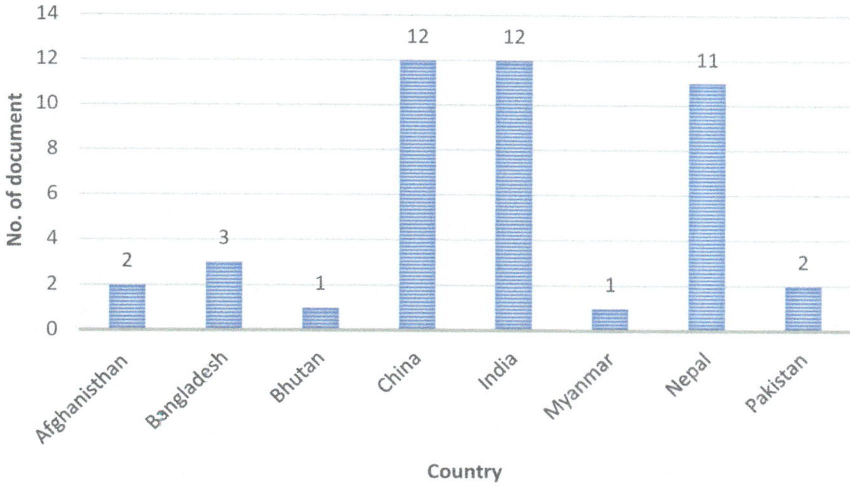

Fig. 2.4 No. of literature on EbAs across the HKH country

risks and threats across sectors were considered important. For instance, about one-fourth of the EbA activities are focused on assessing risks related to flood, disasters and ecology and their management (Fig. 2.3).

What are the similarities and differences in EbA across the HKH countries?: Country-wise EbA.

The literature are unevenly distributed across the region (Fig. 2.4). Altogether 44 literature are focused on EbA in the region. Among the countries, China (12) and India (12) have the highest focus on EbA, followed by Nepal (11). Pakistan, Bhutan and Myanmar have limited EbA literature. Across the countries, ecosystem conservation and livelihood diversification are the major focus for all EbAs. In China, other EbAs are focused on restoration of degraded lands, risks and vulnerability assessment, urban floods and agrobiodiversity with a particular focus on food security. While mainstreaming in policy and plans, Eco-DRR, traditional knowledge, smart city and groundwater recharge are focused in India. In Nepal, activities on capacity building, community-based disaster management, reforestation and crop diversification are other EbAs. Similarly, the socio-ecological resilience of agro-pastoral community in Afghanistan, sustainable fishery in Bangladesh, Eco-DRR in Myanmar and flood management in Pakistan are some of the additional activities. Considering the threats considered for EbA across the region, we also assessed the types of threats.

What types of hazard risks are considered in EbA in the region?

Altogether, ten different types of hazards have been reported for EbA in the Hindu Kush Himalaya (see Fig. 2.5). The analysis shows climate variability and change as the top hazard for EbA in the region, by about 38%. This is followed by floods

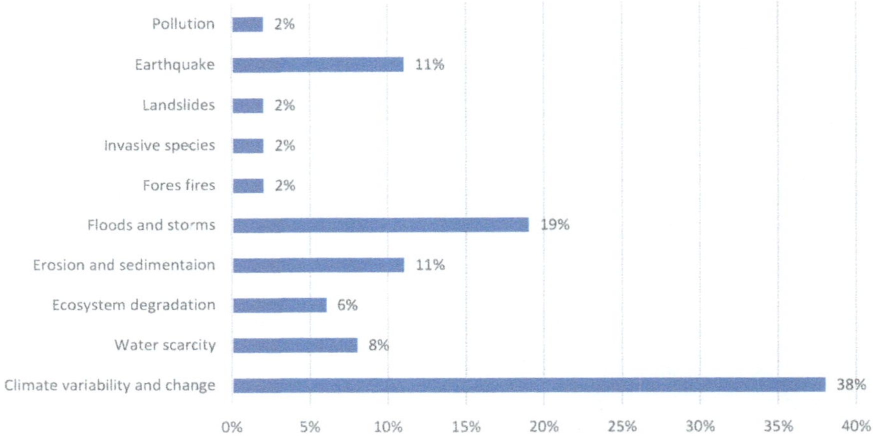

Fig. 2.5 List of hazards reported in EbA literatures, in percentage with regards to total number of reported hazards

and storms (19%), earthquakes (11%) and erosion and sedimentation (11%). Water scarcity (8%) and ecosystem degradation (6%), pollution, landslides, invasive species and forest fires are also threatening the region.

2.3.2 Progress of Countries Towards EbA in the Region

The review of sixth national reports to CBD of HKH countries is the basis of analyzing the progress made by each country towards EbA. The Aichi targets analyzed included the targets 5,7,10,11,13,14 and 15 as these targets are closely linked to EbA (as discussed in Sect. 2.2). For each target assessed, most of the HKH countries show good and fair progress to EbA (see Fig. 2.6). China in particular shows good progress towards all the targets related to EbA, followed by India, Bhutan and Nepal. Bhutan is 'very good' for target 11 as the country has exceeded the target of protected areas coverage. India has also been 'very good' for protected areas coverage and 'good' for all other EbA related targets. Similarly, Nepal exceeding its protected areas coverage shows very good progress in target 11 and report 'good' in other targets. Afghanistan and Myanmar show good to fair progress in all targets related to EbA. Pakistan shows low progress in targets 10 and 14 but reported 'good' for other EbA related targets. This indicates that the countries are trying their best in terms of increasing their resilience to varied hazards including climate change.

The progress is also 'good to fair' for most of the targets (see Fig. 2.7). Target 15, in particular, focusing on restoration for resilience to climate change is 'good' followed by target 11 on protected areas coverage. Except Pakistan, target 14 focusing on

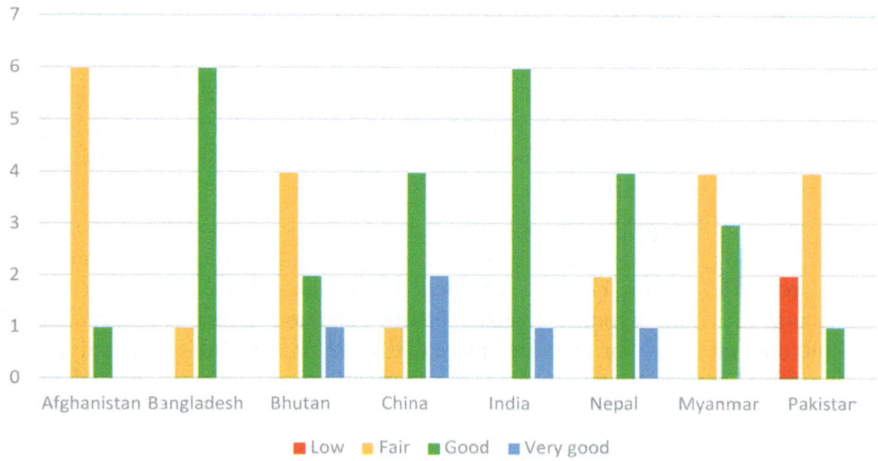

Fig. 2.6 Progress of countries towards Aichi targets related to EbA. Values are shown in percentage of reports identifying levels of progress for each target based on sixth National Reports to CBD

Fig. 2.7 Progress on Aichi targets linked to EbA in the Hindu Kush Himalaya

increasing the capacity of ecosystems to provide services for sustainable livelihoods is 'good' for all the countries.

2.3.3 Benefits of EbA in the Region

Besides the analysis of progress by HKH countries for Aichi targets related to EbA, we also analyzed the EbA benefits. The analysis shows significant benefits to both people and ecosystems arising from EbAs. Some of the notable benefits are thematically described below:

1. Human wellbeing: Livelihoods support and its diversification is the core element of EbA in the region. EbAs contribute to the overall wellbeing of people with increased economic benefits. The benefits are through selling of vegetables, honey, cardamom products, banana, bamboo and non-timber forest products of the communities. For instance, Increased water availability and equal access to water, increased tourism, improved agricultural productivity through agro-forestry practices and better air and water quality through plantation in the abandoned areas and the river banks are reported to contribute to better health of people.
2. Capacity building and networking: Capacity building is also one of the most important components of EbA. Capacity building in terms of enhancing marketing skills, access to markets, networking with stakeholders and knowledge about value chains of products is important for enhancing economic benefits. Similarly, knowledge to manage natural resources sustainably and managing community-based institutions are the focus of the EbAs. Besides, the leadership skill of community people is important for many community-based adaptation initiatives. The activities are the reflection of the type of hazards in the countries. The capacity building of deprived communities has been prioritized to improve their livelihoods and hence build resilience.
3. Traditional knowledge: Recognition of indigenous communities and their traditional knowledge for ecosystem management and resilience building have been promoted in EbAs. Many EbAs have particular arrangements for marginalized communities, especially the poor and indigenous people. For instance, the Chepangs (an indigenous[2] community of Nepal) are being provided with support to promote their agriculture productivity. Similarly, the feminization of homestay businesses has been prioritized for women groups and female-headed households.
4. Improved resilience: Crop diversification, improved irrigation systems and promoting water recharging systems are some of the priorities for building resilience. Activities are also focused on promotion of strong and cost-effective means such as agroforestry practices, minimizing flood risks and agrobiodiversity to deal with the impacts of changing climate and improve sustainability against extreme weather events, pests, weeds and other hazards.
5. Ecosystem health and functions: The benefits to ecosystems, in terms of restoring, managing and enhancing the capacity of ecosystems to withstand

[2] The indigenous ethnic community is a tribe/community native to a particular area with its own mother tongue, traditional culture and egalitarian social structure. They do not fall under the conventional Hindu hierarchical caste structure (GoN 2009).

different stressors, as well as provide services for local communities are the benefits arising from EbAs. The analysis showed notable improvement of *ecosystem health and its functions* through EbA. For instance, the land rehabilitation interventions such as restoration of forest and river through green belt, gulley control in the Panchase Mountain area protected land from degradation and conserved more than 50 hectares of land (UNDP 2015).

6. Restoration: Ecosystems conservation and management have been the foundation of many EbAs in the region. Degraded ecosystems are managed through restoration and sustainable land management. For instance, UNEP's project planted 500,000 seedlings for forest enhancement in Nepal (IUCN 2020). The ecosystem restoration is reported to improve ecosystem health and its functions such as soil and nutrient management, biodiversity and genetic diversities, surface accretion, carbon sequestration, water quality improvement and reducing surface water run-off during storms. For instance, wetland protection and rehabilitation through EbA in India is reported to increase water storage potential, mitigate floods and wise use of water during droughts (Dhyani et al. 2018). Moreover, water quality improvement, groundwater recharge, reducing surface water run-off during storms are also important.
7. Disasters risk reduction: EbA has always been playing an important role in disaster risk reduction. Low-cost and low-tech approaches are especially feasible in the mountainous areas (see Box 1). The restored ecosystems are an integral part of protecting infrastructure and enhancing human security, acting as natural barriers and hence mitigating the impact of and aiding recovery from many extreme weather events and disasters. Biodiversity and genetic diversities were reported to be conserved through an EbA approach. This could help biodiversity especially engendered species to adjust to changing climatic conditions. The successful implementation of EbAs controls soil erosion and provides an alternative livelihood option. For example, an intervention by IUCN Nepal in western Nepal has implemented bio-engineering techniques for 'eco-safe roads' concept to control the landslide. This project also provided different plant species for the protection of slope land erosion and now communities are harvesting broom grass for use as fodder and for sale and are earning 20,000 NPR per year, per kilometre (Monty et al. 2017).

Box 1: Landslide Early Warning Systems for ECO-DRR in Nepal

Landslide occurrence in the Himalayas is a complex and common phenomenon causing many lives and properties. Landslide mitigation using hard engineering techniques such as retaining wall, drainage, rock bolting and iron netting are common examples in the region. However, these techniques are not always feasible due to high cost and topographic challenges, i.e. steep terrain, high altitude, strong monsoon, inaccessibility of areas. Therefore, Landslide Early Warning System (LEWS) is one of the best non-structural mitigative measures in those terrains where the local population at risk will be benefitted by increasing their awareness and enhancing their preparedness. The history of LEWS in Nepal is not so long. Water Induced Disaster Prevention Technical Centre (DPTC), Government of Nepal installed rain gauge, Piezometer, moving pegs, Tiltmeter, extensometers in the Km 19 landslide along the Kathmandu-Trishuli road, central Nepal in 1993 for the landslide monitoring. Some regional LEWS based on rainfall threshold is installed in the Nepal Himalaya but these systems did not consider the threshold based on physical movement of the landslide (JICA 2009; Dahal and Hasegawa 2008). One of the successful examples can be taken from Thapa and Adhikari (2019) where they have explained how Government of Nepal has installed a community-based LEWS. Some of the major features and mechanisms are described below:

The community-based low-cost and low-tech has considered all three parameters: rainfall, displacement and soil moisture content. This system was set up in the Mehele landslide, Dolakha, Nepal (N 27° 43′ 22.54″; E 86° 03′ 49.11″; 1952 m) on 28th May 2018. The system consists of Arduino Mega controller, flash memory of 256 KB, SRAM 8 Kb, EEPROM 4 KB, Click Speed 16 MHz, Click Speed 19 MHz and an LCD 16 × 2 display with 50 W power supply. The system set up a threshold of 60 mm in 24 h or cracking increases equal or greater than 30 cam and moisture content in the soil exceed more than 60%. The LEWS worked perfectly when the landslide occurred at 11 pm on 23rd August 2018. The local community heard the siren and prepared for evacuation. Altogether 495 people from 117 households benefited from this system and most of them are from marginalized population. This LEWS has set up a landmark in the Nepal Himalaya to implement the ECO-DRR approaches for the rural-mountainous communities.

2.3.4 Challenges of EbA in the Hindu Kush Himalaya

The implementation of EbA in the region is challenged by varied issues across sectors and scales. One of the major challenges is the **limited coordination and cooperation** within the government and among different stakeholders. For instance, limited cooperation was reported between two state governments in India (Delhi government and Uttar Pradesh government) during the implementation phase (Singh et al. 2013). In this regard, the **multi-stakeholder engagement** including local communities and private sectors is quite important. Equally important is environmental governance in the region which has been regarded weak in terms of managing institutions and stakeholders. The limited capacity of nodal institutes is one of the challenges for better environmental governance. Similarly, limited regional cooperation among the countries to address hazards and disasters that often cross borders is important.

Climate and water-induced disasters can cross borders requiring collective action across countries and communities for disaster resilience (Molden et al. 2017). Effective cooperation and consultation with different sectors across the scales are of utmost importance. However, limited consultation with multi-stakeholders is challenging for EbA measures. Many EbA projects have failed to ensure participation of relevant stakeholders. For example, the top-down selection of beneficiaries for afforestation projects in Bangladesh has ignored the specific needs of local communities promoting equity and justice issues. As EbA is multi-dimensional in nature, the **integration of different disciplines and working with different sectors** is a challenge. Lack of coordination amongst DRR, climate change and natural resource management policies and interventions are some of the observed barriers for EbA effectiveness in the region.

EbA should involve integrated sectoral approaches, including all relevant sectors such as forestry, fisheries, agriculture and water resources that help the community adapt to climate change and other disasters. The ecological risks and societal vulnerabilities are often analyzed in isolation and without necessarily linking ecological, social and economic aspects that exacerbate vulnerabilities. EbA approaches are often narrowly focused, failing to integrate conservation and development goals. For example, the Tidal River Management project in Bangladesh aimed to trap sediment within polders and restricted fishing and farming for the local communities who are dependent on their subsistence livelihoods (Saroar et al. 2019).

The EbA measures are **donor-driven and often lack funds** for implementation, monitoring and management after project period. As such, major projects are focused on mainstreaming EbA in national policies and plans for sustainability. However, the limited awareness, funding and capacity to integrate EbA in policy and plans are the challenges for mainstreaming EbA. Most of the stakeholders are unaware of the EbA concept, its benefits and adequate understanding of the local conditions. For instance, the private sectors are not aware of the importance of floodplain and wetland ecosystems which has been challenging for EbA counterparts to initiate conservation efforts. Limited knowledge and capacities to apply environmental tools for DRR and mainstream into development planning are other challenges for EbA in the region.

2.4 Discussions

While EbA is gaining worldwide recognition in climate change adaptation and resilient community development (Nalau et al. 2018), the progress in its widespread use is still slow, particularly in the HKH (Bourne et al. 2016). The current EbA activities in the region cover a variety of topics related to process, goal, objective and outcomes. Several EbA projects have a focus on the integration of traditional knowledge, capacity building and mainstreaming of the EbA approach in government policies. Ali Shah et al. (2019) reported that EbA is part of the activities of a majority of smallholder farmers in Pakistan and they suggest its wider application through non-government organization effort in fostering farmer-to-farmer information sharing and appropriate government policy support.

A community-based approach coupled with enhancing social capital and institutional building is crucial for the success and sustainability of EbA (ibid). Participatory plant breeding and community-supported agriculture project in China has applied EbA, which emphasizes four key issues: (i) effectiveness for human societies, (ii) effectiveness for the ecosystem, (iii) financial and economic effectiveness and (iv) policy and institutional issues (Song et al. 2015; Reid and Zhang 2018). The promotion of organic agriculture at national and community scale especially in Bhutan, India and Nepal is also EbA focused. These frameworks and approaches can be useful for other ongoing and future EbA projects across the region.

Climate-induced hazards qualifying EbA are manifold of which floods and storms rank the top. Part of HKH is the hotspots for natural disasters, floods and glaciers lake outburst in particular, with immense impacts on people's lives, livelihoods and economies at a transboundary level (Yusuf and Francisco 2009) since the region is home to several transboundary landscapes and rivers including Ganges, Bramhaputra and Meghna (Mirza 2011). In August 2017, floods and landslides caused by torrential monsoon rains affected almost 41 million people and killed over 900 in Bangladesh, India and Nepal combined (UNOCHA 2017). River basin approach with upstream and downstream linkages would help the countries and communities deal with such disasters. Similarly, the mountain ecosystems are extremely vulnerable to climate change with impacts on people and their livelihoods (Sharma et al. 2009). This calls for a transboundary, regional and inter-country coordinated and collaborative effort. To promote transboundary initiatives, cross-learning and joint undertakings of EbA activities are important. It will require fair and transparent dialogue, followed by cooperation and coordination among the countries at the policy level. Countries can also cooperate on the documentation and exchange of case studies and good practices from the region. This can be further enhanced by north–south and south-south cooperation on science, technology and innovation for better disaster resilience. For instance, regional cooperation among the upstream and downstream countries can be strengthened to maximize benefits such as irrigation and hydropower, while minimizing adverse risks of floods and landslides. It can integrate scientific, economic, social and ecological knowledge to support decision-making (Molden et al. 2017). While there are some similarities in approaches, goals and outcomes, there is differentiated progress in EbAs across countries. Therefore, cross-country learning, among other approaches, can play a pivotal role in building capacity and scaling up good practices. Capacity building, including cross-country learning, is necessary to address the following gaps: (i) inadequate technical knowledge and capacity regarding the designing and implementation of projects; (ii) lack of capacity to plan, (iii) limited technical capacities within DRR sector on implementing natural resource management strategies; and (iv) limited understanding, research and public awareness of the benefits of EbA. The Paris Agreement gives due emphasis on equity to ensure vulnerable groups, communities and ecosystems receive a priority in climate change actions (UNFCCC 2015), so it is important to duly consider gender equality and social inclusion during capacity building and project implementation in future.

When looked through an Aichi Target lens, the majority of countries have made either fair or good progress in implementing EbA related targets. Moreover, all the targets linked with EbA are also moderately good in terms of achieving set goals. Since the Fifteenth Meeting to the Conference of the Parties of the Convention on Biological Diversity (CBD CoP15) is approaching, it would be good to raise the gaps in achieving targets for different countries so that the post-2020 biodiversity framework duly consider important issues that remained poorly attempted so far. The study highlighting the challenges and barriers aims to contribute to the post-2020 biodiversity framework.

Awareness about EbA among government officials and development partners is equally important to accelerate the integration of the approach into policies and strategies. The future of EbA planning, implementation and evaluation will require a combination of "top-down" and "bottom-up" approaches to ensure sustainable outcomes across various levels of implementation. Moreover, an integrated approach to social protection, DRR and climate change adaptation is necessary (Schipper and Pelling 2006; Mercer 2010; Guha-Sapir et al. 2013; Kundzewicz et al. 2014; Bakker and Duncan 2017) with its clear linkage with the nature-based solutions (Cohen-Shacham et al. 2016), which will require an inter-disciplinary and multi-institutional approach. Vertical and horizontal cooperation and collaboration are key to fostering inter-disciplinary and multi-intuitional approaches.

Research work is also limited in the region. So, it is important to take stock of research work conducted in the past and identify research gaps. Future studies are essential to examine how existing EbAs are contributing to relevant national and global initiatives and what further actions are needed to improve efficiency and effectiveness. This will give insights on the ways of integrating EbA activities in different strategies, goals and sectoral approaches. Future work is necessary to evaluate the benefits of EbA projects by assessing the contribution of EbA on adaptive capacity, resilience and reduced vulnerability of human beings in the face of climate change along. It is also essential to examine the co-benefits and impact of EbA on restoring, maintaining or enhancing the capacity of ecosystems in producing useful services and combating climate change impacts (Reid and Zhang 2018). Equally important is to appraise cost-effectiveness, economic viability along social, institutional and political issues pertaining to effective EbA implementation (Seddon et al. 2016a; Reid et al. 2017; Reid and Zhang 2018).

2.5 Conclusion

We systematically reviewed EbA literature in the Hindu Kush Himalaya. EbA practices in the HKH are focused on ecosystem conservation, integration of indigenous knowledge, capacity building and mainstreaming EbA and disaster management. This approach has been applied to address climate change, floods, drought and landslides, however, the goal and outcomes of this approach differ from country to country. For example, China and India are comparatively advanced in capacity building and restoration of ecosystems, while Nepal is ahead in equity issues. Nevertheless, the majority of countries have made fairly good progress in implementing EbA related targets, to achieve their set goals. The benefits of EbA are multi-dimensional. They improve the resilience of ecosystems and communities by enhancing people's wellbeing and maintaining ecosystem's health. EbAs are particularly important in diversifying livelihoods of people and building their capacity to deal with disasters and hazards. However, limited cooperation among the countries, coordination across different sectors and short-lived donor-driven projects are some of the challenges. More efforts are required to improve the socio-ecological resilience of the countries by creating awareness, effective mainstreaming of EbA in national and regional plans and policies. As such, we argue to have an integrated approach encompassing different sectors and disciplines with clear linkages with the nature-based solution for effective and impactful EbA in the region. This would require vertical and horizontal cooperation and collaboration among different stakeholders. Regional cooperation among the countries to share knowledge, best practices and technology is required for disaster resilience. The future research could focus on analysis of contribution of existing EbAs and their effectiveness in improving the resilience of socio-ecological systems.

Acknowledgements ICIMOD gratefully acknowledges the support of its core donors: the Governments of Afghanistan, Australia, Austria, Bangladesh, Bhutan, China, India, Myanmar, Nepal, Norway, Pakistan, Sweden and Switzerland. **Disclaimer:** The views and interpretations in this publication are those of the authors. They are not necessarily attributable to ICIMOD and do not imply the expression of any opinion by ICIMOD concerning the legal status of any country, territory, city or area of its authority, or concerning the delimitation of its frontiers or boundaries, or the endorsement of any product.

Annex

Annex 1: Links between Aichi Targets and EbA and Eco-DRR

Strategic goal	Aichi target	Link to EbA and Eco-DRR
B. Reduce direct pressures biodiversity and promote sustainable use	Target 5: By 2020, the rate of loss of all natural habitats, including forests, is at least halved and where feasible brought close to zero and degradation and fragmentation is significantly reduced	Forests and coastal vegetation can serve as a protective buffer from extreme events

(continued)

(continued)

Strategic goal	Aichi target	Link to EbA and Eco-DRR
	Target 7: By 2020 areas under agriculture, aquaculture and forestry are managed sustainably, ensuring conservation of biodiversity	DRR is a core element of sustainability for forestry and agriculture; forests serve as a protective buffer from erosion and landslides
	Target 10: By 2015, the multiple anthropogenic pressures on coral reefs and other vulnerable ecosystems impacted by climate change or ocean acidification are minimized, so as to maintain their integrity and functioning	Coral reefs can be effective in protecting against coastal hazards, such as by reducing wave energy
C: Improve the status of biodiversity by safeguarding ecosystems, species and genetic diversity	Target 11: By 2020, at least 17 per cent of terrestrial and inland water areas and 10 per cent of coastal and marine areas, especially areas of particular importance for biodiversity and ecosystem services, are conserved through effectively and equitably managed, ecologically representative and well connected systems of protected areas and other effective area-based conservation measures and integrated into the wider landscapes and seascapes	Protection of ecosystems, which allows them to keep providing services that are important for adaptation and disaster risk reduction, even beyond the boundaries of the protected area
	Target 13: By 2020, the genetic diversity of cultivated plants and farmed and domesticated animals and of wild relatives, including other socio economically as well as culturally valuable species, is maintained and strategies have been developed and implemented for minimizing genetic erosion and safeguarding their genetic diversity	Reduces risks of climate change affecting food security and livelihoods

(continued)

(continued)

Strategic goal	Aichi target	Link to EbA and Eco-DRR
D: Enhance the benefits to all from biodiversity and ecosystem services	Target 14: By 2020, ecosystems that provide essential services, including services related to water and contribute to health, livelihoods and wellbeing, are restored and safeguarded, taking into account the needs of women, indigenous and local communities and the poor and vulnerable	Ensures provisioning of essential ecosystem services, including those underpinning DRR
	Target 15: By 2020, ecosystem resilience and the contribution of biodiversity to carbon stocks has been enhanced, through conservation and restoration, including restoration of at least 15% of degraded ecosystems, thereby contributing to climate change mitigation and adaptation and to combating desertification	Resilient ecosystems are a key component of DRR, e.g. restoration of coastal vegetated ecosystems contributes to mitigation, adaptation and disaster risk reduction through shoreline stabilization

Source Lo (2016)

References

Bakker MHN, Duncan JA (2017) Future bottlenecks in international river basins: where transboundary institutions, population growth and hydrological variability intersect. Water Int 42(4):400–424

Brink E, Aalders T, Ádám D, Feller R, Henselek Y, Hoffmann A, Ibe K, Matthey-Doret A, Meyer M, Negrut NL, Rau AL, Riewerts B, von Schuckmann L, Törnros S, von Wehrden H, Abson DJ, Wamsler C (2016) Cascades of green: a review of ecosystem-based adaptation in urban areas. Glob Environ Chang 36:111–123. https://doi.org/10.1016/j.gloenvcha.2015.11.003

Castleberry A, Nolen A (2018) Thematic analysis of qualitative research data: Is it an easy as it sounds? Curr Pharm Teach Learn 10(6):807–815. https://doi.org/10.1016/j.cptl.2018.03.019

CBD (2010) Strategic plan for biodiversity 2011–2020 and the aichi targets. CBD Secretariat, Montreal Canada. https://www.cbd.int/doc/strategic-plan/2011-2020/Aichi-Targets-EN.pdf

CBD (2020) Sixth national reports to convention on biological diversity. Secretariat of the Convention on Biological Diversity, Montreal, Canada

Chaudhary S, McGregor A (2018) A critical analysis of global ecosystem services (*paristhiki sewa*) in Nepal. Land Use Policy 75:364–374. https://doi.org/10.1016/j.landusepol.2018.03.024

Chaudhary S, McGregor A, Houston D, Chettri N (2015) The evolution of ecosystem services: a time-series and discourse-centred analysis. Environ Sci Policy 54:25–34. https://doi.org/10.1016/j.envsci.2015.04.025

Chong J (2014) Ecosystem-based approaches to climate change adaptation: progress and challenges. Int Environ Agreement 14(2014):391–405

Cohen-Shacham E, Walters G, Janzen C, Maginnis S (eds) (2016) Nature-based Solutions to address global societal challenges. Gland, Switzerland: IUCN. xiii + 97p

Dahal RK, Hasegawa S (2008) Representative rainfall thresholds for landslides in the Nepal Himalaya. Geomorphology 100(3–4):429–443. https://doi.org/10.1016/j.geomorph.2008.01.014

Deo A, Garner D (2014) Tropical cyclone activity over the Indian ocean in the warmer climate. In: Mohanty UC (ed) Monitoring and prediction of tropical cyclones in the Indian Ocean and climate change. Springer, New Delhi, pp 72–80

Dhyani S, Lahoti S, Khare S, Pujari P, Verma P (2018) Ecosystem based disaster risk reduction approaches (EbDRR) as a prerequisite for inclusive urban transformation of Nagpur City, India. Int J Disaster Risk Reduction 32:95–100. https://doi.org/10.1016/j.ijdrr.2018.01.018

GoN (2009) Guidelines for Community Forestry Development Programme Community Forest. Ministry of Forests and Environment, Government of Nepal, Kathmandu

Guha-Sapir D, Hoyois P, Below R (2013) Annual disaster statistical review 2012: the numbers and trends. CRED, Brussels

IPCC (2014) Climate change 2014: synthesis report. Contribution of Working Groups, I, II and III to the Fifth Assessment Report of the Intergovernmental Panel on Climate Change, IPCC, Geneva, Switzerland. [Core Writing Team, R.K. Pachauri and L.A. Meyer (eds.)]. IPCC, Geneva, Switzerland, 151p

IUCN (2020) Ecosystem-based approaches to climate change adaptation. International Union for Conservation of Nature (IUC), Gland, Switzerland. https://www.iucn.org/theme/ecosystem-management/our-work/ecosystem-based-approaches-climate-change-adaptation

JICA (2009) The study on disaster risk management for Narayangharh-Mugling highway. Final report, vol I, Summary. Japan International Cooperation Agency (JICA) and Nippon koei co., Ltd., pp 4–11. http://open_jicareport.jica.go.jp/pdf/11922242_01.pdf. Accessed on 02 Aug 2019

Jones HP, Hole DG, Zavaleta ES (2012) Harnessing nature to help people adapt to climate change. Nat Clim Chang 2(7):504–509

Kundzewicz ZW, Kanae S, Seneviratne SI, Handmer J, Nicholls N, Peduzzi P, Mechler R, Bouwer LM, Arnell N, Mach K, Muir-Wood R, Brakenridge RG, Kron W, Benito G, Honda Y, Takahashi K, Sherstyukov K (2014) Flood risk and climate change: global and regional perspectives. Hydrol Sci J 59(1):1–28. https://doi.org/10.1080/02626667.2013.857411

Lo V (2016) Synthesis report on experiences with ecosystem-based approaches to climate change adaptation and disaster risk reduction. In; Secretariat of the Convention on Biological Diversity. Retrieved from https://www.cbd.int/doc/publications/cbd-ts-85-en.pdf

MEA (2005) Linking ecosystem services and human wellbeing. In: 3, C. (ed)

Mercer J (2010) Disaster risk reduction or climate change adaptation: are we reinventing the wheel? J Int Dev 22(2):247–264

Mirza MMQ (2011) Climate change, flooding in South Asia and implications. Reg Environ Change 11:95–107

Mittermeier RA, Robles-Gil P, Hoffmann M, Pilgrim JD, Brooks TB, Mittermeier CG, Lamoreux JL, Fonseca GAB (2004) Hotspots revisited: earth's biologically richest and most endangered ecoregions. CEMEX, Mexico City, Mexico, p 390

Molden D, Sharma E, Shrestha AB, Chettri N, Pradhan NS, Kotru R (2017) Advancing regional and transboundary cooperation in the conflict-prone Hindu Kush Himalaya. Mountain Res Develop 37(4). https://doi.org/10.1659/MRD-JOURNAL-D-17-00108.1

Monty F, Murti R, Miththapala S, Buyck C (eds) (2017) Ecosystems protecting infrastructure and communities: lessons learned and guidelines for implementation. IUCN, Gland, Switzerland, x + 108p

Nalau J, Becken S, Mackey B (2018) Ecosystem-based adaptation: a review of the constraints. Environ Sci Policy 89:357–364. https://doi.org/10.1016/j.envsci.2018.08.014

Ojea E (2015) Challenges for mainstreaming ecosystem-based adaptation into the international climate agenda. Curr Opin Environ Sustain 14:41–48. https://doi.org/10.1016/j.cosust.2015.03.006

Reid H, Adhikari A (2018) Ecosystem-based adaptation: strengthening the evidence and informing policy. International Institute for Environment and Development, Inn Road, London

Reid H, Zhang Y (2018) Ecosystem-based approaches to adaptation: strengthening the evidence and informing policy. Research results from the Participatory Plant Breeding and Community Supported Agriculture project, China. IIED Project Report. IIED, London

Reid H, Seddon N, Barrow E, Hicks C, Hou-Jones X, Kapos V, Rizvi AR, Roe D, Wicander S (2017) Ecosystem-based adaptation: question-based guidance for assessing effectiveness. IIED, London

Saroar MM, Rahman MM, Bahauddin KM, Rahaman MA (2019) Ecosystem-based adaptation: opportunities and challenges in coastal Bangladesh. In: Huq et al. (eds) Confronting climate change in Bangladesh. The Anthropocene: Politik, Economics, Society and Science, vol 28. https://doi.org/10.1007/978-3-030-05237-9_5

Schipper L, Pelling M (2006) Disaster risk, climate change and international development: scope for, and challenges to, integration. Disasters 30(1):19–38. https://doi.org/10.1111/j.1467-9523.2006.00304.x

Seddon N, Reid H, Barrow E, Hicks C, Hou-Jones X, Kapos V, Rizvi AR, Roe D (2016a) Ecosystem-based approaches to adaptation: strengthening the evidence and informing policy: research overview and overarching questions. IIED, London. Available at http://pubs.iied.org/G04045/

Seddon N, Hou-Jones X, Pye T, Reid H, Roe D., Mountain D, Rizvi AR (2016b) Ecosystem-based adaptation: a win-win formula for sustainability in a warming world? IIED Briefing paper. IIED, London

Shah SIA, Zhou J, Shah AA (2019) Ecosystem-based adaptation (EbA) practices in smallholder agriculture; emerging evidence from rural Pakistan. J Cleaner Prod 218:673–684

Sharma E, Chettri N, Tse-ring K (2009) Climate change impacts and vulnerability in the Eastern Himalayas. ICIMOD, Kathmandu, p 27

Sharma E, Molden D et al (2019) Introduction to the Hindu Kush Himalaya Assessment. In: Wester P, Mishra A, Mukherji A, Shrestha A (eds) The Hindu Kush Himalaya Assessment. Springer, Cham

Singh S, Nair SS, Gupta AK (2013) Ecosystem services for disaster risk reduction: a case study of wetland in East Delhi Region, India. Global J Human Social Sci Geogr Geo-Sci Environ Disaster Manage 13(4):37–47. Retrieved from http://socialscienceresearch.org/index.php/GJHSS/article/view/646

Song Y, Zhang Y, Song X, Buckley L (2015) Emerging biocultural innovations for climate resilience in southwest China. SIFOR qualitative baseline study. IIED, London

Swiderska K, King-Okumu C, Monirul Islam M (2017) Ecosystem-based adaptation: a handbook for EbA in mountain, dryland and coastal ecosystems. IIED, London

Thapa PS, Adhikari BR (2019) Development of community-based landslide early warning system in the earthquake-affected areas of Nepal. J Mt Sci 16(12):2701–2713

Timalsina R, Songwathana P (2020) Factors enhancing resilience among older adults experiencing disaster: a systematic review. Australas Emergency Care 23(1):11–22. https://doi.org/10.1016/j.auec.2019.12.007

Triyanti A, Chu E (2018) A survey of governance approaches to ecosystem-based disaster risk reduction: Current gaps and future directions. Int J Disaster Risk Reduction 32:11–21

UNDP (2015) EbA annual progress report. United Nations Development Programme Nepal, Kathmandu, Nepal. Accessed at: https://www.np.undp.org/content/nepal/en/home/operations/projects/closed-projects/environment---energy/eba/home.html

UNFCCC (2015) Adoption of the Paris Agreement. Conference of the Parties twenty-first session, Paris, 30 Nov to 11 Dec 2015. http://bit.ly/2GGWXXa

UNOCHA August 21, 2017. https://reliefweb.int/sites/reliefweb.int/files/resources/Nepal%20Flood%20Sitrep%2021%20August%202017.pdf

Wester P, Mishra A, Mukherji A, Shrestha ABS (eds) (2019) The Hindu Kush Himalaya Assessment—mountains, climate change, sustainability and people. Springer Nature Switzerland AG, Cham. https://doi.org/10.1007/978-3-319-92288-1

Yusuf AA, Francisco HA (2009) Climate change vulnerability mapping for Southeast Asia. Economy and Environment Program for Southeast Asia (EEPSEA), Singapore. Available at https://www.idrc.ca/sites/default/files/sp/Documents%20EN/climate-change-vulnerability-mapping-sa.pdf

Zolch T, Wamsler C, Pauleit S (2018) Integrating the ecosystem-based approach into municipal climate adaptation strategies: the case of Germany. J Cleaner Prod 170:966–977. https://doi.org/10.1016/j.jclepro.2017.09.146

Chapter 3
Evaluation of Ecosystem-Based Approaches for Disaster and Climate Risk Resilience and Policy Perspectives in Pakistan

Muhammad Barkat Ali Khan, Atta-ur Rahman, and Rajib Shaw

Abstract Pakistan is a country with a diverse topography, biodiversity and it is exposed to a range of disasters and climate risks. This chapter discusses the spatial pattern and trends of ecological zones, forest cover and mangroves along the coastal ecosystem and its role in minimizing the risks of coastal hazards. In mountainous areas, consistent degradation of forest ecosystem has increased the risk of soil erosion, siltation in dams, accelerated the flood frequency, modified the micro-climate and posed serious threats to blue–green infrastructure. The fertile agricultural land that was once the food basket land for the local population, is no more meeting the local demand. Primarily, the farmland converted into a built up environment. In changing climate scenario, most of the rivers in Pakistan on which life depends are recharged by springs, rainwater and melting of snow/glaciers are under constant pressure. In urban areas, the increase in population and soil sealing have put tremendous pressure on groundwater aquifers and calls for sustainability in blue–green infrastructure. In conclusion, the chapter suggests approaches for sustainable utilization of blue–green infrastructure in ecosystem-based disaster, climate risk resilience and sustainable development.

Keywords Ecosystem · DRR · Policy · Climate risk · Resilience

3.1 Introduction

Pakistan is located in South Asia and it shares borders with four countries and Arabian Sea. Pakistan is bordering in the north with China, to the east with India, to the south with the Arabian Sea and Indian Ocean and to the west with Afghanistan and Iran. The country extends from 23° 35′ to 37° 05′ North latitudes and 60° 50′ to 77° 50′

M. B. A. Khan
Department of Geography, Higher Education Department, Peshawar, KP, Pakistan

A. Rahman (✉)
Department of Geography, University of Peshawar, Peshawar, Pakistan

R. Shaw
Keio University, Fujisawa, Japan

© The Author(s), under exclusive license to Springer Nature Singapore Pte Ltd. 2021

M. Mukherjee and R. Shaw (eds.), *Ecosystem-Based Disaster and Climate Resilience*, Disaster and Risk Research: GADRI Book Series,
https://doi.org/10.1007/978-981-16-4815-1_3

East longitude (Khan 2006). The total area of the country is 796,095 km^2; based on area, it is the 33rd largest country in the world. The population of Pakistan is sharply growing, according to the census conducted in the year 2017, the total population of the country was 207 million (GoP 2017a) and ranked the 5th most populous country in the world. The average yearly growth rate of population during 1998–2017 census was recorded as 2.4%, which is the highest in urban (2.7%) and low in the rural areas (2.23%; GoP 2017a). The country boasts of rich cultural heritage, being home to several ancient civilizations and cultures ranging from sites of Neolithic period in the Stone Age to the Indus valley civilization in the Bronze Age (Khan 2006). The region has witnessed numerous rulers and dynasties including Greeks, Hindus, Arabs, Mongols, Afghans and Sikhs. In the nineteenth century, it became part of British Empire and later on, Pakistan came into being in August 1947 (Rahman et al. 2015).

The topography of this unique country is highly diverse and includes some of the world's tallest mountain ranges and peaks above the snow line on the one hand and home to some of the world's famous glaciers on the other (Rahman et al. 2015). Coniferous forests envelop the country below the tree line with diverse flora and fauna extending to the sea coast (GoP 2005). There are also vast and fertile plains and deserts in the provinces of Sindh and Punjab. Millions of years ago, South Asia including the present territory of Pakistan was part of a landmass called Gondwanaland, which included Antarctica, Australia and Africa. It is estimated that about 130 million years ago South Asia broke off from the parent landmass and about 20 million years ago reached its present position. After this, a geomorphic process took place which has produced two distinctive regions in Pakistan. The first one is the northern and western highlands produced by the mountain-building movements by the collision of Indian plate with Eurasian plate during the Tertiary Period. As a result, the mountain ranges of Karakorum, Himalayas and Hindukush were formed, whereas the physiographic unit "the Indus plain" was formed during the Quaternary Period by the deposition of sediments in shallow bays by the Indus River and its tributaries (Khan 2006).

The climate of Pakistan is diverse including extremely cold areas with snow covered mountain peaks as well as hot and arid areas in central part of the country (GoP 2017b). Pakistan has four distinct seasons, Pre-monsoon (Hot), from March to June; Monsoon (hot and humid) from July to mid of September; post-monsoon, from mid of September to October and cold from November to February. During summer season the relative humidity differs from 25 to 50% and is very hot. Diurnal temperature in summer season remains 40 °C and above in plain areas. During winter the average temperature ranges from 4 to 20 °C, the temperature drops below freezing point in the northern parts of the country. There is a big difference in the range of low and high temperature; where the mercury falls to −27 °C in the north (at Skardu) during winter and rises to 52 °C in the south-central parts in summer season (ADRC 2015). The varied arid subtropical climate of the coastal region of Pakistan receives about a mean annual rainfall of 100–200 mm. The coastal areas of Pakistan receive less rainfall; the area does not receive or meagre precipitation during monsoon and winter rainfall due to western depressions. The coastal region in Pakistan especially in the Indus Delta is having inlets of sea water into the coastal side which makes a creek. The Indus Delta has 17 major creeks, several minor creeks and wide mudflats, which occupy approximately 600,000 hectares (ha) spreading from Korangi Creek in the north to Sir Creek in the South (Amjad and Jusoff 2007).

The vast diversity in topography, climate and soil characteristics in Pakistan provides a huge assortment in natural vegetation, water resources and distribution of population. There are also differences in the approach of human activities towards utilization of natural resources especially water resources. The diversity in topography, climate and natural vegetation that extends to the sea coast and provide habitat to a range of different flora and fauna with unique eco systems, are the hallmarks of Pakistan that makes her diverse and distinctive from other regions of the world. The numerous traditional practices ranging from ancient irrigation techniques to the modern canal systems enables the country to produce different varieties of vegetables, fruits, cereal crops and cash crops round the year. Apart from the variety in the natural vegetation including alpine forests, coniferous forests, deciduous forests, riverine forests and arid forests, there is also diversity in agro forestry and irrigated plantation. In Pakistan, the population is unevenly distributed, the rate of urbanization is high and the country is also prone to a variety of natural hazards and impacts of climate change (Shaw et al. 2016). In this regard, the policy makers and administrative setup are devising policies and programmes to preserve the diverse eco systems and mitigate the natural hazards and impacts of climate change.

3.2 Agro-Ecological Zones of Pakistan

Physically, Pakistan can be divided into northern and north-western mountains, plateau (Potowar and Baluchistan plateau) and fertile Indus floodplain drained by a dense network of irrigated infrastructures (Khan 2006). Pakistan has a wide diversity in terms of topography, rainfall, temperature, perennial streams and rivers, snow clad mountains, vast deserts, humid mountains and irrigated and unirrigated agricultural lands. Keeping in view the significance of Agro-ecological zones, the Pakistan Agricultural Research Council has divided the country into twelve major agro-ecological zones (Fig. 3.1). The agro-ecological zones includes wet mountains in the north, western dry mountains, northern dry mountains, Sulaiman piedmont, southern irrigated plains, sandy desert (a), sandy desert (b), norther irrigated plain (a), northern irrigated plain (b), northern dry mountains, Indus deltaic plain, dry western plateau and a vast rainfed area. Each agro-ecological zone has typical topographic and climatic characteristics. These agro-ecological zones have several socio-economic and environmental challenges.

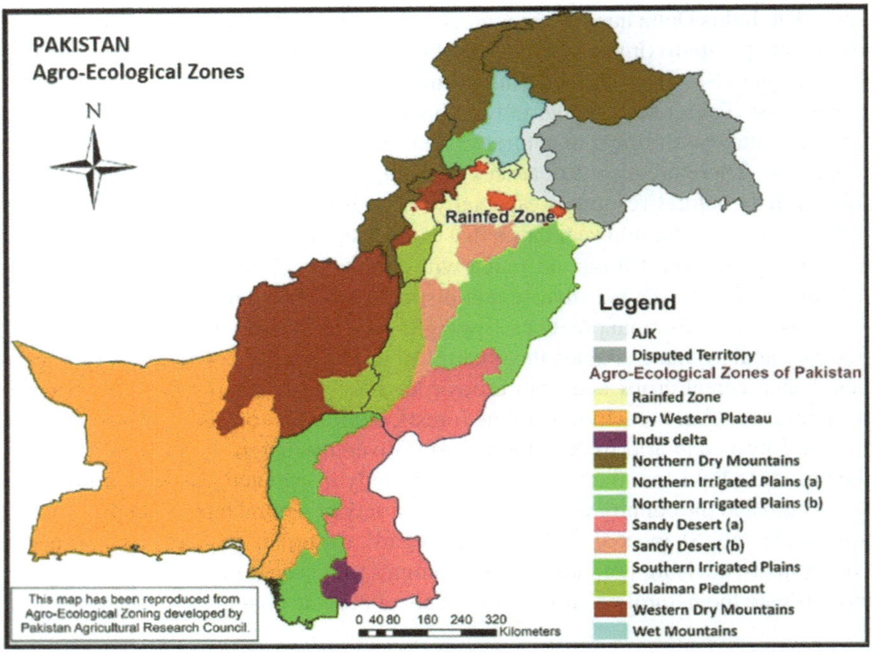

Fig. 3.1 Agro-ecological zones of Pakistan modified after Mahmood et al. (2019a, b)

3.3 Forest Ecosystem, DRR, Climate Risk and Policy

Forest Ecosystem plays a pivotal role in disaster risk reduction, climate change risk reduction, stimulating hydro-climate resilience and enhancing sustainability of blue–green sectors. Forest resources and forest ecosystems are vital for balanced environmental conditions, including water resources, soil cover, flora and fauna and non-timber forest products. According to the last natural assessment conducted in 2004, Pakistan is ranked as a country having only 5% of its land covered by forests (GoP 2015). In Pakistan, the major forest types include coastal mangrove forests, riverine forests, subtropical scrub forests, moist temperate conifer forests, dry temperate conifer forests and irrigated plantations, which modify the micro-climate, a source of water vapours and check the water runoff and wetlands in the region. Pakistan has recently initiated a billion tree tsunami afforestation project in the Khyber Pakhtunkhwa and at national level 10 billion tree tsunami afforestation in Pakistan to address Paris Agreement, Sendai Framework for Disaster Risk Reduction and SDG 13 Climate Action.

In Pakistan, some of the world's distinctive highland forests are also found including Juniper, Cedrus Deodar, Oak and Chilghoza pine (*Pinus gerardiana*) forests. Ecosystem services of forests are having significant impacts on physical and mental health of human beings including medicinal plants, water management, mitigating soil erosion, siltation in dams, retarded landslides over the fragile slopes, climate resilience safety nets and controlling infectious diseases. In this regard, establishing Pakistan forest institute at Peshawar and framing National forest policy, National Rangeland Policy and National Climate Change Policy of Pakistan are some of the glaring initiatives taken by the Ministry of Climate Change to address forest ecosystem, DRR and Climate Risk Resilience in the country. Forests also help in alleviating poverty by providing remarkable livelihoods opportunities for the forest dwelling communities. Pakistan is having big diversity in climate, topography and soil with meagre forest resources and the rate of deforestation is 4.6% per annum which is second highest in the world (Khan and Khan 2009). Pakistan is also considered as it witnessed impacts of climate change in the shape of droughts of 1900–2000 (Anon 2007).

In Pakistan, with the passage of time due to rapid population increase, lack of alternate sources for fuelwood and timber, forests in all provinces are under severe pressure particularly in Gilgit Baltistan and Khyber Pakhtunkhwa. The estimate of deforestation at the national level is 27,000 ha per year which primarily occurs in private and communal natural forests (GoP 2015). The deteriorating land, depletion of biodiversity and uncontrolled deforestation in the major catchment areas is also negatively impacting crop production and quality of water at outlets. In the low-lying and coastal areas especially close to the rivers, deforestation intensifies floods and expedites sea water intrusion causing huge economic losses. Natural hazards and anthropogenic factors collectively result in loss of forest biodiversity in terms of shrinking number of flora and fauna species.

3.3.1 Spatial Pattern of Mangrove Forests, DRR and Climate Risk Resilience

Globally, mangroves ecosystem is considered as the important resource that mitigates the coastal hazards and a major source of livelihood earnings for the coastal communities (Rahman et al. 2017). In the changing climate scenario, increasing trend of hydro-meteorological disasters and calls for boosting-up of climate risk resilience, the role of mangrove ecosystem cannot be ignored. The coastline of Pakistan along Arabian Sea reaches up to 1046 km in length (Khan 2003) with numerous patches of threatened ecosystems of mangroves managed by dedicated authorities.

The coastline of Pakistan extended in east from Indian coast at Rann of Kutch to as far as Gawatar 'the Iranian coast' in the west (Rahman et al. 2017). The coast of Pakistan is extended over two provinces including Sindh and Baluchistan includes 250 km in Sindh province and 800 km in Baluchistan province. Mangroves are found in sparse to dense patches in the coastal region of Pakistan. Within the total covered area, 97% is in Sindh province while the 3% lies in Baluchistan province (Rahman et al. 2017). The mangrove forests in Pakistan are mostly found in the arid climatic region and the forest is reliant upon fresh water of Indus and Hub rivers (Barkati and Rahman 2005).

The Indus Delta provides a good habitat for mangrove forests. In Pakistan, the Indus Delta covers almost 600,000 ha area, the delta consists of 17 major creeks, several minor creeks and vast mudflats (Amjad and Jusoff 2007). Mangroves are found extensively in the Indus Delta along the south-eastern coastline for about 240 km along Arabian Sea (Saifullah 1982). In the country, almost 95% of mangroves are found in the Indus deltaic marshes of the Sindh province. Apart from Indus Delta, the other areas in Pakistan with mangrove forests are Kalmat Khor, Sandspit, Sonmiani (Miani Hor) and Jiwani. The Indus Delta is considered as the second largest mangrove forests in the sub tropics (Giri et al. 2015). The mangrove forests that cover estuaries in the Indus delta are found between the levels of mean high water spring tides and mean sea level. The mangrove trees are sometimes partially submerged under water while sometimes it is completely above sea level, the inundation differs due to differences in the water level according to the tidal cycle. The roots and lower stems remain submerged under water during high tides, while at low tides for several hours the mangroves remain dry (without water cover).

During spring tides the water reaches the outer peripheries on elevated ground, while during neap high tides the lower extents of the mangrove marshes are mostly under water. This condition of rise and fall of sea level due to tides create a fluctuating environment in the mangrove covered areas. The roots are immersed with saline water of Arabian Sea during high tides. During monsoon season, the river discharge increases due to torrential rainfall and the season bring more fresh water with silt and reduces salinity. The movement of water during rains, surface runoff and tides also affect soil and water by altering nutrients, oxygen levels and temperature.

In Pakistan, there are eight species that dominate the mangroves ecosystem including Avicenna Marina, Aegiceras corniculatum, Rhizophora apiculate, Ceriops roxburghiana, Ceriops tagal, Sonneratia caseolaris, Bruguiera conjugate and Rhizophora mucronata. The most dominant mangrove species in Pakistan is Avicennia Marina. During the year 1984, the mangrove in the Indus Delta was considered as the fifth largest area under mangrove forest in the world (Snedaker 1984). In winter seasons, the valuable birds migrate from Siberia to the coastal region of Pakistan. In Pakistan, it is an important destination for animals and provides a good habitat for migratory birds.

The estimates of the total area covered by mangrove forests differ in different publications. We can see the difference, as according to FAO the total covered area was 347,000 ha (FAO 2005), in the same year according to IUCN it was reported 86,727 ha (IUCN Pakistan 2005) and 108,058 ha in the same year by Mangrove for the Future (MFF) Pakistan, an international organization working in eleven countries (Beresnev et al. 2016). In the year 2015, it was reported as 283,280 ha by the Forest Department Government of Sindh (GoS 2015), while according to FAO during the same year the estimate is 95,000 ha (FAO 2020). According to the current official data of Forest Department, Government of Sindh Pakistan, mangrove forests of Indus delta cover an area of almost 600,000 ha, established an important ecosystem in the coastal deltaic region Mangroves in the Indus delta are exceptional in being the largest

Table 3.1 Area under Mangrove forests, modified after GoS (2020)

Organization	Area under control
Sindh Forest Department	280,470 ha (693,000 acres)
Sindh Board of Revenue	255,130 ha (630,000 acres)
Port Qasim Authority	64,400 ha (159,000 acres)

arid climate mangroves in the world. The forests are almost totally dependent upon the discharge of fresh water from River Indus and a small quantity of fresh water from local sources. The mangroves in the Indus delta are being managed by three different organizations, including Sindh Forest Department, Port Qasim Authority and Sindh Board of Revenue. The detail of area under control of these three organizations is shown in Table 3.1.

Mangrove forests are very important for environment and livelihoods of local people, the local population depend upon the resources produced and sustained by these vigorous ecosystems. Mangrove ecosystems are also known as places of high biological productivity and provide habitat and nurseries for a diversity of marine organisms. Mangroves are important for trapping sediments during runoff which mitigate erosion, also act as carbon sinking points, protect coasts from storms, floods and cyclones; therefore it stabilizes the coastal areas. Mangroves provide habitat to a variety of edible fishes and shrimps and also provide other resources especially firewood to the people living close to mangroves. Therefore, the mangroves are also playing an important role in providing livelihoods sources to the coastal communities. According to the estimates of IUCN, 1 ha of mangroves in a year can produce 100 kg fish, 25 kg shrimp and 15 kg crab (IUCN 2005).

According to the figures of Zoological Survey of Pakistan, there are 98 species of fish documented in the mangroves of Indus delta. Due to the availability of nutrients rich food, most of the fish species are attracted by mangrove swamps due to rise of water during high tides. The mangrove forests of Indus delta is a good example of mixture of terrestrial and aquatic organisms in an area. The mangrove forests of Indus delta also provide habitat to waterfowls, migratory birds, such as pelicans, flamingos, cormorants, cranes. These birds migrate to the area in flocks of thousands for staying, feeding and breeding every year from November to February. In the last decade, it was being explored that degradation of habitat resulted in shrinking the number of migratory birds. Similarly, the dolphins which live and thrive in the Indus delta mangroves have also declined due to marine pollution. The Indus Delta provides sustenance to approximately 200,000 people continuously, living in the vicinity of Mangroves of Indus delta (GoS 2020).

The forest cover of mangroves was sharply reduced in the time period of eight years from 1992 to 2000 (Fig. 3.2). It was being analyzed through LANDSAT satellite data, the forest cover of mangroves decreased from 122,028 ha in 1992 to less than 73,000 ha in the year 2000, showing a decrease of more than 50,000 ha (Rahman et al. 2017). The factors behind the depletion of mangroves forest cover were recorded as a decrease in the flow of fresh water, cutting mangrove forest for fuelwood, intensive camel grazing, urban expansion, nutrients deficiency in the coastal soil, coastal

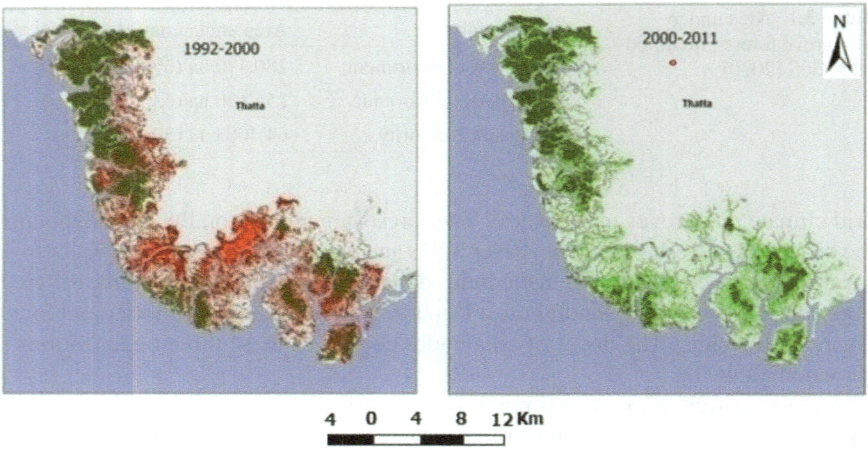

Fig. 3.2 Mangrove forest cover along Indus Delta 1992–2011 modified after MoCC (2014)

water pollution including industrial and domestic means (Barkati and Rahman 2005). Due to this sharp depletion, there were impacts on the increase of coastal hazards, depletion of marine ecology including fisheries. Therefore, government of Pakistan prioritized to restore and increase mangrove forest cover along the coastal region. Due to these efforts, the mangrove forest cover was increased to 250,000 ha during the year 2002 and was ranked as the sixth largest mangrove forest in the world (Saifullah and Rasool 2002). In another study by IUCN in 2005 during the year 1932 the mangrove forest cover was 604,870 ha in the present day Pakistani territory, which was reduced to 440,000 ha in 1986 and then it was further shrunk to 160,000 ha in 1992 and further decreased to 86,000 ha in 2005 (IUCN 2005).

The analysis revealed that there is change in spatial pattern and trends in the area under mangrove forest cover in the time frame 1992 to 2000 and 2000 to 2011 (Fig. 3.2). The analysis further indicated that a sharp decrease in the forest cover was recorded during 1992–2000. In the eight year span, the situation related to depletion of mangrove forest in the Indus delta was alarming. In Fig. 3.2, the red colour on the map is showing the deforested land. After 2000, due to some efforts by the government, the forest was regenerated and an increase was observed in the mangrove cover.

Depletion in the mangrove forests has environmental and socio-economic impacts. Decrease in mangrove forest reduces production of commercial fish products, accelerate the susceptibility of coastal hazards, silting up the navigational channels, increases coastal erosion and loss of biodiversity.

3.3.2 *Spatial Pattern of Coniferous Ecosystem, DRR and Climate Risk Resilience*

The Himalayas, Karakoram and Hindu Kush region lies to the north and north-western part of Pakistan. It hosts thousands of glaciers, snow clad mountains and springs that recharge the Indus river system. These mountains are the treasury for the local communities and a place of beautiful alpine meadows, alpine forests, range-lands and lush green coniferous forests. These coniferous forests are found in the mountainous region of north and north-western part of the country and major source of livelihood earnings. In Pakistan, coniferous forests are more prone to deforestation due to its commercial significance and high rewarding timber are obtained from these softwood forests. These coniferous forests modify the local climate and provide ample amount of water vapour to the atmosphere, which later on became a source of precipitation and recharge river system.

A study was conducted by Ahmad et al. (2012) found that a decrease in the coniferous forest cover has been recorded from 1992 to 2010 and estimated as 40,000 ha in a year. In Pakistan, Khyber Pakhtunkhwa and the then Northern Areas are important for thick coniferous forests, but the rate of deforestation is also highest in these areas. Table 3.2 shows the details of changes in the coniferous forest cover. The analysis further revealed that the decrease is highest in the areas having thick coverage of coniferous forests. The major depleted areas reported from Khyber Pakhtunkhwa province and the province of Gilgit Baltistan. There is an overall decrease of 440,000 ha in the area covered by coniferous forests in the timeframe from 1992 to 2010 (Table 3.2). However, during the stated period, an increase in the coniferous forest cover has been reported from the province of Punjab and the state of Azad Jammu and Kashmir (AJK). This increase is attributed mainly to the regeneration and decrease in deforestation. However, after billion tree tsunami project

Table 3.2 Coniferous forest cover change from 1992–2010 modified after Ahmad et al. (2012)

Province/territory	Status of vegetation cover (000 ha)					Rate of annual change % (Base year 1992)
	1992	1997	2001	2005	2010	
Khyber Pakhtunkhwa	940	805	858	840	845	−0.56
Punjab	30	34	34	38	41	1.22
Baluchistan	42	39	35	32	30	−0.6
Northern Areas (Gilgit Baltistan)	660	312	318	301	285	−0.08
Sindh	No conifer forests in Sindh province					
Azad Jammu and Kashmir	241	289	267	288	272	0.71
Total forests	1913	1479	1512	1499	1473	−1.27

followed by 10 billion tree tsunami project, a massive move towards the regeneration and afforestation has made and recorded positive implications on forest coverage.

Coniferous forests are very important for providing habitat to a variety of species of flora and fauna. A study conducted in Malam Jabba valley of Swat district of Khyber Pakhtunkhwa Pakistan, reveals that the coniferous forests in the area are providing habitat to 90 plant species of ethnobotanical importance. These species including 84 species of angiosperms, three species of gymnosperms, 3 species of pteridophytes and one species of fungi. The valley is famous for thick coniferous forests and providing habitat to plant species including 71 of medicinal use, 20 of fodder, 14 of wild fruit, 18 of fuel wood, nine of furniture and agriculture tools, ten of vegetable, nine of thatching, fencing and hedges, two of religious use, two poisonous, two evil eye and also four honey bee species. Another study conducted related to environment-vegetation relationship of Coniferous Dominating Forests (CDF) in the Indus Kohistan area of northern Pakistan. There is a diversity of habitats for plants, birds and animals species in the thick coniferous forests.

These forest resources are having far-reaching ecological and socio-economic implications on climate change. Apart from conventional role of forests, worldwide it has also contributed revenue of US$ 100 billion, annually and at the same time, it is providing trillions of dollars in ecosystem services. According to estimates, global forests consume 3 billion tonnes of carbon each year and also provide habitat for approximately 65% of all terrestrial species (Pedlar and McKenny 2017).

Swat valley is located in the north of Pakistan and falls in Khyber Pakhtunkhwa province. The valley is famous for its scenic beauty, forest cover and plenty of fresh water resources. A research work conducted for appraisal of deforestation in Swat valley and the Landsat images for the years 2011 and 2016 were analyzed and showed a sharp decrease in the forest cover in different areas of the valley. According to the study, during this time period about 11 km^2 area is converted from forest to barren land and an estimated 9985 km^2 area of forest cover was degraded in district Swat. It also results in the depletion of water resources especially in the southern part of the valley (Owais and Siddique 2019).

Climate change is also causing loss of biodiversity and resulting in shifting of forest areas northwards to the high altitude and cooler places (Khan and Ahmad 2015). As the heatwaves are increasing, it will result in forest fires and unpredictable excessive rainfall causing damages to regeneration in the forest and plantation areas (Saeed et al. 2017; Dawood et al. 2020). Changes in the ecology can also change the species composition in the forests. The changes in the pattern of climate can also result in different pests and diseases (Saifullah 2017). In this regard, the Climate change policy, National Forest Policy and Disaster Management Act have specifically highlighted the conservation of forest cover as a DRR and climate risk reduction strategies.

3.3.3 *Forest Ecosystem, DRR and Policy Measures*

In Pakistan, to address and overcome issues and challenges related to forest ecosystem and enhancing forest resources the Ministry of Climate Change is framing policies from time to time (GoP 2015). The National Forest Policy 2015 has a three pronged approaches for the solution of issues related to forests, climate change and Disaster Risk Reduction:

i. Conservation of the existing forests by restricting deforestation and promoting forest conservation.
ii. Initiating mass afforestation programmes with particular focus on communal lands with participation of all sections of the society.
iii. Fulfilling international requirements and opportunities by ensuring capacity building, legal reforms and scientific planning.

For achieving the objectives of enhancing forest cover and adopting the approach, following are some of the key policy measures framed by the federal government of Pakistan.

i. Mass afforestation programmes are to be organized by each of the provinces and other administrative units including Islamabad Capital Territory and the state of Azad Jammu and Kashmir (AJK). In the present circumstances at national level on average a target of 70–80 million tree saplings is planted annually, which is inadequate to meet even domestic demand for wood. Therefore, in collaboration with all provinces and territories, the federal government must undertake long term mass afforestation programme through relevant national organizations. The process must include investment in communal, Guzara forests and private forests by Government, donor agencies and private funding.
ii. Conventionally forestry was an isolated wood producing sub sector, with the passage of time it is being transformed into an integrated subject with cross cutting themes influencing a number of economic sectors including water, energy, agriculture, non-timber forest products, health, tourism, housing and. It is therefore essential to incorporate forestry with the development policies and programmes of these sectors at the levels of planning and implementation.
iii. Establishing ecological corridors or networks consist of core areas, buffer zones and interconnecting corridors. The core areas include existing or new protected areas which are surrounded by buffer zones that need to be interconnected. The process will contribute to sustainability through holistic approach.
iv. Regulating commerce and trade including inter-provincial timber movement. In the existing set up the provincial governments, Azad Jammu and Kashmir and Gilgit Baltistan have the legal decree to manage forest resources, checking unlawful tree cutting, controlling trade and movement of timber within the jurisdictions. According to article 151 of the Constitution and Federal Legislative List, the Federal Government is having authority to regulate inter-provincial coordination on forest related matters, including inter-provincial trade of timber and non-timber forest products and export and import of wood. In Pakistan,

the rate of deforestation and forest degradation is high; some of the areas are more notorious for it like Kohistan district of Khyber Pakhtunkhwa and Diamer district of Gilgit Baltistan. In these areas, most of the valuable conifer forests are either communal or privately owned. These forests are the only source of earning livelihoods for local communities through commercial sale of wood. For not cutting forest trees there is no mechanism from government to provide incentives to right holders and lawful owners of these forests. According to the agreement of United Nations Framework Convention on Climate Change (UNFCCC), any action that reduces carbon emission by minimizing deforestation and forest degradation is eligible for incentives.

v. For protecting existing forests and enhancing forest cover the government shall undertake all possible options including Payment of Ecosystem Services (PES), retiring the rights to forests and purchase of privately owned forests and entitling it for sustainable utilization by existing and coming generations to stop the trend of deforestation in particular from communal and private forests. The government should also protect the rights of forest owners and right holders and the process should be done with their consent. The government should promote and encourage integrated approach to forests, wildlife and biodiversity management. In Pakistan, there is diversity in forest ecosystem and it spread in mountains, terrestrial and marine wetlands and deserts. The wetlands in Pakistan apart from fishes provide habitat to different species of fauna and unique aquatic flora. The forests and related ecosystems are managed by more than one provincial governmental department such as the departments of Forest, Wildlife and Fisheries. Unfortunately, there is a lack of integrated approach in management; each department has its own system of management and implementation of policies in isolation from other closely related departments. According to the recent approaches, there is a dire need for ensuring holistic or all-inclusive ecosystem approach for better achievement of conservation and sustainable utilization.

vi. All provinces and territories shall develop updated forest monitoring systems according to the globally accepted standards restructuring forest management including policies, planning and laws in accordance with national priorities and international standards (GoP 2015).

The federal government also framed a detailed implementation mechanism for the policies related to forest ecosystems.

3.4 Agriculture, Food Security Nexus with DRR, Climate Risk and Policy

Pakistan is a country with highly diversified environmental conditions; there are 12 agro-ecological zones, where livestock mixed farming and more than 35 types of crops are practised (Rahman et al. 2016). Pakistani farmers are facing difficulties to compete in international markets, they are lacking advanced technologies due to which the production is comparatively low, the other causes include, high production cost, large international stock build up and reduced international prices. With the passage of time, there is a decrease in the contribution of agriculture sector in the GDP of Pakistan. In the 1960s the share of agriculture to the GDP was 46% which is decreased to 25% in the year 2016. In the year 2016 still 45% of the population is engaged in agriculture activities but the share in GDP is reduced to 25% (Asim and Akbar 2019). Agriculture continue to be the largest contributor to GNP, but its relative share dropped from 52% in 1951 to 25.1% in 2000 and then 22.5% in 2004 (Khan 2006).

3.4.1 Agriculture Sector, Yield and Production

In Pakistan, different cereal and cash crops are cultivated both in Rabi and Kharif seasons. In Cereal crops, the most important are wheat, rice and maize, while in cash crops the most important are cotton and sugarcane (Table 3.3).

In Pakistan, the most important cereal crop is wheat, with the passage of time the area and production is on increase but due to rapid population increase and low yield as compared to developed countries mostly the country is not self-sufficient in wheat production (Khan 2006). The province of Punjab is most important for wheat cultivation and production. The maize is cultivated mostly in the mountainous region or where the soil characteristics and irrigation water are not sufficient for rice cultivation. The area was not being increased but the production is slightly increasing due to improved farming practices. After wheat the important cereal crop is rice, which is mostly cultivated in the provinces of Punjab and Sindh. There is a little increase in the production of rice from 2011 till 2014. Pakistan is self-sufficient in the production of rice and the surplus rice is being exported.

Keeping in view the present trends, Pakistan would require taking measures to bring reforms in its production systems. With the passage of time area under rice and sugarcane crops will be decreasing due to the cultivation of other economically high yield crops, such as fruit trees, vegetables, pulses, oilseed, soybean and fodder.

Table 3.3 Pakistan Important crops, modified after Pakistan bureau of Statistics, 2020 (area in 000 ha, production in 000 tonnes, cotton production in bales 375 lbs each)

Crops	2011–12		2012–13		2013–14	
	Area	Production	Area	Production	Area	Production
Wheat	8649.8	23,473.4	8660.2	24,211.4	9199.3	25,979.4
Rice	2571.2	6160.4	2308.8	5535.9	2789.2	6798.1
Maize	1087.4	4338.4	1059.5	4220.1	1168.5	4944.2
Cotton	2834.5	13,595	2878.8	130,307.7	2805.7	12,768.9
Sugarcane	1057.5	58,397	1128.8	63,749.9	1172.5	67,460.1

Some more problems which result in low income of farmers in the rural areas are exploitative market practices and lack of storage and value addition facilities. In Pakistan, there are potentials to utilize each part of land productively to contribute to the improvement of livelihoods and opulence of the rural population of Pakistan. Most of the country is arid and semi-arid and through different irrigation techniques the lands in arid regions including Thar, Cholistan, Southern and western parts of Khyber Pakhtunkhwa, Thal and arid coastal belt, can be cultivated and utilized for specific and innovative farming practices by the provision of irrigation facilities.

In Pakistan, adequacy and spatial distribution of Rabi crops are grown in different parts of the country (Fig. 3.3). In Pakistan, Rabi crops including wheat, barley, tobacco, onion and tomato. In cereal crops, wheat is the important Rabi crop of Pakistan. In the southeast, within Sindh province, the area is producing excessive Rabi crops. The reason behind that is plenty of irrigation water and fertile soil being deposited by river Indus. Apart from that there is an extensive region in the central part of the country mostly consist of Punjab province, remaining parts of Sindh, a coastal strip in Baluchistan and a small area in central Khyber Pakhtunkhwa are producing adequate Rabi crops. Apart from that rest of the country mostly including

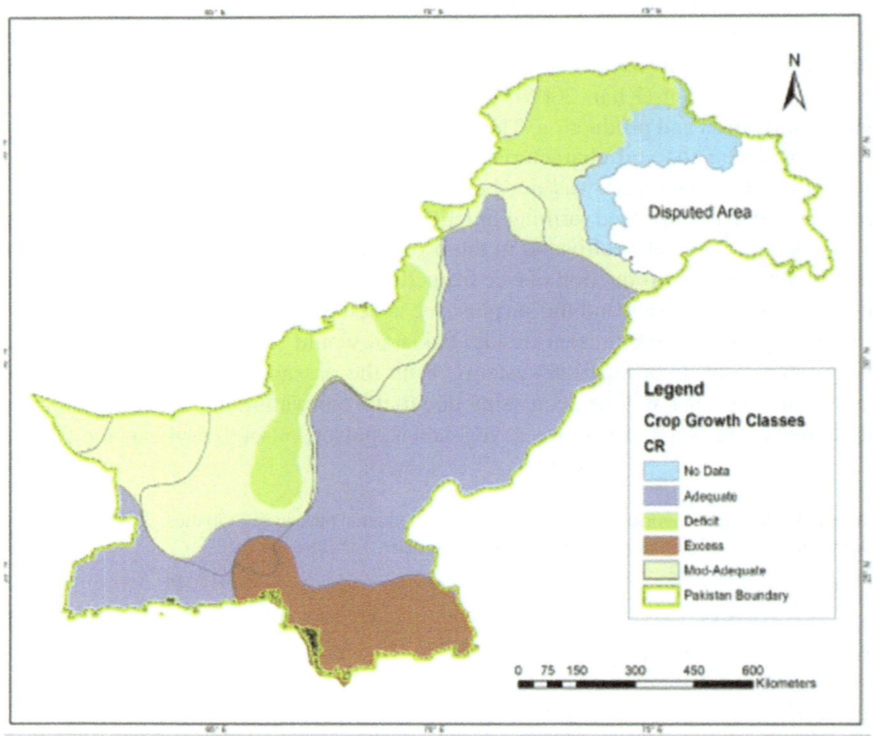

Fig. 3.3 Pakistan crop growth classes (Rabi) modified after MoCC (2014)

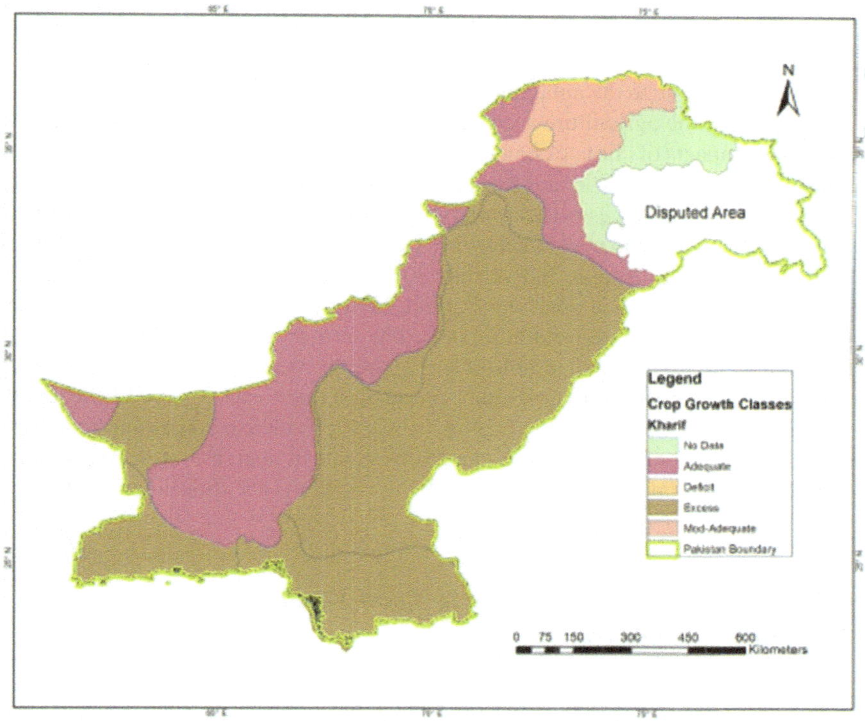

Fig. 3.4 Pakistan crop growth classes (Kharif) modified after MoCC (2014)

Baluchistan and Khyber Pakhtunkhwa provinces are mild adequate or deficit in the production of Rabi crops.

In Pakistan, the Kharif crops including rice, maize, cotton and sugar cane, in cereal crops rice is the most important Kharif crop followed by maize. Figure 3.4 is showing Kharif crops in Pakistan, in most of the areas especially Punjab, Sindh, central parts of Khyber Pakhtunkhwa and parts of Baluchistan is having excessive production of Kharif crops. Therefore the country is exporting Kharif crops especially rice. Apart from that, Fig. 3.4 is showing a strip towards west including areas of Baluchistan and Khyber Pakhtunkhwa which is producing adequate Kharif crops. In the northern mountains due to topography and soil characteristics, the area is mild adequate in the production. The reason behind the excessive and adequate production is that the Sindh and Punjab due to fertile soil and improved irrigation facilities are producing more rice, while the mountainous areas of Khyber Pakhtunkhwa and Baluchistan are producing excessive or adequate maize during Kharif season.

3.4.2 Impacts of Climate Change on Agriculture

Pakistan is also prone to the antagonistic impacts of climate change; besides impacts on other sectors, the agriculture sector is also prone to the consequences of climate change. The impacts of climate change that will most likely affect agriculture include increase in the temperature in arable areas, changes in the rainfall pattern, variability of Monsoon winds, changes in the availability of irrigation water, extreme conditions like floods, droughts, cold waves, heat waves, locust, severe water stressed conditions in the arid and semi-arid areas. In Pakistan, 80% of the farming community including small landholders and such conditions affect them by reducing agriculture productivity and damaging their agricultural lands. The farmers in arid regions, mountains and coastal areas are most vulnerable to such extreme events. The impacts of climate change result in reduction in agriculture production due to shortening of the length of growing season increase in the heat at critical reproductive stages and increased water requirements of crops. These conditions result in decrease in the agriculture production in the arid and semi-arid regions by about 6–18% (Saifullah 2017).

3.4.3 Policy Measures Related to Food Security and Agriculture

In Pakistan, there is a federal division named the ministry of food security and research and there are several institutions working under the division like Federal Seed Certification and registration department, Agriculture Policy Institute, department of plant protection, animal husbandry department and animal quarantine department etc. Then there are provincial departments including agriculture research, agriculture extension and livestock departments. There are detailed policies related to food availability, diversification for income and nutrition, agriculture inputs, agriculture mechanization, livestock, fisheries, poultry and food accessibility etc. (GoP 2017c). The subsectors are covered in the broader policy related to agriculture and food security, the policy is addressing the subsectors and is being managed in an integrated way by the concerned departments.

The main operational responsibility will be at provincial level, but effective policy implementation will require the participation of various institutions, ministries and departments at federal level. The new policy directions will additionally require drawing in government institutions, including agriculture and agriculture extension departments, livestock and dairy development, irrigation, on-farm water management, fisheries, forestry on the agriculture side at provincial level. At other departments including education, health and women's development and social welfare at the sub provincial level and particularly at the grass-root level including district and union council level. The policy will also require working with the private sector including academia, NGOs/CSOs and farmers' organizations (GoP 2017c).

3.5 Land and Water Resources and Its Link with DRR and Climate Risk

In Pakistan,natural resource management is a challenge due to degradation of natural resources, including ground water depletion, soil health, rapid withdrawal of water resources and grassland degradation. The rate of urbanization is high and the land use is changing from fertile agricultural lands to build up an environment of urban areas for residential and other purposes. The main driver for the rural transformation is rural to urban migration; the people migrate to urban localities for jobs, business and amenities. The high population growth rate results in increased demand for food which also makes the water resources under stress. With the passage of time, the per capita availability of water has hugely decreased from 5000 m^3 in 1950s to less than 1000 m^3 in 2017. In the river discharges from a total of 142 million acre feet (MAF) nearly 104 MAF is diverted to irrigation canals and only 57 MAF reaches the farm head. The figure shows high loss of irrigation water for productive use in agricultural lands. The canal water is also supplemented with groundwater abstraction of about (50.3 MAF), there are more than one million tube wells in the country and it makes water availability at farm head of about 108 MAF. In the process, about 27 MAF is lost in field applications and only 81 MAF of water is used for crop consumption for the requirement of 102 MAF. The excessive use of tube well water has resulted in reduction of ground water and results fluctuations in water table. It also results in interloping of saline water in rock aquifers of ground water. From the sources of hill torrents about 19 MAF water is unused, there are potentials to harvest hill torrent water. It can result in bringing seven million hectares area under cultivation mostly in Baluchistan province which is deficient in agriculture production. The lands which can be cultivated due to harvesting of hill torrent water cover 67%, in Baluchistan followed by Khyber Pakhtunkhwa 13%, Punjab 8%, Sindh 8% and the federally administered areas 4% (GoP 2017c).

There are also some traditional means of irrigation in the rural areas of Pakistan especially in the provinces of Punjab, Sindh and Baluchistan. These traditional methods of irrigation include Karez irrigation and left irrigation methods. Karez irrigation is practised in Baluchistan province in the arid areas where the depth of water table varies due to slope and rock characteristics. The water is pitched from a high altitude point along a slope in an underground tunnel with a series of vertical tunnels used for digging underground tunnels and later on for repair and cleaning. The length of underground tunnel from source to command area is generally 1–2 km and reaches maximum of 10 km, the underground tunnel opens into land surface in the downslope areas, where mostly a storage tank is being constructed for irrigation water storage. The site selection for Karez is done by local well experienced elders, while the digging is being done by skilled workers specially trained for making Karez. Left irrigation is another traditional and oldest method of irrigation in Pakistan; it is mostly practised in the rural areas of the provinces of Punjab and Baluchistan. The earliest method of left irrigation was to pitch water in a bucket fixed with rope from shallow wells and ditches by hand. With the passage of time improvements were made to the left irrigation system including, *Dhenkli*, *Charsa* and finally Persian well. In *Dhenkli* or *Shaduf* a movable wooden stick with weight at one side and water carrying bucket at another side was fixed to a pole. The pole was fixed close to a ditch with water or well or river. After that, the system was further improved by using animals and *Charsa* was made. In *Charsa* there was a pole with pulley and a rope with bucket towards well and the next end is tied to a donkey for pitching water. The system was further improved by making Persian wheel locally known as *Rahat*, in which two wheels perpendicular to each other, an animal mostly bull or donkey and a series of bucks in the vertically fixed wheel. This system is still in practice in patches in the provinces of Punjab and Sindh.

According to National Food Security Policy 2017, irrigation sector including water availability and use face different challenges including lack of approved National Water Policy. Rapid population increase result in shrinking water resources especially fresh water resources and diminishing agricultural land. The water storage is inadequate and sedimentation in the water reservoirs further deteriorates the problem of water shortage. The high irrigation water losses results in low water availability and decreased agriculture productivity. The rainwater in the hill torrents is untapped and cannot be used productively for irrigation purposes. The energy resources in the country including electricity and petroleum are diminishing; there is shortage of electricity and cannot fulfil the present demand and the cost of petroleum resources is high according to the purchasing power. There is a lack of institutional arrangements and framework for regulating groundwater management which results in quarrying of groundwater aquifers. The intrusion of saline water is deteriorating the quality of ground water and making it saline. The violation of Indus Water Treaty by India by constructing water reservoirs on the rivers allocated to Pakistan under the treaty is also creating problems for the water and irrigation sector of Pakistan (GoP 2017c).

3.5.1 Impacts of Climate Change on Water Resources

Climate change is having negative impacts on water resources, due to glacial melt, unpredictable and uncertain pattern of rainfall results in shortage of irrigation water, increase in floods and soil erosion. Increase in the temperature will result in increase in the evapo-transpiration and will result in increased water demand of crops up to 10–30%. In the world, the glaciers are receding faster due to global warming. Pakistan is home to world's famous glaciers in the Himalaya Karakurum Hindukhush (HKH) region, including Siachin, Biafo, Hisper and Batura glaciers. Shrinking glaciers will have severe consequences for the sustainable water supply of the country. According to researches within the next two or three decades, glacial melt is projected to increase flooding. As the glaciers recede, it will also result in decreased river flow during normal conditions. Apart from that, the people of mountainous areas are becoming more prone to natural hazards due to increase of glacial lake outburst floods (GLOFs) which are increasing due to climate change (Saifullah 2017).

3.5.2 Water Versus DRR, Climate Risk and Policy Measures

According to national water policy 2018, following are the strategic priorities and planning principles related to water sector in Pakistan for addressing water related issues.

i. Due to seepage, the water loss is very high and in the Indus River system, almost 50% of canal water does not reach the farms. These canals are not lined and the seepage can be reduced by at least one third if these canals are lined. It will result in mitigating wastage of irrigation water and enhancing agriculture productivity.
ii. Storage of water is important for mitigating impacts of climate change on water resources. In Pakistan, the pattern of rainfall is not uniform and mostly the wet years with more precipitation are followed by dry years with very low rainfall. Therefore storing the water in wet years can result in having enough water in the dry years. For storage of water and irrigation projects, the policy recommends developing national master plan to focus on storage, floods, urban areas, arid areas, irrigation and tariff rationalization. There is a great potential to construct small and medium size dams and enhancing the lifetime of existing dams by remodelling.
iii. In Pakistan, there is a dire need for new technologies, including preparation of inventory of water resources through remote sensing and GIS and precise monitoring of irrigation water distribution.
iv. Sustainable water resource development is having close relationship with renewable energy. The water reservoirs including large, medium and small dams not only produce power but also deliver water for irrigation and other human needs. A large number of tube wells can be operated through solar energy where the water table is not so deep, it will provide additional water. The policy also proposes to use solar energy for day time use of desalination of sea water especially in the coastal areas of Baluchistan.
v. The policy proposes integrated water resource management (IWRM), in the world, the system is changing from sectoral to more integrated approach. Under IWRM the interests of areas including upstream and downstream can be secured against contamination and mining. The second point is to prolong the life of water reservoirs by managing the watershed and catchment. The policy also proposes to strengthen institutional and management capacities at all levels to properly adopt IWRM.
vi. The policy proposes the federal government to play primary role in assisting protocols to guarantee the efficient and sustainable utilization of ground water, waste water management and industrial uses. The sectors of water security, food security and energy security are closely interlinked; therefore the regulatory framework must deal with all the associated issues holistically including waste water treatment, ground water pollution and WASH related activities (GoP 2018).

vii. The policy also consists of detailed planning principles and implementation arrangements for the proposed points in the water policy 2018.

3.6 Urbanization, DRR and Climate Risk

The urban population is on increase throughout the world. It increased from 746 million in 1950 to 4 billion in 2016 (UN 2014, 2016; Table 3.4). In Pakistan, the urban population is also rapidly growing. In the year 1951, the total population of Pakistan was 33.74 million with urban population share of 6.019 million. During the year 2016, the total population was 195.40 million with urban population of 77.93 million. The percentage of urban population in the year 1951 was only 17.8%, which increased to 39.9% in the year 2016 (Khalid and Khalid 2017).

Due to increase in the urban population, the urban land cover is also rapidly increasing and results in increase in the soil sealing (Rahman et al. 2019). The soil sealing results in creating problems to the water resources; it reduces the seepage of water to aquifers and results in increase in surface runoff, which can cause lowering of water table and encourage urban floods. In Pakistan, the urban population in most of the cities is growing in an unplanned way and apart from other environmental problems it is also resulting into increase in the impermeable surfaces and results in soil sealing. Therefore, the water fluctuations are considerably affected by surface sealing caused by the rapid urban growth and expansion.

In a research study of Peshawar city district of Pakistan, the city population is rapidly growing just like other cities of Pakistan, due to that the abstraction of groundwater is also increasing for fulfilling needs of fresh water supply. The population of the study area was 1.113 million in the year 1981. The population of the study area increased to 3.8 million in the year 2016. The analysis shows that the sealed surfaces have increased, in the year 1981 the total built up area was 4635 ha making 3.7% of the total area, which is increased to 20,451 ha which is 16.27% of the total area. The same trends can be observed in all the cities of Pakistan, the population is increasing,

Table 3.4 Pakistan trend in urban population (in million) modified after Khalid and Khalid (2017)

Year	Total population	Urban population	Rural population	Percentage of total population	
				Urban	Rural
1951	33.74	6.01	27.72	17.8	82.2
1961	42.88	9.65	33.22	22.5	77.5
1972	65.30	17.33	48.72	26.5	73.5
1981	84.25	23.84	60.41	28.3	71.5
1998	130.85	42.14	88.12	32.5	67.5
2008	162.37	57.32	105.06	35.3	64.7
2016	195.40	77.93	117.48	39.9	60.1

it is resulting into increase in the built up area which increases the sealed surface. Such conditions results in modifications in ground and surface water. It result in decrease in the ground water, due to which the water table goes downward and sometimes the quality of water is also affected by the seeping of polluted water in the cracks and unconsolidated rock material. Due to soil sealing, the rate of surface water runoff also increases and results in the risk of urban flooding (Rahman et al. 2019). In Pakistan,due to high land value in the cities encroachment towards streams and rivers is also common, which results in making the population more vulnerable during floods (Mahmood et al. 2019a, b).

In Pakistan, the provincial governments are managing the urban areas and policy units are established like Urban Policy Unit of Khyber Pakhtunkhwa and Urban Unit of Punjab. There are also city management authorities for urban areas, like Karachi Development Authority, Lahore Development Authority and Islamabad Development Authority. In the present circumstances however, there is a lack of policy related to urban areas in the federal government but the government is in the process to formulate a detailed policy for managing cities and urban localities.

3.7 Natural Hazards, DRR, Climate Risk and Policy

Pakistan is prone to different types of hazards including geo-tectonic and hydro-meteorological hazards (Rahman 2015). The northern part of Pakistan and Azad Jammu and Kashmir are vulnerable to natural hazards including, earthquakes, landslides, floods, glacial lake outburst floods (GLOFs) and drought (Rahman and Shaw 2015). The southern part of the country is arid and semi-arid including Indus plain which is exposed to floods, river erosion, drought and pest attacks. The coastal region of Pakistan is exposed to hazards including cyclones, hydrological drought and storm surges, while some coastal areas in Sindh province along Indus River are also exposed to river floods. During heavy rains especially in the monsoon season, the country's big cities are prone to floods. These hazards make the country prone to serious dangers, it can be judged by the figures, that 6073 people were killed and more than 8,989,631 people were affected in the period between 1993 and 2002 (World Disaster Report 2003). During the year 2005, these figures were outnumbered by the devastating earthquake, 7.6 on Richter scale, killing more than 73,000 people and affecting more than 3.5 million in northern Pakistan. During the year 2010, a massive flood affected most parts of the country and proved to be a worst natural disaster. Apart from that the manmade disaster also traumatized the society, environment and the economy. These anthropogenic hazards include civil and communal conflicts, urban fires, nuclear and radiological mishaps and oil spills.

Table 3.5 is showing the details of earthquakes and floods since 2010, before that the country was hit by the notorious earthquake of 2005. Pakistan is prone to different types of natural hazards; the most dominant of these are floods and earthquakes. The floods can be considered as the most frequent and affecting big number of population. The floods also affect agricultural lands, crops, orchards and

Table 3.5 Pakistan major natural hazards since 2010 modified after ADRC (2015)

Year	Nature of disaster	Deaths	Injured	Population affected
2015	Earthquake	272	2123	–
2015	Flood	238	232	1.6 million
2014	Flood	367	673	2.5 million
2013	Earthquake	386	816	–
2013	Flood	333	173	1.5 million
2012	Flood	571	2902	4.8 million
2011	Flood	520	1180	9 million
2010	Flood	1750–2200	–	20 million

forests. Apart from river induced floods the northern mountainous areas experience flash floods during summer season. Pakistan is also situated in an active seismic zone making it more prone to earthquake hazards. There are different faults and thrusts which are formed due to the collision of Indian plate with Eurasian plate and making the country vulnerable to earthquakes.

3.7.1 Earthquake Resilience and Policy

Pakistan is part of the Indian plate and situated close to the point of collision between Indian and Eurasian plates. The country is prone to tectonic earthquakes, which mostly originates in the mountains of Hindukush, Karakorum or Himalaya. In Pakistan, the level of resilience can also be considered as low because of rugged topography in the north, poor roads infrastructure and shortage of health and logistic facilities. The early warning system and evacuation plan is also not effective in the remote and mountainous areas of the country, while the rescue system is also lacking advanced skills and equipment. However, the people of Pakistan are more proactive to help each other during any unanticipated situation. In the northern and western mountains, the earthquake is followed by land sliding due to unconsolidated rock material on mountain slopes. These hazards especially land sliding results in casualties and closure of roads and bridges and make the people less resilient to the hazard.

There are numerous faults and thrusts going through Pakistan especially in the Hindukush ranges in the western and northern part of Pakistan, which is making Pakistan prone to earthquakes. These faults and thrusts including Chaman fault, Main Karakorum Thrust, Himalayan Frontal Thrust etc. These faults and thrusts are developed due to the collision of Indian Plate and Eurasian Plate and the earthquakes in Pakistan are of tectonic origin. In October 2005 the country was hit by the worst earthquake and the aftershocks were continued for few days. Figure 3.5 is showing seismic hazard zones in Pakistan, there are four patches in west, northwest, northeast

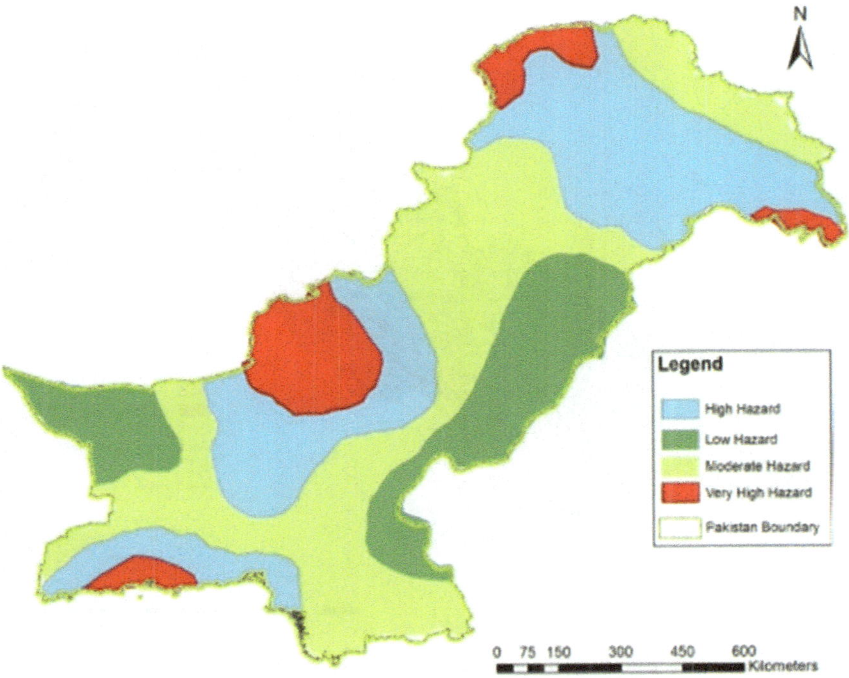

Fig. 3.5 Pakistan seismic Hazard zones modified after MoCC (2014)

and southwest of Pakistan; they are classified as very high hazard zones. It includes parts of Baluchistan, Azad Jammu and Kashmir (AJK) and Khyber Pakhtunkhwa. Around these very high hazard zones are the zones of high hazard, they are also situated in Baluchistan, Khyber Pakhtunkhwa and AJK (Fig. 3.5). Apart from that in the central part of the country, the zone of moderate hazard is extended to the south and west and there is also a strip in the extreme north. In the eastern part of the country along the border with India and the western part consist of Chaghi Baluchistan are considered as the zones of low hazard related to earthquakes.

3.7.2 *Flood Risk, Resilience and Policy*

Flood is the most frequent hazard in Pakistan, apart from seasonal floods which occur almost every year especially in the mountainous areas of the country was hit by several rivers induced floods in the last two decades including the worst flood of the year 2010. In Pakistan, the mountain areas are prone to flash floods especially during monsoon season and high snowmelt during summer. The central and southern

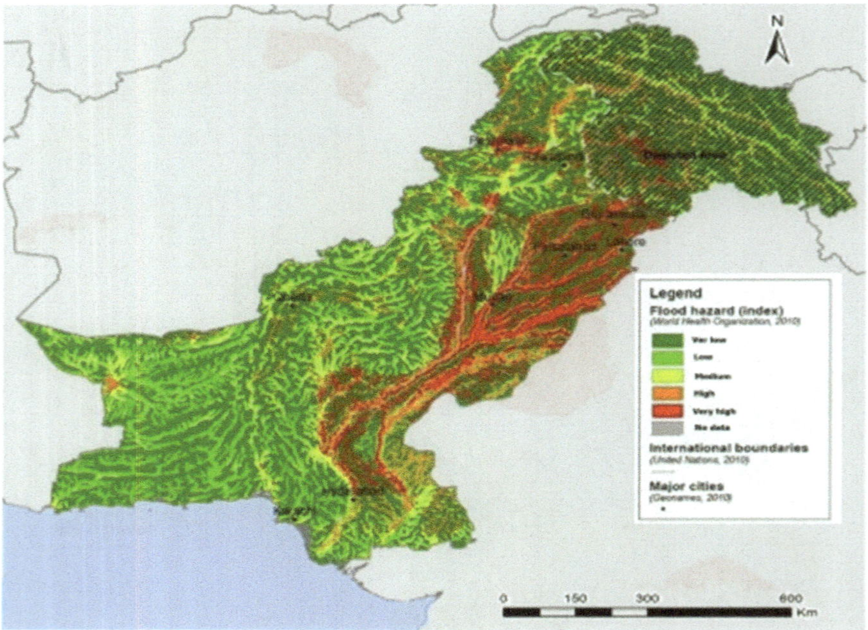

Fig. 3.6 Pakistan Flood Hazard Distribution modified after MoCC (2014)

parts of the country including some urban localities are prone to river induced floods in the same season.

Figure 3.6 is showing distribution of floods in Pakistan. The areas shown red along the rivers are highly prone to floods; these areas are situated along the rivers and streams in the Indus plain and northern mountains. The areas around very high flood hazards are considered as high flood hazard zones, in the Indus plain these areas situated in the active flood plains. Then there are the areas of moderate flood hazard around the high hazard zone. The regions of low and very low hazard zones are situated mostly in the western extreme northern part of the country. In the west the rainfall is low and the mountains are not having glaciers, therefore the areas are not prone to floods, while in the northwest Chitral is also arid and there are no risks of floods.

3.7.3 Climate and Disaster Risk Management System

In Pakistan, there are potentials to construct small and medium water reservoirs, which will mitigate the risk of flood and on the other hand, can provide more water for irrigation and other purposes. There are three administrative levels of governance related to disaster management; National Disaster Management Authority (NDMA) on federal level as central point, the provincial Disaster Management Authority (PDMA) in respective provinces is the central point for disaster risk management and District Disaster Management Authority (DDMA) for management on the district levels (GoP 2010, 2012). In case of any unforeseen event, the responder will be DDMA for rescue and rehabilitation. In case of more severity of the hazards, the PDMA and NDMA will intervene. For formulation of policies on national level National Disaster Management Commission (NDMC) was established and Provincial Disaster Management Commission (PMDC) was established headed by Chief Minister. The Deputy Commissioner of each district is heading the DDMA of respective districts at grass-root level. In the mechanism, NDMA is having lead role to formulate plans, strategies and programmes and to provide technical directions to the national and provincial stakeholders. While PDMA provide technical guidance to the district level and DDMA focus on rescue and timely response to natural hazards (ADRC 2015).

Pakistan is a disaster-prone country due to climatic extremes, geo-physical conditions and high degree of vulnerability and exposure. The earthquake 2005 and flood 2010 and 2011 have revealed the vulnerability of Pakistani society and economy to natural hazards. There is a range of hydro-meteorological, geo-physical and biological hazards, including floods, earthquakes, avalanches, droughts, cyclones, landslides, pest attacks, tsunamis, glacial lake outburst floods and epidemic pose risks to Pakistan.

National disaster risk reduction (DRR) policy was prepared in the year 2013 by the federal Ministry of Climate Change and National Disaster Management Authority (NDMA; GoP 2013). The policy provides a general guiding framework for addressing the high level of proneness to natural hazards. The policy covers both natural and anthropogenic hazards. The policy aims to promote measures on priority basis to mitigate vulnerability to hazards and measures to ensure development processes and programmes for enhancing resilience. The policy also provides guidelines both for DRR and related development plans and programmes to strengthen attention on priority issues. The policy is based on a detailed analysis of existing related documentation including policies, assessments, plans and frameworks. The foundations of the current DRR policy focus on the priority actions of the UN Hyogo Framework for Action (HFA) and are within National Disaster Management Act, 2010. The policy decentralizes authorities with clear roles and responsibilities at national, provincial and district levels for the effective implementation of DRR strategies and Climate risk resilience. The policy also promotes consultations among different stakeholders of district, provincial and federal government as well as civil society and development partners.

3.8 Strategies in Addressing Climate Change, Disaster Resilience and Policy

During the record of natural calamities in the last two decades, Pakistan is one of the topmost ten countries in the world, which are highly affected by climate change. However, the country's per capita greenhouse gases emissions is one of the lowermost in the world. In recent years, the country is frequently faced with unanticipated events caused by climate change, including climate-induced phenomena like glacial melt, seasonal floods, sea water intrusion, droughts and desertification. The sufferings due to the result of climate change are much more in the country than its total greenhouse gases emissions.

Pakistan is having great diversity in topography, climate and natural vegetation. The population of the country is sharply increasing there is unplanned built up and the settlements are encroaching towards mountain slopes, agricultural lands and river banks. Such conditions are making the country more prone to natural hazards and lack sustainable utilization of blue–green infrastructure. Pakistan is a party to the United Nations Framework Convention on Climate Change (UNFCCC), 1992; and Kyoto Protocol 1997.

Pakistan also started the process for evaluating the Paris Agreement 2015, which illustrates that climate change characterizes a very essential and possibly irreversible risk to the planet earth and human and therefore is a global challenge and require the extensive teamwork by all countries. The country has adopted the following policy measures for increasing resilience to natural disasters and climate change and paving the way for sustainable development.

3.8.1 Pakistan Climate Change Act, 2017

In Pakistan, a bill was passed in the national assembly to meet Pakistan's responsibilities under international conventions related to climate change and to address the effects of climate change. The Act describes "climate change" as a change in the climatic pattern which is caused by considerable changes in the concentration of greenhouse gasses as a direct or indirect result of human activities and which is in addition to natural climate change that has been observed (GoP 2017).

3.8.1.1 Formation of Climate Change Council

According to this Act, a climate change council was formed, the council was headed by the Prime Minister or the person nominated by the Prime Minister. The members including federal minister, chief ministers of all the provinces, provincial ministers dealing with environment, Chairman, National Disaster Management Authority (NDMA), Chairman Pakistan Climate Change Authority, Secretary of the Division owed with the subject of climate change, who shall also be the Secretary of the Council.

Following are the functions of the council.

(a) Coordinate and supervise implementation of this Act.
(b) Monitor the execution of international agreements related to climate change.
(c) Guide, supervise and coordinate, mainstreaming of climate change miseries into decision-making by Federal Government's and Provincial Government's ministries, divisions, departments and agencies, for creating favourable conditions for holistic climate resilient and climate-compatible processes in the development of different segments of the economy.
(d) Approve and regularly monitor execution of integrated adaptation and mitigation policies, programmes, strategies, plans, projects and other measures which are framed by the authority to achieve Pakistan's roles under the international conventions and agreements related to climate change and sustainable development goals (SDGs).
(e) Monitoring effective implementation of National Adaptation Plan and its integral constituent provincial and local adaptation action plans. Regular monitoring of the interventions related to National Appropriate Mitigation Action Framework and National Communications submitted to the concerned Secretariat of the United Nations Framework Convention on Climate Change (UNFCCC).
(f) Approval of procedures and guidelines for the security and protection of renewable and non-renewable energy resources, habitats, species and bio diversity that is harmfully affected or susceptible to climate change.
(g) Consider the National Climate Change Report for related activities and give timely proper guidelines thereon.

3.8.1.2 Formation of Climate Change Authority

According to Climate Change Act 2017, a Climate Change Authority was established with technical people as members. The Chairperson and members shall be professionals, scientists, academicians, industrialists, serving or retired government servants, agriculturists or technocrats. There is a specified list of functions of the authority in the act, the authority will be responsible for framing policies related to climate change and will also formulate mechanism for its implementation. The other functions also include preparing appropriate adaptation and mitigation projects for submission to national and international institutions for fundraising, including clean development mechanism, green climate fund, global environmental facility and adaptation fund. The functions also include preparing projects and initiating activities for financing under the mitigating emissions from deforestation and forest degradation and conducting needs assessments and preparing climate change technology action plan. The authority is also liable to frame and convey procedures for the protection and management of renewable and non-renewable energy resources, habitats, species and biodiversity in general which are negatively affected or prone to the vagaries of climate change.

The climate change authority was planned to be responsible for all the functions related to climate change including coordination with international organizations, putting forward recommendations to the government and monitoring and implementation of different initiatives and projects related to climate change. According to the act the government also established a climate change fund which will be utilized by the authority for different activities and functions.

3.9 Conclusion

Pakistan is a country with great diversity in topography, climate, natural vegetation and anthropogenic environment. The population is sharply increasing making it the fifth most populous country in the world. Due to inappropriate practices related to natural resources the country is facing problems of meagre forest resources and other ecosystems. Due to the rapid increase in population the problem of food insecurity is also increasing with the passage of time. The settlements are encroaching towards fertile agricultural lands due to unplanned built up. The urban population is sharply growing due to rural urban migration and the cities are growing mostly in an unplanned way. The growth of cities also results in soil sealing having impacts on ground water and surface runoff. Pakistan is also prone to different natural hazards including geo-tectonic and hydro-meteorological hazards. The country was worse affected during earthquake of 2005 and flood in 2010. Pakistan is also ranked among the top ten countries which are prone to the impacts of climate change. In Pakistan,due to north–south expansion of the country, diversity in topography and precipitation there is diversity in ecosystems. The diversity in forest related ecosystems play a important role in disaster and climate risk resilience. The mangrove forests in the coastal areas protect the coastal areas from saline water and erosion by sea waves. In the vast plains of Punjab and Sindh province, the riverine forests and irrigated plantation also mitigate soil erosion and positively contribute to air quality. The dry forest in the arid regions of Pakistan including Baluchistan, Thar Desert and Cholistan Desert provide opportunities for the local people to live and to have livelihoods sources. In the mountainous areas of northern Pakistan there are coniferous forests which make the fragile slopes to be stable and also mitigate soil erosion. Apart from enhancing disaster resilience, these diverse forests provide habitat to plants and animal species.

In Pakistan, there are also potentials to enhance food security, water security and renewable energy resources. For taking maximum advantage of the available resources and mitigating impacts of climate change and natural disasters, the government of Pakistan is framing policies and procedures on a need basis. In the past, due to political instability and sometimes law and order situations the long term policies and plans were mostly not achieved. Now the government established different organizations on federal and provincial levels and there are detailed policies and implementation arrangements for natural resources management, food security and disaster risk reduction.

Pakistan is facing challenges related to natural resources management, the country is also prone to natural hazards but there are potentials of proper management for development and preserving and enhancing the diverse ecosystems. The government is making policies and programmes for mitigating natural hazards and impacts of climate change. In Pakistan, there are potentials related to integrated natural resource management and renewable sources of energy. Being a country with exceptional diversity in topography, climate, natural vegetation and anthropogenic environment, Pakistan has evoked interest not only for Geographers worldwide but also environmental scientists, botanists and zoologists. The phenomenal growth in population makes it the fifth most populous country in the world. Due to inappropriate and unscientific practices related to the utilization of natural resources, the country is facing problems of depleting forest resources and destruction of unique ecosystems. The alarming and uncontrolled increase in population coupled with the problem of food insecurity is also raising head gradually and slowly. Human settlements are encroaching upon fertile agricultural lands as unplanned and unscientific methods of construction are in progress. The urban population is growing at alarming level in the process of contiguous rural urban migration. New cities are proliferating everywhere in an unplanned and unsupervised way with little consideration for the welfare of future generations. This unprecedented growth of cities also results in soil sealing with deliberating impacts on ground water and surface runoff. The country is also vulnerable to different natural hazards and disasters including geo-tectonic and hydro-meteorological hazards. The country was hit hard by the colossal earthquake in 2005 and devastating floods in 2010. Pakistan is also on the list of the top ten countries which are severely affected by the rising global warming and continuous climate change that may eventually reach calamitous proportion if not contained immediately.

It was found from the analysis that Pakistan can enhance food security, water security and renewable energy resources if these resources were effectively managed. In order to avail maximum advantages from the available resources and reduce the deteriorating impacts of climate change and natural disasters, the government of Pakistan is framing new policies and procedures on a need basis. The long term policies and plans could not be implemented audaciously keeping in view the consistent political instability and deteriorating law and order situation in the country. Now the government has established different organizations on federal and provincial levels with detailed policies, line of actions and implementation arrangements for natural resources management, food security and disaster risk reduction.

Apart from encountering issues related to natural resources management, Pakistan is also on the hit list of innumerable natural hazards but proper management, development and utilization of remarkable natural resources can enhance the overall infrastructure and preserve the uniquely diverse ecosystem. The current government seems serious in its stance and is devising policies and programmes to combat and mitigate natural hazards in consideration of its impacts upon climate change within as well as outside the country. Pakistan offers amazing potentials related to integrated natural resource management and renewable sources of energy. It's high time to develop visionary plans, take urgent initiatives, raise awareness and implement the long awaited projects in the interest of the country as well as the world at large.

References

Ahmad SS, Abbasi QA, Jabeen R (2012) Decline of conifer forest cover in Pakistan: a GIS approach. Pak J Bot 44(2):511–514

Amjad AS, Jusoff K (2007) Mangrove conservation through community participation in Pakistan: the case of Sonmiani Bay. Int J Syst Appl Eng Dev 1(4):75–81

Anonymous (2007) Intergovernmental panel on climate change (IPCC). In: Working group, vol 3. Cambridge University Press, Cambridge

Asian Disaster Reduction Centre (ADRC) (2015) Country Report Pakistan, ADRC, Kobe, Japan

Asim H, Akbar M (2019) Sectoral growth linkages of agricultural sector: implications for food security in Pakistan. Agric Econ Czech 65(6):278–288

Barkati S, Rahman S (2005) Species composition and faunal diversity at three sites of Sindh mangroves. Pak J Zool 37(1):17–31

Beresnev N, Phung T, Broadhead J (2016) Mangroves related policies and institutional framework in Pakistan, Thailand, and Vietnam. Mangroves for the future (MFF). FAO and IUCN

Dawood M, Rahman A, Ullah S, Mahmood S, Rahman G, Azam K (2020) Spatio-statistical analysis of rainfall fluctuation, anomaly and trend in the Hindu Kush region using ARIMA approach. Nat Hazards 101(2):449–464

FAO (2005) Trends in forest ownership, forest resources tenure and institutional arrangements: a case study from Pakistan. Food and Agriculture Organization of the United Nations

FAO (2020) FAO website. http://www.fao.org/forestry/112650f977bbb5c6a591b2924c6443ef171d08.pdf

Giri C, Long J, Abbas S, Murali RM, Qamer FM, Pengra B, Thau D (2015) Distribution and dynamics of mangrove forests of South Asia. J Environ Manage 148:101–111

Government of Pakistan (GoP) (2005) National environmental policy. Ministry of Environment, Islamabad

Government of Pakistan (GoP) (2010) National disaster management act. Government of Pakistan, Islamabad

Government of Pakistan (GoP) (2012) National climate change policy. Ministry of Climate Change, Government of Pakistan, Islamabad

Government of Pakistan (2013) National disaster risk reduction (DRR) policy. Ministry of Climate Change and National Disaster Management Authority (NDMA), Islamabad

Government of Pakistan (2015) Economic survey of Pakistan. Federal Bureau of Statistics, Islamabad

Government of Pakistan (2017a) Population Census Report, Pakistan Census Organization, Bureau of Statistics Islamabad

Government of Pakistan (2017b) Pakistan Climate Change Act 2017. Ministry of Climate Change, Islamabad

Government of Pakistan (GoP) (2017c) National food security policy. Ministry of Food Security and Research, Government of Pakistan, Islamabad

Government of Pakistan (2018) National Water Policy, Ministry of Water Resources. Government of Pakistan, Islamabad

Government of Pakistan (2020) Agriculture statistics, important crops. Bureau of Statistics Government of Pakistan, Islamabad

IUCN (2005) Mangroves of Pakistan—status and management. IUCN, Pakistan

Khalid NA, Khalid F (2017) Pakistan: a study of geographical environment, economy and human resources. Azeem Academy Publishers, Lahore, Pakistan
Khan FK (2003) Geography of Pakistan. Oxford University Press, Karachi
Khan FK (2006) Pakistan: geography, economy and people. Oxford University Press, Karachi
Khan SM, Ahmad H (2015) Species diversity and use patterns of the alpine flora with special reference to climate change in the Naran, Pakistan. In: Climate change impacts on high-altitude ecosystems. Springer, Cham, pp 155–175
Khan SR, Khan SR (2009) Poverty–deforestation links: evidence from Swat, Pakistan. Ecol Econ 68:2607–2618
Mahmood S, Rahman A, Shaw R (2019) Spatial appraisal of flood risk assessment and evaluation using integrated hydro-probabilistic approach in Panjkora River Basin, Pakistan. Environ Monit Assess 191:1–15
Mahmood N, Arshad M, Kächele H, Ma H, Ullah A, Müller K (2019) Wheat yield response to input and socioeconomic factors under changing climate: evidence from rainfed environments of Pakistan. Sci Total Environ 688:1275–1285
MoCC (2014) Ministry of climate change, environmental Atlas of Pakistan. Ministry of Climate Change, Islamabad
Owais SM, Sadique S (2019) (2019) Appraisal of Deforestation and Forest Degradation in District Swat, Pakistan. Int J Econ Environ Geol 10(3):07–15
Pedlar JH, Mckenny DW (2017) (2017) Assessing the anticipated growth response of northern conifer populations to a warming climate. Sci Rep 7:43881
Rahman A (2015) Effectiveness of the disaster risk management system in Pakistan. Arab World Geogr 18(1–2):124–138
Rahman A, Samiullah, Shaw R (2017) Fragile mangroves and increasing susceptibility to coastal hazards in Pakistan. In: DasGupta R, Shaw R (eds) Participatory mangrove management in a changing climate. Disaster risk reduction (methods, approaches and practices). Springer, Tokyo
Rahman A, Shaw R (2015) Disaster resilience: generic overview and Pakistan context. In: Rahman A, Khan AN, Shaw R (eds) Disaster risk reduction approaches in Pakistan. Springer Publishing Co., Tokyo, pp 53–75
Rahman A, Khan AN, Shaw R (2015) Disaster risk reduction approaches in Pakistan. Springer, Tokyo
Rahman A, Farzana RG, Shaw R (2016) Flood disasters and land use planning in Swat Valley, Eastern Hindu Kush. In Banba M, Shaw R (eds) Land use management in disaster risk reduction. Springer, Tokyo
Rahman A, Khan A, Haq N, Samiullah, Shaw R (2019) Soil sealing and depletion of groundwater in rapidly growing Peshawar City District, Pakistan. In: Ray B, Shaw R (eds) Urban drought. Disaster risk reduction (methods, approaches and practices). Springer, Singapore
Saeed F, Almazroui M, Islam N, Khan MS (2017) Intensification of future heat waves in Pakistan: a study using CORDEX regional climate models ensemble. Nat Hazards 87:1535–1647
Saifullah SM (1982) Mangrove ecosystem of Pakistan. In: The third research on mangroves in Middle East, Japan Cooperative Centre for the Middle East, vol 137, pp 69–80
Saifullah SM, Rasool F (2002) Mangroves of MianiHor lagoon on the north Arabian Sea coast of Pakistan. Pak J Bot 34(3):303–310
Saifullah (2017) Climate change impact on agriculture of Pakistan—a leading agent to food security. Int J Env Sci Nat Resour 6(3)
Sendai framework for disaster risk reduction 2015–2030, adopted at the third United Nations world conference on disaster risk reduction held from 14 to 18 March 2015, in Sendai, Miyagi, Japan
Shaw R, Rahman A, Surjan A, Parvin GA (2016) Urban disasters and resilience in Asia. Elsevier, New York
Sindh Forest Department (SFD) (2015) Sindh Forest Department. Department website. http://www.snv.org/project/mangroves-and-markets
Sindh Forest Department (SFD) (2020) Sind Forest Department. Department website. https://sindhforests.gov.pk/index.php

Snedaker SC (1984) Mangroves: a summary of knowledge with emphasis on Pakistan. In: Haq BU, Milliman JD (eds) Marine geology and oceanography of Arabian Sea and Coastal Pakistan. Van Nostrand Reinhold, New York, pp 255–262

United Nations (2014) World urbanization prospects. Population Division of the Department of Economic and Social Affairs of the United Nations Secretariat, New York

United Nations (2016) The world's cities in 2016: data booklet (ST/ESA/SER.A/392). Population Division of the Department of Economic and Social Affairs of the United Nations Secretariat, New York

US Geological Survey (2020) Dawn news Pakistan, 12 Apr 2020. https://www.dawn.com/news/1215636

World Disaster Report (2003) International federation of red cross and red crescent societies. Geneva, Switzerland. ISBN 92-9139-092-5

Chapter 4
Ecosystem-Based Approaches and Policy Perspectives in Nepal

Shobha Poudel, Bhogendra Mishra, and Rajib Shaw

Abstract Mountain ecosystems are particularly vulnerable to climate change since increasing temperatures and disruptive precipitation patterns have led to floods, droughts and other natural disasters. Ecosystem-based approaches to disaster risk resilience (Eco-DRR) and Adaptation (EbA), is a nature-based method for climate change adaptation (CCA). That can reduce the vulnerability of the ecosystem to extreme events enhancing sustainability in various sectors, including but not limited to agriculture, forestry, energy and water. Similarly, by increasing the resilience of vulnerable communities, EbA helps countries to meet the goals of the Sendai Framework for Disaster Risk Reduction. This study reviewed the existing Eco-DRR/EbA approaches and its integration into policy and planning in Nepal. Literature suggests that EbA approaches (1) enhance community adaptive capacity or resilience, (2) help ecosystems to produce goods and services for local communities and (3) is financially and economically viable in Nepal. However, EbA is not in mainstream for CCA so far in the country. Existing policies, institutional and political obstacles are the major challenges for the effective implementation, despite EbA has a high potential in Nepal. Policymakers should bring it into the mainstream of development that could make significant progress in mitigating the climate impact at local, provincial and national scales.

Keywords Climate change · Ecosystem-based disaster risk resilience · Ecosystem-based adaptation · Policy · Nepal

---

S. Poudel (✉) · B. Mishra
Science Hub, Balaju, Nepal

Policy Research Institute, Kathmandu, Nepal

R. Shaw
Keio University, Fujisawa, Japan

© The Author(s), under exclusive license to Springer Nature Singapore Pte Ltd. 2021

M. Mukherjee and R. Shaw (eds.), *Ecosystem-Based Disaster and Climate Resilience*, Disaster and Risk Research: GADRI Book Series,
https://doi.org/10.1007/978-981-16-4815-1_4

4.1 Introduction

The mountain ecosystem of Nepal is the source of various ecosystem goods and services such as clean water, mountainous food and habitat for the mountainous animal etc. (Poudel and Shaw 2016; Spehn et al. 2010). The mountain environment is fragile to change, which increases vulnerability in the region in the context of climate change (Houet et al. 2010; Poudel et al. 2020). The number of natural hazards such as landslides and floods are more frequent which cause considerable human and economic losses every year (Poudel and Shaw 2015, 2016). Natural and climatic factors, such as heterogeneous topography can induce such hazards. These hazards can claim a number of lives and enormous losses of resources every year (Table 4.1). Human interventions also exacerbated these natural risk factors. In addition to that climate-induced disasters are expected to increase in number and intensity in the future (Mishra 2011; Poudel et al. 2017; Poudel and Shaw 2016). To minimize the risk from these disasters, the disaster risk reduction (DRR) system in Nepal relies mainly on response and rehabilitation rather than preparedness and mitigation (Nepal et al. 2018). The climate change adaptation (CCA) measures are still not in the mainstream of development. Ecosystem-based measures for Disaster Risk Resilience (Eco-DRR) or adaptation (EbA) have not been widely planned and implemented albeit they have a high potential for mitigating potential risk. As well as it contributes to sustainable development because of their mutual benefits. Hence, ecosystems should be incorporated as a part of climate-compatible infrastructure. It should be treated as an integral part of facilities and equipment that played a vital role to make economic resilience with climate change and its induced disasters (Emerton et al. 2016).

EbA is one of the measures to adapt to the adverse effects of climate change and induced disasters by using biodiversity and ecosystem services. It aims to reduce vulnerability and improve the resilience of ecosystems and people considering the potential vulnerability under the climate change scenario (Mensah et al. 2011). The application of EbA spans through the diverse ecosystems, geography, sectors and stakeholders across the countries around the world (UNDP 2015). Forest conservation and sustainable management to prevent rainfall-induced disasters such as landslides with more intense rains; restoration of degraded wetlands to protect against increasing floods are examples of EBA measures. As well as sustainable management of grassland to protect against floods and soil erosion; and agroforestry can contribute against changing climate enables production (Secretariat of the Convention on Biological Diversity 2009).

Eco-DRR is a well accepted tool for the DRR in a sustainable way (Renaud et al. 2013). As Eco-DRR has a number of co-benefits over the conventional engineering-based solution, it is recognized as a no-regret strategy (Daigneault et al. 2016; Renaud et al. 2016). Andrade et al. (2011) suggested to take in account when implementing EbA: (i) increase multi-stakeholder engagement; (ii) incorporate flexible operational and management structures; (iii) operating at several geographical areas; (iv) increasing the co-benefits of development and conservation goals; (v) rely on the

Table 4.1 Damages and losses of disaster incidents in Nepal in 2018

Damage and losses of disaster incidents										
S. No.	Incident	No. of incidents	Death			Injured	Affected family	Houses destroyed		Estimated loss (NPR)
			Male	Female	Total			Partial	Complete	
1	Floods	418	128	55	183	61	16,196	14,424	286	60,944,400
2	Landslides	483	96	65	161	182	1083	149	328	191,662,000
3	Lightening	432	87	72	160	551	618	23	14	14,687,000
4	Fire	3973	74	76	150	557	6027	549	3234	6,422,638,013
5	Cold wave	48	26	22	48	0	48	0	0	–
6	High Altitude	45	37	4	41	6	46	0	0	–
7	Heavy Rainfall	342	8	22	30	84	538	193	252	89,415,160
8	Animal Terror	141	14	8	22	69	280	136	8	4,390,150
9	Wind storm	254	7	12	19	84	1527	763	301	51,447,998
10	Boat Capsize	10	9	7	16	9	27	0	0	–
11	Epidemic	22	11	4	15	1881	420	0	0	–
12	Snake Bite	18	7	7	14	5	18	0	0	–
13	Snow storm	2	10	0	10	0	10	0	0	–
14	Avalanche	1	1	0	1	0	1	0	0	–
15	Hailstone	3	0	0	0	0	127	2	0	457,000
16	Others	189	54	45	99	150	289	69	10	3,181,599
Total		6381	569	399	968	3639	27,255	16,308	4433	6,838,823,320

Source The Government of Nepal—Ministry of Home Affairs (MoHA) (2019) best available evidence based on indigenous knowledge; (vi) enhancing the resilience of the most vulnerable people and (vii) ensure the active participation of socially excluded groups. Climate change is a cross-cutting issue and should be incorporated into different policy fronts. In this study, the integration of Ecosystem-based approaches within policies has been carried out especially in three areas: (1) Integration of international policies into ecosystems and natural resource management; (2) Analysis of National Adaptation Program of Action (NAPA) and national policies from the perspective of Ecosystem-based approaches (3) How water resource and forests sector have been integrated into the mainstream of development, climate change and sectoral policies in Nepal. There are several international policies that support Ecosystem-based approaches for climate change adaptation and disaster risk reduction.

4.2 Ecosystem-Based Approaches for Reducing Risks and Adapting to Climate Change

The main aim of EbA and Eco-DRR is the sustainable management, conservation and restoration of ecosystems to combat with consequences of climate change or to minimize the impacts of natural disasters (Munang et al. 2013; Renaud and Murti 2013). Both EbA and Eco-DRR intersect each other in making use of properly managed and well-functioning ecosystems, their goods and services, to support vulnerable communities to cope with the changing climate and natural disasters. These EbA and Eco-DRR incorporate the mixed approaches that connect naturally with scientific approaches (Van Bohemen 2012; UNEP 2015). Besides, they also have a common purpose to integrate multiple disciplines and multiple sectors. It seems Nepal had a proper balance of actions across sectors, where the forest sector (31%) covered most of the actions of EbA, agriculture covered 27% and water and biodiversity had (21%) (18%) respectively (Fig. 4.1; Table 4.2) (Ikkala 2011). Most of the actions in Nepal, Peru and Uganda on ecosystem-based approaches were under biodiversity conservation or under the natural resource management sectors of forests, agriculture and water. In Peru and Uganda, there were several actions in the tourism sector, related to wildlife and ecosystem conservation for tourism as another source of livelihood amid climate change whereas there was only one action in the energy sector in Nepal.

With the aim of providing benefits in poverty reduction and rural employment, as well as for climate change adaptation, community forest and multi-stakeholder and multi-actor participation in governance and management of forests have been promoted. Ecosystem-based approaches in the forest sector have taken into account of conservation, sustainable management, agroforestry and non-timber forest product (NTFP) management for adaptation. Forests and biodiversity come together in many plans and policies, which shows that many of the recognized forest adaptation approaches are ecosystem based. Additionally, several biodiversity conservation programs are concentrated particularly on forest ecosystems and its surroundings.

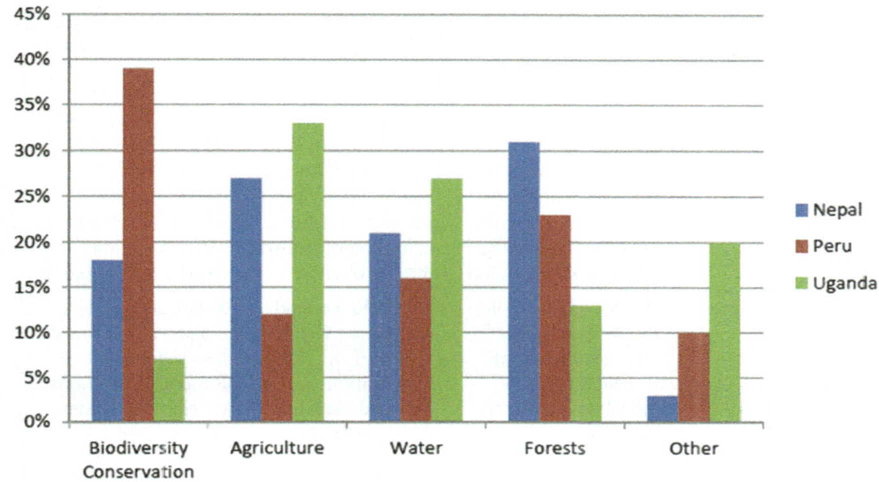

Fig. 4.1 Measures taken on ecosystem-based approaches to adaptation in different sectors in Nepal, Peru and Uganda. *Source* UNDP (2015)

Some examples are area increment of wetlands to minimize flood risks and forest restoration in the slopes to minimize the risk of landslides (Brink et al. 2016; Lange et al. 2019). Several researches prove the appropriateness and comparative advantages of Eco-DRR and EbA measures (Doswald and Estrella 2015; Renaud et al. 2016). Among them, some studies indicate that, besides their advantages in lower costs and sustainability, they can be the alternative sources of local livelihoods. As well as it plays an imperative role to minimize the risk of climate change impacts if they are well planned and focus on implementation (Daigneault et al. 2016; Renaud et al. 2013, 2016). Ecosystem-based approaches to adaptation and disaster risk reduction are interlinked in many cases; they can reduce vulnerability and enhance the resilience to both natural and human-induced disasters.

4.3 Integrating EbA and Eco-DRR into Planning and Policy

Ecosystem-based approaches to adaptation and disaster risk reduction are promoted by leading international development agencies. One of them is the Sustainable Development Goals (SDGs) adopted by the UN General Assembly. The SDGs associated to DRR are: developing cities for all that can integrate all the stakeholders, safe and resilient to disasters and sustainable (SDG 11), taking immediate action to cope with unprecedented disasters and minimize the adverse impacts (SDG 13). Similarly, protect and use of oceans, seas and marine resources in a sustainable way (SDG 14) and sustainably manage forests, combat desertification, reverse of degraded land and

Table 4.2 Examples of how three Eco-DRR projects in mountain countries of Nepal, Peru and Uganda integrated Community based-DRR goals and addressed four recommended principles

	Nepal	Peru	Uganda
Actions	Wetlands, pond, spring, forest restoration; women leased abandoned land to plant broom grass; gabion walls; roadside stabilization	Traditional water canal restored; land taken out of domestic grazing for vicuña grazing; wetland management; animal fibre production	Hydrological gravity flow schemes and reforestation; conservation agriculture; improved water retention; riverbank restoration
Benefits environment DRR	Water conservation; improve degraded land. Reduction of floods, fires, landslides, drought impacts	Grass and wetland restoration; vicuña (wildlife) conservation Decrease landslides, floods, fire risk, drought impacts	Water and soil conservation; forest loss reduction Reduction of floods, erosion, landslides, drought impacts
Principles governance	Strengthened institutional capacity for community management; used Panchase Protected Forest structure; worked with forest groups and women's groups. Informed Forest Management Plan	Strengthened water, pasture committees; formed new committees and plans; worked in No Yauyos Chochas Landscape protected area; contributed to the regional government's climate change strategy	Community water user groups formed Mt Elgon Conservation Forum brought together up/downstream actors for joint planning and decision making; worked with government extension
Capacity	Training promoted EbA; reduced water conflicts and diseases; enhanced incomes and employment	Capacity building for livestock and vicuña management; income from vicuña and tourism; value chain development increased	Increased income from crops; enhanced social capital; communities less reliant on food aid
Knowledge equity	Restored traditional water storage ponds Activities fit with women's schedules and workloads; included women in different castes	Restored a forgotten traditional water management model. Communication training focused on women, youth and elders, as under-represented groups	Farmer to farmer exchanges, peer to peer learning. Less time spent finding water and collecting firewood, especially for women

Source Klein et al. (2019)

reduce biodiversity loss (SDG 15). EbA and Eco-DRR can be promoted through effective incorporation into policy and practice in addition to the suggested goals related to sustainable development, poverty reduction and biodiversity conservation. This should be integrated into the policy cycle from policy planning to policy evaluation.

Top-down or bottom-up approaches are applied to integrate EbA and Eco-DRR most effectively. Meaningful participation of local people and indigenous communities, as well as practitioners, plays an imperative role in policymaking processes. Several countries have included EbA and Eco-DRR into their national plans and strategies. They have integrated EbA and Eco-DRR even in National Biodiversity Strategies and Action Plans (NBSAPs) under the CBD, disaster management plans, development policy and drought relief policy, NAPAs under the United Nations Framework Convention on Climate change (UNFCCC). NAPA under the UNFCCC supports the Least Developed Countries (LDCs) in prioritizing the tasks to cope with emerging climate change adaptation needs.

Many LDCs have identified the importance of ecosystems to support people to adapt to climate change in their spotlighted NAPA projects. Fifty-six percentage of high priority projects from NAPA had significant natural resource components (Reid et al. 2009). Some of the countries such as Cape Verde, Eritrea, Sudan, Solomon Islands and Vanuatu have NAPA projects that had a robust natural resource component, while each NAPA considered at least one project with a natural resource component. Pramova et al. 2010) identified the incorporation of EbA into the NAPAs and found that 68% of the NAPAs have mentioned the ecosystem services. It was estimated that around 42% of NAPA projects incorporated ecosystem restoration as an adaptation component Stucki and Smith (2011).

Climate change was not in mainstream development while formulating most sectoral policies. While looking into the climate change and sustainable water management policies in 9 countries, the biodiversity policies and to some extent water policies were not updated for many years. These policies were prepared before realizing the climate change impacts (Pittock 2011). Though most of the sectoral policies that did not consider the EbA are formulated two decades ago. Some recently formulated policies, including Nepal's Strategic Vision for Agricultural Research (NARC 2010) has not taken into account the EbA. A few Ecosystem-based approaches to adaptation-related activities are already being practised in Nepal. Forests and livelihoods project that is operationalized currently has mainstreamed adaptation. A new program on EbA in Mountain Ecosystems, supported by the German Federal Ministry for the Environment, Nature Conservation and Nuclear Safety (BMU), is being carried out in Nepal, Peru and Uganda (Fig. 4.1; Table 4.2). Several countries have been initiated and implementing activities on ecosystem-based approaches to adaptation. Some of them are related to prioritizing the NAPAs activities, however, it shows that these highly prioritized projects have not received the funding adequately. The NAPA highly prioritized project, integrated management of agriculture, water, forest and biodiversity sectors in Nepal has applied several ecosystem-based approaches to adaptation. Ecosystem-based approaches often come together with other approaches for adaptation for example integration of blue, green and grey infrastructure.

Institutional arrangements and structures play an imperative role to enhance coordination among the sectors. As the government body has coordinating powers over line ministries, it is better to lead the cross-sectoral issues like EbA and Eco-DRR. Capacity building of the local, provincial and federal governments should be mainstreamed for Eco-DRR/EbA activities. As well as for the effective implementation of Eco-DRR and EbA approaches, awareness among different stakeholders could be raised and technical skills need to be developed in Nepal.

4.4 Ecosystem-Based Approaches to DRR and Adaptation in the International Policy

The Rio Conventions have recognized the importance of ecosystems for adaptation. This type of platform provides policy guidance to the countries to get information on how to implement and promote action related to sustainable management, conservation and restoration of ecosystems that are practised at the international level (Lange et al. 2019). These types of conventions can also be opportunities for identifying and advancing the joint effort in the delivery of adaptation. The first set of decisions addresses the need for ecosystems to be adapted amid climate change, which is the objective of the UNFCCC as well. Incorporation of climate change in countries' strategic plan to managing ecosystems and maintain their resilience is mentioned in the CBD mandate. The UNCCD (2008–2018) aimed to enhance the resilience of impacted ecosystems to climate change. The emphasis of the second set of references is enabling people to adapt to the impacts of climate change from ecosystem management and the goods and services provided by it. The UNFCCC Cancun Adaptation Framework asks Parties to act on adaptation through sustainable management of natural resources while CBD defines Ecosystem-based Approaches to Adaptation. The UNEP/UNDP/IUCN partnership program on Ecosystem-Based Adaptation in Mountain Ecosystems funded by the German Government also emphasizes this kind of adaptation.

The two major global initiatives, the Sendai Framework for Disaster Risk Reduction and the Sustainable Development Goals (SDGs) acknowledge the EbA and Eco-DRR in 2015. It also shows the increasing importance as well as the benefits of EbA and Eco-DRR. The Sendai Framework for Disaster Risk Reduction was approved in Sendai, Japan, in March 2015 that replaced the Hyogo Framework for Action (UN 2015a). And In September 2015 Sustainable Development Goals has been adopted by the UN General Assembly (UN 2015b). A new consensus to address climate change was done in Paris in December (UNFCCC 2015). These agreements have recognized the importance of ecosystem and ecosystem services to climate change adaptation, mitigation, disaster risk reduction and sustainable development. The application of ecosystems for DRR, CCA and sustainable development has increased and ideas such as EbA, Eco-DRR, Nature-based Solutions, Green Infrastructures, Working with Nature etc. have come out or been advanced during the last twenty years. This type of acknowledgement has eased the increased implementation of Eco-DRR/EbA projects. However, there is variation in the definition and concepts of ecosystem-based approaches (Renaud et al. 2016).

4.5 Ecosystem-Based Approaches to DRR and Adaptation in National Plans and Policies in Nepal

The significant role of ecosystems and ecosystem services for development has been emphasized in various sectoral policies without integrating it with adaptation. Some of the plans and policies such as the Agricultural Perspective Plan (1996), Conservation Strategy (1988), the Forest Sector Policy (2000), the Biodiversity Strategy (2002) and the Wetlands Policy (2003) in Nepal. All these policies emphasize the interrelation between ecosystems, biodiversity and human wellbeing, but do not integrate to climate change. The NBSAPs hardly recognize the relationship between biodiversity conservation, land degradation and desertification and adaptation. Nepal has already developed Local Adaptation Plans of Action (LAPAs) and NAPA, which provide a good opportunity for integrating ecosystem-based approaches to DRR and adaptation. And National Adaption Plan (NAP) formulation is ongoing. Several national and sectoral policies in the country do not integrate ecosystem-based approaches to adaptation.

4.5.1 National Adaptation Plan (NAP)

The NAP has been regarded as a pathway to identify the need for medium and long-term adaptation and to develop and implement strategies. This plan would build on the lesson learned from the NAPA process that was developed to address the short-term adaptation needs. Nepal has already pointed out that the NAP should take into account the vulnerable ecosystems and cross-sectoral approaches that should align with existing plans such as NAPA, National biodiversity strategies action plans (NBSAPs) while submitting its views on the NAP. The NAPs that is being developed under the UNFCCC framework could adopt the cross-sectoral, ecosystem-based approach for the CCA.

4.5.2 National Adaptation Program of Action (NAPA)

The National adaption program of action (NAPA) is developed through a comprehensive analysis of the climate change induced vulnerability and risk assessment. This includes development of a long list of adaptation options and identifies the most promising one through a set of prescribed steps such as cost-benefit analysis among others. Some of them were ecosystem-based adaptions. In NAPA, four of such options included the ecosystem-based among nine. The adaptation options are often cross-sectoral, multi-stakeholders and multi-criteria process that is likely to be the ecosystem-based approaches. Therefore, the NAPA actions were selected based on a set of given criteria, which include potential to support the livelihood of rural part of the country and easy to implement. The focus was given to the cross-sectors approaches that could have mutual benefit in the range of sectors. The NAPA develop the project profile on community-based adaptation through integrated management with the ecosystem-based adaptation was the key that supports strengthening the local institutions and develop the early warning systems.

4.5.3 Nepal Climate Change Policy

Nepal Climate Change Policy aims to reduce the vulnerability to climate change enhancing the climate change adaptation capacity of local people to utilize the maximum products and services provided by natural resources and their efficient management (Climate Change Policy 2011). All three levels of government, federal, province and local should incorporate ecosystem-based approaches as important elements of climate change policy. National Development Plans also acknowledge the imperative role of ecosystems in supporting people to adapt to climate change. This indicates that ecosystem management and adaptation are taken as part of economic prosperity, poverty alleviation and sustainable development. Nepal has an integrated ecosystem-based approach to adaptation plans of water and forests. Overall national development is guided by these types of cross-sectoral documents, considering the ecosystem-based approaches as a mainstream development is significant. All countries should have built and structured the climate change and adaptation policies to complement one another to get the synergetic effect between priorities, objectives, visions and actions. The country's development plans are also lined up with the objectives around these approaches. The National Environmental Policies do not link adaptation and ecosystems. The National Environmental Policy of Nepal (1993) is too old and does not speak about climate change. Focus must be given at national level to take the advantage of adaptation in the nation's plan and implementation of commitments under all Rio Conventions.

4.5.4 National Strategy for Disaster Risk Management

In Nepal, disasters Management Strategies do not make interrelation between disaster risk management and ecosystems or adaptation. Nepal's National Strategy for Disaster Risk Management neither mentions adaptation, nor considers ecosystem management as a means to enhance resilience to disasters. On one hand, most of the policies related to natural and environmental disaster risk reduction are formulated or derived from international discussions and frameworks. While new policies were also formulated aftermath of the disasters that occurred in the country, especially after the Nepal earthquake 2015. As a reaction to this event, the government of Nepal has formulated and amended several disaster risk reduction policies that establish responsibilities at the national, state and municipal level. Nepal recently established National disaster risk reduction and management authority (NDRRMA) in 2019. This legislation aimed to change disaster response from disaster risk prevention and mitigation in disaster risk management. However, risk professionals argue that this is the proper way; there is a big problem in the implementation of risk prevention and mitigation strategies. National authorities claim that this is mainly due to the lack of enough knowledge. Disaster practitioners are mainly trained for post-disaster responses like search and rescue; and relief. They are only trained on technical safe-guarding, not in risk prevention and mitigation approaches. The political will and financial availability highly determine the effectiveness of the NDRRMA.

4.6 The Potential and Limitations of EbA and Eco-DRR

Several literatures depict that ecosystem-based approaches in Nepal have promising results, however, these kinds of measures are not taken into account. Consequently, it is bounded either in a specific location or in some piloting sites (Renaud et al. 2016; Ikkala 2011). Therefore, their full potential is not well explored how beneficial for resilience building and to reduce vulnerability. Nevertheless, Nepal government makes an effort for conservation and restoration of the forest sector at the national level. Specifically, with the implementation of the Mountain EbA program in the Panchase region of Nepal (Figs. 4.2 and 4.3). It shows that Mountain EbA has a direct contribution to minimize the disaster risk and climate change impacts. Additional information and concrete data on climate change impact and disaster risk could help to label these efforts as EbA and Eco-DRR measures.

A pilot project such as Panchase is also key to gaining experiences and makes a success story of EbA and Eco-DRR that can help to mainstream these measures on a national level. Nonetheless, to reduce the vulnerability and enhance the resilience of the landscape, a bunch of programs are needed which include but not limited to massive reforestation for regulating the regional climate and land restoration (Ikkala 2011; UNDP 2015). Forests can be used to protect from landslides and rockfalls and need to increase the buffer zone areas to reduce the loss from flooding.

While developing and designing the green or habitat corridors, wildlife protection should also be taken into consideration in a way that they simultaneously aligned with DRR activities. Nevertheless, this fosters a harmonious relationship between environmental planning authorities and disaster prevention.

Ecosystem-based measures that are still not recognized by the decision-makers is one of the major problems for its implementation. Difficulties in determination of cost and benefit ratio and assessing the effectiveness are major challenges in the short

Fig. 4.2 Wetland in Panchase area Nepal. *Source* Ikkala (2011)

Fig. 4.3 Village in Panchase area Nepal. *Source* Ikkala (2011)

run. Commitment of the local population is required as well as awareness-raising campaigns and environmental education should be conducted for the effective implementation of the ecosystem-based measures. These awareness campaigns should be conducted in a way that enhances community perception for better engagement (Lange et al. 2019).

There is not a proper monitoring and evaluation system and mechanism for the appropriate environmental impact assessment. In this regard, Panchase can be considered as one of the key pilot cases to upscale the EbA for the proper implementation and long-term monitoring and evaluation. Changes in land-use patterns sometimes may raise obstacles to manage floodplains areas, agricultural land and residential areas since floodplains are massively used for agriculture and settlements. Hence, shifting the residential areas and agricultural pocket areas to create retention areas may cost high and could generate major social and economic conflicts. Generally, EbA and Eco-DRR seem to be beneficial in the context of Nepal.

The ongoing CCA and DRR projects that are based on ecosystem-based services can contribute to make its widespread acceptance in the region.

4.7 Conclusion

EbA and Eco-DRR play imperative roles to reduce the vulnerability to climate change and its induced disasters for a country like Nepal. Therefore, ecosystem management, conservation and restoration should take into account to increase socio-ecological resilience and support people to adopt the EbA. Nepal has incorporated EbA and Eco-DRR into its climate change and adaptation policies. However, it is not enough to know the full potential of the EbA and Eco-DRR in the country's economic growth, poverty reduction and national development.

Ecosystem-based approaches to adaptation have not been well addressed in several natural resources management related policies even if they acknowledge the imperative role of ecosystem management for human wellbeing and development. The commitments made on all Rio conventions should be taken in high priority for the implementation and focus must also be given at the national level to take the benefits of adaptation in the planning. The existing NBSAPs and UNCCD NAPs also have not integrated adaptation. While formulating the environmental and sectoral policies climate change was not mainstreamed on the development agenda. On the other hand, these policies have not been updated for more than two decades. First of all, these policies need to be updated.

As discussed in the earlier section, several constructive developments have also taken place on the policy level. Several international agreements have highlighted the facts and mentioned the critical role of ecosystems for DRR and CCA. Sendai Framework for Disaster Risk Reduction (SFDRR) (UN 2015a) and Sustainable Development Goals (SDGs) have highlighted the role of ecosystems (UNFCCC 2015). Similarly, the Paris Agreement has also mentioned the environmental or ecosystem integrity (UNFCCC 2015). In the twelfth conference of the Parties in 2015, the Convention on Biological Diversity has also emphasized the importance of ecosystem-based solutions for CCA and DRR. Similarly, they have made a decision in the Ramsar Convention on Wetlands adopted resolution XII.13 on "wetlands and disaster risk reduction" (Renaud et al. 2016). Studies show that there is a possibility of interlink among major international agreements and Eco-DRR/CCA. The importance of ecosystem-based solutions for DRR and CCA is increasing globally. Ecosystem-based approaches are important in every sector at the local unit for development and resources management. Therefore, it is crucial to develop as a component of mainstreaming in the local and sub-national level, which is responsible to plan and deliver the services in the field. Multi-stakeholders' engagement in planning and implementing EbA and Eco-DRR actions create the ownership among the different parties. These approaches can build on people's participation, cost-effective and easy to implement and can contribute significantly to local livelihoods.

Agriculture, water, forests and biodiversity conservation sectors have mainly integrated actions on ecosystem-based approaches to adaptation. Nevertheless, other sectors that focus on natural resource management, such as energy, health and tourism can also be benefited from ecosystem-based approaches to adaptation. And it is urgent to explore such opportunities. Hard infrastructure and institutional strengthening can also come together with ecosystem-based approaches to adaptation in many ways. Therefore, ecosystem-based approaches should be taken into account as part of broader, overall adaptation strategies. Research on ecosystem-based approaches to adaptation is highly recommended for future study. Ecosystem-based approaches should be linked with scientific cases to assess the vulnerability and climate change impact that support evidence-based decision making and action on the field level. Multi-stakeholder institutional arrangements are needed for better integrated, cross-sectoral policy formulation that is imperative to ensure the delivery of effective, sustainable solutions which enhance the resilience of both ecosystems and human being to climate change.

References

Andrade Á, Córdoba R, Dave R, Girot P, Herrera FB, Munroe R, Oglethorpe J, Paaby P, Pramova E, Watson E (2011) Draft principles and guidelines for integrating ecosystem-based approaches to adaptation in project and policy design: a discussion document. Turrialba, Costa Rica

Brink E, Aalders T, Ádám D, Feller R, Henselek Y, Hoffmann A, Ibe K, Matthey-Doret A, Meyer M, Negrut NL (2016) Cascades of green: a review of ecosystem-based adaptation in urban areas. Glob Environ Chang 36:111–123. https://doi.org/10.1016/j.gloenvcha.2015.11.003

Climate Change Policy (2011) Climate change policy Nepal

Daigneault A, Brown P, Gawith D (2016) Dredging versus hedging: comparing hard infrastructure to ecosystem-based adaptation to flooding. Ecol Econ 122:25–35. https://doi.org/10.1016/j.ecolecon.2015.11.023

Doswald N, Estrella M (2015) Promoting ecosystems for disaster risk reduction and climate change adaptation: opportunities for integration. United Nations Environment Programme, Geneva

Emerton L, Huxham M, Bournazel J, Kumara MP (2016) Valuing ecosystems as an economic part of climate-compatible development infrastructure in coastal zones of Kenya and Sri Lanka. In: Ecosystem-based disaster risk reduction and adaptation in practice. Springer, pp 23–43. https://doi.org/10.1007/978-3-319-43633-3_2

Houet T, Loveland TR, Hubert-Moy L, Gaucherel C, Napton D, Barnes CA, Sayler K (2010) Exploring subtle land use and land cover changes: a framework for future landscape studies. Landsc Ecol 25:249–266. https://doi.org/10.1007/s10980-009-9362-8

Ikkala N (2011) Ecosystem based approaches to adaptation in national policy: a case study from Nepal, Peru and Uganda

Klein JA, Tucker CM, Steger CE, Nolin A, Reid R, Hopping KA, Yeh ET, Pradhan MS, Taber A, Molden D, Ghate R, Choudhury D, Alcántara-ayala I, Lavorel S, Müller B, Grêt-regamey A, Boone RB, Bourgeron P, Castellanos E, Chen X, Dong S, Keiler M, Seidl R, Thorn J, Yager K (2019) An integrated community and ecosystem-based approach to disaster risk reduction in mountain systems. Environ Sci Policy 94:143–152. https://doi.org/10.1016/j.envsci.2018.12.034

Lange W, Sandholz S, Viezzer J, Becher M (2019) Ecosystem-based approaches for disaster risk reduction and climate change adaptation in Rio de Janeiro State, pp 345–359

Mensah A, Deeb A, Grünwaldt AH (2011) Making the case for ecosystem-based adaptation: building resilience to climate change. United Nations Environment Programme. UNEP

Mishra B (2011) Analysis and forecasting of snow cover using ANN in Kaligandaki Basin, Nepal. A thesis submitted in partial fulfillment of the requirements for the degree of Master of Science in Remote Sensing and Geographic Information Systems, Asian Institute of Technology School of Engineering and Technology, Thailand

Munang R, Thiaw I, Alverson K, Liu J, Han Z (2013) The role of ecosystem services in climate change adaptation and disaster risk reduction. Curr Opin Environ Sustain 5:47–52. https://doi.org/10.1016/j.cosust.2013.02.002

NARC (2010) NARC's strategic vision for agricultural research (2011–2030). Meeting Nepal's Food and Nutrition Security Goals through Agricultural Science and Technology

Nepal P, Khanal NR, Pangali Sharma BP (2018) Policies and institutions for disaster risk management in Nepal: a review. Geogr J Nepal 11:1–24. https://doi.org/10.3126/gjn.v11i0.19546

Pittock J (2011) National climate change policies and sustainable water management: conflicts and synergies. Ecol Soc 16

Poudel S, Shaw R (2015) Demographic changes, economic changes and livelihood changes in the HKH Region. In: Krishna Nibanupudi H, Shaw R (eds) Mountain Hazards and disaster risk reduction. Springer Japan, Japan, pp 125–137. https://doi.org/10.1007/978-4-431-55242-0

Poudel S, Shaw R (2016) The relationships between climate variability and crop yield in a Mountainous environment: a case study in Lamjung District, Nepal. Climate 4:13. https://doi.org/10.3390/cli4010013

Poudel S, Funakawa S, Shinjo H (2017) Household perceptions about the impacts of climate change on food security in the mountainous region of Nepal. Sustainability 9:641. https://doi.org/10.3390/su9040641

Poudel S, Funakawa S, Shinjo H, Mishra B (2020) Understanding households' livelihood vulnerability to climate change in the Lamjung district of Nepal. Environ Dev Sustain 1. https://doi.org/10.1007/s10668-019-00566-3

Pramova E, Locatelli B, Brockhaus M, Fohlmeister S (2010) Ecosystem-based adaptation in the national adaptation programmes of action (NAPAs). In: National climate change adaptation research facility (NCCARF) 2010 climate adaptation futures conference. Gold Coast, Australia.

Reid H, Alam M, Berger R, Cannon T, Milligan A (2009) Community-based adaptation to climate change, participatory learning and action. International Institute for Environment and Development, London, UK

Renaud FG, Murti R (2013) Ecosystems and disaster risk reduction in the context of the Great East Japan Earthquake and Tsunami: a scoping study Report to the Keidanren Nature Conservation Fund. UNU-EHS

Renaud FG, Sudmeier-Rieux K, Estrella M (2013) The role of ecosystems in disaster risk reduction. United Nations University Press

Renaud FG, Sudmeier-Rieux K, Estrella M, Nehren U (2016) Ecosystem-based disaster risk reduction and adaptation in practice. Springer

Spehn EM, Rudmann-Maurer K, Körner C, Maselli D (2010) Mountain biodiversity and global change. Global Mountain Biodiversity Assessment

Stucki V, Smith M (2011) Integrated approaches to natural resources management in practice: the catalyzing role of national adaptation programmes for action. Ambio 40:351–360

The Government of Nepal—Ministry of Home Affairs (MoHA) (2019) Nepal disaster report
UN (2015a) Sendai framework for disaster risk reduction 2015–2030
UN (2015b) Sustainable development goals (WWW document). https://www.un.org/sustainabledevelopment/news/communications-material/. Accessed 7 May 2020
UNDP (2015a) Making the case for ecosystem-based adaptation: the global mountain EbA programme in Nepal, Peru and Uganda
UNEP (2015b) Promoting ecosystems for disaster risk reduction and climate change adaptation: opportunities for integration
UNFCCC (2015) Adoption of the Paris agreement (WWW document). https://unfccc.int/resource/docs/2015/cop21/eng/l09r01.pdf. Accessed 7 Apr 2020
Van Bohemen H (2012) (Eco)system thinking: ecological principles for buildings, roads and industrial and Urban areas. In: Sustainable urban environments. Springer, pp 15–70

Chapter 5
Ecosystem-Based Approaches and Policy Perspective from India

Shweta Bhardwaj and Anil Kumar Gupta

Abstract Ecosystem-based approaches to climate change adaption (EbA) and disaster risk reduction (Eco-DRR) utilise the opportunities created by sustainable management, conservation and restoration of ecosystem assets and services. These are not just cost-effective and flexible but can also deliver multiple benefits. These approaches have been widely recognised and accepted in the backdrop of increasing disaster risks which are known to be exacerbated due to environmental changes, viz. climate change, land use changes and ecosystem degradation. Integration of ecosystem-based interventions into developmental planning and actions can significantly contribute towards achieving developmental and economic efficiency. Over the years, there has been an evolution in India's policy frameworks across various strategically important thematic areas and sectors, which has offered number of suitable pathways for mainstreaming of these ecosystem-based approaches into developmental planning and practice. Present paper makes a diagonal review of ecosystem-based strategies and interventions for disaster risk reduction and climate change adaptation through the existing policy environment across key themes and sectors. A systematic discussion of key policies for their present strengths, opportunities and customization required to foster integration of ecosystem approaches, is presented with detailed case analysis of climate change, disaster management and sectoral developmental policy contexts in India.

Keywords Ecosystem-based approaches · Disaster risk reduction · Climate change adaption · Sustainable development · Developmental policies

5.1 Development in India: An Overview

Since independence, India has achieved remarkable economic advancement which can be well reflected through growth that India has witnessed across various spheres including its national income and agricultural sector which has helped India in

S. Bhardwaj · A. K. Gupta (✉)
National Institute of Disaster Management (Ministry of Home Affairs), Sector 29, Rohini, Delhi 110042, India

© The Author(s), under exclusive license to Springer Nature Singapore Pte Ltd. 2021

M. Mukherjee and R. Shaw (eds.), *Ecosystem-Based Disaster and Climate Resilience*, Disaster and Risk Research: GADRI Book Series,
https://doi.org/10.1007/978-981-16-4815-1_5

becoming self-reliant in food production, diversify its industrial base (steel, power, fertiliser etc.) and make advancement across its critical economic infrastructures including energy, transport, communication etc. India, in these years, has also made substantial efforts in advancing social development across the country, there are a number of commendable policies, programmes and schemes launched by the government to promote education, health, livelihood and social security. Despite these efforts, India's social development particularly in terms of overall poverty and inequality reduction remains very limited. India positions at 129 out of 189 countries as per its human development index value, assessed based on long-term progress made by countries across three dimensions which include: access to knowledge, decent living standard and long healthy life (UNDP 2019a, b). In contrast to this, India is also one of the fastest growing market economies in the world and the third largest economy by GDP in the Asian continent (IMF 2017).

While world looks at India as an eminent power, India holds a huge responsibility of meeting the developmental needs of its growing population within its environmental boundaries. In a country like India, development is an imperative; however, growing population and rapid urbanisation poses a serious threat to sustainability of India's developmental growth. Increased consumption demands are placing serious constraints on available resources such as fossil fuels, land, water, forest, minerals etc. Along with these challenges climate change which is often argued to be a developmental problem rather than solely being an environmental issue has increased our exposure towards extreme weather events; leaving us more vulnerable towards disasters. A report, '*EnviStats India*' (Govt. of India 2018), highlights some of the worrisome statistics on the state of India's natural capitals viz. land, water and forest. The report mentioned about the increased extraction of groundwater which projects the unsustainability and future availability of the resource; reduction in area under forest cover (except from a very few states which showed the increase) and increase in built-up areas is witnessed at the cost of farmlands; and there are also evident effects of climate change with declining areas under snow and glaciers. All these trends are alarming and highlights the ecological footprints of human development. Shortcomings in development or ill-planned development most often shape up the disaster risks in the society and these disasters have potential to undo years of hard earned developmental gains. Disasters can set back the economic and human developmental progress of any country by destroying the vital social infrastructure/resources, disrupting the economy and impacting the local livelihoods.

5.2 Disaster, Climate Change and Development Linkages: Indian Context

The unique geo-climatic condition of India makes it vulnerable to a number of natural hazards such as floods, earthquakes, cyclones, drought etc. Almost 58.6% of landmass of the country is prone to earthquake (moderate to high intensity); 12% of the landmass is vulnerable to floods and 68% of cultivable area in the country is prone to droughts (GOI 2011). India's vulnerabilities to disaster risk are further driven by a number of socio-economic factors including poverty, inequality, growing populations, vulnerable livelihoods and developmental priorities of the growing economy. In India, on an average, direct losses from disasters cost up to 2% of the country's GDP and almost 12% revenues of the central government (World Bank 2003). Floods are the greatest source of annual losses in the country accounting for $7 billion losses every year (UNISDR 2015). During 1980–2017, India experienced 278 flood events that caused about $58.7 billion of economic loss and affected more than 750 million people across the country (UNISDR & CRED 2018). These estimations of human and economic cost of disasters mainly include the immediate and direct losses incurred from the disasters which include deaths, injuries, destruction caused to buildings/infrastructure etc. Most often, the indirect impacts of disasters on communities remain understated or unaccounted.

Disaster risk pervades through vulnerabilities and poverty is one of the most influential drivers of socio-economic vulnerabilities in India. Poverty is a multi-dimensional problem which encompasses a lack of access to education, health, sanitation, safe livelihood, food and nutritional security, safe drinking water and other essential resources and services. Poverty is considered a cause as well as a consequence of disaster risk (Wisner et al. 2004). Building resilience of the communities by strengthening their capacities is what is required to address the disaster risks and vulnerabilities. Very often gains achieved through poverty reduction and developmental advancements helps in reducing disaster risk. Poverty risk reduction and developmental programmes related to livelihood, social protection, food security etc. also help in achieving disaster risk reduction goals by reducing vulnerabilities of poor and enhancing their resilience.

Climate change poses an additional threat to India's development. Climate change directly impacts development by affecting the climate sensitive developmental sectors (agriculture, water, energy etc.), which indirectly result in socio-economic consequences (migration, loss of livelihood, poverty etc.) borne by communities. Climate change tends to affect the frequency, spatial extend and intensity of extreme events and would result in unprecedented extreme climatic and weather-related events in future; thereby excreating the disaster risk. Just like for disasters, unsustainable development remains the underlying cause for climate change as well. Therefore, the developmental pathways most often determine the degree of vulnerabilities of natural and socio-economic systems towards climate change.

Development which is considered as a major driver of disaster and climate change risk can create excessive stress on natural ecosystems and resources; accumulate risk in urban areas due to unplanned rapid developmental activities and can exacerbate socio-economic inequalities (UNISDR 2015). Developmental policies and strategies can play a critical role in addressing disaster and climate change concerns. Figure 5.1 illustrates the linkages between development, disasters and climate change. The relationship between development, disasters and climate change and provides the rationale for integration the disaster risk reduction and climate adaption strategies into developmental planning, which have been supported by many international policy

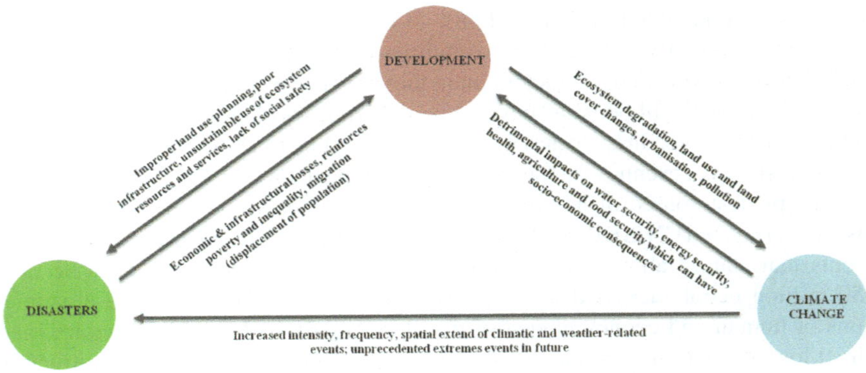

Fig. 5.1 Development, disaster and climate change linkages

frameworks including Sendai Framework for Disaster Risk Reduction (SFDRR), Sustainable Developmental Goals(SDGs) and United Nation Framework Convention on Climate Change (UNFCCC).

Risk informed development planning can address underlying disaster and climate change vulnerabilities, which can be achieved through management of environment and natural resources; promotion of social and economic developmental measures and proper land use planning. Integrating disaster risk reduction measures into ongoing developmental policies and programmes are viewed as possible entry points for mainstreaming Disaster Risk Reduction (DRR) and Climate Change Adaption (CCA) into developmental planning. There are a number of national programmes and schemes which can act as potential entry points for mainstreaming disaster and climatic concerns at national and sub-national levels (Gupta et al. 2016). These developmental programmes can enhance resilience of communities towards disaster and climatic risks by ensuring access to sustainable livelihood and basic services like healthcare, education, sanitation, water supply etc.; thereby, reducing vulnerabilities and enhancing capacities of communities to withstand the disaster and climate change impacts. Table 5.1 provides a list of some of the key ongoing government initiatives across different developmental areas in India.

5.3 Ecosystem-Based Approaches to Disaster Risk Reduction and Climate Change

The goods and services provided by the ecosystem are the basis of existence of human civilisation and its well being. Conversely, humans also have an important role in defining the well being of ecosystems and their services. Human actions have substantially changed the functioning and structures of ecosystems. These changes made to ecosystems have resulted in substantial gains in terms of human well being

Table 5.1 National developmental programmes and schemes across different themes (India)

Themes	Programmes
Health and sanitation	National Rural Health Mission (NRHM) Integrated Child Development Services Swastha Bharat Mission Swacch Bharat Mission
Infrastructure/ Rural or Urban development	Atal Mission for Rejuvenation of Urban Transformation Pradhan Mantri Gram Sadak Yojana (PMGSY) Smart Cities Mission
Livelihood	Mahatma Gandhi National Rural Employment Guarantee Act (MGNREGA) National Rural Livelihood Project (NRLP)
Education	Mid-Day Meal Sarva Shiksha Abhiyan
Agriculture	Pardhan Mantri Krishi Sinchayee Yojana Pardhan Mantri Fasal Bima Yojana Soil Health Card Scheme

and economic growth, achieved at the cost of degraded ecosystem services. There are evidences that suggest the degradation of ecosystems have resulted in increased risk of nonlinear changes in ecosystems which can have potential impacts on well beings of humans and these harmful effects of degraded ecosystems is disproportionately borne by the poor, who are directly dependent on these services and are the most vulnerable to any changes in these services (MEA 2005). These changes in ecosystems can put serious pressure on the supply of critical ecosystems goods and services, having severe consequences such as loss of livelihoods, health impacts and local migration of population.

The interlinkages among ecosystems, disasters and climate change are multifaceted and complex. Ecosystem degradation often results in reduced capacities of ecosystems to effectively deliver critical goods and services, leaving communities dependent on them more vulnerable towards disaster risks and impacts of climate change (Gupta et al. 2013). Ecosystems themselves, act as natural buffers towards modulating the impact of natural hazards and climate change on human systems, while the services provided by them are also of critical importance in disaster risk reduction and climate change mitigation and adaption. Therefore, maintaining healthy ecosystems is considered as a key requirement for building resilience of natural and human systems towards the impacts of disasters and climate change (IUCN 2013; Munang et al. 2013). Climate change affects disaster risks by exacerbating climate and weather-related hazards. Disaster risk which is assessed on the basis of three components: frequency and severity of hazard; number of people and assets exposed to hazard and their vulnerabilities, can be effectively addressed by adopting a sustainable approach towards ecosystem management. Sustainable management of ecosystems can effectively address each of the three components of disaster risk as well as climate change and its impacts. Ecosystem-based approach allows strategic management of ecosystems and its services which can effectively contribute towards disaster risk reduction and climate change mitigation. Ecosystem-based Disaster Risk Reduction (Eco-DRR) and Ecosystem-based Climate Change Adaption (EbA) are two such approaches which emphasise on role of ecosystems in effectively managing the risk associated with disasters and climate change respectively.

5.4 Ecosystem-Based Disaster Risk Reduction (Eco-DRR) and Ecosystem-Based Adaption (EbA)

Both Ecosystem-based Disaster Risk Reduction (Eco-DRR) and Ecosystem-based Climate Change Adaption (EbA) have lots of similarities as the governing principle behind both the approaches remains the same, which advocates for sustainable management, conservation and restoration of ecosystems and their services. The only divergence that comes among these two is in the specific objectives that are being addressed by these two concepts, where one specifically deals with disaster risk reduction while other addresses climate change and its impacts. In contrast to conventional approaches towards disaster risk reduction and climate change adaption, ecosystem-based approaches are cost-effective approaches, capable of providing multiple economic, environmental and social benefits; and can help in enhancing resilience of communities at local and national levels towards disasters and climate change. These approaches deliver these multiple co-benefits irrespective of the uncertainty of disaster occurrence or potential climate change impacts, these approaches are therefore considered as a 'no-regret' investment (Renaud et al. 2013).

Ecosystem-based disaster risk reduction (Eco-DRR) put forward such an approach for reducing disaster risk through sustainable management, conservation and restoration of ecosystems and their services while promoting sustainable and resilient development (Estrella and Saalismaa 2013). Similarly, Ecosystem-based Adaption (EbA) consider sustainable management, conservation and restoration of biodiversity and ecosystem services as the fundamental part of overall adaptation strategy towards climate change and its impact; focuses on maintaining and increasing the resilience of ecosystems and communities, while reducing their vulnerabilities towards adverse impacts of climate change (SCBD 2009). Climate change is recognised as the driver of present and future disaster risk and in this regard, an integrated Eco-DRR approach can also effectively address CCA concerns. However, to do so, Eco-DRR needs to adopt a broader perspective of sustainable development evolved from knowledge, experiences and expertise from three domains: environment, disaster management and climate change (Renaud et al. 2013).

Eco-DRR and EbA utilise ecosystem-based management strategies, tools and principles for addressing disaster and climate concerns. These tools and strategies provide opportunities for maintaining and enhancing ecosystems and their services while building resilience of ecosystems and communities towards disasters and climate change. Some of these strategies include environmental impact assessment tools; integrated risk and vulnerability assessments; protected area management; integrated ecosystem management tools (integrated watershed management, integrated coastal zone management, integrated forest management, sustainable dryland management etc.); and community-based natural resource management (traditional or indigenous knowledge, ecosystem-based sustainable livelihood practices etc.) (PEDRR 2010; SCBD 2009).

Box 1: Case-study example for Ecosystem-based Disaster Risk Reduction

Protected Area Management for Disaster Risk Reduction using traditional community knowledge: A case-study of Manas Biosphere Reserve, India

Manas Biosphere Reserve in Assam, India, is a designated UNESCO world heritage site. It is one of the most diverse biodiversity hotspots in the country characterised by a range of forested hills, tropical evergreen forest and alluvial grassland providing critical habitats for several endangered species. The Subankhata Reserve Forest located on the eastern side of Manas Biosphere Reserve comprises of rocky terrains and course changing Himalayan Rivers and is prone to floods and soil erosions (landslides) during heavy rainfalls.

The indigenous tribal community living in the biosphere reserve through an old traditional technique of channelizing and regulating water resources (Himalayan Rivers), popularly known as *Dong Bundh System*, reduced soil erosion and flood risk in the region. In this technique, micro-check dams are constructed along the river using local resources such as timber, bamboo and boulders. The technique contributed significantly towards disaster risk reduction by successfully mitigating flood and landslides hazard risk (due to soil erosion) in the lower catchment areas and also ensured availability of water for drinking and irrigation purposes in the otherwise water-deficient regions; thereby, also reducing socio-economic vulnerabilities of the communities.

Source IUCN (2016), Ekaratne and Vidanage (2013)

Box 2: Case-study example for Ecosystem-based Climate Change Adaption

Enhancing Climate Resilience of India's Coastal Communities: A GOI-UNDP Project

The project aims at enhancing resilience of vulnerable coastal communities towards climate change and its impact through an integrated ecosystem centric and community-based approach for coastal management. The project is being implemented across coastal districts for the state of Andhra Pradesh, Maharashtra and Odisha. The three broad objectives of the project include: adopting a co-management approach for enhancing resilience of coastal and marine ecosystems and services; promoting climate-resilient coastal livelihood; and strengthening governance and institutional framework for integrated coastal management. The key activities undertaken during the project include:

- Conducting coastal vulnerability; developing decision-support adaptive planning tool and preparing site-specific restoration guidelines
- Community-based restoration and conservation of coastal ecosystems
- Promotion of climate-resilient livelihoods and enterprises; strengthening market-linkages
- Improving local capacities for supporting climate-adaptive livelihoods through awareness on climate change and impacts, role of ecosystem; building and strengthening village-level capacities of CBOs, NGOs, SHGs etc.
- Conducting training on climate-adaptive agriculture and aquaculture for communities and knowledge exchange programmes; providing financial and technical support to communities for up taking these practice
- Developing network of institutions/organisations for promoting knowledge exchange

Source UNDP (2019a, b)

5.5 Policy Perspective from India

The effectiveness of ecosystem-based approaches in addressing disasters risk and climate change by increasing the social and the ecological resilience in a cost-effective manner is well evident. However, most of these evidences are anecdotal and project-derived, reflecting lack of integration of these approaches across national/regional policies, planning and practices. These approaches can be easily scaled up through their effective mainstreaming into national developmental policies and practices supported by proper institutional and financial frameworks. However, ensuring effectiveness and efficiency of these policies on ground, can only be achieved by developing coherence across multi-level policy planning, implementation and financing across multiple sectors such as agriculture, tourism, forestry, infrastructure etc.

In India, there are several national policies, plans and strategies which support the mainstreaming of ecosystem-based approaches. However, most of these policies/strategies do not explicitly deal with ecosystem-based approaches but include them indirectly, illustrating the linkages between policies/strategies and these approaches. These approaches are often implemented within broader strategies across different policy issues. For example, there are policies related to environment, climate change, disaster management and sectoral development which have direct and indirect references to both these approaches.

The following section discusses some of key policies across climate change, disaster management and different developmental sectors analysing their linkages with ecosystem-based approaches (Eco-DRR and EbA).In most of these policies, there are different types of strategic instruments that have been employed for mainstreaming ecosystem-based approaches as a part of overall strategy and for achieving the desired policy targets.

5.5.1 Climate Change

5.5.1.1 National Action Plan for Climate Change (NAPCC)

NAPCC is the foundational climate policy in India, released in 2008. NAPCC reflects the country's vision of environmentally sustainable development. While addressing the climate change concerns, this strategic framework provided the directional shift in the developmental pathways; promoting strategies to achieve developmental objectives delivering multiple co-benefits so as to address climate change concerns simultaneously. The eight missions that form the core of NAPCC are: National Solar Mission, National Mission for Enhanced Energy Efficiency (NMEEE), National Mission for Sustainable Habitat, National Water Mission, National Mission for Strategic Knowledge on Climate Change, National Mission for Sustainable Agriculture, National Mission for Green India and National Mission for Sustaining the Himalayan Ecosystem (NMSHE). These missions represent the multi-dimensional integrated long-term strategy advocated by national action plan for achieving key objectives in context of climate change while advancing overall development. Later on, three more missions were added to NAPCC which are National Wind Mission, National Mission on Impact of Climate Change on Human Health and National Mission on Waste to Energy. For each of these missions, there are different nodal ministries responsible for implementing and monitoring. Ecosystem-based approaches have been widely recognised as a part of the overall framework and strategies across various NAPCC missions for realising the mission objectives. Table 5.2 analyses some of the missions under NAPCC and their linkages with ecosystem-based approaches.

5.5.1.2 State Action Plan for Climate Change (SAPCC)

NAPCC very well recognises the role of sub-national authorities in successfully achieving the national objectives and priorities for addressing climate change and its impacts. SAPCC is representative of strategic approach aimed at mainstreaming climate change concerns within developmental planning. In India, state governments are expected to play a key role in advancing climate change adaption and mitigation strategies at state level by incorporating the climate change concerns into local governance and adopting climate sensitive developmental policies, programmes and investments (MoEFCC 2017). SAPCC addresses the state-specify climate change

Table 5.2 References to ecosystem-based approaches across missions under NAPCC

Mission	References to Ecosystem-based approaches
National Mission for Sustainable Habitat (GoI 2007) Aims at promoting understanding of climate change, its adaption and mitigation through energy efficiency and natural resource conservation Mission provide implemented strategies for climate change mitigation and adaption across various sectors which include: Energy efficiency, urban transportation, water supply and sewerage, municipal solid waste management, urban stormwater management and urban planning No separate fund is allocated to this mission as the mission is being implemented through 4 flagship missions which are 1. Atal Mission on Rejuvenation and Urban Transformation (AMRUT) 2. Swachh Bharat Mission 3. Smart Cities Mission 4. Urban Transport Programme	• Creating educational programmes for students, engineers, urban planners and other professionals on green buildings focussing on regional solutions • Green Demonstration projects across key locations in the country • Consumer awareness programmes on benefits (economical and environmental) of green buildings • Increasing green covers in urban • Reducing heat island effects • Developing green belts as envisioned in NUHHP • EIA of master plans and other infrastructure projects
National Water Mission (GoI 2008) The mission was established with the aim of conserving water, reducing wastage and ensuring equitable distribution of water (cross and within states) through integrated water resource development and management The mission envision five goals: 1. Establishing comprehensive database on water in public domain and assessing the impact of climate change on water resources 2. Promoting citizen and state actions for conserving, augmenting and preserving water 3. Focussing on vulnerable and over-exploited areas 4. Increasing water efficiency 5. Promoting basin level integrated water resource management	• Developing inventory of wetlands • Reassessment basin-wise water situation • Promotion of traditional system of water conservation • Promotion of water purification and desalination techniques • Promotion of participatory irrigation management • Intensive programme in over-exploited areas on groundwater recharge • Incentives for recycling of water (including wastewater) • Promoting water efficient techniques and technologies • Incentive for the use of efficient irrigation practices • Promoting mandatory water audits (including those for drinking water purposes) • Incentives as awards for water conservation and efficient use • Guidelines for use of water for different purposes (irrigation, drinking etc.) with respect to basin situation • Planning principle for integrated water resources development and management

(continued)

Table 5.2 (continued)

Mission	References to Ecosystem-based approaches
National Mission for Sustainable Agriculture (GoI 2010a) Mission aims at promoting sustainable agriculture ensuring food security, equitable access to food resources and enhanced livelihood and economic stability at national level The mission focuses on four functional areas viz. Research and development; Technologies, products and practices; Infrastructure and capacity building while synergizing modern technology and research with traditional knowledge and agricultural heritage Key components under the missions: • Rainfed Area Development (RAD) • On-Farm Water Management (OFWM) • Soil Health Management (SHM) • Climate Change and Sustainable Agriculture Monitoring, Modelling and Networking (CCSAMMN)	• Promoting biotechnology • Research and promotion of higher carbon pathways in low-carbon plants (increase effective use of CO_2, reducing GHG emission) • Conservation of indigenous genetic resources and agricultural heritage • Promotion of water efficient irrigation techniques • Development of mechanism for integrated rainwater, surface and groundwater management • Improved agronomic activities (soil treatment, efficient use of water, increased soil carbon storage etc.-increase carbon sequestration and carbon soil sink) • Soil conservation, bio-fertilisers • Optimum Land use • Promotion of SRI (System of Rice intensification) • Supplementing income from off-farm activities • Promoting Crop diversification • Promoting Crop-livestock fisheries farming system
National Mission for Green India (GoI 2010b) Mission aims at protecting, restoring and enhancing the forest cover in the country and responding to climate change through adaption and mitigation measures Key objective of the mission includes: 1. Increased forest/tree cover 2. Improved/enhanced ecosystems services like carbon sequestration and storage; hydrological services and biodiversity; provisioning services (fuel, fodder and timber and non-timber forest produces) 3. Increased forest based livelihood income Mission lays emphasis on landscape approach, unlike conventional afforestation program. The landscape which are large contiguous areas of forest/non-forest land at different scale/level provide opportunities to meet targets of both national and state policy Landscape are identified on basis of biophysical and socio-economic parameters often co-terminus with micro/milli watershed	Mission focuses on multiple ecosystem services including biodiversity, water, biomass, conserving mangroves, wetlands and other such important habitats along with carbon sequestration as co-benefit. Thus establishing a holistic view of greening Mission set out targets for each of the following activities: • Enhanced and increased forest cover and ecosystem services through adoption of • Eco-restorations/afforestation, agroforestry and social forestry for Enhanced tree cover in urban and peri-urban areas • Restoration of wetlands and enhancing their ecosystem services • Improved fuel-use efficiency and promoting alternative energy sources in project area households • Enhanced community livelihood

(continued)

Table 5.2 (continued)

Mission	References to Ecosystem-based approaches
National Mission for Sustaining the Himalayan Ecosystem (NMSHE) (GoI 2010c) Mission aims at developing sustainable national capacities to assess the health status of Himalayan ecosystem and enabling policy bodies in their functions so as to provide support to states in Indian Himalayan regions ineffective implementation of actions supporting sustainable development The mission attempts at addressing the following issues conversing Indian Himalayan ecosystem which includes: • Himalayan glaciers and associated hydrological consequences • Biodiversity conservation and protection • Wildlife protection and conservation • Traditional knowledge of societies and their livelihoods • Issues related to sustenance of Indian Himalayan ecosystem Key actions include: 1. Continuous monitoring of the ecosystem and data generation 2. Glaciology research 3. Generation of bio-geo database and ecological modelling for Himalayas 4. Prediction of socio-economic and climate change scenarios 5. Vulnerability assessment 6. Identification of desirable adaption policies for improving regional sustainability 7. Sustainable forestry 8. Strengthening of Regional cooperation 9. Enhanced implementation of guidelines for priority actions	• Promoting Sustainable agriculture and food security – Awareness campaigns on agro-biodiversity conservation, food security, legal rights and bio-resources – Encouraging Adoption of integrated pest management and nutrient supply techniques • Plantation systems in Himalayas – Promoting plantation of native/endemic species • Sustainable forestry – Promoting joint forestry management and participatory mechanisms for long-term management and sustainable use of ecosystem resources • Sustainable urbanisation in mountain habitats through – Proper town planning – Solid waste management – Promotion of sustainable pilgrimage – Ecologically sustainable commercial and adventure tourism – Green road construction – Enhancing water security – Building environmental awareness

Adapted from: Bhardwaj et al. (2020) and various Government of India documents, guidelines and websites.

challenges and needs while aligning them with national priorities as outlined under NAPCC. SAPPC supports the overall developmental objectives of states as it advances climate-adaptive strategies into existing programmes and policies being implemented at state level. In this regard, a common framework with sufficient flexibility was provided to sub-national authorities (states and UTs) for helping states/U.Ts to prepare their respective SAPCC.

Most of the states have already prepared their SAPCC, by making use of consultancy (technical assistance) provided by different development organisations. Formulation of SAPCC is considered as an important milestone towards decentralisation of India's national policy on climate change. However, implementation of SAPCCs on the ground still remain very weak due to lack of enabling environment in presence of lack of coordination between the centre and the state; lack of mainstreaming across different line departments (agriculture, environment, forest etc.) and limited financial resource and technical capacity requisite for implementation (Kumar 2018; Gogoi 2017).

5.5.1.3 Intended Nationally Determined Contributions (INDCs)

In 2015, India submitted its INDCs to United Nations Framework Convention on Climate Change (UNFCCC) under Paris Agreement, setting out targets for achieving reduction in the country's GHG emissions. India's INDCs encompass the country's commitment to embrace low-carbon pathways for achieving its developmental objectives. The INDCs are based on national programmes and policies promoting clean, renewable energy, enhanced energy efficiency, green transportation, enhanced carbon sink (increased forest and tree cover), low-carbon and resilient developments(urban centres), abating pollution and promoting resource efficiency (waste to wealth). As per a report (GOI 2015), India's INDCs proposal focuses on following key areas:

- Sustainable lifestyle
- Clear economic development
- Increasing the share of non fossil fuel based electricity
- Reducing emission intensity of gross domestic product
- Technology transfer and capacity building
- Enhancing carbon sink
- Mobilising finance
- Adaption.

India's plans to achieve following quantified INDC targets:

1. Lowering its GDP emission intensity by 33–35% by 2030 as compared to 2005 levels
2. Achieving 40% increase in total cumulative generation of energy from fossil-free energy sources by 2030
3. Creating additional carbon sink of 2.5 to 3 billion tons by increasing tree and forest cover by 2030

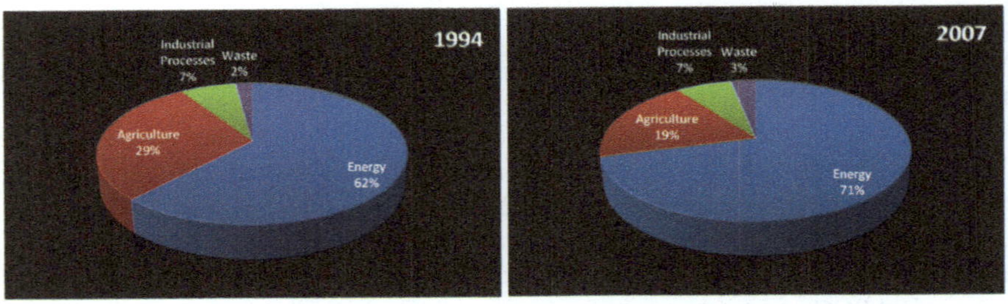

Fig. 5.2 Sectoral Distribution for Green House emissions in India for the years 1994 and 2007. Adapted from MoEFCC (2012)

Central to India's INDCs objectives remain the reduction in GHGs emissions and enhancement of the country's carbon sequestration capacity; hence, these two have influenced the country's response towards achieving INDCs targets. In India, energy sector is the largest contributor of GHGs with a total contribution of 71% for the year 2007. As illustrated in Fig. 5.2, it has been witnessed that the shares of GHG emission from energy sector have increased from 62 to 71% between the years 1994 and 2007 (MoEFCC 2012). Therefore, reduction in GHG emission from energy sector is of critical importance for India to achieve its INDCs targets. In this regard, India is making constant efforts towards decarbonisation of its power sector, enhancement of energy efficiency and management and reduction of its sectoral energy demands by introducing energy efficient technologies. NAPCC and its missions are the main driver of India's INDCs efforts (UN and RIS 2016; Pahuja et al. 2014). Missions under NAPCC particularly—National Solar Mission, National Mission on Sustainable Habitat and National Mission on Energy Efficiency focus on measures to contain GHG emissions. While the Mission for Green India and Mission for Sustainable Himalayan Ecosystems focus on increasing the country's carbon sequestration capacities. Table 5.3 provides a list of some of the key policy instruments that can significantly contribute to realising INDCs goals.

5.5.2 Disaster Management

5.5.2.1 National Disaster Management Act, 2005 and National Disaster Management Policy, 2009

Disaster management in India has evolved from a relief centric approach to a multidimensional holistic and pro-active approach for dealing with disasters and the two major policy responses that helped in advancing this paradigm shift in India's disaster management approach are: National Disaster Management Act, 2005 and National Disaster Management Policy, 2009. These two policy responses helped in establishing the much needed comprehensive framework for management of disasters in

Table 5.3 Existing Policy instruments that can contribute in realising INDCs goals

Instruments type	Existing tools
Missions under NAPCC	• National Solar Mission • National Mission for Enhanced Energy Efficiency (NMEEE) • National Mission on Sustainable Habitat (NMSH) • National Mission for a Green India • Mission for Sustainable Himalayan Ecosystems • National Renewable Mission (can be taken up as a new national action plan)
Fiscal	• Coal tax • Subsidy cuts on fossil fuels
Regulatory	• National Wind-Solar Hybrid Policy • National Policy on Biofuels • Green highways policy • National Tariff Policy 2006 • National Electricity Policy • National Rural Electrification Policy • Energy Conservation Building Code

the country. The National Disaster Management Act, 2005, provided the legal framework for disaster management and mandates establishment of three-tier institutional arrangement in the country at national, state and district levels while clearly defining the role, responsibilities and structures of different agencies. The act also provided for financial provision for disaster management in the country. National Disaster Management Policy, 2009 prepared in line with National Disaster Management Act, 2005 emphasised on adopting an holistic and integrated approach for disaster risk management by focussing on each component of disaster management continuum. The new approach as laid down by DM act and policy aimed at promoting culture of preparedness, prevention, mitigation and response at all levels and also focussed on mainstreaming disaster management across developmental planning and processes (NDMP,2009).

National Disaster Management Policy, 2005 very well recognises the role that environment plays in effective management of disaster risk; based on which one of the main objectives of DM policy encourages the adoption of mitigation measures based on traditional wisdom, technology and environmental sustainability. NDMP clearly mentions about environmental degradation being one the main driver of vulnerabilities towards disaster risk along with others such as climate change, increasing population, industrialisation etc. The policy advocates for environmentally sustainable development; encouraging restoration of ecological balance and sustainable development; and incorporating them wherever necessary in disaster management plans. The policy mentions about the need for restoring balance of ecosystems such as forests, coastal areas, islands etc. across urban, agricultural and industrials environments. The policy further promotes a holistic approach towards land use planning by directing concerned authorities (ministries and departments) to formulate alternate land use plans for different geographic and administrative areas based on environmental and hazard data analysis.

Apart from National Disaster management act and policy, there are disaster specific guidelines provided by the government for management of different disasters. These guidelines very well capture and promote the ecosystem-based approaches for reducing disaster risks and vulnerabilities. Table 5.4 analyses strategies across some of the key disaster guidelines and highlight their linkages with Eco-DRR.

The information provided in the table is taken from different disaster guidelines provided by the Government of India.

5.5.3 Sectoral Policies

Ecosystem-based approaches are often promoted as a part of strategies to deal with challenges associated with different developmental sectors and thus indirect or direct reference these approaches are often found across several sectoral policies. This integration of ecosystem-based approaches into sectoral policies provides an opportunity for up-scaling these solutions to local levels, ensuring their effective implementation on ground. Table 5.5 analyses key Indian policies across some of strategically important development sectors and their linkages with ecosystem-based approaches.

5.6 Conclusion

It has been clearly evident from the analysis of national plans and programmes that there is a wide scope for ecosystem-based approaches to get easily mainstreamed into the number of interventions across various sectors. Despite wide international recognition, the uptake of ecosystem-based approaches for disaster risk reduction and climate change adaption remains very limited at regional/local levels. Some of the key challenges and opportunities associated with up-scaling of these approaches can be summarised as follow:

- Most of these interventions are implemented on project or pilot demonstration basis. There remains a lack of consideration of these approaches within developmental planning and practices due to limited information and knowledge about the efficacy of these approaches in addressing disaster risk, climate change and sustainable development concerns. For addressing this specific constraint of informed decision making, it is required to create a strong science-policy evidence-base so as enhance the understanding of social and economic values of these ecosystems based solutions among stakeholders, while contextualising the country specific disaster, climate change and developmental needs and challenges.

Table 5.4 References to ecosystem-based approaches across different disaster guidelines

Disasters guidelines	References to ecosystem-based approaches (Eco-DRR)
Flood (NDMA 2008a, b)	1. **Prevention, preparedness and mitigation** Eco-friendly Structural measures and non-structural measures including • Identification of Natural detention basin • Channel improvements • Desiltling/dredging of rivers • Drainage improvement • Diversion of flood water • Catchment area treatment/afforestation • Anti-erosion measures • Flood plain zoning • Flood proofing- by adopting flood resilient design for flood shelters, settlements • Integrated Water Resource management 2. **Regulation and enforcement framework supporting** • Wetlands: Conservation and Restoration • Watershed Management including Catchment Area Treatment and Afforestation
Drought (NDMA 2010a, b)	3. **Prevention, preparedness and mitigation** • Judicious use of surface and groundwater • Cloud seeding in drought prone regions • Micro-irrigation systems-promoting crop diversification through drip/sprinkle irrigation systems • Water conservation, storage structures and management • Afforestation with bio-diesel species 4. **Capacity development** Building capacities across: • Natural resource management with focus on water conservation, watershed development and revival and creation of water bodies • Groundwater recharge management alternatives • Use of micro-irrigation and supplementing irrigation through harvested water • Agroforestry, farm forestry and dryland horticulture 5. **Developing Community participation in drought mitigation measures such as rainwater harvesting etc**

(continued)

Table 5.4 (continued)

Disasters guidelines	References to ecosystem-based approaches (Eco-DRR)
Landslides and snow avalanches (NDMA 2009)	1. **Landslide risk treatment** • Landslide remediation practices-drainage improvement measures, soil/debris removal works, river training work and buttress filling works • Providing protective covering to treated slope by practising afforestation or biotechnical methods 2. **Important area for Research and development includes** • Quantification of environmental degradation, anthropogenic impact, agricultural produce, loss of land, livelihood and tariff delays
Cyclones (NDMA 2008a, b)	1. **Coastal zone management** • **Conservation of coastal wetlands** – Conversation of coastal wetland – Analysing geomorphic characteristics of the coastal ecosystems such as beaches, sand dunes and bars etc. and taking up measures for integrated coastal management – Supporting coastal aquaculture – Implementing coastal zone regulation as per coastal zone notification • **Coastal resources management** – Delta water management(involving proper drainage and regulation of aquaculture even beyond CRZ – Initiatives for best land use practices will be encouraged through implementation of policies and incentives for afforestation for safeguarding ecological and income security incentives for reclamation o wasteland and degraded forest land encouraging agroforestry, organic farming, environmentally sustainable cropping patterns and efficient irrigation techniques • **Mangrove conservation and plantation** • **Shelterbelt plantation** • **Coastal flood plain management** • **Groundwater resource development through groundwater recharge schemes, adopting roof water or rainwater harvesting techniques** • **Effective protection against coastal erosions** 2. **Community-based disaster management** • Sensitising communities about the importance of different coastal resources, construction of shelters etc., encouraging community ownership and involving communities in management and maintenance of coastal environment • Encouraging communities for adapting traditional techniques for sustainable land use and reclamation of wasteland/degraded forest • Spreading culture of ecosystem and biodiversity conservation through school education

(continued)

Table 5.4 (continued)

Disasters guidelines	References to ecosystem-based approaches (Eco-DRR)
Tsunami (NDMA 2010a, b)	1. **Structural Mitigation Measures** • **Tsunami mitigation** Coastal villages can be safeguarded from the impacts of tsunami by adopting a range of soft solutions including – Developing sand dunes with sea weeds or shrubs – Raising the ground level (above the design water level) with natural beach sand for rehabilitating the entire coastal village – Development of coastal forest (green belt) – Periodically degrading the inlets and associated water bodies, so as to absorb the influx due to Tsunami – Adopting natural beach nourishment so as to create a steep beach face – Creating sandy ramps at close intervals along the coast – Establishment of mangroves plantations as coastal defence against tsunami – Construction of seawall in conjunction with natural bio-shields, subject to topography and bathymetry • **Protecting and strengthening of fragile seafronts, coastal resources and lifeline structures** – Creating an inventory of coastal natural resources and exiting built environment – Assessing their vulnerabilities – Prioritising of environments based on their vulnerability – Developing measures for their protection and strengthening 2. **Regulation and Enforcement** • **Implementation of coastal zone regulation as per CRZ notification** • **Land use** – Coastal ecology to be protected and strengthened – Coastal habitats to be planned in low hazard zone – Conserving and protecting mangrove plantation – Encouraging casuarinas, bamboo and other shelterbelt plantations – Coastal geomorphic features(beaches, sand dunes etc.) to be protected as they act buffer against coastal hazards • **Ensuring efficient land use practices through** – Policies and incentives for afforestation – Encouraging adoption of traditional, sustainable land use promoting reclamation of wasteland and degraded forest – Encouraging agroforestry, organic farming, environmentally sustainable cropping patterns and efficient irrigation techniques – Funding for greenbelt creation and conservation of mangroves • **Plan for conservation and restoration of mangroves and shelterbelts plantation** • **Selection of species and efforts for community involvement** – Selection of species based on factors such as biodiversity, tidal amplitude, soil adaptability, maturity characteristics and enrichment of species diversity – Encouraging community involvement and beneficiary orientation nursery programmes crucial for regeneration of forest cover and coastal shelter belt • **Funding support for spread of mangroves and shelterbelts**

Table 5.5 References to ecosystem-based approaches across various sectoral policies

Sector	Policy	References to ecosystem-based approaches
Water	***National Water Policy*** (GoI 2012) The policy aims at understanding the existing situation of water resources and their management in the country; proposing a legal and institutional framework for effective water resource planning, development and management Key areas related to water resources management as addressed by NWP-2012 include: 1. Water framework law 2. Use of water 3. Adaptation to climate change 4. Enchaining water available for use 5. Demand Management and Water Use Efficiency 6. Water Pricing 7. Conservation of River Corridor, Water Bodies and Infrastructure 8. Project planning and implementation 9. Management of Flood and Drought 10. Water supply and sanitation 11. Institutional arrangement 12. Transboundary River 13. Database and Information Systems 14. Research and training needs	• Increased storage capacities in various forms including soil moisture, lakes, ponds, small and large reservoirs • Planning and managing water resource structures (dams, flood and tidal embankment etc.) incorporating possible climate change strategies • Assessment and periodical review of availability of water resources and its usage in various sectors • Addressing the over-exploitation of groundwater through improved technology and techniques of water use, incentivising efficient use of groundwater and promoting community-based management of aquifer • Inter-basin transfers after assessing the environmental, social and economic impacts • Recycle and reuse of water • Encouraging and incentivising water saving irrigation techniques and methods including aligning cropping pattern to water resource availability, microirrigation, reducing evaporation-transpiration etc • Restoring and managing encroached and diverted water bodies and drainage channels • Preventing industrial effluents, residual fertiliser and chemical water from reaching groundwater • Effective maintenance and management of water resources infrastructure • Planning water resource projects considering the social and environmental aspects • Time-bounded Environmental and investment clearance • Greater emphasis should be laid on rehabilitation of natural drainage systems • Evolving different agricultural strategies for effective management of drought through scientific local research • Encouraging rainwater harvesting and de-salinization for increasing water availability in urban and industrial areas

(continued)

Table 5.5 (continued)

Sector	Policy	References to ecosystem-based approaches
Energy	***National Electricity Policy*** (GoI 2005) Key objectives of policy includes: • Access to electricity • Availability of power • Reliable and quality power supply at reasonable price • Increase in per capita availability • Minimum lifeline consumption • Addressing Financial gap and commercial viability of sector • Protecting consumer interest Key issues to be addressed by policy: Rural Electrification, Generation; Transmission; Distribution Recovery of Cost of services &Targeted Subsidies; Technology Development and Research and Development (R&D); Competition aimed at Consumer Benefits; Financing Power Sector Programmes Including Private Sector Participation; Energy Conservation, Environmental Issues; Training and Human Resource Development; Cogeneration and Non-Conventional Energy Sources; Protection of Consumer interests and Quality Standards	• Promoting Environmental Impact Assessment(EIA) and implementing Environment Action Plan (EPA) • Catchment area treatment for hydroprojects to be encouraged and monitored • Setting up of coal washeries and utilising fly ash as per environmental guidelines • Ensuring compliance with environmental norms and standards during operations of power plants
Urban Development	***National Urban Housing and Habitat Policy*** (GoI 2007) The policy aims at providing "affordable housing for all" with special emphasis on urban poor. The key objectives of policy include: • Urban planning • Affordable housing • Increasing flow of funds • Spatial incentives • Increasing supply of land • Special provision for vulnerable groups • Generating employment • Promoting Public–Private Partnership • Establishing and managing information systems • Promoting healthy environment	Promoting healthy environment • Special attention to be paid to housing in coastal areas to protect and conserve fragile ecology • Special emphasis will be given on 'green lungs' of city such as of parks, zoos, botanical gardens, social forestry and green belts; keeping a significant proportion for them in master plans • Protection of water bodies particularly keeping floodplains areas free from construction and encroachment • Developing special area development plans with fragile ecological characteristics on basis of EIA; taking care of all environmental concerns at planning stage itself

(continued)

Table 5.5 (continued)

Sector	Policy	References to ecosystem-based approaches
Agriculture	***National Agriculture Policy*** (GoI 2000) The policy aims at promoting sustainable agriculture by making optimal utilisation of land, water, forest and other natural resources The broad objectives of the policy are: • Realising the untapped growth potential of the sector • Strengthening rural infrastructure for faster development of the sector • Accelerating growth of agro-business • Creating employment opportunities in rural areas • Securing fair standard of living for farmers and agricultural workers • Discouraging migration from rural to urban areas • Facing sectoral challenges that arise due to globalisation and economic liberalisation	• Utilising wastelands for agriculture and afforestation purposes • Promoting inter-cropping and multi-cropping practices for increasing cropping intensity • Developing long-term sustainable plan for rainfed agriculture through watershed approach • Emphasised on rational utilisation of ground and surface water • Involving farmers and landless labourers into developmental programmes (pastures/forestry) by offering incentives (financial) or entitlements (trees or pastures) • Development of rainfed and integrated horticulture, floriculture, aromatic and medical plants, beekeeping and sericulture for expanding food supply, exports and employment opportunities in rural areas • Diversifying agriculture, Development of animal husbandry, poultry, dairying and aquaculture • Promoting sustainable agriculture through integrated marine and inland fisheries approach Supporting integrated pest management • Promotion of selective and eco-friendly farm mechanisation

Adapted from: Bhardwaj et al. (2020) and various Government of India documents, guidelines and websites

- Committed efforts by the national government are essential for mainstreaming these approaches into relevant sectors which can be easily done through developmental plans, programmes and strategies. Providing proper roadmaps for mainstreaming and implementing these solutions, with specific guidelines for stakeholders particular at regional/local levels can only help in translating these plans into on-ground actions.
- Identifying and ensuring stable financial mechanisms remains the major challenge in supporting these approaches particularly for developing nations. With very limited avenues of funding, the investments in these approaches are usually considered an extra burden on public funding and have to compete with other developmental priorities. Also, with the much longer temporal scale required for ecosystem-based solutions for implementation and delivering benefits, the choice of conventional engineered solutions are most often considered over these approaches for providing immediate benefits and solutions.
- For successful implementation of these cross-disciplinary approaches on ground, it is very important to ensure multiple levels of governance and cross-sectoral integration so as to ensure strategic linkages across national and sub-national ***levels*** with regard to planning, financing, implementation, monitoring and evaluation of these ecosystem-based solutions by effectively engaging various ***stakeholders*** across various ***sectors*** (agriculture, tourism, health etc.).
- Involving local stakeholders is also very important to ensure long-term sustainability and benefits of these approaches. Local communities most often are the direct users of ecosystems and their services; their indigenous knowledge about local ecosystems can provide valuable contributions in planning effective ecosystem based solutions for addressing local needs and challenges. Therefore, it is important to ensure representation of local stakeholders and indigenous people including local government authorities, community organisations, local leaders and others.

References

Bhardwaj S, Gupta AK, Dhyani S, Thummarukuddy M (2020) Nature-based solutions entry points through sectoral policies, strategic instruments and business continuity. In: Dhyani S et al (eds) Nature based Solutions: science, innovations and strategies in South Asia. Springer Nature

Ekaratne K, Vidanage S (2013) Small grants, large gains : lessons from MFF Small Grant Facility Projects in Sri Lanka (2011–2013). IUCN

Estrella M, Saalismaa N (2013) Ecosystem-based disaster risk reduction (Eco-DRR): an overview. In: Renaud FG, Sudmeier-Rieux K, Estrella M (eds) The role of ecosystem in disaster risk reduction. United Nations University Press

Gogoi E (2017) India's state action plans on climate change: towards meaningful action. Oxford Policy Management, New Delhi

GoI (2000) National Agricultural Policy. Government of India, New Delhi

GoI (2005) National electricity policy. Ministry of Power, Government of India, New Delhi. Accessed at https://powermin.nic.in/en/content/national-electricity-policy

GoI (2007) National urban housing and habitat policy. Ministry of Housing & Urban Pverty Alleviation, Government of India, New Delhi. Accessed at https://nhb.org.in/Urban_Housing/HousingPolicy2007.pdf
GoI (2008) National Water Mission under National Action Plan on Climate Change. Ministry of Water Resources, Government of India. Accessed at http://mowr.gov.in/sites/default/files/Mission_Doc_Vol22880755143_0.pdf
GoI (2010a) National mission for a green India. Ministry of Environment and Forests, Government of India. Accessed at http://www.indiaenvironmentportal.org.in/files/green-india-mission.pdf
GoI. (2010b). National mission for sustainable agriculture. Department of Agriculture and Cooperation, Ministry of Agriculture, Government of India. Accessed at http://agricoop.nic.in/sites/default/files/National%20Mission%20For%20Sustainable%20Agriculture-DRAFT-Sept-2010.pdf
GoI (2010c) National mission for sustaining the Himalayan Ecosystem under National Action Plan on Climate Change (NAPCC). Department of Science and Technology, Ministry of Scince and Technology, Government of India, New Delhi. Accessed at http://www.knowledgeportal-nmshe.in/Pdf/NMSHE_MissonDocument.pdf
GOI (2011) Disaster management in India. Ministry of Home Affairs. Government of India
GoI (2012) National water policy. Ministry of Water Resources, Government of India, New Delhi. Accessed at http://mowr.gov.in/sites/default/files/NWP2012Eng6495132651_1.pdf
GOI (2015) India's intended nationally determined contribution. Press Information Bureau, Government of India
Gupta AK, Nair SS, Singh S (2013) Environmental legislation for disaster risk management. National Institute of Disaster Management & Deutsche Gesellschaft für internationale Zusammenarbeit GmbH (GIZ)
Gupta AK, Singh S, Katyal S, Chopde S, Wajih SA, Kumar A (2016) Training manual on climate resilient and disaster safe development—process framework. NIDM, GEAG, ISET, CDKN
https://ndma.gov.in/images/guidelines/droughtguidelines.pdf
IMF (2017) Promoting inclusive growth. The International Monetary Fund (IMF)
IUCN (2016) Regional assessment on ecosystem-based disaster risk reduction and biodiversity in Asia. IUCN
Kumar V (2018) Coping with climate change: an analysis of India's state action plans on climate change. Centre for Science and Environment, New Delhi
MEA (2005) Millennium ecosystem assessment: ecosystems and human well-being: synthesis. Island Press, Washington, DC.
MoEFCC (2012) India Second National Communication to the United Nations Framework Convention on Climate Change. Ministry of Environment, Forests & Climate Change.Government of India
MoEFCC (2017) Towards evolution of a framework for the preparation of state level climate change strategy and action plan. Ministry of Environment, Forest and Climate Change, Government of India
Munang R, Thiaw I, Alverson K, Liu J, Han Z (2013) The role of ecosystem services in climate change adaptation and disaster risk reduction. Curr Opin Environ Sustain 47–52
NDMA (2008a). National Disaster Management Guidelines: management of cyclones. National Disaster Management Authority, Government of India, New Delhi. Accessed at https://ndma.gov.in/images/guidelines/cyclones.pdf
NDMA (2008b) National Disaster Management Guidelines: management of floods. National Disaster Management Authority, Government of India, New Delhi. Accessed at https://ndma.gov.in/images/guidelines/flood.pdf
NDMA (2009) National Disaster Management Guidelines: management of lanslides and snow avalanches. National Disaster Management Authority, Government of India, New Delhi. Accessed at https://nidm.gov.in/pdf/guidelines/new/landslidessnowavalanches.pdf
NDMA (2010a) National Disaster Management Guidelines: management of drought. National Disaster Management Authority, Government of India, New Delhi. Accessed at http://agricoop.nic.in/sites/default/files/Manual%20Drought%202016.pdf

NDMA (2010b) National Disaster Management Guidelines: management of tsunami. National Disaster Management Authority, Government of India, New Delhi. Accessed at https://ndma.gov.in/images/guidelines/ndmaguidelinesmanagementofsunamis.pdf
Pahuja N, Pandey N, Mandal K, Bandyopadhyay C (2014) GHG mitigation in India: an overview of the current policy landscape. Working Paper. World Resources Institute, Washington, DC
PEDRR (2010) Demonstrating the role of ecosystem-based management for disaster risk reduction. Partnership for Environment and Disaster Risk Reduction
Renaud FG, Sudmeier-Rieux K, Estrella M (2013) The relevance of ecosystems for disaster risk reduction. In: Renaud FG, Sudmeier-Rieux K, Estrella M (eds) The role of ecosystems in disaster risk reduction. United Nations University Press
SCBD (2009) Connecting biodiversity and climate change mitigation and adaptation: Report of the second ad hoc technical expert group on biodiversity and climate change. Technical Series No. 41. Secretariat of the Convention on Biological Diversity, Montreal
UN & RIS (2016) India and sustainable development goals: the way forward. United Nations and Research Information systems of developing counties
UNDP (2019a) Enhancing climate resilience of India's Coastal Communities. UNDP
UNDP (2019b) Human Development Report 2019: beyond income, beyond averages, beyond today: inequalities in human development in the 21st century. United Nations Development Programmme, New York
UNISDR & CRED (2018) Economic losses, poverty & disasters: 1998–2017. UNISDR & CRED
UNISDR (2015) GAR 2015: making development sustainable: the future of disaster risk management. United Nations Office for Disaster Risk Reduction (UNISDR)
Wisner B, Blaikie P, Cannon T, Davis I (2004) At risk: natural hazards, people's vulnerability and disasters. Routledge, London

Chapter 6
Ecosystem-Based Approaches and Policy Perspectives: Towards an Integrated Blue–Green Solutions in Vietnam

Thi My Thi Tong and Ngoc Huy Nguyen

Abstract Vietnam's international economic integration is strongly promoted through various forms, following a roadmap for the proper adoption of international principles and standards of the global economy and market. The country is constantly striving to move from a brown economy to a blue and green economy, for which the ecosystem-based approaches have been employed as critical factors. The National Vision for Socio-Economic Development to 2020 has emphasised on nature-based solutions for environmental protection, creating an essential basis for policies and practices emerging areas of interest in ecosystem-based approaches. Despite the promulgation of laws and decrees, ratification of international agreements, and participation in international initiatives, policy reviews show limited entry points for mainstreaming ecosystem-based solutions in decisions and practices. This chapter provides an analysis of the policy and practices of ecosystem-based approaches with 14 case studies across Vietnam. They were classified into different categories depending on the nature of solutions and level of blue–green integration. The shortcomings and challenges facing those models are also analysed in order to contribute to the efforts of bringing opportunities for further integration of blue–green infrastructure at both national and local levels.

Keywords Ecosystem-based approach · Blue–green integration · Structural measures · Climate and risk resilience · Vietnam

6.1 Introduction

Vietnam has a rich biodiversity. The significant ecosystems include terrestrial ecosystems (forest ecosystems), wetland ecosystems and marine ecosystems. The forest ecosystem has the highest species diversity and is home to many species of wild animals, plants and microorganisms. Vietnam has 30 types of natural wetlands,

T. M. T. Tong (✉)
Vietnam Institute of Economics, Vietnam Academy of Social Sciences, Hanoi, Vietnam

N. H. Nguyen
Oxfam International, Hanoi, Vietnam

© The Author(s), under exclusive license to Springer Nature Singapore Pte Ltd. 2021
M. Mukherjee and R. Shaw (eds.), *Ecosystem-Based Disaster and Climate Resilience*, Disaster and Risk Research: GADRI Book Series,
https://doi.org/10.1007/978-981-16-4815-1_6

belonging to three groups of wetlands: inland wetland (19 types), coastal wetland (11 types) and artificial wetland (9 types), of which the Mekong River Delta and the Red River Delta have the largest wetland areas. With a long coastal line, Vietnam's marine ecosystems are the basis for maintaining fish reserves of over 5.3 million tons and can meet about 47% of domestic protein needs annually.

However, a long coastline together with diverse topography and variable climate causes the country one of the most climate hazard-prone countries in the world (Garschagen et al. 2016). Vietnam was ranked as 16th most at risk by the World Risk Index (2011) and as 13th most vulnerable by the Climate Change Vulnerability Index (2016) and worsened from ninth (2017) to sixth (2018) on the global vulnerability ladder by the Global Climate Risk Index 2020. According to the Ministry of Natural Resource and Environment (MONRE) (2016), the country experienced annual increased temperatures, at the same time with fluctuated precipitation. During the period 1958–2014, the average annual temperatures increased by 0.62 °C. Meanwhile, precipitation patterns increase by 15–20% in the wet season and decrease by 10% in the dry season in the same period, meaning that, while summers will be drier, rainy periods will be wetter (MONRE Vietnam 2016). These changes in both temperature and precipitation are leading to more uncertain and extreme climatic events (GoV 2015). Annually, the country is affected by a range of hydrometeorological and climatological hazards, such as sea-level rise and storms along the coast, highlighting the critical importance of wetland and marine ecosystems (including mangrove and coastal forests) (EM-Dat Database 2019), particularly sea-level rise and storms along the coast, highlighting the critical importance of wetland and marine ecosystems (including mangrove and coastal forests).

With the diverse ecosystems distributed throughout the country, the application of nature-based solution in climate change adaptation (CCA) and disaster risk reduction (DRR) is being considered as appropriate approaches for Vietnam in the context of increasing adverse impacts. Simultaneously, Vietnam has the majority of the population whose livelihoods depend essentially on the services that natural ecosystems provide, highlights the need for the country to prioritise nature-based measures, particularly in highly exposed areas such as coasts and deltas. In Vietnam, Government and society always emphasise on the role of the forest, mostly coastal protection forest in strengthening people's livelihood as well as reduce their exposure to climate risks. As such, terrestrial forest and mangrove protection has been given high priority in policies and plans. The second *National action plan for climate change response (2010–2015)* has emphasised on nature-based solutions, creating an essential basis for policies and practices emerging in areas of interest in ecosystem-based approaches in dealing with climate change and disaster risks. However, current CCA and DRR in Vietnam are still predominantly human-built or rely on "grey" infrastructure, which is associated with adverse environmental and social side effects.

This chapter provides an analysis of the policy and practices of ecosystem-based approaches with case studies in different areas in Vietnam. The policy analysis was conducted to see the extent to which natural and ecological solutions (green solutions) and water management solutions (blue solutions) are linked to climate change adaptation. Results from policy review show that the Government is prioritising terrestrial and coastal forests for enhancing climate resilience and is constantly striving to employ more ecosystem-based solutions for promoting climate and disaster risk resilience in Vietnam. There is a shift from infrastructural to natural solutions revealed in related policies issued after 2015. However, there are still little entry-points for the integration of ecosystem-based solutions in practices, yet no mention of the integration of "blue" or water resource management. The case studies go beyond mere description to include a degree of analysis, which maps out 14 case studies in different categories based on their primary targets and actions to apply ecosystem-based approaches in practices. Challenges and remains from case studies will be analysed in more detail to contribute to efforts of bringing opportunities for further integration of nature-based solutions at both national and local levels.

6.2 EbA Reflected Through Policies on Climate Change and Disaster Response in Vietnam

This section provides a list of policies related to climate change and disaster response in Vietnam (Table 6.1) and classified policies in different levels (from 1 to 4) based on their implications to nature-based solution (Table 6.2).

6.2.1 Climate Change Response Policies

The Vietnamese Government has shown strong commitment to the fight against climate change since the early 1990s. Vietnam ratified the United Nations Framework Convention on Climate Change (UNFCCC) in 1994 and the Kyoto Protocol (KP) in 2002, approved to implement the UN Convention United Nations Against Desertification (UNCCD) in 1998 and the Hyogo Framework for Action in 2011. Furthermore, Vietnam pledged to contribute to the Paris Climate Agreement (PA) in 2016.

In order to help with the implementation of UNFCCC and KP at the national level, Vietnam set up the National Steering Committee in 2011. This Committee has already completed and submitted the Initial National Communication (NDC1) (2003), the Second National Communication (NDC2) (2010), and the Initial Biennial Update Report (2014), and the first Intended Nationally Determined Contributions (INDC) presented at UNFCCC COP 21 in Paris (2015). These reports reflect the latest efforts of Vietnam in responding to climate change. Reviewing the changes of climate change adaptation solutions through these reports will see an evident change in response solutions in the direction of "hard" to "soft", increasing of natural-based solutions. The adaptation measures proposed for water resource management

Table 6.1 List of reviewed policies

Acronym	Name of documents	Year	Document no	References
Climate change response policies				
NTP-RCC	National target program to respond to climate change	2008	Decision 158/QD-TTg	(GoV 2008)
NS-RCC	National strategy on responding to climate change	2011	Decision 2139/QD-TTg	(GoV 2011)
NAP-RCC 1	National action plan on responding to climate change in the period 2012–2020	2012	Decision 1474/QD-TTg	(GoV 2012b)
NTP-RCC 2	Target program to respond to climate change and green growth	2016	Decision 1670/QD-TTg	(GoV 2016b)
Resolution 24	Proactively responding to climate change, strengthening natural resource management and environmental protection	2013	Resolution 24/NQ-TW	(GoV 2013)
SP-RCC	Support program to respond to climate change	2014		(GoV 2012a)
NAP-PA	Plan of implementation of the Paris agreement on climate change in 2021–2030	2016	Decision 2053/QĐ-TTg	(GoV 2014)
Resolution 120 MKD	The strategy of sustainable development of the Mekong Delta to adapt to climate change to 2050	2016	Resolution 120/NQ-CP	(GoV 2016a)
NAP-MKD	National action program to implement Resolution 120/NQ-CP on sustainable development of the Mekong Delta to adapt to climate change	2019	Decision 417/QD-TTg	(GoV 2019b)
NAP-PA/MARD	Plan of implementation of the Paris agreement on climate change in the 2021–2030 period of MARD	2020	Decision 891/QD-BNN-KHCN	(GoV 2020)

(continued)

Table 6.1 (continued)

Acronym	Name of documents	Year	Document no	References
NAP-RCC 3	National action plan on responding to climate change in the period 2012–2030, vision to 2050	2021	Decision 1055/QĐ-TTg	(GoV 2020)
Disaster response policy				
O-FSPC	Ordinance on prevention and control of floods and storms	1993	Ordinance 9-L/CTN	(GoV 1993)
NS-DRPM	National strategy for natural disaster prevention, response and mitigation to 2020	2007	Decision 172/2007/QD-TTg	(GoV 2007)
NAP-DRPM 1	National action plan for natural disaster prevention, response and mitigation to 2020	2019	Decision 649/QD-TTg	(GoV 2019a)
P-CBDRM	Community awareness-raising and national community-based disaster risk management project	2009	Decision 1002/QD-TTg	(GoV 2009)

in NDC1 mainly focused on structural measures such as building and upgrading dams, reservoirs and irrigation systems. For coastal management, NDC 1 presented three strategic options including full protection (which requires all dike higher and strengthen coastal management), adaptation (which focus on build "adaptive" infrastructures and transfer to suitable farming techniques) and withdrawal (which focuses on resettlement, moving houses and infrastructures from threatened areas). As such, a nature-based solution was not employed. According to NDC2, the majority of adaptation measures during 2005–2010 still focused on structural solutions, including building new hydrologic and hydropower reservoirs, dams, and dykes, structural embankments for flood protection, and saltwater intrusion barrier. It aimed to upgrade 2700 km of the sea and river dykes following new standards corresponding to rising sea levels and construct an additional 585 km of dykes to protect critical economic areas. INDC provides a holistic approach to the use of coastal forests for resilience, which states that CCA must be carried out in a focused manner and respond to urgent, immediate impacts and potential long-term impacts. INDC also priorities the importance of improving the quality of coastal forests.

Vietnam's commitment to climate change response is highlighted in the national action plans and decisions and has been mainstreamed in socio-economic development plans.

Table 6.2 Policy mapping

Policy documents	Level 1	Level 2	Level 3	Level 4	EbA-related terms[a]			Main objectives/tasks related to EbA
					"Natural"	"Eco"	"Forest"	
Climate change response policies								
NTP-RCC 1					1	1	3	
NS-RCC					6	5	20	Natural resource protection is one of the key objectives for climate change response
NAP-RCC 1								Coastal mangroves and headwater protection forests
NTP-RCC 2					2	0	3	Planting and restoring 10,000 ha of coastal mangroves, watershed protection forests to adapt to climate change, absorb 2 million tons of CO_2 per year and create stable livelihoods for the people
NQ24					13	8	26	Prioritise programs to reduce greenhouse gas emissions through efforts to combat deforestation, forest degradation and create livelihoods for communities
NAP-PA					0	2	3	Integrating climate change adaptation based on ecosystems and communities, through developing ecosystem services, conserving biodiversity, using indigenous knowledge, giving priority to the most vulnerable communities

(continued)

Table 6.2 (continued)

Policy documents	Level 1	Level 2	Level 3	Level 4	EbA-related terms[a]			Main objectives/tasks related to EbA
					"Natural"	"Eco"	"Forest"	
Resolution 120 MKD					22	27	26	Respect the laws of nature, in accordance with the actual conditions, avoid harsh interference in nature; choose a natural, environment-friendly and sustainable adaptation model with the motto of proactively living with floods, inundation, brackish water and saltwater
AP-MKD					20	34	15	Building mangrove forests to protect the system of river and sea dikes, and develop ecological livelihoods
NAP-PA/MARD					1	14	39	Integrating EbA into forest management and sustainable fisheries models
NAP-RCC 2					16	30	27	Strengthening the resilience and enhancing adaptive capacity of ecosystems through science and technology and awareness is one of the key objectives
Disaster response policy								
O-FSPC					0	4	7	
NS-DRP					4	1	15	Afforestation and watershed protection

(continued)

The first national program on climate change in Vietnam, the *National Target Program to Respond to Climate Change* (NTP-RCC) issued in 2008, stressed the need to mainstream climate change responses into social and economic development and develop feasible action plans for CCA and mitigation. The program was established in three phases: inception phase (2009–2010), implementation phase (2011–2015) and development stage (after 2015). During the first phase, the Ministry of Agriculture

Table 6.2 (continued)

Policy documents	Level 1	Level 2	Level 3	Level 4	EbA-related terms[a]			Main objectives/tasks related to EbA
					"Natural"	"Eco"	"Forest"	
AP-DRPM					0	0	3	Planting watershed protection forests and coastal protective forests to mitigate flash floods and river banks or coastlines erosion

Level 1 Policy implies nature-based solution, though does not mention precisely about EbA
Level 2 Policy mentions about EbA or related terms but not employ EbA solutions/actions/projects
Level 3 Policy employs EbA solutions/actions/projects
Level 4 Policy embraces the integration between water resource management and EbA
[a]EBA activities in Vietnam mainly focus on (1) watershed protection and afforestation and (2) protection and development of coastal protective forests in vulnerable areas. Because the current EbA solutions in Vietnam focus on watershed and mangrove ecosystems, the research uses the keywords "natural", "ecological" and "forest" when reviewing EbA through various policies in VN

and Rural Development (MARD) was assigned to design and be responsible for the application of an ecosystem-based approach to adapt to climate change in the vulnerable coastal areas with particular focus on protection forest. Simultaneously, MARD was responsible for ensuring water security for irrigation systems, the safety of sea dikes and reservoirs. This creates a favourable condition for MARD to be the focal point for the integration of blue into green solution in adapting to climate change in the later periods. In the second stage, the first pilot project on ecosystem-based adaptation (EbA) was approved to carry out in Quang Nam and Ben Tre, which put efforts towards restoring and protecting mangroves in order to combat coastal erosion under the impacts of climate change.

It was until 2013 when "ecological balance" was first mentioned in a supreme document on climate change response, the Resolution 24-NQ/TW *on proactively responding to climate change, strengthening natural resource management and environmental protection* (Resolution 24) adopted by the 7th Central Party Executive Committee. Together with this document, the Government emphasises the role of strengthening capacity to proactively respond to climate change based on available resources, especially natural ecosystems.

In 2014, the Government augmented the *Support Program to respond to climate change* (SP-RCC), promoted activities to cope with impacts caused by climate change and reduced GHG emissions. Its main objectives are to integrate climate change policies into socio-economic plans and mobilise funds for climate change response. The Program prioritised 61 projects, of which 8 out of 61 projects were based on natural solutions, which accounted for 10% of the total budget.

The *National Strategy on Responding to Climate Change* (NS-RCC) for 2011–2020 recognises that climate change brings significant risks to food security and agricultural development, human health, natural resources and ecosystem function, and overall sustainable development. The strategy first mentioned strengthening the adaptive capacity of natural systems in its main objectives. It links the EbA to community-based adaptation (CbA) while emphasising the importance to mobilise the participation of socio-economic components to cope with climate change effectively. In this stage, although still prioritising adaptation, the Government also put efforts towards mitigation together with a timeline for developing a low-carbon economy.

The shift towards nature-based solutions in Vietnam can be seen clearly with the updated versions of the *National Action Plan on Responding to Climate Change* (NAP-RCC) through three periods: NAP-RCC 1 during 2012–2015, NAP-RCC 2 during 2016–2020, and NAP-RCC 3 during 2021–2030. The first NAP-RCC 1 outlines a total of 63 priority tasks for the 2012–2020 period, of which ten prioritised projects for the 2012–2015 period. Most projects still follow a "hard" approach, with a focus on infrastructure development, such as dams and sluices. There are only two tasks on forest protection aiming to reduce GHG emissions, and none is related to EbA. Updated in 2015, the NAP-RCC 2 (2016–2020) highlighted the importance of the natural system (aside from the human system) in the overall objectives, and *natural capital becomes the key trend in sustainable economic development* with a shift towards a low-carbon economy. The targets to 2020 include 42 projects on coastal mangrove planting and watershed protection. The expected outcome is planting and restoring 10,000 hectares of coastal mangroves and headwater protection forests in order to adapt to climate change. However, the proposed solutions are still mainly structural-based, including the construction of lakes and dams with a capacity of 100 million m^3, upgrade of the saltwater–freshwater control systems, building an additional 200 km of dykes and embankment in the vulnerable areas. The *National Action Plan for Climate Change Response for the period 2021–2030* (NAP-RCC 3) is expected to be adopted in 2020. Unlike NAP-RCC 1 and 2 (mainly focused on infrastructure solutions), the NAP-RCC 3 plan focuses on natural and ecological solutions. One of the key goals of NAP-RCC 3 is *to improve the adaptability of natural ecosystems and biodiversity to the impacts of climate change through strengthening the management of ecosystems and biodiversity, at the same time, enhance the resilience of natural ecosystems, protect and conserve biodiversity under the impacts of climate change and sea-level rise*. The plan identifies tasks for six field groups: (i) agriculture; (ii) natural disaster prevention; (iii) environment and biodiversity; (iv) water resources; (v) infrastructure; and (vi) other fields (public health, labour—society, culture—sport—tourism). There are 10 out of 63 main tasks designed to enhance the capacity of natural and ecological systems, and at the same time, take advantage of ecological services in response to climate change. The plan also emphasises the strong connection between EbA and CbA with the focus on developing and replicating EbA and CbA models, while strengthening the participation of local communities in biodiversity monitoring, conservation and management.

To advance Vietnam's commitments towards the 2015 Paris Climate Agreement, the GoV has approved the *Implementation Readiness Plan of the Paris Agreement* (NAP-PA) in October 2016. The plan is divided into two main phases: the "readiness" phase (before 2020) and "implementation" phase (2021–2030). There are in total 68 main tasks, and many of these activities promote the implementation of EbA projects, using indigenous knowledge, and intricate linkages between the EbA and CbA approaches. In this document, it is the first time that "EbA" was mentioned in full-term and as the critical approach specifically to CCA. *The Action plan on implementing the Paris Agreement by MARD* (NAP-PA/MARD) in the 2021–2030 period was issued in March 2020. The plan shows the consistency in the views with that of NAP-PA, one of the vital priority strategies for climate change response that mentioned in the document is "harmony with nature". MARD was assigned to lead and participate in the implementation of 41 out of 68 tasks under 03 levels: mandatory, priority and encouragement, belonging to 05 target groups: (i) reduce emissions greenhouse (GHG); (ii) climate change adaptation; (iii) prepare resources; (iv) establish a transparent system; and (v) develop and finalise policies and institutions. Of which, EbA activities are categorised in the "priority and encouragement" group. Specifically, task No. 35 in the plan refers to the implementation of integrated EbA in forestry and aquaculture through the development of ecosystem services and biodiversity conservation. Accordingly, it is expected that EbA will be integrated into forest development and sustainable fisheries models. Nevertheless, there is no mention of the integration of water management into a nature-based approach or a green system.

At the regional level, the GoV issued the *Resolution 120/NQ-CP on sustainable development of the Mekong Delta (MKD) to adapt to climate change to 2050* (Resolution 120 MKD) in 2016 with an effort to transform MKD into *a more developed region with advanced social organisation, higher per capita income, the proportion of eco-agriculture and high-tech agriculture accounts for over 80%, forest cover reaches over 9% (compared to the current baseline of 4.3%), critical natural ecosystems are preserved and developed*. The Resolution highlights natural solutions, emphasising the vital role of protecting natural ecosystem systems for sustainable development. In the document, the term "natural" is mentioned 18 times, and "eco" is mentioned 27 times. Its views on CCA and DRR are *respected for the laws of nature, following functional conditions, avoid rough intervention in nature; choose a model that is adaptable, environmentally friendly and sustainable with the motto of actively living with floods, inundation, brackish water and saltwater*. More importantly, the strategic direction emphasises on taking water resources as the core element and the basis for strategic development planning and policies for the whole region. In this way, the integrated planning of water resources management will help the "blue" or "water source" element be integrated throughout EbA or green projects.

Following Resolution 120, the *National action plan to implement Resolution 120 in MKD* (NAP-MKD) was issued to provide specific actions in adaptation to climate change towards sustainable development in MKD. Ecological solutions are applied throughout the strategy as well as the NAP-MKD. The term "natural" is mentioned 17 times, "eco" is mentioned 34 times. Opinions on CCA based on ecosystems are made clear: *consistent with natural conditions, biodiversity, culture, people and natural laws, focusing on biodiversity conservation, the ecological environment of the region*. At the same time, the plan encourages (i) investment for embankments using mangrove planting areas, and development of mangrove forests to protect sea and river dike systems; and (ii) vary ecological livelihoods associated with forests, which aims to shift livelihoods from relying on freshwater ecosystems to brackish and saltwater ecosystems in order to adapt to climate change and sea-level rise.

At the local level, all provinces have approved the *Provincial Action Plan to implement PA* (PAP-PA), which employs EbA through the development of ecosystem services and biodiversity conservation. Until now, 100% of the provinces have their own PAP-PA. However, most of PAP-PA just mentioned generally but did not include any specific actions on how to apply EbA in local conditions, even in some provinces with large areas of the natural ecosystem, high biodiversity, but high vulnerability to climate change and sea-level rise such as Hai Phong (Decision 3337/QD-UBND), Ca Mau (Plan 03/KH-UNBD), Quang Nam (Decision 3462/QD-UBND). There were only very few provinces that propose specific adaptation measures at localities; i.e. Hoa Binh Province proposed to develop community-based ecological village adaptation to climate change for the area of Hoa Binh reservoir (Plan 152/KH-UBND); Quang Tri and Tra Vinh Province proposed to develop models of ecotourism, eco-friendly tourism, community tourism and cultural tourism linking environmental protection and biodiversity conservation in mountainous and coastal areas (Plan 5244/KH-UBND and Plan 45/KH-UBND).

6.2.2 Disaster Response Policies

Parallel to its commitment to climate change response, Vietnam is also a member of several international commitments on DRR, including Hyogo Framework for Action (HFA) 2005–2015, Sendai Framework for Disaster risk reduction (SF-DRR) 2015–2030, ASEAN Agreement on Disaster Management and Emergency Response (AADMER) and UN General Assembly Resolution 68/211 in 2013. This helps to promote the development of DRR activities in recent years.

With a long history of fighting against natural disasters, DRR is a long-standing strategy and an integral part of national socio-economic development in Vietnam. Vietnam has developed a complete legal policy system to settle the implementation of DRR from the central to the local level. The first *Ordinance on Prevention and Control of Floods and Storms* was put in circulation in 2013, laying the foundation for the development of DRR policies and activities in Vietnam. The Ordinance emphasises the impact of floods and storms on human life as well as the ecological environment. Therefore, the first Article in the Ordinance prescribed flood and storm prevention as *the activities of preventing, combating and overcoming the harmful consequences of floods and storms in order to mitigate losses of human life and production and to limit adverse impacts on the ecological environment.*

The *National Strategy for Natural Disaster Prevention, Response and Mitigation* (NS-DPRM) to 2020 was issued in 2007, predating the NS-RCC (2008). The strategy proposed 51 solutions for DRR, including improving the legal and institutional system (9), non-structural measures (34) and structural measures (8). There are 2 out of 34 priority non-structural measures with EbA approaches including (i) planting and protection of watershed forests and (ii) planting of breakwater trees to protect dykes implemented by MARD.

In the same year, the Government approved the *National Community-Based Disaster Risk Management* (CBDRM) Project at Decision 1002/QD-TTg, brought more focus on communities and emphasised on the linkage between CbA and EbA while highlighted the critical role of ecosystem services for DRR, as well as the role of communities in protecting the watershed forest and coastal forest areas.

As such, compared to CCA strategy, DRR policies applied natural base solutions earlier, more in number and larger in scale.

6.3 Integration of EbA Through Sectors' Policies

Natural-based solutions have been prioritised and promoted in many national strategies and policies The *Sustainable Development Strategy 2011–2020* highlighted the role of a balance between economic growth, environmental protection and social progress. Accordingly, response to climate change in vulnerable coastal areas needs to focus on social advance and equality, national resources and environment protection, sociopolitical stability, simultaneously.

The new *Planning Law* (2017) remarked a new way in planning, which targeted towards integrated and sustainable planning. On one hand, it enforces the integration of climate change and environmental considerations for future planning. On the other hands, it aims to bridge the gap of planning activities in different sectors, leading to an agreement in planning between the central and local levels and bringing more transparency to the overall planning process.

The integration of EbA into the overall planning activities is also repeated in the *Master Plan and 5-Year Plan for the Vietnam Marine Economic Sustainable Development Strategy to 2030, with a vision to 2045* issued by the Government in March 2020. Accordingly, the planning of economic zones and coastal urban areas needs to be carried out sustainably, followed by an ecosystem-based approach and smart adapting to climate change.

The strategy of the Government for natural resource management is embodied in several documents, legislation and programs. The most important is the *Vietnam National Environment Action Plan* (August 1995), which demonstrates a strategic framework for forest and ecosystem conservation in both terrestrial and wetland areas and for watershed protection and management. A *Biodiversity Action Plan* (1997) also sets out a ranking of special-use forests (parks and wildlife sanctuaries) in terms of biodiversity values and extent of the threat. The amended *Law on Environment* was issued in 2014 includes a chapter on responding to climate change. The *10th National Assembly passed the Law on Water Resources* (LWR) (No. 08/1998/QH10) in 1998. The law stipulates the management, protection, development and utilisation of water resources. Apart from the general provisions, the LWR provides provisions on the prevention, control and mitigation of the consequences of flooding, inundation and other adverse effects caused by water.

Forest Management in Response to Climate Change
With a large area of forests and biodiversity, the Vietnamese Government considers forests an essential and valuable ecological resource for socio-economic development, especially in response to climate change. Forests in Vietnam are considered as a significant sink for carbon and play an essential role in adapting to climate change through environmental functions such as erosion control and water circulation assurance (CIFOR 2014). The Government has made its effort in issuing many critical legal documents to make use of forest resources for the implementation of CCA and DRR, especially in remote and upland areas where the population are poor and ethnic minorities, and in coastal areas where impacts of climate change are severe (UNDP/GEF SGP 2017). In Vietnam, coastal forests are classified primarily as protection forests, and most of the coastal forestlands follow the direct management of the Government. Vietnam has 29 provinces and cities with coastal mangrove forests and tidal zones along from Mong Cai to Ha Tien province. Of which: the Northern coastal region has five provinces of Quang Ninh, Hai Phong, Thai Binh, Nam Dinh and Ninh Binh. The Central Coast region has 14 provinces from Thanh Hoa to Binh Thuan. Southeastern and the Southwest coastal areas have ten provinces of Ba Ria Vung Tau, Dong Nai, Ho Chi Minh city, Ben Tre, Tien Giang, Tra Vinh, Soc Trang, Bac Lieu, Ca Mau and Kien Giang.

Many laws and regulations on forest resource use and protection have been inaugurated. The first and most important document is *Forestry Resource Protection and Development Act* issued in 1991, has defined three main forest land types (a) protection forests (watersheds and wetlands); (b) special-use forests (protected areas); and (c) production forests. To strengthen forest protection from national to local levels, a number of decrees had been allotted and outlined the management principles of each forest type, i.e. the separation of forest protection from production, clarification of the role of forest enterprise, distribution of forest land to local villagers, etc.

Since 1992, the Government has initiated the 327 Program. This is essentially a national program on expanding and protecting fallow or bare land. According to a report of World Bank (1997), this program used up US$45–50 million annually in the period of 1992–1995, yet it was criticised for too broad scope, which reduces the investment impacts. In 1995, the 327 Program was adjusted by the Decision 556/TTg to be more focused on protective forests and special-use forests, with the application of household-based agroforestry. This project is a precursor to the *Five Million Hectare Reforestation Program,* which was approved by the Decision 661/QĐ-TTg; therefore, the project is referred to as Project 661. The Project 661 was considered as a key socio-economic program, which had goal to plant 5 million hectares of new forests and protect existing forests in the period 1998–2010 in order to increase the forest cover of Vietnam to 43% by 2010. However, the Project 661 had been criticised as waste of money and low investment impacts (VTC 2008).

Another important program is the Decree 776 Program, which provides infrastructure and planning support for allocating land to establish new farm holdings adjoining the coastal full protection zones so that joint protection with the forest protection institutions can be brought about. Many other socio-economic programs have also been opened since 1992. These include Program 120 for employment promotion, programs to fight against hunger and poverty, including nutrition, literacy, and subsidies for the provision of essential goods. However, these programs have mainly focused on remote districts and have not been realised in connection with forest protection.

In 2008, the Government started the successful *Mangrove Restoration and Development Program 2008–2015* in increasing coastal mangrove coverage. However, the current program was designed to apply a monoculture-planting regime, and survival rates for mangrove forests supported under the program are only 50%. The reason is due to the lack of low seedling quality, limited protection for seedlings in the early stages of growth, less diversity in species selection, inappropriate planting methods and inadequate community engagement for maintenance and monitoring. The statistics from Government indicate that 62% of the existing mangrove forests in Vietnam are newly planted and monoculture. Lack of community consultations and awareness-raising among residents in terms of the benefits of mangrove reforestation has led to low levels of community engagement in some areas.

In 2015, the Government approved the *Program for coastal forest protection and development in response to climate change* (Decision 120/QĐ-TTg) for the 28 coastal provinces of Vietnam. Its view stresses the importance of coastal forest management and protection as an urgent task to prevent natural disasters and cope with climate change and sea-level rise. This program is of high priority and is specifically cited as a cornerstone for adaptation action in Vietnam's first INDC (GoV 2015). The program aims to promote the protective function of coastal forests to mitigate natural disasters, protect dike systems, infrastructure and conserve biodiversity. The targets include (i) the protection of existing coastal forest areas of 310,965 ha, and (ii) establishment of new plantations of 46,058 ha, increasing the coastal forest coverage from 16.9% (as of 2014) to 19.5% by 2020. Decree 119/2016/ND-CP on management, protection and development policies for coastal forests identifies the importance of these forests for responding to climate change. Accordingly, the allowable activities include (1) forest protection, afforestation, non-timber forest product plantation, agroforestry production and aquaculture; and (2) construction of sedimentation, coastal erosion protection, coastal mangrove restoration, ecotourism and forest environmental services. The importance of forest in responding to climate change is also reflected in the *National Target Program on sustainable forestry development* (Decision 886/2007/QD-TTg). The program has a goal to improve the quality of forests to meet the requirements of natural disaster mitigation, protecting the ecological environment, effectively coping with climate change and rising sea levels.

6.4 EbA Implementation Projects in Vietnam

The change in the policy of prioritising natural solutions has promoted the formation and development of a series of EbA projects in Vietnam. Many examples of natural risk management approach being used as a cost-effective and environmentally sound means of reducing risk to climate-related hazards, especially in vulnerable coastal regions. Initially, the Government of Vietnam focused only on afforestation, including the development of watershed and mangrove protection forests, breakwaters, wind and sand protection forests for coastal protection and erosion control. In recent times, solutions related to mangrove protection have been dominant and have been combined with other options to diversify livelihoods for people, increasing people's participation in protecting mangroves and coastal wetlands. The application of EbA approach in Vietnam has brought about positive results such as increasing the area and growth of mangroves, creating corridors for coastal protection, and also contributing to improving livelihoods from sustainable exploitation of forest and non-timber forest products.

Vietnam has many examples of successful EbA in various locations from northern to central and southern areas, though mostly at a small scale. EbA projects in Vietnam are mainly implemented under the auspices and technical and financial sponsorship of international organisations, non-governmental organisations such as the World Bank (WB), United Nations Development Programme (UNDP), Japan International Cooperation Agency (JICA), The Federal Ministry of the Environment, Nature Conservation and Nuclear Safety (BMU). The pilots mainly strived to reform current challenges by using advanced technologies and planting methods. The improved approaches include broadening and tailoring the application of mangrove varieties to local conditions, transforming planting techniques, employing tools such as bamboo breakwaters to encourage young plant growth, and intensifying soil quality and maintenance of young forests. Some pilots have also explored to more actively engage communities in planning and sustaining, and in the site selection, creating favourable conditions for local livelihoods, such as aquaculture, and mangroves can coexist.

This study reviews 14 projects in natural resource management whose objectives targeted climate change either directly or indirectly (Box 6.1). Results show that the majority of reviewed EbA projects targeted the poor or ethnic minorities group and mainly implemented in coastal and central mangrove regions. Among the sort of ecosystems, mangrove forest and terrestrial forest are the most frequently addressed with 8 out of 14 projects targeted at mangrove forests. The main goals of these projects are not only to adapt with climate change but also to strengthen the carbon sink, while aim at reducing erosion, land loss, and the impacts of disasters such as typhoons, cyclones, storms and sea-level rise. These so-called EbA projects have adopted technical measures such as reforestation, restoration, rehabilitation or improved varieties and land, focusing on improving livelihoods through the provision of small credit funds or financial assistance and small infrastructure. In the next section, the projects will be numbered and named in the order as set out in Box 6.1 (P1–P14). The analysis of 14 projects is presented in the Table 6.3.

Table 6.3 Review of 14 EbA projects

	Ecosystem types	Projects	Donors/implementers	Regions	CCA/DRR Objectives		EbA links						WRM links	References
					Direct	Indirect	P/R	L	B	G/F/P	CB/AR	Others		
1.	Wetlands	Coastal Wetlands protection and development project (CWPDP) (1999–2006)	WB/PMU	Coastal/Northern		Y	Y	Y	Y	Y	Y	Y (small infrastructure)	Y	(World Bank 1999)
2.	Mangrove forest	Mangrove plantation and disaster risk reduction in the period of 1994–2015 (MP-DRR)	IFRC/VNRC	Coastal/Northern	Y		Y	Y			Y		N	(IFRC 2011a)
3.	Terrestrial forest	Forest sector development project (2004–2015) (FSDP)	WB/PMU	Central/sloping land		Y	Y		Y	Y			N	(World Bank 2018)

(continued)

Table 6.3 (continued)

	Ecosystem types	Projects	Donors/implementers	Regions	CCA/DRR Objectives		EbA links						WRM links	References
					Direct	Indirect	P/R	L	B	G/F/P	CB/AR	Others		
4.	Terrestrial forest	The Greater Mekong Subregion Biodiversity Conservation Corridor Project (2011–2015) (BCC) Integrating biodiversity conservation, adaptation to climate change and sustainable forest management in the Trung Truong Son landscape (2016–2019)	ADB/MONRE (VEA) and PPC	Central/Upland	Y		Y	Y	Y			Y	N	(ADB 2010; Quang Nam PPC 2019; Thua Thien Hue PPC 2019)

(continued)

Table 6.3 (continued)

	Ecosystem types	Projects	Donors/implementers	Regions	CCA/DRR Objectives		EbA links						WRM links	References
					Direct	Indirect	P/R	L	B	G/F/P	CB/AR	Others		
5.	Mangrove forest	Mangrove restoration in a degraded Peri-Urban site in Central Vietnam (2012–2015)	Rockefeller/ISET Vietnam	Coastal Central Urban		Y	Y	Y				Y		(Tuyen and Tyler 2017)
6.	Terrestrial forest	Protection forests restoration and sustainable management project (2012–2021)	JICA	Upland Central		Y	Y	Y	Y					(JICA 2013)

(continued)

Table 6.3 (continued)

	Ecosystem types	Projects	Donors/implementers	Regions	CCA/DRR Objectives		EbA links						WRM links	References
					Direct	Indirect	P/R	L	B	G/F/P	CB/AR	Others		
7.	Mangrove forest	Improving the resilience of vulnerable coastal communities to climate change-related impacts in Vietnam (2015–2021)	GCF/UNDP	Coastal Central	Y		Y					Y (housing)		(UNDP 2015)
8.	Terrestrial forest	Sustainable forest management and biodiversity as a measure to decrease CO_2 emissions	KfW/GFA	Upland Northern	Y					Y		Y		

(continued)

Table 6.3 (continued)

	Ecosystem types	Projects	Donors/implementers	Regions	CCA/DRR Objectives		EbA links						WRM links	References
					Direct	Indirect	P/R	L	B	G/F/P	CB/AR	Others		
9.	Mangrove forest	Strategic mainstreaming of ecosystem-based adaptation (EbA) in Vietnam	BMU/GIZ	Coastal Central	Y		Y	Y			Y	Y (policy integration)		(ISPONRE and GIZ 2016, 2018)
10.	Mangrove forest	Promoting ecosystem-based adaptation through reforestation and sustainable use of mangroves (MAM1: 2012–2016) Scaling up Mangrove EbA in the Mekong Delta and (MAM2: 2016–2020)	BMU/SNV	Southern coastal area	Y		Y	Y		Y	Y			

(continued)

Table 6.3 (continued)

	Ecosystem types	Projects	Donors/implementers	Regions	CCA/DRR Objectives		EbA links						WRM links	References
					Direct	Indirect	P/R	L	B	G/F/P	CB/AR	Others		
11.	Mangrove forest	Mekong Delta Integrated Climate Resilience and Sustain Livelihoods Project (2016–22)	WB/MARD		Y		Y	Y		Y			Y	(World Bank 2016)
12.	Mangrove forest	Forest Sector Modernization and Coastal Resilience Enhancement Project (2017–2023)	WB/MARD	Coastal Northern to central	Y		Y			Y		Y (physical structure)		
13.	Dune and Mangrove forest	Ecosystem-based Adaptation in the North Central Coast of Vietnam: restoration and co-management of degraded dunes and mangroves 2018–2021	BMU/UNIQUE forestry and land use GmbH/MARD	North Central Coast	Y		Y	Y	Y		Y			(BMUB 2020)

(continued)

Table 6.3 (continued)

	Ecosystem types	Projects	Donors/implementers	Regions	CCA/DRR Objectives		EbA links						WRM links	References
					Direct	Indirect	P/R	L	B	G/F/P	CB/AR	Others		
14.		Coastal Management in the Red River Delta (2020–2027)		Coastal Northern	Y		Y	Y						

P/R Forest plantation/restoration, replantation/reforestation
L Livelihood diversity for community
EP/B Biodiversity enhancement
G/F Provide grant or fund or payment (PES)
O Other measures such as the focus on small infrastructure

Box 6.1: 14 Reviewed Projects of EbA in Viet Nam

1. ***Coastal Wetlands Protection and Development Project (CWPDP) (1999–2006)***: funded by World Bank aims to reduce coastal erosion and building community resilience through re-establishing the coastal mangrove wetland ecosystems and protect their aquatic nurturing and coastal protection functions sustainably.
2. ***Mangrove plantation and disaster risk reduction in the period of 1994–2015 (MP-DRR)***: was a multi-phase project operated by Vietnam Red Cross (VNRC) with support of the International Federation of Red Cross and Red Crescent Societies. The program aims at building up local resilience to natural hazards and climate change through the provision of support to forest plantation and management, awareness-raising and capacity building for selected communities and the local authorities. This program has been implemented in eight coastal (Quang Ninh, Hai Phong, Thai Binh, Nam Dinh, Ninh Binh, Thanh Hoa, Nghe An, Ha Tinh) and two upland provinces (Vinh Phuc, Hoa Binh) in northern Vietnam.
3. ***Forest Sector Development Project(2004–2015) (FSDP)***:

4.1. ***The Greater Mekong Subregion Biodiversity Conservation Corridor Project(2011–2015) (BCC)***: was funded by ADB with the long-term goal is to establish a system of biodiversity corridors across Vietnam, Lao PDR and Cambodia, to ensure sustainable forest ecosystem services and climate change adaptation across the Central Truong Son region. The focus was on Quang Nam, Quang Tri, and Thua Thien—Hue provinces to rehabilitate and preserve the construction of the ecosystem in the region, producing livelihood benefits to local communities.

4.2. ***Integrating biodiversity conservation, adaptation to climate change and sustainable forest management in the Trung Truong Son landscape (2016–2019)***: is funded by the Global Environment Fund via ADB grant. VEA/MONRE is assigned to deploy project at three provinces: Quang Nam, Quang Tri and Thu Thien Hue. The project aims to improve management and maintenance of the ecological integrity of the protected areas and surrounding areas, to ensure sustainable forest ecosystem service and adaptation to climate change in the Trung Truong Son area that benefits for the livelihoods of local communities and contribute to the economic growth of the area.

5. ***Mangrove Restoration in a Degraded Peri-Urban Site in Central Vietnam (2012–2015)***
6. ***Protection forests restoration and sustainable management project (2012–2021)*** was funded by JICA with objectives focused on sustainable protection and management of protection forests; restoration and preservation of biodiversity and poverty reduction in the mountainous regions.

7. ***Improving the resilience of vulnerable coastal communities to climate change-related impacts in Vietnam (2015–2021)*** was funded by GCF and implemented by UNDP to increase the resilience of vulnerable coastal communities to climate change-related impacts in Vietnam through robust mangrove coverage to provide a natural buffer between coastal communities and the sea.
8. ***Sustainable Forest Management and Biodiversity as a Measure to decrease CO_2 emissions***: was funded by KfW and implemented by GFA aimed to promote biodiversity in the Northern mountainous forest ecosystem, contributing to climate change adaptation and livelihood support for the local population.
9. ***Strategic Mainstreaming of EbA in Vietnam(2014–2019)***: was funded by the German Government to integrate the EbA strategy in the land use planning and development planning, systematically integrate EbA into national adaptation policies and pilot EbA approach at the local level.
10. ***Promoting Ecosystem-Based Adaptation through Reforestation and Sustainable Use of Mangroves*** (MAM1: 2012–2016) and ***Scaling up Mangrove EbA in the Mekong Delta*** (MAM2: 2016–2020) funded by BMU supports mangrove restoration and protection in the Mekong Delta in Vietnam, while strengthening the livelihoods and resilience of smallholder shrimp farmers and their families.
11. ***Mekong Delta Integrated Climate Resilience and Sustain Livelihoods Project*** (2016–22): The project's development objectives are to strengthen means for climate-smart planning and advance climate resilience of land and water management practices in some provinces of the Mekong Delta in Vietnam.
12. ***Forest Sector Modernisation and Coastal Resilience Enhancement Project*** (2017–2023): funded by World Bank and is being implemented by MARD. The objective is to improve coastal forest management in the most vulnerable provinces (including Quang Tri, Thanh Hoa, Nghe An, Ha Tinh, Quang Binh, Thua Thien Hue, Quang Ninh and Hai Phong). These provinces have about 400 km of coastline (accounted for 12% of Vietnam's total coastline). Its goal is to support the Government of Vietnam to achieve its higher objective of using coastal forests for boosting resilience to climate change.
13. ***EbA in the North Central Coast of Vietnam: Restoration and co-management of degraded dunes and mangroves*** 2018–2021: funded by BMU and implemented by UNIQUE aims at illustrating the technical probability of particular EbA interventions using indigenous and endemic tree species for restoring the protective functions of degraded dunes and estuaries in the region.

14. ***Coastal Management in the Red River Delta*** (2020–2027): The project aims to increase the resilience of local communities and coastal ecosystems to climate change by integrated coastal zone management, coastal hazard risk management (e.g. dikes), climate-resilient communities and livelihood systems studies.

6.4.1 Return Overweighs Cost in EbA Projects

Most projects show financial efficiency when comparing results of plantation and cost of reduced damages with the investment budget (comparing plantation area with investment budget). It was estimated that planting and protecting 12,000 ha of mangroves in Vietnam cost approximately $1 million but reduced the costs of sea dyke maintenance by $7.3 million per year (Turnbull et al. 2013). The P3 estimates of the economic rate of return exceeded the opportunity cost of capital. At completion, the economic rate of return for plantations was 13.2%, compared to 17.0% at appraisal. The P2 reported that the mangroves planted by VietNam Red Cross (VNRC) represent 4.27% of all mangroves in Vietnam and almost a quarter of those in the eight program provinces. During 2006–2010, P2 had cost US$8.88 million, with which 9462 ha of forest, including 8961 ha of mangroves, was created in 166 communes, protecting about 100 km of sea dyke (IFRC 2011a). The last phase of P2 (2011–2015) focused on the sustainable care and protection of existing mangroves rather than on area expansion. Since then an additional 107.6 ha of mangroves, 5.0 ha of the coastal protective trees (casuarina and acacia) and 20.6 ha of the upland protective trees had been planted in seven communes out of 74 coastal program communes with mangroves, contributing significantly to climate change adaptation, awareness-raising as well as providing livelihood support.[1] The project provides essential evidence on how mangroves contribute to enhance local livelihoods and help mitigate climate change and disaster impacts. It is estimated that the project affects around 350,000 direct beneficiaries and around 2 million indirect beneficiaries who are now better protected by mangroves and other trees (IFRC 2011a). Comparing the damage caused by similar typhoons before and after the intervention, it found that damages to dykes have been reduced by USD 80,000 to USD 295,000 in the studied communes, and these savings represent less than the costs for mangrove planting. The report ascribes annual probabilities to significant disasters and calculates the overall avoided risk up until 2025 (up to USD 37 million in one commune). Besides,

[1] This part was synthesis mainly from three reports of the MP-DRR program: (1) *Breaking the waves: Impact analysis of coastal afforestation for disaster risk reduction in Vietnam* (IFRC 2011a); (2) *Planting protection: Community-based Mangrove Reforestation and Disaster Preparedness Program, 2006–2010* (IFRC 2011b); and (3) *Final evaluation of the mangrove plantation and disaster risk reduction in the period of 2011–2015* (Asian Management and Development Institute 2015).

the report finds that the program's protective benefits over its costs in each of the communes. Of which, assets located between mangroves and dykes (shrimp farms, boats) benefit the most (IFRC 2011a). A large proportion of beneficiaries in mangrove communes ascribe a positive impact of the program on their income. The bamboo plantation has also shown to bear the high potential to raise incomes; however, the overall impact has been relatively minor mainly due to the small lots allocated to each planter. During the first and second stages (1994–2010), the direct economic benefits (e.g. from aqua product collection, honey bee farming) are also found to be substantial, although much smaller than protective benefits. Direct economic benefits are found to be between USD 344,000 and USD 6.7 million in the communes studied. Of which mangroves have also led to an increase of yield from aquaculture product collection (e.g. shells, oysters) by 209–789%, providing more income for coastal communities, in particular its poorer members. By far the most significant benefit identified concerns the mangroves' carbon value. Extrapolating from locally conducted research on accumulated carbon and CO_2 absorption capacity, the result of the project finds that the mangroves planted by VNRC will have absorbed at least 16.3 Mio t CO_2 by 2025. Assuming a price of USD 20 per t of CO_2 emissions and having applied the discount rate of 7.23%, this represents a value of USD 218.81 Mio.

6.4.2 EbA Targets the Poor Communities and Ethnic Minority Groups

Most of the projects target some of the poorest districts in upland or coastal areas, whose exposure to climate extremes is high. The P3 targeted communes with high poverty rates (World Bank 2018). The P4 contributed to poverty reduction and improvement of ethnic minorities in poor mountainous districts of the project area. The primary beneficiaries of the project are those who were living in remote areas with high rates of poverty and ethnic minority groups. The P7 prioritised vulnerable families identified as poor by the Government and highly disaster-exposed people. P8 benefited 184,000 local people whose livelihoods were dependent mainly on the forest resources, of which 76% are ethnic minorities. However, despite the advantages of EbA options as more available to the poor than other structural adaptation measures, the paradox is that poor and ethnic minority groups often lack inadequate capacity and resources to take good care of the forest. This was cited as one of the main difficulties of EbA projects. The final report of P9 shows that the lack of capacity and resources of poor households is one of the barriers to the effectiveness of the project. Training costs during project implementation increased more than initially planned, but the training effect was not as expected (ISPONRE and GIZ 2016, 2018).

A report of UNDP in 2013 pointed out that it was complicated to establish mangrove areas and rate of established forest in Vietnam was only about 50% due to low seedling quality, lack of protection of seedlings from the physical effects in the early stage of growth, and lack of species diversity selection and suitable planting methods in each specific site. With that, it was highly recognised that community knowledge and skill, as well as experience in mangrove restoration or replantation, is fundamental. However, due to their limited access to knowledge and resources, the poor often lack the skills and ability to meet technical requirements for mangrove planting.

6.4.3 *Management Approach Shifted from Top-Down to Bottom-Up and Co-management in EbA Projects*

In the past, the project management unit (PMU) has usually been set up at the central level for coordination and administration of the project at the local level. Most of the projects were implemented in a top-down approach, where project activities were designed by practitioners based on initial assessment results with contribution from national or international experts. However, this style has confronted many troubles in implementation. The shortcoming of the top-down approach seems visible enough for the donors and implementers to employ decentralised approach and co-management. Most of the projects established after 2010 applied a decentralised approach with the establishment of local project management units (PPMUs) and participation from grassroots organisations, research institutes and private sector in different project implementation stages. Valuable lessons have been learnt from management approaches to forest protection and rural development under many projects.

The P1 was designed following the Government's emphasis on rural development and targeted towards raising agricultural productivity, enhancing investment in research, generating employment, reforming state-owned enterprises and improving the management of natural resources (especially reforestation) (World Bank 1999). The objectives of the P3's were aligned with the Government of Vietnam's Socio-Economic Development Strategy 2011–2020 and Socio-Economic Development Plan 2011–2015, which supported for establishing smallholder plantations on barren hillsides to raise the productivity of these plantations made environmental sense, as did measures to strengthen the capacity of special-use forest management boards to protect biodiversity (World Bank 2018).

For the P2, in addition to the establishment of the National Project Steering Committee (NPSC) and its secretariat, it set up a Project Field Coordination Unit (PFCU) based at the regional level to assist different provinces in the project area and a PPMU in each province to help with decentralising implementation down to the local level (IFRC 2011a). One of the key factors that decided the success of the P2 is that the program is consistent with governmental legislation for mangrove and coastal forestry and DRR. The project integrates EbA and CbA from the early stages. When CBDRM's program (Decision 1002) was launched in 2009, the program involved

Disaster Management Centre (DMC), and at the same time signed an agreement with MARD to support carrying out Decision 1002 in at least 1000 communes, and provide technical support to carry out the Decision in 6000 communes nationwide. VNRC at all levels took this opportunity to participate actively in DRR activities at the local level, contributing to improving the capacity of local Government and people to reduce the risk of disasters and respond to climate change. This project is highly appreciated in terms of efficiency, not only effectively using the funding in mangrove planting, but also creating an effective management system, and sustainable livelihoods for the people. The earlier phases of the project mainly focused on afforestation and increased its area in the planting of bamboo along river dykes, and coastal and riverbank stretches. The planting component aims to protect dykes and communities from various hazards such as typhoons, storms and floods, as well as to provide additional income to poor community members. The following phases of the project focus on building management systems, improving care and management skills for officials and people. As emphasised by the project, creating an effective management system and sustainable livelihoods for the people is a prerequisite in determining the effectiveness and sustainability of the project.

The P4 follows a bottom-up approach, by way of the initial consultation in the communes, and focuses on beneficiaries considered an indispensable part of the project. During the project's preparation phase, the participation and consultations with multi-stakeholders were held to confirm the beneficiaries' priorities, their contributions and commitments regarding labour, local materials, land and safety corridors (Quang Nam PPC 2019; Thua Thien Hue PPC 2019).

To overcome the shortcomings of the top-down and the bottom-up approaches, a co-management solution was introduced and lately applied by some projects. However, in Vietnam, the application of co-management in protection mangrove forest is limited due to the policy perspective, the forest tenure law defines all mangroves as "protected" forests, which cannot be accessed or used. Meanwhile, co-management, in its nature means that local communities participate in and benefit from management decisions for sustainable resource use. Results from the P5 and the P13 showed that simply adopting consistent processes for co-management did not secure consistent outcomes, yet local leadership and commitment to mangrove forest management and sustainable use was a significant contributing factor (BMUB 2020; Tuyen and Tyler 2017).

In fact, despite a decentralised approach applied in the field, the management role of commune authorities was not clarified, thus limiting their participation in the monitoring process, yet just focus solely on guidance for the implementation. The consequence is a delay in the implementation compared to the original plan. Also, the lack of close supervision of local authorities (both at villages and communal levels) with the community forest protection groups makes the group's activities less active and regular (ISPONRE and GIZ 2018).

6.4.4 CbA is Complementary to the Successfulness of EbA

Evidence from the projects' review showed that the involvement of the local community in the preparation and implementation is critical for EbA. These lessons were reflected in the P1 project's design, especially for the vulnerable (poorer) communities where pilot participatory rural appraisal exercises in developing commune action plans have started (World Bank 1999). Community involvement in patrolling reserves, monitoring inventory and abstaining from illegal tree felling and poaching was also highly recognised in the P3. It is said that communities hold the essential role in protecting the resources and mainly contributing to the project's value.

On the other hand, the lack or delay of a community's engagement leads to unpredictable consequences. Another example from the P5 proved that due to lack of a proper process of community consultation and assignation, initial local consensus and commitment in Nhan An Village (Quy Nhon City) was broken down due to people want to earn more from raising a much bigger volume of oysters than agreed in the contract, which in turn, leading to loss of all newly planted seedlings of mangrove. Besides, the project also experienced high losses of seedlings due to not follow community advice on the selection of mangrove species (Tuyen and Tyler 2017).

6.4.5 Sustainability of the EbA Projects

The sustainable outcome is one of the significant issues among EbA projects. Many reasons are affecting the sustainability of the EbA project, the results from the review of 14 projects show that the lack of guiding and supporting policies for EbA, as well as the lack of monitoring and evaluation measures is the two main factors leading to the project's unsustainability.

For the P2, the sustainability of the project is still limited due to the lack of legal frameworks that direct the maintenance and management of a sustainable project area. Therefore, despite that the P2 was recognised as a successful case study, it did not receive direct and official financial support as well as legal support (for the care and protection of mangroves) from the Government to expand the planting and protection of mangroves. The P2 project is one of the clearest examples of the effectiveness of the EbA strategy in Vietnam, the mangrove afforestation, in particular, has also been a highly efficient way to reach its protective, direct economic and ecological benefits. However, policies to manage and support the implementation of the EbA strategy at the local level need to be finalised to ensure project sustainability.

The omission of the specific EbA policy also leads to limitations of the EbA project's sustainability as it causes the absence of appropriate support as well as appropriate monitoring and evaluation measures for EbA activities. Lack of supportive policies will make EbA dependent on financial and technical resources from external sources, such as in the case of the P2, VNRC has done very well, but due to financial dependence on external support sources, they were unable to replicate EbA activities further. Lack of monitoring and evaluation measures will result in the effectiveness of the project being invisible and unrecognised, or underestimated, thus scaling up or continuing to carry out EbA activities are often overlooked. In case of the P9, the implementation of the EbA takes a long time to ensure the achievements of the goals and results set out, so the evaluation of project implementation results is limited, the effectiveness of the project has not been seen, affecting the continued implementation or expansion of the project. Besides, the lack of a plan in the P9 to monitor and evaluate the effectiveness of the project after finishing the project also affects the sustainability of the project because EbA activities are usually significant long after the project is completed.

Furthermore, the lack of supporting policy also causes a lack of incentives for local stakeholders and investors to protect and restore coastal forests for long-term benefits. When considering investing in the protection of coastal forests, especially for the case of Vietnam, coastal forests are mainly protection forests, and local stakeholders face a trade-off between short-term benefits from exploitation/harvesting and longer-term public goods regarding response to climate change. It means that these stakeholders should receive more incentives for their trade-off for long-term management and protection of coastal forest systems. Results from the P2 recognised the formal protection of mangroves and the commitment of the Government as well as a definite sense of local ownership as essential factors for a more sustainable project. The messages from the projects to stakeholders are that mangroves, bamboo and casuarina trees cannot be seen as being there for good but rather require long-term work on protection, future planning and awareness (IFRC 2011a).

On the other hand, there are also examples of success related to sustainability. For the P3, since the project ended, an additional 3457 ha of plantations in the provinces of Thua Thien Hue and Quang Nam and 293 ha in Quang Ngai Province were FSC-certified. The management modalities introduced by the project gained positive impacts and are expected to be further improved. In addition, there are 24,500 project households that kept investing in afforestation after harvesting the trees from the first rotation. They continued to maintain and expand afforestation with the technical and financial support from the Vietnam Bank for Social Policy (World Bank 2018).

6.5 Conclusions

Ecosystem-based adaptation is a proven management approach to natural hazards that protects ecosystems, enhances the resilience of vulnerable groups, supports the targets set by the 2030 Agenda, the Paris Agreement and the Sendai Framework and offers a more flexible and adaptive solution to disaster risk than grey infrastructure within an uncertain climate. Despite these multiple benefits, ecosystem-based approaches are still rarely implemented in practice. There is growing recognition of the importance of ecosystem-based approaches for adaptation to climate change—it is a cost-effective measure that has multiple benefits and can overcome many of the drawbacks of more common engineering adaptation options.

Currently, Vietnam has no separate and specific policy for the application of ecosystem-based; however, the integration of EbA through various sectors and in CCA and DRR encourages and enhances the application of the nature-based solution. The role of ecosystems, sustainable management of natural resources to maintain benefits as well as enhance the adaptability of the community is increasingly recognised. Resolution No. 24, the NS-RCC, NAP-RCC and AP-MKD emphasises the need to maintain and restore ecosystems in response to climate change in Vietnam. These important documents and policies demonstrate the strong commitment of the Party and the Government in the fight against climate change in parallel with conservation and management of natural resources. Analysis from the EbA-related policies showed a shift from hard to soft solutions, especially with strategies and policies formed after 2015. Results from a review of policy also pointed out that the role of MARD as a coordinating role to integrate the EbA measures among different sectors is still limited. MARD's AP-RCC has not presented a road to integrate EbA through the cross-sectoral actions and programs related to forest ecosystems, coastal ecosystems, freshwater ecosystems. At the provincial level, PPAP has integrated the EbA strategy. However, the integration was limited at a very general level, and few provinces have made specific action on EbA activities, which are based on local socio-economic and natural conditions. Also, local authorities typically lack the legislative backing and resources to provide valuable incentives to developers to implement EbA and issues over ongoing institutional arrangements remain barriers.

Although the commitment and policy to support the implementation of EbA has been developed and initially developed, the implementation of EbA in Vietnam is still limited. Governmental efforts are trying to integrate ecosystem-based approaches in climate change response. However, the practical level goes behind the plan, due to many reasons, especially the lack of specific support policies and monitoring and evaluation mechanisms for EbA projects. Lack of specific direction from the Government on the implementation of EbA caused the formation of EbA projects that relied on related documents and policies in the field of climate change response and ecosystem management. However, the policy interpretation of each organisation is often different and based on the initial evaluation (which is mostly done by independent consultants with different specialities). This causes EbA implementation difficult and delayed, strongly affecting projects' efficiency. The P2 is a typical example, due to the lack of institutional arrangement, the role of local authorities was vague in terms of critical tasks such as directions, consult, monitoring and evaluation of the project's implementation process.

Another challenge of Vietnam in managing and approaching the EbA method is that the national policy was created and developed following a state-led mechanism (all management is concentrated in the hands of the state). The formation of EbA projects in Vietnam is very top-down, expressed through the change of the number of projects in line with the policy of the central Government, but less based on local practical conditions. Most of the projects emphasised the "in compliance with" the central governmental policies. Previously, mangrove restoration projects were used to concentrate in northern and central Vietnam. Furthermore, recently, together with Resolution 120, mangrove restoration in southern Vietnam, especially the Mekong River Delta, has received more impetus. Sustainability is also one of the significant issues of EbA projects. The review of 14 projects shows that even projects rated as highly effective are facing difficulties in replicating and maintaining the sustainability of projects.

References

ADB (2010) Greater Mekong Subregion biodiversity conservation corridors project (Vietnam component). Retrieved from https://www.adb.org/sites/default/files/project-document/62555/40253-02-reg-pam-02.pdf

Asian Management and Development Institute (2015) Final evaluation of the mangrove plantation and disaster risk reduction in the period of 2011–2015. Hanoi, Vietnam. Retrieved from https://www.rcrc-resilience-southeastasia.org/wp-content/uploads/2016/01/MP-DRR-final-report-26Oct.pdf

BMUB (2020) Ecosystem-based adaptation on the northern central coast of Vietnam: restoration and co-management of degraded dunes and mangroves. Retrieved from https://www.international-climate-initiative.com/en/details/project/ecosystembased-adaptation-on-the-northern-central-coast-of-vietnam-restoration-and-comanagement-of-degraded-dunes-and-mangroves-18_II_160-3013

CIFOR (2014) Summary after three years of implementation of the policy on payment for forest environmental services in Vietnam. Hanoi, Vietnam

EM-Dat Database (2019) Disaster statistics. Retrieved from https://www.emdat.be/emdat_db/

Garschagen M, Hagenlocher M, Comes M, Dubbert M, Sabelfeld R, Lee YJ (2016) World risk report 2016. Bündnis Entwicklung Hilft and UNU-EHS. Retrieved from https://collections.unu.edu/view/UNU:5763

GoV (1993) Ordinance on prevention and control of floods and storms—ordinance 9-L/CTN (in Vietnamese). Hanoi, Vietnam

GoV (2007) National strategy for natural disaster prevention, response and mitigation to 2020—decision 172/2007/QD-TTg (in Vietnamese). Hanoi, Vietnam

GoV (2008) National target program to respond to climate change—decision 158/2008/QD-TTg (in Vietnamese). Hanoi, Vietnam

GoV (2009) Community awareness-raising and national community based disaster risk management project 2009—decision 1002/QD-TTg (in Vietnamese). Hanoi, Vietnam

GoV (2011) National strategy on climate change—decision 2139/QD-TTg (in Vietnamese). Hanoi, Vietnam

GoV (2012a) Decision no. 1443/TTg-QHQT of the Prime Minister approving the list of priority projects under SP-RCC program (in Vietnamese)

GoV (2012b) National action plan on responding to climate change in the period 2012–2020 (in Vietnamese)—decision 1474/QD-TTg. Hanoi, Vietnam

GoV (2013) Proactively responding to climate change, strengthening natural resource management and environmental protection—resolution 24/NQ-TW (in Vietnamese). Hanoi, Vietnam

GoV (2014) Support program to respond to climate change (in Vietnamese). Hanoi, Vietnam

GoV (2015) Intended nationally determined contribution of Vietnam. Hanoi, Vietnam. Retrieved from https://www4.unfccc.int/sites/ndcstaging/PublishedDocuments/VietNamFirst/VIETNAM%27SINDC.pdf

GoV (2016a) Strategy of sustainable development of the Mekong Delta to adapt to climate change to 2050—resolution 120/NQ-CP (in Vietnamese). Hanoi, Vietnam

GoV (2016b) Target program to respond to climate change and green growth—decision 1670/QD-TTg (in Vietnamese). Hanoi, Vietnam
GoV (2019a) National action plan for natural disaster prevention, response and mitigation to 2020—decision 649/QD-TTg. Hanoi, Vietnam
GoV (2019b) National action program to implement resolution 120/NQ-CP on sustainable development of the Mekong Delta to adapt to climate change. Hanoi, Vietnam
GoV (2020) Plan of implementation of Paris agreement on climate change in the 2021–2030 period of MARD 2020—decision no. 891/QD-BNN-KHCN (in Vietnamese). Hanoi, Vietnam
IFRC (2011a) Breaking the waves: impact analysis of coastal afforestation for disaster risk reduction in Vietnam. Hanoi, Vietnam. Retrieved from https://www.preventionweb.net/files/globalplatform/entry_bg_paper~mangroveimpactreportfinallowapril2011.pdf
IFRC (2011b) Planting protection: community based mangrove reforestation and disaster preparedness programme, 2006–2010
ISPONRE and GIZ (2016) Strategic mainstreaming of ecosystem-based adaptation (EbA) in Vietnam: participatory Identification of EbA measures for piloting in Ha Tinh
ISPONRE and GIZ (2018) Ecosystem-based adaptation: coastal protective forest in Hoa Binh Village, Quang Hung Commune, Quang Binh Province, Vietnam
JICA (2013) JICA's support to forestry and nature conservation in Vietnam. Vietnam. Retrieved from https://www.jica.go.jp/vietnam/english/office/others/c8h0vm0000cydg8v-att/sector_03_01.pdf
MONRE Vietnam (2016) Climate change and Sea level rise scenarios for Vietnam. Retrieved from https://www.preventionweb.net/files/11348_ClimateChangeSeaLevelScenariosforVi.pdf
MONRE Vietnam (2019) Submitted to the Prime Minister for approval the National Plan for climate change adaptation for 2021–2030, with a vision to 2050 (in Vietnamese). Electronic Portal of the Ministry of Natural Resources and Environment of Vietnam. Hanoi, Vietnam, 29 Nov 2019. Retrieved from http://www.monre.gov.vn/Pages/trinh-thu-tuong-chinh-phu-phe-duyet-ke-hoach-quoc-gia-thich-ung-voi-bien-doi-khi-hau-giai-doan-2021---2030,-tam-nhin-den-nam-2050.aspx
Quang Nam PPC (2019) Greater Mekong subregion biodiversity conservation corridors. Vietnam Component (BCC Project) Quang Nam Province. Vietnam
Thua Thien Hue PPC (2019) Greater Mekong subregion biodiversity conservation corridors. Vietnam Component (BCC Project) Thua Thien Hue Province. Vietnam. Retrieved from https://www.adb.org/sites/default/files/project-documents/40253/40253-023-smr-en_10.pdf
Turnbull M, Sterrett CL, Hilleboe A (2013) Toward resilience: a guide to disaster risk reduction and climate change adaptation. In: Catholic relief services—United States conference of catholic bishops. Practical Action Publishing Ltd., UK. https://doi.org/10.1080/12294659.2013.10805272
Tuyen NP, Tyler S (2017) Mangrove restoration in a degraded peri-urban site in central Vietnam: variable success in different villages. Hanoi, Vietnam. Retrieved from https://docs.wixstatic.com/ugd/558f8a_c71ea0274268463088752692a165788f.pdf
UNDP (2015) Improving the resilience of vulnerable coastal communities to climate change-related impacts in Vietnam. Hanoi, Vietnam. Retrieved from https://www.undp.org/content/dam/vietnam/docs/Publications/FP-UNDP-281215-5708.pdf
UNDP/GEF SGP (2017) Improve livelihoods associated with forest protection and development—practical and policy recommendations. Hanoi, Vietnam. Retrieved from https://data.vietnam.opendevelopmentmekong.net/vi/dataset/c-i-thi-n-sinh-k-g-n-v-i-b-o-v-va-phat-tri-n-r-ng-th-c-t-va-khuy-n-ngh-chinh-sach
VTC (2008) Project of 5 million hectares of forest: throwing money into space? (in Vietnamese). Retrieved from https://vtc.vn
World Bank (1997) Forest protection and rural development project. Hanoi, Vietnam. Retrieved from http://documents1.worldbank.org/curated/en/246311468779401261/pdf/multi-page.pdf
World Bank (1999) Coastal wetlands protection and development project. Hanoi, Vietnam. Retrieved from http://documents.worldbank.org/curated/en/633501468779402279/pdf/multi-page.pdf

World Bank (2016) Mekong Delta integrated climate resilience and sustain livelihoods project (2016–22). Retrieved from http://documents.worldbank.org/curated/en/840701467996680631/pdf/PAD1610-PAD-P153544-IDA-R2016-0110-1-Box394889B-OUO-9.pdf
World Bank (2018) Vietnam: forest sector development project. Washington, DC. Retrieved from http://documents.worldbank.org/curated/en/767121522168227379/Vietnam-forest-sector-development-project

Chapter 7
Turning Blue, Green and Gray: Opportunities for Blue-Green Infrastructure in the Philippines

Noralene Uy and Chris Tapnio

Abstract Nature-based solutions represent a critical concept that harnesses natural systems to provide essential services for disaster risk reduction and climate change adaptation. As a nature-based solution, blue-green infrastructure takes advantage of nature's innate ability to substitute for or strengthen infrastructure systems by preserving, enhancing, or restoring a natural system's elements to build high quality, resilient and lower-cost infrastructure. The chapter describes how ecosystem-based disaster risk reduction, ecosystem-based adaptation, and blue-green infrastructure are implemented in the Philippines, including the policies that support them, the status of implementation, and through a case study in Polillo, Quezon, Philippines. Findings show that despite policies in place to support and advance the mainstreaming of nature-based solutions in the country, the environment's cross-cutting nature as a sector makes enforcement and implementation of programs, plans, and activities extremely challenging. Implementing nature-based solutions in the Philippines has so far been undertaken as a response to environmental challenges. More than being reactive, a proactive focus on nature-based solutions for prevention, mitigation, and rehabilitation is needed. The science and evidence for blue-green infrastructure would need to be strengthened to inform decision-making better, gain political commitment at all levels, secure funding and private sector engagement, and ultimately advance its implementation.

Keywords Blue-green infrastructure · Nature-based solutions · Ecosystem-based disaster risk reduction · Ecosystem-based adaptation

---

N. Uy (✉)
Department of Environmental Science, Ateneo de Manila University, Loyola Heights, Quezon City 1108, Philippines
e-mail: nuy@ateneo.edu

C. Tapnio
Yale School of the Environment, Yale University, Prospect St., New Haven, CT, USA

© The Author(s), under exclusive license to Springer Nature Singapore Pte Ltd. 2021
M. Mukherjee and R. Shaw (eds.), *Ecosystem-Based Disaster and Climate Resilience*, Disaster and Risk Research: GADRI Book Series,
https://doi.org/10.1007/978-981-16-4815-1_7

7.1 Introduction

Ecosystems provide a range of ecosystem services and goods that support livelihoods and human well-being. Ecosystem functions and processes that underpin the provision of these ecosystem services are challenged by environmental and human-induced disruptions that exacerbate socio-ecological systems' vulnerability (MEA 2005). Natural ecosystems have been shown to reduce vulnerability to natural hazards and extreme climatic events by providing essential ecosystem services such as storm protection, green coastal defenses, and regulating water and nutrient recharge, contributing to resilient outcomes. Over the years, the case for investing in ecosystems has been increasingly supported through mainstreaming in disaster risk reduction and climate change adaptation.

As the latest addition to the lexicon of ecology, nature-based solutions (NbS) have become a buzzword that is used as an umbrella term for ecosystem-related approaches such as ecosystem-based adaptation (EbA), ecosystem-based disaster risk reduction (Eco-DRR), blue-green infrastructure (BGI), ecosystem management and ecological engineering, among others. The International Union for the Conservation of Nature (IUCN) defines NbS as "actions that protect, manage, and restore natural capital in ways that address societal challenges effectively and adaptively while simultaneously providing human well-being and biodiversity benefits." The World Wide Fund for Nature describes NbS for the climate as "ecosystem conservation, management and/or restoration interventions intentionally planned to deliver measurable positive climate adaptation and/or mitigation benefits that have human development and biodiversity co-benefits managing anticipated climate risks to nature that can undermine their long-term effectiveness" (WWF undated). NbS represents a critical and urgently significant concept that harnesses natural systems to provide essential services (e.g., wetlands for flood mitigation; mangroves to reduce the impact of waves, storm surge, and coastal erosion) (World Bank 2019). It can provide a cost-effective and flexible approach with multiple benefits for disaster risk reduction and environmental management such as buffers from climate-related impacts; resource base for critical economic sectors such as fisheries and tourism; resilient water-related services and can increase the resilience of vulnerable communities that depend on natural resources through sustainable livelihoods (GFDRR undated).

As a nature-based solution, EbA aims to maintain and increase the resilience and reduce the vulnerability of ecosystems and people in the face of the adverse effects of climate change by harnessing biodiversity and ecosystem services through conservation, sustainable management, and restoration of ecosystems as part of an overall adaptation strategy (IUCN 2017). IPCC (2014) provides examples of EbA options, including wetland and floodplain conservation and restoration; biological diversity conservation; afforestation and reforestation; mangrove conservation and rehabilitation; green infrastructure; ex-situ conservation and seed banks; community-based natural resource management; and adaptive land use management, among others. Eco-DRR, on the other hand, is "the sustainable management, conservation, and restoration of ecosystems to reduce disaster risk, to achieve sustainable and resilient development" (Estrella and Saalismaa 2013). It entails combining ecosystem management with disaster risk reduction strategies, such as early warning systems and preparedness planning, for disaster prevention and recovery, thereby reducing the impact of disasters on people and communities (Sudmeier-Rieux et al. 2019). When these nature-based interventions are synergized with gray infrastructure to form the so-called "hybrid" solutions, these are referred to as blue-green infrastructure (BGI), green infrastructure, bioengineering, or next-generation infrastructure.

Blue-green infrastructure is starting to gain momentum as successful examples have demonstrated a "triple-win" with benefits for the environment, economy, and people. BGI takes advantage of nature's innate ability to substitute for or strengthen infrastructure systems by preserving, enhancing, or restoring a natural system's elements to build high quality, resilient and lower-cost infrastructure. In this way, combining biological systems with technological and engineering structures such as dams, levees, reservoirs, treatment systems, and pipes can result in enhanced system performance and protected populations (Browder et al. 2019). These natural building materials are integrated into infrastructure design to increase their (i) strength over time; (ii) flexibility and capability to absorb ground movement and energy from water and wind; (iii) ability to recover from damage (ADB 2020).

7.1.1 Principles, Criteria, and Standard for NbS

In efforts to deepen understanding of nature-based solutions and provide an operational framework for their application, international organizations have identified principles and approaches and standards to guide implementation of all types of NbS interventions and build momentum in both policy and practice (Table 7.1). Because the ecosystem approach is a central concept in the work programs of the Secretariat of the Convention on Biological Diversity (CBD), it provided operational guidance through the 12 Principles of the ecosystem approach and its implementation guidelines (CBD 2004). Cohen-Shacham et al. (2016) proposes eight principles to form the basis of NbS implementation. To promote the application of a common set of qualification criteria and standards in the context of implementing the Paris Agreement and Nationally Determined Contribution commitments and informing national adaptation planning processes, FEBA (2017) developed an assessment framework for designing, implementing, and monitoring EbA measures by proposing a set of three elements and five qualification criteria, and 20 quality standards and indicators. Recently, the IUCN released the Global Standard for NBS to recognize the need to have a user-friendly framework for the verification, design, and scaling up of NbS (IUCN 2020). The Standard seeks to answer the need for clarity and precision on using the concept, have a common understanding of its application, and develop a shared vision of the desired outcomes. The Standard consists of eight criteria with corresponding 28 indicators. These principles, criteria, and standard have common themes and focus, which demonstrate an increasing understanding of the concept as

Table 7.1 Principles, criteria, and standard for nature-based solutions

CBD (2004)	**12 Principles of ecosystem approach** Principle 1: The objectives of management of land, water, and living resources are a matter of societal choice; Principle 2: Management should be decentralized to the lowest appropriate level; Principle 3: Ecosystem managers should consider the effects (actual or potential) of their activities on adjacent and other ecosystems; Principle 4: Recognizing potential gains from management, there is usually a need to understand and manage the ecosystem in an economic context. Any such ecosystem management program should: (a) Reduce those market distortions that adversely affect biological diversity; (b) Align incentives to promote biodiversity conservation and sustainable use; and (c) Internalize costs and benefits in the given ecosystem to the extent feasible; Principle 5: Conservation of ecosystem structure and functioning, to maintain ecosystem services, should be a priority target of the ecosystem approach; Principle 6: Ecosystems must be managed within the limits of their functioning; Principle 7: The ecosystem approach should be undertaken at the appropriate spatial and temporal scales; Principle 8: Recognizing the varying temporal scales and lag-effects that characterize ecosystem processes, objectives for ecosystem management should be set for the long term; Principle 9: Management must recognize that change is inevitable; Principle 10: The ecosystem approach should seek the appropriate balance between, and integration of, conservation and use of biological diversity; Principle 11: The ecosystem approach should consider all forms of relevant information, including scientific and indigenous and local knowledge, innovations and practices; and Principle 12: The ecosystem approach should involve all relevant sectors of society and scientific disciplines

(continued)

Table 7.1 (continued)

Cohen-Shacham et al. (2016)	**Eight principles for NbS** 1. NbS embrace nature conservation norms and principles; 2. They can be implemented alone or in an integrated manner with other solutions to societal challenges (e.g., technological and engineering solutions); 3. NbS are determined by site-specific natural and cultural contexts that include traditional, local and scientific knowledge; 4. They produce societal benefits fairly and equitably, in a manner that promotes transparency and broad participation; 5. NbS maintain biological and cultural diversity and the ability of ecosystems to evolve; 6. They are applied at a landscape scale; 7. NbS recognize and address the trade-offs between the production of a few immediate economic benefits for development and future options for the production of the full range of ecosystems services; and 8. They are an integral part of the overall design of policies and measures to address a specific challenge
FEBA (2017)	**Three elements and five criteria that help answer the question: Is the approach EbA or not?** Element A: EbA helps people adapt to climate change Criterion 1. Reduces social and environmental vulnerabilities Criterion 2. Generates societal benefits in the context of climate change adaptation Element B—EbA makes active use of biodiversity and ecosystem services Criterion 3. Restores, maintains, or improves ecosystem health Element C—EbA is part of an overall adaptation strategy Criterion 4. Is supported by policies at multiple levels Criterion 5. Supports equitable governance and enhances capacities
IUCN (2020)	**Global Standard for NbS** Criterion 1: NbS effectively address societal challenges; Criterion 2: Design of NbS is informed by scale; Criterion 3: NbS result in a net gain to biodiversity and ecosystem integrity; Criterion 4: NbS are economically viable; Criterion 5: NbS are based on inclusive, transparent, and empowering governance processes; Criterion 6: NbS equitably balance trade-offs between the achievement of their primary goal(s) and the continued provision of multiple benefits; Criterion 7: NbS are managed adaptively, based on evidence; and Criterion 8. NbS are sustainable and mainstreamed within an appropriate jurisdictional context

operationalized over the years. These include ecosystem integrity, scale, co-benefits, adaptive governance, and participation, among others.

In the following, the chapter describes how NbS, particularly Eco-DRR and EbA, are implemented in the Philippines, including the policies that support it, the status of implementation, and a case study to demonstrate the focus of projects. It provides a discussion and some insights into opportunities for scaling up BGI in the country. Finally, it presents the conclusions and some recommendations for advancing NbS and BGI in the Philippines.

7.2 Nature-Based Solutions in the Philippines

7.2.1 Implementing NbS in the Philippines

The Philippines is endowed with a rich environment and natural resources, and therefore has a long history of environmental management and natural resources conservation. Recognizing that ecosystem-based approaches play a critical role in disaster risk reduction, climate change adaptation, and environmental management, the Philippine government has developed policies to support and advance the mainstreaming of NbS, Eco-DRR, and EbA in programs, projects, and activities in these sectors as described in Table 7.2.

Despite policies in place to support and advance the mainstreaming of NbS in the country, the environment's cross-cutting nature as a sector makes enforcement and implementation of programs, plans, and activities extremely challenging. On the one hand, several national-level agencies are central to the discussion concerning the environment, such as the DENR, Climate Change Commission, Department of Science and Technology, and Department of Human Settlements and Urban Development. They often work in siloes and implement activities not in coordination with other agencies and stakeholders. Also, local government units play a critical role in enforcement but not many local chief executives champion and prioritize actions for the environment nor undertake implementation in sync with the national government and the participation of relevant actors. On the other hand, competing priorities and interests of different stakeholders, with implication to budget and spending, frequently side-tracked efforts to protect, restore, and rehabilitate the environment to build resilience.

Nevertheless, the uptake of NbS in the Philippines, particularly Eco-DRR and EbA, is increasing over the years due to the policy support mentioned above and technical assistance from development partners promoting ecosystem-based approaches. Exploring and advancing NbS is an important priority due to the country's high vulnerability to climate and disaster risks. A study by Losada et al. (2017) shows that mangroves provide the most protection for frequent lower intensity storms (e.g., 1-in-10-year storm events) in the Philippines. More than USD1.6 billion damages can be averted for more catastrophic events (e.g., 1-in-25-year storm events) due

Table 7.2 Relevant policies on NbS in the Philippines

Policy	Year of approval/implementation	Description
Republic Act 9729 (Climate Change Act of 2009)	2009	Recognizes the need for national and sub-national government policies, plans, programs, and projects to be founded on sound environmental considerations and the principle of sustainable development
National Framework Strategy on Climate Change	2010	Advocates for the integrated ecosystem-based management approach, which acknowledges the interrelationships across the country's ecosystems and strengthens the integrity of decision-making processes toward the formulation of comprehensive adaptation strategies from the ridge to reef
Republic Act 10121 (Disaster Risk Reduction and Management Act of 2010)	2010	Adopts a comprehensive and integrated disaster risk reduction program within the context of environmental protection and poverty reduction to minimize the socioeconomic and environmental impacts of disasters and climate change
National Disaster Risk Reduction and Management Plan	2011	Identifies disaster risk reduction and management- and climate change adaptation-sensitive environmental management as an outcome where disaster risk reduction is mainstreamed in environmental policies and plans, including for the management of land use, natural resource, and climate change
National Climate Change Action Plan (2011–2028)	2011	Focuses on ecological and environmental stability to achieve the outcome: protection and rehabilitation of critical ecosystems and restoration of ecosystem services

(continued)

Table 7.2 (continued)

Policy	Year of approval/implementation	Description
Intended Nationally Determined Contribution to the United Nations Framework Convention on Climate Change	2015	Supports the management of climate- and disaster-resilient ecosystems as a priority measure for them to contribute to increased resilience to climate change
Department of Public Works and Highways Design Guidelines, Criteria and Standards Volumes 3–6 (2015)	2015	Incorporates resilient design and standards that allow ecosystems to adapt naturally for climate change adaptation, disaster resilience, and sustainability of public buildings and other related structures, water engineering projects, and highways and bridges
Green Building Code	2015	Enhances environmental protection by promoting resource use efficiency and environmentally-friendly designs and technologies in buildings
Department of Environment and Natural Resources (DENR) Administrative Order No. 2016-26	2016	Seeks to effectively manage the country's coastal and marine ecosystems to increase their ability to provide ecological goods and services and ensure food security and climate and disaster resilience under the Coastal and Marine Ecosystems Management Program (CMEMP)
DENR Mainstreaming Ecosystem-based Adaptation Program	2016	Aims to integrate EbA in all programs, projects, and activities among its bureaus and services from national to sub-national and the community as DENR's contribution to the implementation of the Philippine INDC concerning the Paris Agreement

(continued)

Table 7.2 (continued)

Policy	Year of approval/implementation	Description
Philippine Development Plan (2017–2022)	2017	Contains more aggressive strategies to rehabilitate and restore degraded natural resources and protect fragile ecosystems. The interventions are to be designed following a sustainable area development framework and a ridge-to-reef approach so that the interdependence of the different ecosystems is considered. Even in urban areas, green spaces will be expanded
Cabinet Cluster on Climate Change Adaptation, Mitigation, and Disaster Risk Reduction Roadmap (2018–2022)	2018	Targets outcomes on (i) adequate supply of clean air, water, and other natural resources by addressing effective management of forest areas, terrestrial and marine protected areas, urban green spaces, wetlands, and water bodies, and developing market-based instruments for enhanced ecosystem-based management and (ii) resilience of critical infrastructure by improving design and construction, and increasing investments in disaster-resilient infrastructure
Harmonized National Research and Development Agenda (2017–2022)	2017	Sets the research priorities for disaster risk reduction and climate change adaptation, including (i) observation and monitoring networks for the weather, climate, geologic and oceanographic processes; (ii) modeling and simulation for improved forecasting; (iii) hazard, vulnerability, and risk assessments of socio-ecological systems; and (iv) technology development and application to adapt to climate change in all sectors, including food, water, health, environment, businesses, infrastructure, and settlement

(continued)

Table 7.2 (continued)

Policy	Year of approval/implementation	Description
National Disaster Risk Reduction and Management Plan 2020–2030	2020	Targets NbS as an output with activities such as (i) conduct of natural resources and ecosystem accounting, valuation, stocktaking, inventory, and assessment, (ii) behavior of carrying capacity studies and research on terrestrial, coastal, and marine habitats and resources, (iii) mainstreaming of ecosystem values into national and local development planning, including ecosystem services indicators, monitoring systems and databases, (iv) protecting intact habitats, restore and rehabilitate degraded ecosystems/habitats (e.g., mangroves, wetlands, forest, coral reefs), and improve urban green spaces, (v) promoting sustainable integrated area development, and participatory environmental governance, (vi) promoting sustainable community resource-based enterprises and livelihood programs, and (vii) formulating and implementing the Integrated River Basin Management and Development Plan. It highlights the need for mainstreaming BGI, including enhancing understanding of its contribution to disaster resilience, especially among at-risk groups and stakeholders

to the presence of mangroves. It is estimated that one hectare of mangroves in the Philippines provides approximately USD 3200/year of direct flood reduction benefits. Another study by Seriño et al. (2017) in areas affected by Typhoon Haiyan estimates that the average cost of saving a life by retaining the remaining mangrove vegetation amounts to as much as USD 302,000 while the estimated reduction in compensation for totally damaged houses is around USD 53,000. Thus, mangroves are considered to have the best defense against storms and coastal flooding when combined with built infrastructure. Consequently, mangrove reforestation is a common Eco-DRR and EbA initiative in the country.

In recent years, the management of watersheds, protected areas, and river basins have placed greater emphasis on putting ecosystems and ecosystem services at the center. Forests, floodplains, and soils can contribute to a clean and reliable water supply and protect against floods and drought. For this reason, the 2018 United Nations World Water Development Report highlights the essential role that NbS plays in the achievement of the Sustainable Development Goals. The project, "Ecosystem-based management and application of ecosystem values in two river basins in the Philippines (E2RB)," implemented by DENR in 2019–2023, aims to improve a fragmented water governance regime. Also, it seeks to use ecosystem services values as a basis for private sector buy-in for them to support the financing of conservation and protection measures that will maintain ecosystem services and protected areas.

Similarly, the "Emerging Champions for Biodiversity Conservation and Improved Ecosystem Services (ECBCIES)" project implemented by DENR in 2011–2015 sought to improve the management of local ecosystems, particularly watersheds, and their associated ecosystem services. Indigenous technologies and solutions have also been promoted through projects such as "Increasing the Reduction of Hazards and Threats through Preparedness and Alternative Indigenous Technologies Along the River Systems of Allah and Safali Rivers (I-RePAIR)" implemented by the Foundation for the Philippine Environment in 2017. The project demonstrated bamboo as an indigenous technology for flood protection and the adoption of a participatory process in rehabilitating around 21 km of degraded riverbanks using the Naci Dike Technique to keep the riverside farming communities safe from floods.

Although BGI is at its very early stages in the country, it holds great promise as demonstrated by the Green, Green, Green program of the Department of Budget and Management (DBM) launched in 2017 with a budget totaling over USD 52 million under the 2018 national budget. 'Green, Green, Green' is a special assistance program that aims to make the Philippines' 145 cities more livable and sustainable by developing public open spaces. Specifically, it aimed to (DBM undated):

1. Enrich open spaces and create a sustainable and livable urban environment through turfing, landscaping, and green space architecture;
2. Establish forest parks, conservatory, and botanical gardens;
3. Transform streetscapes, such as the installation of eco-friendly street furniture and fixtures and shading;
4. Augment connectivity and accessibility, such as the construction of eco-friendly bike lanes and walkways; and

5. Enhance green infrastructure through tree planting, construction of bioswales, and pervious surfaces.

Design proposals submitted by a total of 143 cities were reviewed following a set of criteria, including accessibility, the usability of spaces, ecosystem services, and administrative management. Among the 68 funded proposals are the following: (i) Masbate City Mangrovetum, a mangrove area with an elevated walkway to increase appreciation of the mangroves, including the species that use mangroves as their habitat; (ii) Puerto Princesa Balayong Nature Park, which is designed to provide ecological services, and serve as a refuge for insects and birds, and exotic forest flora and native plants; (iii) Public Park of Lamitan, Basilan, a public space to improve social and cultural connectivity among citizens thereby promoting peace, safety, and security within an area of conflict; (iv) Franklin Bridge in Dagupan, which aims to develop an integrated area consisting of an esplanade, a plaza and a boat terminal that will promote coastal resource management, and address climate issues in the city (Palma 2018).

The application of BGI requires looking at infrastructure challenges from a different perspective and developing innovative techniques for green infrastructure. This new way of doing business often accounts for its slow advancement, especially in a developing country like the Philippines. Browder et al. (2019) identify the challenges in integrating green infrastructure to various sectoral plans as follows:

- Assessment of technical performance and its interaction with gray infrastructure;
- Engagement of different types of stakeholders, which can be time-consuming and costly and requires new skill sets;
- Lack of enabling conditions and policies for financing and implementing green infrastructure;
- Lack of compilation of good practices and lessons learned from green infrastructure projects; and
- Lack of comprehensive scientific knowledge and data to inform green designs.

Overall, mainstreaming Eco-DRR, EbA, and BGI in national and local development and investment plans remains a challenge and continues to be slow despite the growing recognition of its multiple benefits to many sectors in the country.

7.2.2 Rehabilitating Ecosystems and Renewed Attention to Environmental Protection

In recent years, significant environmental challenges sounded alarm bells and brought renewed attention to the importance of ecosystem-based management in the country. The Typhoon Haiyan, Manila Bay, and Boracay rehabilitation experiences, which triggered the revisiting of policies and systems related to environmental protection of different sectors at all levels, are discussed briefly in the following.

Typhoon Haiyan and Mangrove Rehabilitation

Super Typhoon Haiyan struck the Philippines in 2013, resulting in many lives lost, homes and infrastructure destroyed, agriculture damaged, forest and wildlife devastated, and other socioeconomic impacts. Communities living along the coast were vulnerable due to the loss of natural buffers such as mangroves and coral reefs, which could have provided defense against storm surges and strong winds. Areas that have intact mangrove forests incurred less damage (Seriño et al. 2017) such as the municipality of General MacArthur in Eastern Samar, where not a single death was reported due to the protection of mangroves, which were found to have the highest values in terms of species richness, the diversity index (H′), and evenness (Delfino et al. 2015), and a well-executed evacuation plan (Madarang 2018).

Large areas of mangrove forest have been damaged by the typhoon, with many species dead and rotting, consequently becoming toxic and harming organisms within the system (ZSL 2017). As part of recovery work, mangrove rehabilitation was promoted and undertaken by various organizations. The DENR earmarked around USD 8 million to support efforts to replant the affected coastal areas in Leyte and Samar with mangroves Stringer and Orchard (2013). Most of the funds were channeled through cash-for-work programs where communities were engaged in mangrove replanting activities in heavily affected coastal regions. The Zoological Society of London embarked on a marine protected area and mangrove management project in Bantayan Island, Cebu, which included carrying out biological and social surveys to understand the habitat and local resource use with the participation of the communities (ZSL 2017). In Concepcion, Iloilo, Conservation International implemented green-gray solutions to restore coral and marine biodiversity, including mangrove habitats, while introducing alternative livelihood options and encouraging sustainable fisheries (Conservation International undated).

Due to extensive mangrove restoration in Typhoon Haiyan's aftermath, activities that were not supported by science were unsuccessful and led to unsustainable, non-typhoon resistant, and non-culturally specific mangrove replanting efforts. Many of these efforts fail to establish self-sustaining mangrove forests that are sizeable, diverse, and functional due to weak or non-participation of the community, inappropriate or mono-species planting, and poor choice of location (Global Resilience Partnership 2018). This underlines the importance of using science and ground evidence to inform ecosystem planning and management.

Manila Bay Rehabilitation

The Manila Bay area covers eight provinces, four of which are coastal (Bataan, Bulacan, Cavite, and Pampanga), and 178 local government units in three regions: National Capital Region, Region III (Central Luzon), and Region IV-A (Calabarzon). Its coastline measures around 190 km. Seventeen major river systems are draining to Manila Bay with a drainage area of 1994 km^2 or 199,400 ha. Several areas declared as protection zones under the National Integrated Protected Areas System Act of 1992 can also be found in the Manila Bay area.

A Supreme Court Mandamus on Manila Bay (G.R. 171947-48) was issued on 18 December 2016 ordering 13 government agencies to clean up, rehabilitate, and preserve Manila Bay, and restore and maintain its waters to Class SB level to make them fit for contact recreational activities (e.g., bathing, swimming, skin-diving), and fisheries as spawning areas for milkfish, and the commercial propagation of shellfish (DENR undated).

The improvement of water quality, protection of shoreline features, and conservation of biological resources are among the priorities based on the 2005 Operational Plan for the Manila Bay Coastal Strategy Manila Bay Environmental Management Project (2005). Several factors account for the slow rehabilitation of the Manila Bay, including its broad coverage, and around 233,000 informal settler families that reside along its waterways, directly discharging wastes to the water because they are not connected to sewerage treatment facilities. Apart from water pollution and solid wastes, other issues identified in the Manila Bay are overexploitation of resources, siltation and sedimentation, habitat degradation, natural and man-made hazards, and multiple resource use conflict. DENR has established indicators for government agencies to target waste management, relocation of informal settler families along the watershed, and reviving biodiversity. Besides, a Manila Bay Sustainable Development Master Plan is being formulated under the National Economic and Development Authority, which focuses on five areas: improved water quality, ecosystem protection, upgrading of informal settlements, disaster risk reduction and climate change adaptation, and inclusive growth (NEDA 2020). Many laud the environmental benefits of the long-overdue clean-up but recognize that it will take years or even decades to save Manila Bay (Rey 2019).

Boracay: Business Versus Environment

The controversy that erupted in 2018 surrounding the findings of coliform contamination in the resort island of Boracay, several times dubbed as the best beach in the world, is not new. As early as 1997, water quality issues in the paradise island have been identified, and a resolution process to address this was undertaken (Almodiel 1999). The closure and redevelopment of this prime tourist destination, which attracts more than a million tourists and an estimated USD 772.5 million in revenue every year (Santos 2018), in April 2018 came about following a resolution from the National Disaster Risk Reduction and Management Council recommending the declaration of a State of Calamity in the Boracay islands due to an environmental disaster, and the implementation of a temporary closure for purposes of rehabilitation (NDRRMC 2018). Investigation and validation done by the Department of Interior and Local Government (DILG), DENR, and Department of Science and Technology reveal a high level of environmental degradation due to beach erosion, the disappearance of wetlands, high concentration of fecal coliform, and pollution caused by improper waste management and encroachment of protected areas by illegal structures. Unsustainable tourism was identified as a root problem that created interrelated environmental issues on habitat degradation and coral reef destruction, water quality, green tides or algal blooms, governance, and the island's carrying capacity (Aranza 2018).

The media contributed to sensationalizing the island's environmental issues (Canoy et al. 2020) and brought heightened attention to the important role of politics and governance in the implementation, monitoring, and enforcement of environmental laws and development controls. Having battled the recurring issues on wastewater and solid waste management for two decades, local environmental groups in Boracay called for a moratorium on new and expansion of establishments and strict adherence to environmental laws.

Although the rehabilitation was welcomed, it came at a high cost to people's livelihoods resulting in loss of revenues and jobs mostly from the tourism sector with potential economic loss projected at USD 427 million-USD 1.7 billion in total output, and USD 144–573 million in salaries (Reyes et al. 2018). Despite the negative economic impacts of the closure, it was deemed necessary to ensure public safety and public health and protect biological resources and diversity, and ecosystem functions and processes of the islands.

Following the closure, a significant number of hotels and restaurants were ordered to close for violating local environmental laws, three casinos have been permanently shut down, and establishments built within the 30-m shoreline easement have been demolished. The DILG has filed legal complaints against 17 government officials over the neglect of Boracay, citing laxity in issuing building permits and failure to address illegal development activities on the island.

7.2.3 *Village-Level Green-Gray Management Strategy for Coastal Protection in Polillo, Quezon, Philippines: A Case Study*

Small island communities in the Philippines bear the brunt of coastal hazards, including typhoons, storm surges, intense monsoon season, flooding, and rainfall-induced landslides. Sea level rise has increased coastal flooding and land loss due to erosion. This has set a domino effect that reduced fish stocks due to temperature-induced shifts in adult fish distributions and reproduction cycles. The loss of coastal defense, marine habitat, and biodiversity coupled with increased coral bleaching and collapse of coral reefs from ocean acidification and runoff has caused reduced fish populations and, consequently, loss of livelihoods. Climate-related hazards impact different key sectors in island municipalities and provinces, including agriculture, water and water systems, energy, coastal ecosystems, infrastructure and services, and human health.

The study location is Barangay (village) Libjo in Polillo, a small island municipality in the province of Quezon in the Philippines. Based on the Comprehensive Development Plan of Polillo, most of its 17 coastal barangays are dependent on coastal resources, with fisheries and fishery-related livelihoods among the primary sources of income (Fig. 7.1). Due to its geographical location facing the Pacific Ocean, Polillo is exposed to various hazards, notably typhoons that cause surges

Fig. 7.1 Mangroves along the sedimentary plain segment of the coast of Barangay Libjo (left); mangroves in the flat hard rock segment (right). These support other floral and faunal species

and tidal waves. These climate-related hazards directly impact the communities along the coastal villages of Polillo and their livelihoods. Coastal flooding and storm surges have damaged coastal defenses, including the green systems (i.e., mangrove stands, coral reefs, seagrass areas, tidal flats) and hard engineering structures such as seawalls, dikes, and groins.

The literature points to the insufficiency of traditional, hard engineering structures (gray infrastructure) to protect populations from climate-induced disasters (Browder et al. 2019). Aside from being costly and only provides a temporary solution to a long-term problem that is likely to worsen with climate change (Acclimatise News 2019), these structures may even contribute to further erosion and sediment transport by changing wave and energy regimes (Uy and Shaw 2012). This study seeks to validate existing risks and vulnerabilities to climate-induced hazards of the community in Libjo utilizing participatory and community-based approaches combined with a web-based tool, the Coastal Hazard Wheel (CHW), and propose a green-gray management strategy to address disaster and climate risks.

Participatory Community Risk Assessment

The study utilized Participatory Community Risk Assessment (PCRA) in an attempt to fill the gaps caused by the lack of inventory or inadequate assessment of damage and losses from climate-induced hazards done at the community level since these activities are undertaken by the Municipal Disaster Risk Reduction and Management Officer with other municipal government officials and not with the community. The conduct of PCRA was significant in promoting collaborative research and analysis with community groups as a basis for formulating effective community interventions.

The PCRA examined the communities' vulnerability and adaptation strategies through participatory action research methods (i.e., focus group discussion, key informant interview, life history, and community risk mapping). The community risk mapping in PCRA is critically revealing the ecological-environmental (i.e.,

Hazard	Elements At Risk				Vulnerabilities
	Physical	Economic/Financial	Social	Environmental/Natural	
Typhoon/Storm surge: HIGH Level of threat	• Houses made of light materias located near the shoreline • Fishing related infrastructures • Human safety • Water resources and infrastructure • Road networks ans bridges • Power and communication lines • Infrastructure and other barangay facilities	• Fish production • (Majority of community dependent on fishing as major source of livelihood) • Agricultural production (i.e. copra) and livestock raising	Social services such as health, food, education and other basic services are at risk	• Mangroves • Seagrasses • Coral reefs • Fish sanctuaries • Agricultural lands	• Houses are made of light materials • Houses located with No Build Zones less than 40m along the coastline • Residents are dependent on the water transportation in transporting prime commodities from the mainland • Insufficient food supplies coming from the mainland during typhoon.
More intense monsoon season: HIGH Level of threat	• Houses made of light materias located near the shoreline • Fishing related infrastructures • Human safety • Water resources and infrastructure • Road networks and bridges • Power and communication lines • Infrastructure and other barangay facilities	• Fish production • (Majority of community dependent on fishing as major source of livelihood) • Agricultural production (i.e. copra) and livestock raising	Social services such as health, food, education and other basic services are at risk	• Mangroves • Seagrasses • Coral reefs • Fish sanctuaries • Agricultural land	• Houses are made of light materials • Houses located with No Build Zones less than 40m along the coastline • Residents are dependent on the water transportation in transporting prime commodities from the mainland • Insufficient food supplies coming from the mainland during typhoon.
El Nino/La Nina: HIGH Level of threat	• Water resources and infrastructure • Food security • Livestock	Agricultural production (i.e. copra) and livestock raising	Social services such as health, food, education and other basic services are at risk	Agricultural lands	• Highly dependent on rain since crops are rain-fed • There are no available irrigation facilities • Water sources cannot be sustained during prolonged dry season • Increase incidence in pests and diseases during dry season
Flooding/Flash flood: LOW level of threat	• Houses near coastline, riverbanks • Human safety • Water resources and infrastructure • Road networks and bridges • Power and comminication lines • Infrastructure and other barangay facilities	Agricultural production and livestock raising	Social services such as health, food, education and other basic services are at risk	Agricultural lands	• Agricultural production areas are flooded due to overflow of inland bodies of water • 50 Families affected in Sitio Corregidor Putian; 80–100 households affected in Sitio Ilulo during high tide (0.5 meter flood)
Landslide: LOW level of threat	• Water resources and infrastructure • Road networks and bridges • Power and comminication lines • Infrastructure and other barangay facilities	Agricultural production and livestock raising	Social services such as health, food, education and other basic services are at risk	Agricultural lands	

Fig. 7.2 Community risk registry of Barangay Libjo, Polillo

storm/tidal surges; coastal inundation) and social factors (i.e., population increase in settlements along river banks; clogging of canals and creeks with garbage) that create risks. Results show that the coastal communities are at risk from typhoons, intense monsoon, storm surge, and coastal flooding. Through the PCRA, the study produced a community risk registry that summarizes overall risk, which could be used by the local Barangay Disaster Risk Reduction and Management Committee (Fig. 7.2). The study predicts that the study site's identified risks will intensify with the changes in the future climate, as projected by the Philippine Atmospheric, Geophysical, and Astronomical Administration unless appropriate hazard management options are put in place.

Coastal Hazard Wheel

The study utilized the Coastal Hazard Wheel (CHW) 3.0, developed mainly from Rosendahl-Appelquist and Halsnæs (2015)'s work, as an information and decision support system that can be used for multi-hazard assessments and the identification of appropriate management options for a specific coastline. This system successfully addresses the gap in current protocols for coastal hazard assessment and management, which has high requirements for input data and domain expertise (Ramieri et al. 2011). Compared to other systems, the CHW is developed as a multi-tool that can address critical coastal management issues collectively and is directly accessible to coastal decision-makers at all management levels. The CHW generated different coastal hazard maps showing the hazard levels of ecosystem disruption, erosion, flooding, saltwater intrusion, and gradual inundation on the coast of Libjo (Fig. 7.3). It produced a portfolio of options in addressing the different coastal hazards in Libjo.

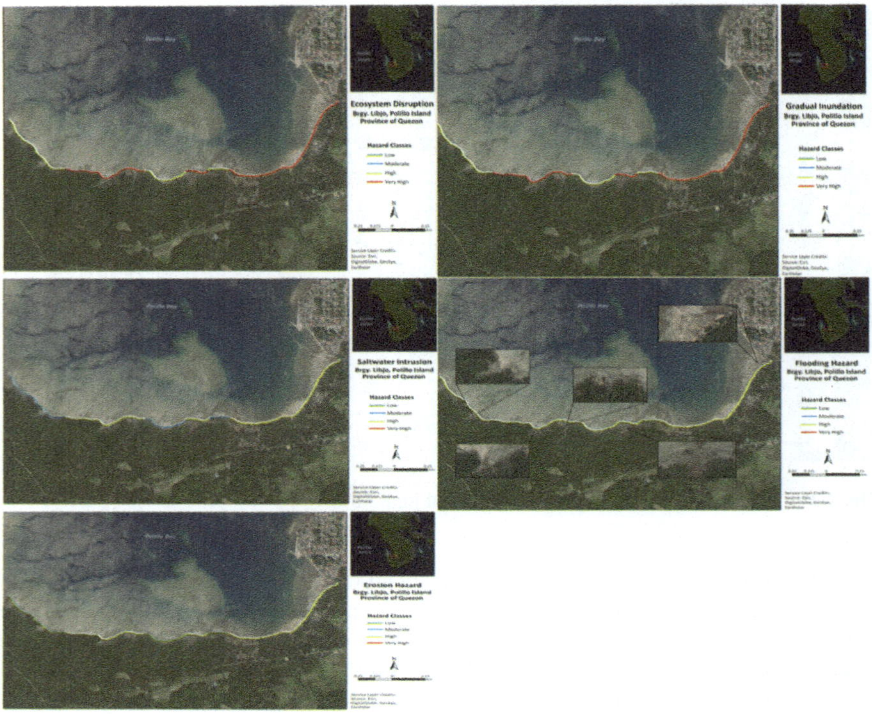

Fig. 7.3 Hazard classes generated in CHW 3.0 for Barangay Libjo, Polillo

Studies from Hallegate (2009), Rosendhal-Appelquist and Halsnæs (2015), Cohen-Shacham et al. (2019), and the nature-based risk management strategy (World Bank 2017) were used in the systematic evaluation of the hazard management options.

Recommendations for a Green-Gray Management Strategy

The study recommended, "accommodation" measures under the "protect-accommodate-retreat-avoid" (PARA) framework (Doberstein et al. 2019) and management options needed for the Integrated Coastal Zone Management Plan. These include coastal zoning and groundwater management for Integrated Water Resource Management. Flood warning systems, floodproofing, and flood shelters were also included since these have been identified and integrated into the Municipal Disaster Risk Reduction and Management Plan of Polillo for 2016–2021. Management options that were not recommended are the use of hard/soft options such as dikes and dune construction due to the high cost of engineering works, availability of workforce and materials, and the possibility of risk transfer. These management options may only be used in conjunction with green infrastructure.

The study highly recommends ecosystem-based management and wetland restoration in a village-level setting as a hazard management strategy for coastal protection.

This will provide the needed technical expertise, institutional policy, and enabling legal framework for the effective conservation, restoration, and sustainable management of the green infrastructure in Libjo to reduce coastal communities' vulnerability and encourage multi-stakeholder support following hazard events. Nature-based solutions to coastal flooding and storm surges in Libjo is recommended to be a community-based, community-centered, and community-led approach that is inclusive (bottom-up), proactive (engages multiple stakeholders), and appropriate (green-gray solutions are suited to the local context). Ideally, incorporating green-gray concepts in the development plans of Polillo should (i) lead to enhanced coastal protection and resilient infrastructure; (ii) provide alternative livelihoods; (iii) build capacities and empower institutions; (iv) protect and sustainably manage the environment; and (v) mainstream green-gray solutions in the plans of the barangays in Polillo.

7.3 Discussion: Opportunities for Scaling Up

Implementing NbS—Eco-DRR, EbA, and BGI—in the Philippines has so far been undertaken as a response to environmental challenges. More than being reactive, a proactive focus on NbS for prevention, mitigation, and rehabilitation particularly on protecting intact habitats, restoring and rehabilitating degraded ecosystems, and improving urban green spaces is needed. This would be important to prevent environmental concerns from intensifying and to mitigate against future impacts of climate variability and change. Stewardship is key therefore; the participation of relevant stakeholders is essential from planning to monitoring and evaluation of NbS. This is critical because local livelihoods are often linked to ecosystems. Moreover, the allocation of financial resources to NbS programs and projects is required to implement the solutions sustainably.

The growing literature and interest in NbS suggest many strategies for scaling these up. Cook (2020) proposes three steps, including raising an understanding of the value of nature, embedding NbS into climate adaptation planning, and encouraging investment in NbS. Strategic entry points for EbA mainstreaming are identified as follows (GIZ 2019):

National level: Embed EbA as a cross-cutting approach in national climate policies and build-in relevant targets, and add EbA-relevant criteria in the national budget allocation planning and the screening process of projects.

Sectoral Policy: Build-in NbS to achieve adaptation objectives of sectoral plans (e.g., water, protected areas, disaster risk reduction), and adopt EbA principles in land use planning for cross-sectoral targets.

Local-level: Consider EbA micro-projects in village action plans to target food security and water availability, and integrate EbA in instruments for municipal development planning processes.

Private sector: Include EbA criteria into proposals of public partnerships with private businesses.

In promoting green and gray integration, Browder et al. (2019) recommend the following:

1. **Legislation**. Encourage policymakers to advocate for green-gray approaches through policies, laws, and regulations;
2. **Regional, master, and land use planning**. Proactively integrate green infrastructure approaches in regional, master, and land use planning.
3. **Advanced methods and tools**. Utilize advanced techniques and tools to analyze the performance of green infrastructure.
4. **Social support**. Prioritize social support for green infrastructure and build long-term coalitions.
5. **Innovative financing**. Take advantage of green infrastructure's characteristics to sell innovative financing approaches.
6. **Partnerships**. Develop collaborative partnerships with approving bodies, civil society organizations, potential co-investors, and technical experts.
7. **Development Partner Support**. Development partners can advance green-gray approaches by building capacity with their organizations, using green-gray assessment tools and strategies in internal processes, investing in performance monitoring, and widely communicating results and experience.

Focusing on the Philippines, USAID (2018) presents some strategies to sustain EbA based on project experience such as (i) use a "learning by doing" approach to advance EbA, (ii) build sustainability through EbA champions, (iii) bring EbA champions together regularly to build capacity, confidence and working relationships, (iv) provide support to develop fair and equitable payment for ecosystem services arrangements, and (v) facilitate site visit exchanges to demonstrate EbA successes. Moreover, GIZ (2018) identifies entry points to mainstreaming EbA in land use planning, sustainable integrated area development strategy, protected area management, climate change adaptation and disaster risk reduction agenda, and national adaptation planning processes.

7.4 Conclusions

Natural ecosystems have demonstrated multiple benefits to people and the environment by reducing climate and disaster risks, contributing to human well-being, and enhancing resilience. However, governments prefer to invest in traditional "hard" engineering solutions, which are costly and do not provide the same range of services as fully functioning ecosystems do while proving to be unsustainable in the long term. Over the years, NbS have increased their evidence base, and its application has gained clarity through the identification of principles and standards by which to operationalize the concept. The renewed focus on ecosystems and ecosystem services in the Philippines provides an opportunity for exploring more insights, solutions, and actions for putting nature to work in harmony with gray infrastructure. Past and ongoing projects on NbS have generated many good practices and lessons learned that can inform future interventions. It is essential now more than ever to make a case for the benefits and cost-effectiveness of ecosystem-based approaches especially BGI. The science and evidence for BGI would need to be strengthened to gain political commitment at all levels, secure funding and private sector engagement, inform decision-making better, and ultimately advance its implementation. A triple-win scenario where economic benefits, environmental conservation, and community empowerment are achieved can be facilitated by well-designed NbS that target disaster and climate resilience. Specific recommendations include:

- Conducting a climate and disaster risk assessment to inform NbS project planning and design;
- Using science, technology, and innovation to identify the appropriate NbS given the local context;
- Coordinated efforts following a whole of government and whole of society approach in implementing NbS initiatives;
- Participation of relevant stakeholders in the planning, implementation, and monitoring of NbS programs and projects;
- Adaptation planning that focuses on investing in the environment, mainly through BGI, to build climate and disaster resilience; and
- Knowledge management to document and compile good practices and lessons learned.

References

Acclimatise News (2019) Life's a beach: nature-based solutions for coastal protection. Acclimatise News [Internet] [Cited 2019 June 6]. Available from: http://www.acclimatise.uk.com/2019/06/06/lifes-a-beach-nature-based-solutions-for-coastal-protection/

Almodiel FC (1999) Boracay: a case study on the use of conflict management to catalyze collaboration in coastal management. Tropical Coasts 6(2)

Aranza RA (2018) Policy recommendation on the rehabilitation of Boracay Island and management of its marine environment. National Academy of Science and Technology Science Advisory (Series 2018 No. 2). NAST, Manila, Philippines, 4 p

Browder G, Ozment S, Rehberger Bescos I, Gartner T, Lange GM (2019) Integrating green and gray: creating next-generation infrastructure. World Bank, Washington DC, USA; World Resources Institute, Washington DC, USA

Canoy NA, Roxas GKT, Robles AMQ, Alingasa APT, Ceperiano AM (2020) From cesspool to fortified paradise: analyzing news media territorial assemblages of rehabilitating Boracay Island, Western Philippines. J Sustain Tour 28(8):1138–1157. https://doi.org/10.1080/09669582.2020.1726934

Cohen-Shacham E, Andrade A, Dalton J, Dudley N, Jones M, Kumar C, Maginnis S, Maynard S, Nelson CR, Renaud FG, Welling R, Walters G (2019) Core principles for successfully implementing and upscaling Nature-based Solutions. Environ Sci Policy 98:20–29. https://doi.org/10.1016/j.envsci.2019.04.014

Cohen-Shacham E, Walters G, Janzen C, Maginnis S (eds) (2016) Nature-based solutions to address global societal challenges. IUCN, Gland, Switzerland

Conservation International (Undated) Green-gray storm shelters: fusing nature and engineering to safeguard communities. Conservation International [Internet]. Available from: https://www.conservation.org/philippines/projects/green-gray-storm-shelters

Cook J (2020) 3 steps to scaling up nature-based solutions for climate adaptation. World Resources Institute [Internet] [Cited 2020 May 21]. Available from: https://www.wri.org/print/66436

Delfino RJP, Carlos CM, David LT, Lasco RD, Juanico DEO (2015) Perceptions of Typhoon Haiyan-affected communities about the resilience and storm protection function of mangrove ecosystems in Leyte and Eastern Samar, Philippines. Clim Disaster Dev J 1(1). http://doi.org/10.18783/cddj.v001.i01.a03

Doberstein B, Fitzgibbons J, Mitchell C (2019) Protect, accommodate, retreat, or avoid (PARA): Canadian community options for flood disaster risk reduction and flood resilience. Nat Hazards. https://doi.org/10.1007/s11069-018-3529-z

Estrella M, Saalismaa N (2013) Ecosystem-based disaster risk reduction (Eco-DRR): an overview. In: Renaud FG, Sudmeier-Rieux K, Estrella M (eds) The role of ecosystems in disaster risk reduction. United Nations University Press, Tokyo, Japan

Global Resilience Partnership (2018) One resilient team: replanting mangroves to combat flood risk in the Philippines. Global Resilience Partnership [Internet] [Cited 2018 September 14]. Available from: https://www.globalresiliencepartnership.org/news/2018/09/14/one-resilient-team-replanting-mangroves-to-combat-flood-risk-in-the-philippines/#:~:text=One%20Architecture%20%26%20Urbanism%2C%20a%20firm,flood%20risk%20in%20the%20Philippines

Hallegate S (2009) Strategies to adapt to an uncertain climate change. Glob Environ Chang 19(2):240–247

Losada IJ, Beck M, Menéndez P, Espejo A, Torres S, Díaz-Simal P, Fernández F, Abad S, Ripoll N, García J, Narayan S, Trespalacios D (2017) Valuation of the coastal protection services of mangroves in the Philippines. World Bank, Washington DC, USA

Madarang CRS (2018) Mangroves are a natural shield against typhoons in the Philippines. Interaksyon [Internet] [Cited 2018 July 30]. Available from: https://www.interaksyon.com/breaking-news/2018/07/30/131382/mangroves-are-a-natural-shield-against-typhoons-in-the-philippines/

Manila Bay Environmental Management Project (2005) Operational plan for the Manila Bay coastal strategy. PEMSEA, Quezon City, Philippines

Palma RAL (2018) Taking a closer look at the Green, Green, Green program. Inquirer Features [Internet] [Cited 2018 September 15]. Available from: https://business.inquirer.net/257289/taking-closer-look-green-green-greenprogram#ixzz6VCyWwMZn

Ramieri E, Hartley A, Barbanti A, Santos FD, Gomes A, Hilden M, Laihonen P, Marinova N, Santini M (2011) Methods for assessing coastal vulnerability to climate change. ETC CCA technical paper 1/2011. ETC CCA, Bologna, Italy

Rey A (2019) Manila Bay rehab: the challenge of cleaning up the nation's waste. Rappler Newsbreak In-Depth [Internet] [Cited 2019 March 16]. Available from: https://rappler.com/newsbreak/in-depth/stilt-houses-manila-bay-rehabilitation-series-part-1

Reyes CM, Albert JRG, Quimba FMA, Ortiz MKP, Asis RD (2018) The Boracay closure: socioeconomic consequences and resilience management. Philippine Institute for Development Studies Discussion Paper Series No. 2018-37. PIDS, Quezon City, Philippines

Rosendahl-Appelquist L, Halsnæs K (2015) The coastal hazard wheel system for coastal multi-hazard assessment & management in a changing climate. J Coast Conserv 19:157–179. https://doi.org/10.1007/s11852-015-0379-7

Santos AP (2018) Business vs. environment—Philippines' Boracay Island faces closure. Deutsche Welle [Internet] [Cited 2018 March 19]. Available from: https://www.dw.com/en/business-vs-environment-philippines-boracay-island-faces-closure/a-43039362

Seriño MN, Ureta JC, Baldesco J, Galvez KJ, Predo C, Seriño EK (2017) Valuing the protection service provided by mangroves in typhoon-hit areas in the Philippines. EEPSEA research report no. 2017-RR19. Economy and Environment Program for Southeast Asia, Laguna, Philippines

Stringer LC, Orchard S (2013) Mangroves, nature's shield against typhoons and tsunami. The Conversation [Internet] [Cited 2013 December 4]. Available from: https://www.theconversation.com/mangroves-natures-shield-against-typhoons-and-tsunami-21051

Sudmeier-Rieux K, Nehren U, Sandholz S, Doswald N (2019) Disasters and ecosystems: resilience in a changing climate—Sourcebook. UNEP, Geneva (Switzerland); TH Köln—University of Applied Sciences, Cologne, Germany

Uy N, Shaw R (eds) (2012) Ecosystem-based adaptation. Community, environment and disaster risk management, vol 12. Emerald Group Publishing Limited, Bingley, United Kingdom

World Bank (2017) Implementing nature-based flood protection: principles and implementation guidance. World Bank, Washington DC, USA

World Bank (2019) Nature-based solutions: a cost-effective approach for disaster risk and water resource management. World Bank Brief [Internet] [Cited 2019 April 10]. Available from: https://www.worldbank.org/en/topic/disasterriskmanagement/brief/nature-based-solutions-cost-effective-approach-for-disaster-risk-and-water-resource-management

[ADB] Asian Development Bank (2020) Bioengineering for green infrastructure. ADB, Manila, Philippines

[CBD] Secretariat of the Convention on Biological Diversity (2004) The ecosystem approach (CBD guidelines). CBD, Montreal, Canada

[DBM] Department of Budget and Management (Undated) DBM launches "Green, Green, Green!" city public open spaces assistance program. DBM [Internet]. Available from: https://dbm.gov.ph/index.php/secretary-s-corner/press-releases/list-of-press-releases/766-dbm-launches-green-green-green-city-public-open-spaces-assistance-program

[DENR] Department of Environment and Natural Resources (Undated) Fact sheet on Manila Bay rehabilitation [Internet]. Available from: https://denr.gov.ph/index.php/2-uncategorised/853-compendium-of-enr-statistics-2016

[DPWH] Department of Public Works and Highways (2015) The Philippine green building code: a referral code of the national building code of the Philippines. DPWH, Manila, Philippines

[FEBA] Friends of Ecosystem-based Adaptation (2017) Making ecosystem-based adaptation effective: a framework for defining qualification criteria and quality standards. FEBA technical paper developed for UNFCCC-SBSTA 46. GIZ, Bonn, Germany; IIED, London, UK; IUCN, Gland, Switzerland

[GFDRR] Global Facility for Disaster Risk Reduction (Undated) Natural hazards: nature-based solutions. GFDRR [Internet]. Available from: https://www.gfdrr.org/en/nbs

[GIZ] Deutsche Gesellschaft für Internationale Zusammenarbeit (2018) Entry points for mainstreaming ecosystem-based adaptation. The case of the Philippines. GIZ, Bonn, Germany

[GIZ] Deutsche Gesellschaft für Internationale Zusammenarbeit (2019) Emerging lessons for mainstreaming ecosystem-based adaptation: strategic entry points and processes. GIZ, Bonn, Germany

[IPCC] Intergovernmental Panel on Climate Change (2014) Climate change 2014: impacts, adaptation, and vulnerability. Part A: global and sectoral aspects. In: Field CB, Barros VR, Dokken DJ, Mach KJ, Mastrandrea MD, Bilir TE, Chatterjee M, Ebi KL, Estrada YO, Genova RC, Girma B, Kissel ES, Levy AN, MacCracken S, Mastrandrea PR, White LL (eds) Contribution of Working Group II to the fifth assessment report of the Intergovernmental Panel on Climate Change. Cambridge University Press, Cambridge, United Kingdom, New York, USA, 1132 p

[IUCN] International Union for the Conservation of Nature (2017) Ecosystem-based adaptation issues brief. IUCN, Gland, Switzerland

[IUCN] International Union for the Conservation of Nature (2020) Global standard for nature-based solutions: a user-friendly framework for the verification, design, and scaling up of NbS, 1st edn. IUCN, Gland, Switzerland

[MEA] Millennium Ecosystem Assessment (2005) Ecosystems and human well-being. Island Press, Washington DC, USA

[NDRRMC] National Disaster Risk Reduction and Management Council (2018) Resolution recommending the declaration of a state of calamity in Boracay Island (Barangays Balabag, Manon-Manoc, and Yapak), Municipality of Malay, Province of Aklan, and the implementation of a temporary closure thereof as a tourist destination for purposes of rehabilitation (NDRRMC Resolution No. 1, s. 2018). NDRRMC, Quezon City, Philippines

[NEDA] National Economic and Development Authority (2020) Manila Bay sustainable development master plan: final master plan report. Version January 2020. NEDA, Pasig City, Philippines

[USAID] United States Agency for International Development (2018) Improving ecosystem management to strengthen resilience to extreme weather in the Philippines: an ecosystem-based management approach. Case study. USAID, Washington DC, USA

[WWF] World Wide Fund for Nature (Undated) What are nature-based solutions for climate change? WWF [Internet]. Available from: https://wwf.panda.org/discover/our_focus/climate_and_energy_practice/what_we_do/nature_based_solutions_for_climate/

[ZSL] Zoological Society of London (2017) Mangrove protection after Typhoon Haiyan. ZSL Conservation [Internet] [Cited 2017 January 4]. Available from: https://www.zsl.org/blogs/conservation/mangrove-protection-after-typhoon-haiyan

Chapter 8
Making Resilience a Reality: The Contribution of Peri-urban Ecosystem Services (BGI) to Urban Resilience

Celeste Norman, Akhilesh Surjan, and Miranda Booth

Abstract The pressure on urban development to meet the needs of growing populations heavily influences spatial planning priorities. Ecosystem-based approaches (EBA) to development, which incorporate blue and green infrastructure (BGI), allow growth to balance ecocentrism with anthropogenic aspirations. This is particularly evident in peri-urban areas (PUA). The ill-defined nature of PUA gives rise to opportunities for ecosystem services to urban centres. However, the current land value favours development models. The disconnect between land value and ecosystem services does not consider the benefits of EBA to urban centres. Law, engineering and planning frameworks result in inflexible responses to changing risk. The planning of PUAs can facilitate beneficial growth strategies for ecosystems services that may include urban farms, allotment gardens and agricultural parks. The adaptability of PUA zones creates opportunities for residents to innovate and sustain their livelihoods as their environment undergoes change from urban pressures. Planning policy is lacking for PUA and as a result development outcomes are poor with ad hoc, developer-led approaches to growth. Strengthening links between urban and rural areas through considered PUA planning creates opportunities for the preservation of natural environments and the capacity of these environments to effectively reduce the negative effects of human development.

Keywords Peri-urban · Ecosystem-based approaches · Urbanisation · Climate change · Planning

---

C. Norman (✉)
Farco Architecture, Brisbane, QLD 4051, Australia
e-mail: celeste@farcoarchitecture.com

A. Surjan · M. Booth
Humanitarian, Emergency and Disaster Management Studies Program, Charles Darwin University, Casuarina Campus, Darwin, NT 0909, Australia
e-mail: akhilesh.surjan@cdu.edu.au

M. Booth
e-mail: miranda.booth@cdu.edu.au

© The Author(s), under exclusive license to Springer Nature Singapore Pte Ltd. 2021

M. Mukherjee and R. Shaw (eds.), *Ecosystem-Based Disaster and Climate Resilience*, Disaster and Risk Research: GADRI Book Series,
https://doi.org/10.1007/978-981-16-4815-1_8

8.1 Introduction

Urban environments only cover less than 2% of the earth's land area but house over 50% of the population and produce 78% of the worlds energy consumption resulting in 60% of the total greenhouse gas emissions (UN Habitat 2016). By 2050 the world population is expected to reach almost 10 billion (United Nations [UN] Department of Economic and Social Affairs 2019). Urban areas will continue to grow placing pressure on supporting ecosystems. Industrialisation has enabled humanity to develop urban areas with exceptional speed; today more than half of the human population live in urban areas and this is projected to increase to 68% by 2050 (UN Development Program [UNDP] 2019). The UN Global Assessment Report (GAR 2019) argues that anthropogenic metabolism will be central to the sustainable development of urban areas as it encompasses,

> the systemic interaction between humans and the environment that consists of the inputs, outputs and stock of materials and energy required to sustain physiological needs for food, air, water and shelter, as well as the products, substances and services necessary to sustain modern human life. (UN Office of Disaster Risk Reduction [UNDRR], 2019, p. 35)

Humanity is moving forward faster than we collectively know and understand how to sustain the current drive for development (Bahadur and Tanner 2014; Davis and Alexander 2015; Tyler and Moench 2012). The conditions that have created and sustain anthropogenic induced climate change need to be addressed as part of a dynamic, multiscale sociological system (Butler et al. 2016; UN 2017). The UN GAR's (2019) definition of anthropogenic metabolism emphasises the interrelationships of people and the environment and acknowledges that the natural resources available to us are finite. As reported in 2010, the worlds cities are consuming 40 million tonnes of raw materials increasing to 90 million tonnes by 2050. This rate of consumption is outside sustainable limits of natural resources (International Resource Panel [IRP] 2018).

One of the many consequences of urbanisation is population migration. Social aspirations, focused on opportunities in urban areas, is a strong driver for rural populations moving into urban areas adopting new lifestyles (Zasada 2001). Compounding and counteracting this trend within mixed demographic of urban dwellers is a migration to the more amenable natural landscapes at the city fringe contributing to urban sprawl (Llausàs et al. 2016). To achieve sustainable development, rural and urban linkages needs to be better managed; by establishing awareness of the ecological benefits that are provided by Peri-Urban Areas (PUA) and rural areas.

Policy and planning reforms are needed to support innovation and diversification of the city fringe (Millennium Ecosystem Assessment 2005; Zasada 2001). Short-term electoral cycles reflect actions with short delivery time frames that typically maintain the status quo. It is with multi-level governance and engagement with the private and civil sectors that transformation will occur.

PUA has capacity to contribute to risk reduction and the overall resilience levels of urban centres. New approaches to planning and managing PUA are needed to counteract the inflexibility of current legal, engineering and planning practices. Planning the expansion of PUAs must include strategies for ecosystem services such as urban farms, allotment gardens and agricultural parks alongside conservation and restoration areas (Mabon et al. 2019). The challenge for PUAs lie in balancing the spatial allocations required for population growth with suitable areas for ecosystem-based services. Planning these zones will require links between urban and city fringes areas to be strengthened, establishing relationships between people and the natural landscape in the pursuit of sustainability.

8.2 Urban Resilience, Strengthening the Rural–Urban Link

Urban resilience is the ability of a city to absorb disturbance while retaining identity, structure, and key processes (Bahadur and Tanner 2014; Tyler and Moench 2012). Developing strategies for urban resilience is complex, balancing social, political, and economic factors (Schipper 2009; Wisner 2016). To achieve urban resilience, the components or systems that make up a city must be constructed so that they can each withstand any stresses imposed on them through adaptability, achieving a holistic resilient system (Renn 2008). Strengthening the rural–urban link through the management of PUAs is central to this process.

8.2.1 Economic Preferencing in Planning

Currently, economic development goals are generally preferred over environmental goals in urban planning, placing urban centres at risk (La Rosa et al. 2014). This preference assumes that human capital can be used to fill any deficit in the environmental systems. For example, coastal and riparian areas are subject to increased flooding events, storm surges and sea level rises. Due to the high land values in these areas, risk mitigation typically results in resistance strategies that bolster a business as usual approach. In contrast, strong sustainability policies acknowledge the hierarchy of dependencies that must be considered in urban planning to achieve sustainable development. According to this perspective, the biosphere is the governing system that sustains all life Using this approach to development assumes more sophisticated strategies that balance human-centred (anthropogenic) economic and social aspirations, with environmental-centred (ecocentrism) aspirations to create, promote and protect urban resilience (Baxter 1996; Morandin Ahuerma et al. 2019).

Weak sustainability policies have led to uncontrolled development at the city fringes resulting in the degradation of local natural systems. These important ecosystem services include air and water purification, flood and drought mitigation, regeneration of soil fertility, moderation of temperature and enhancement of landscape quality (Gupta et al. 2017; La Rosa et al. 2014; United Nations Environment Programme [UN Environment] 2019). The degradation of these natural systems impact the livelihoods of those living in the adjacent urban centres. Development generated risks leave urban areas overly exposed to hazard events which are increasing in severity due to climate change. New climate and environmental conditions can impact global supply chains, making the communities that are connected to this network more vulnerable to shocks, undermining sustainable development. For example, prolonged flooding and drought events impacting agricultural production leading to food insecurity (Filippini et al. 2018).

As cities become more densely populated, their capacity to absorb the adverse effects of hazard events becomes increasingly difficult. This is particularly true for emerging cities in developing countries (World Bank Group 2015). The subsequent migration from rural to urban areas, has sparked concerns related to water quality and poor sanitation, diminished land fertility leading to, poor health and food insecurity. These stressors are exacerbated when populations are concentrated in slums and squatter settlements, typically located within areas that are prone to disaster events (Gupta et al. 2017).

Pro-urban monopolies are further exacerbating the divide between urban and rural communities. The high population density of urban areas sways voting and campaigns in favour of urban communities leaving PUA's underrepresented (Prabhakar et al. 2009; Gupta et al. 2017). These political imbalances continue to leave PUAs vulnerable to uncontrolled or ill-defined development, and a further diminishing of rural and urban interdependencies (La Rosa et al. 2014; Zasada 2001).

Lack of support and the under representation of Peri-Urban [PU] and rural groups can diminish the understanding of the community's vulnerability and reduce resilience. Without adequate planning, cities cannot sustain growth without disrupting livelihoods and their economic, social and environmental systems. This risk is higher in developing nations, where the majority of projected city population growth is expected to occur (UNDP 2019).

8.3 Multifunctional Nature of Peri-urban Space

8.3.1 PUA Adaptability

PUA can be defined by land that is at the city fringes and is captured between urban and rural zones. The exact extent of PUAs however is difficult to define due to its inherently transitional nature (Pahl-Wostl 2009).

The multifunctionality of PUAs has adapted from productionist uses to non-productionist or consumerist offerings (Llausàs et al. 2016; Zasada 2001). Adapting in this way has seen a socio-economic shift from traditional rural activities to high intensity local offerings back to urban centres such as agri-tourism, high intensity and organic farming. The ability of PUA to diversity reflects the community's resilience and willingness to adapt to change. Food production demonstrates a shift that is emerging throughout PUA, integrating various modes that engage consumers by mixing social, cultural and recreational functions within green space (La Rosa et al. 2014). Land yield in this sense is maximised making use of the same site for a variety of activities whilst providing ecosystem benefits such as water purification, flood mitigation, fire breaks, reduction of the urban heat island effect as well as providing a local food system (Gupta et al. 2017; La Rosa et al. 2014; Pauleit et al. 2019).

8.3.2 Challenges Facing PUA

Challenges affecting the development of PUAs is that stakeholders cannot easily conceptualise long-time horizons nor understand systemic causes of vulnerability. As a result, planning is often ad hoc, with symptoms typically identified and incremental responses actioned, while the root causes are neglected (Butler et al. 2016).

In developing nations, uncontrolled development at the city fringe is often occupied by poor and marginalised groups. These groups may be defined by poverty, gender, ethnicity or age and have limited infrastructure and little political representation. The current pressures placed on cities by growth can exceed the capacity to deliver community wide infrastructure such as sewerage and waste management. These stressors can cause health problems for the residents and contribute to the degradation of environmental systems (Alves et al. 2019). For example, sewerage is dumped into natural environments where pollutants seeping into the soil impact runoff from these areas polluting the water supply and contaminating food production (IRP 2018).

In developed countries, challenges are dominated by similar shortcomings in planning legislation and thoughtful zoning. Policymakers typically consider the economic value of land to be higher than the benefits that PUA can provide to the wider urban community (Filippini et al. 2018; La Rosa et al. 2014; Teng et al. 2012). Thoughtless planning solutions that do not allow for a transitional zone between urban centres and rural areas are equally short sighted and reflect a preference for certain and visible gains; mirroring developing nations inability to conceptualise long-term horizons and translate this into sustainable urban planning. While protective zoning of natural areas is one common solution to ensure natural environments are preserved, this approach is considered outside the realm of urban development and thus not an integrated planning strategy. This does little to influence the development of adjacent zones and puts pressure on surrounding areas to compensate for loss of land elsewhere (Zasada 2001).

The effects of population growth and the high economic value of land leads to increased subdivisions in PUAs (Llausàs et al. 2016). In Victoria, Australia, for example, subdivisions in PUA are increasing, as new residents are attracted to the amenity of natural landscapes and proximity to metropolitan areas. The increased popularity of these areas drives the development of infrastructure projects to improve accessibility, which in turn reinforces the areas' desirability and thus population and density increases. This cycle degrades and fragments agricultural and natural landscapes (Mackey et al. 2008; Stagoll et al. 2010). Increased pressure for development in PUA often ignores long standing knowledge of the land and its role and impact due to natural disaster events. This disaster amnesia due to demographic change and population growth can have serious consequences when future events put more people in harm's way. Land development driven by short-term gain limits socio-environmental opportunities and the benefits that these have to people's livelihoods. Sustainable planning considers economic, social and environmental within one system to facilitate co-benefits.

Planning of PUA that supports innovation and resilience requires good understanding of historic local land use data to address the impact of new typologies being developed. By providing an accurate picture of the adaptive land uses within the region value can be refocused on balancing ecosystem services that are both productive and consumer driven. A study undertaken in the PUA in Sydney, Australia, suggests that adaptive and intensive farming techniques do not rely exclusively on the quality of land, indicating that there is an opportunity for greater land use distribution of productive and commercial uses (James and O'Neill 2016) However, due to the ill-defined nature of land uses within the PU, it is unclear the extent to which intensive or urban farming occurs in these areas (James and O'Neill 2016). The practice of intensive farming in PUA is a response to urban pressures, which force land subdivisions and fragment land uses, driving the need for innovation in agricultural practices (Filippini et al. 2018). Industry innovation such as intensive farming highlights the opportunities for new productive land use typologies to be used in PUA where competing land uses, such as housing, are occurring. By reducing the amount of land required, land uses can be reprioritised and incentivised for non-market uses such as conservation and regeneration. In order to facilitate ecosystem-based services the value of land needs to incorporate the value of ecosystem benefits to the wider community by placing economic value on green and blue assets (Cork et al. 2007; Jayakody et al. 2018; Mackey et al. 2008).

8.4 Peri-urban Ecosystem: Threats, Benefits, and Opportunities

8.4.1 Climate Change—Anthropogenic Threats

In the context of climate change and disaster management, adaption and resilience capacities refer to institutional, technological, economic and social capacities to plan and implement programmes of change that could reduce the vulnerabilities and increase the capacities of communities (Prabhakar et al. 2009). Humans are part of a self-constructed social system built to allow different types of people to engage with each other to develop shared values and beliefs. Parallel to this are environmental systems which are interconnected, and reliant systems characterised by complex and unpredictable behaviour consisting of vulnerable and dynamic system states (Bahadur and Tanner 2014; Giebels and de Jonge 2014). Both systems are distinct and to fully appreciate how resilience sits within and is relevant to climate change the two systems need to be connected. Resilience is the conduit for balancing ecological and human services (Bahadur and Tanner 2014).

Technocratic solutions that rely on the abundance of resources are dominating the way humanity is addressing issues of risk. With the destruction and degradation of natural ecosystems to facilitate areas for people to live the surrounding natural areas become increasingly valuable to sustain the health and security of the urban centres (Pauleit et al. 2019). Environmental solutions that focus on the benefits of the natural environment are no regrets solutions due to their low impact, affordability, and environmental trade-offs (Gupta et al. 2017).

Urban environments are built-up areas that contain a density of human occupation that utilises centralised services and infrastructure. Urban environments are categorised by major changes to the natural environment to make way for human needs. Typically, the economic value of land at the city fringes is typically valued in favour of urban expansion over agricultural and non-productive low density uses (La Rosa et al. 2014; Zasada 2001). The cost per unit of development, for example housing, greatly outweighs the feasibility of maintaining unproductive land (La Rosa et al. 2014). The transition of land use generated by urban growth moves outward as PUAs are transformed into urban areas, and rural areas become new PUAs. The expansion of these zones characteristically pushes out agricultural uses and developments are thus then established on fertile land which they have no requirement (Gupta et al. 2017).

8.4.2 Threats to Peri-urban Ecosystems

Due to its large spatial extent, agriculture plays a key role in managing the peri-urban landscape and the social, aesthetic, and environmental functions of urban areas nearby (Zasada 2001). Over time it is common for land uses to be redefined, allowing for different types of activities to be undertaking in a particular zone, for instance agriculture pasture to residential. Land acquisition patterns associated with land use changes often result in large titles being subdivided and individually sold off which causes fragmentation of productive land making it difficult to sustain a traditional rural and agricultural livelihood. Without the allocation of land to ecosystem-based benefits, these zones are left vulnerable to disjointed and fragmented uses with an economic bias for urban development at the cost of environmental sustainability (Jayakody et al. 2018). PU farming has adapted to be multifunctional in its uses, growing multiple types of crops and utilising various aspects of agri-tourism (La Rosa et al. 2014; Jayakody et al. 2018). With the introduction of policy instruments such as carbon trading, farmers can generate income from carbon sequestration that incentivises ecosystem services in the transition of "non-use" land functions into the market space.

Consequences of urban expansion to rural communities can also reside in the form of in situ migration. Examples from China demonstrate how villages remain in place while dense urban areas are built up around them (Shih 2017). This practice removes the original land carers from their traditional roles of maintaining landscapes that once provided ecosystems services to an urban centre (Jayakody et al. 2018). Forced migration can negatively impact livelihoods due to poor consideration of the sustainability of those populations or the consequent cost of displacement. Threats generated by in-migration, urban to rural, exacerbate the need for development and highlights the support of planning legislation to develop these areas (Llausàs et al. 2016; Pahl-Wostl 2009). Trends in PUAs are towards post-productive conversions of land use, which are facilitated by short sighted planning overlays, and market driven development aspirations, that do not consider the cumulative effects of approved developments (Jayakody et al. 2018). As natural landscapes and ecosystems span municipal boundaries, the differences in approaches between governing bodies do little to respect the continuity of ecosystems to cross human-made boundaries (Llausàs et al. 2016).

8.4.3 Opportunities for Adaptive and Multifunctional Peri-urban Areas

Global warming predicts increased frequency and severity of natural hazards (IPCC 2019, UN Environment 2019). The potential increase to 5 degrees above pre-industrial temperatures is set to have catastrophic system failures relating to health, livelihoods, food security, water supply, human security, and economic growth (IPCC 2019).

The convergence of climate change adaption (CCA) and disaster risk reduction (DRR) is an indication of the state of climate emergency of which urbanisation is a driver and focus for threats (UN Environment 2019). DRR and CCA are working towards the survival of human civilisation through establishing and defining sustainability of livelihoods. PUAs have the capacity to respond to these current predictions by integrating natural systems with human activity (La Rosa et al. 2014).

Ecosystem-based approaches (EbA) that inform the appropriate management of PUAs seek to harness the environment to benefit human livelihoods (Dovers and Hussey 2013; Jayakody et al. 2018; Morandin Ahuerma et al. 2019). *This line of thinking enables urban planning to revitalise the urban condition through a mixed mode of green, blue and grey infrastructure* (Alves et al. 2019). *A growing body of literature acknowledges that, developing models to encapsulate and stimulate natural environments will provide urban areas co-benefits beyond physical infrastructure changes. These changes must incorporate stormwater management, air quality improvements, increased biodiversity and wildlife leading to improved health and wellbeing of humans* (Alves et al. 2019; Mabon et al. 2019; Pauleit et al. 2019; Millennium Ecosystem Assessment 2005).

Discussions about the co-benefits of integrating BGI with urban planning have generally focused on the effects on city centres. Studies on the implementation of BGI planning have not necessarily addressed the issue of available land, particularly in developing countries where 90% of urban growth is occurring (UNDP 2019). *This inward view towards development causes conflicts between short-term priorities, such as housing, with long-term goals related to su*stainable development and resilience (Pauleit et al. 2019). *As a result of this disconnect, PUAs are experiencing shifts towards higher density development* (Stagoll et al. 2010). *The types of occupants that are moving there vary and are resulting in different modes of operation including agriculture, rural amenity, pluriactivity, peri-metropolitan, marginalised agriculture and conservation* (Llausàs et al. 2016).

8.4.4 Blue, Green, and Grey Infrastructure

Introducing BGI into cities needs careful consideration. The effectiveness of this new infrastructure rests with their ability to complement existing natural systems. To link new and existing natural corridors together the requirement of functions and services provided by PUA needs to be from an urban consumerist perspective, which acknowledges the environmental dependencies (Pahl-Wostl 2009; Zasada 2011). New types of peri-urban activities have led to industries such as agriculture to transform into multifunctional business models (IRP 2018; SAGE 2005). The multifunctional attributes of peri-urban agriculture have responded to the post-productive, consumption-orientated requirements of urban society but intensifying their land uses to diversify offerings. Offerings include publicly used spaces such as wildlife habitat areas and nature-related recreational activities as well as urban farming and community gardens (La Rosa et al. 2014). These changes create business continuity, encourage social interactions between urban and rural occupants and aid in education with trends like farm to table dining experiences. Industries adapting to PUA are acting as the gateway for urban occupants to reconnect with nature and have an opportunity for co-habitation while maintaining a modern lifestyle.

Agricultural parks are amongst the largest land use typology for PUA however, being the largest typologies by area agricultural parks rely on land that is less fragmented. Agricultural parks are intended to be diverse, multifunctional, and holistic systems that exemplify sustainability principles at the urban–rural interface (SAGE 2005). There is a productive agricultural aspect that is maintained as part of the parks operation as well as providing landscape protection and leisure (La Rosa et al. 2014). The diverse uses within an agricultural park drive agri-tourism and income into the region. Jayakody et al. (2018) found that there is a preference for small scatterings of farmland to be amongst natural settings opposed to large expansive agricultural uses. Such studies are important, because they relate to the strengthening of the urban–rural link in terms of responding to urban consumers and establishing relationships between users and occupants.

Urban farming is concerned with being able to provide an alternate food supply to the local community, which aids in food security and goes a long way in the reduction of food miles (Teng et al. 2012). Filippini et al. (2018) suggest that there is a preference for local food purchasing over organic food options. Nonetheless, there is a gap between food produced in the PUA and the food delivered to local markets despite geographical proximity. There is also a consumer knowledge gap between quality and nutritional value of the food; proximity does not always equate to the healthiest option. Consideration must be given to the environmental context in which food is produced (Teng et al. 2012).

Allotment gardens are a useful and compact way to introduce equitable social activities. They provide a place for communities to come together and make positive contributions to their livelihoods (La Rosa et al. 2014; Zasada 2001). Allotment gardens are flexible in scale and easily inserted into PU and urban contexts. They are often a place for education and knowledge sharing across a broad socio-demographic population (La Rosa et al. 2014).

Maintaining natural environments and landscapes is the most cost-effective solution for land use, because little to no additional infrastructure is required to support functionality (Gupta et al. 2017). Recreational activities such as walking tracks, mountain biking or educational discovery centres can be incorporated into natural landscapes to compliment the tourism aspect (Jayakody et al. 2018; La Rosa et al. 2014).

Ecosystem-based approaches to DRR and CCA also contribute to this PUA land use matrix. Ecosystems-based DRR refers to the use of natural environments or systems to buffer impacts of changing climate, extreme weather events and flooding. The basic objective is to maintain the resilience of natural ecosystems and their services to help communities survive and cope with extreme events (Gupta et al. 2017). The ability of natural environments to mitigate natural hazards at the city fringe is highly valuable to urban centres. When coordinated with urban development, using natural environments, or preserving open space can create potential for multifunctional uses. One example is the use of flooding or tsunami exclusions for agriculture, open space or scenic amenity (Jayakody et al. 2018).

To overcome these concerns, the scope for urban planning must stretch beyond municipal boundaries of cities and address the interdependencies of the surrounding areas at a local level to appreciate the effects of urban expansion into adjacent areas and the negative impact on ecosystem services. EbA operates as multifunctional devices that align with the adaptive nature of PUA (Pahl-Wostl et al. 2013). Just as naturally occurring ecosystems rely on various inputs and outputs, so does human survival. At present, there is a gap between anthropogenic systems and natural systems with an emphasis on anthropogenic resource consumption (IRP 2018; UN Environment 2019). Humanities' reliance on consumption is unsustainable both in the short and long term. EbAs facilitate the bridging of the gap between human and natural ecosystems (Alves et al. 2019).

Urban green infrastructure can build on a long history of green space planning and has its antecedents in theories and applications of greenway and green structure planning, urban and landscape ecology (Pauleit et al. 2019). Initiatives implemented by the city of Fukuoka in Japan provide one of many strong examples of green infrastructure application in the world. The city has included green infrastructure policy since the late 1990s inclusive of large open spaces, green rooves, and green scaping to building facades to mitigate temperature (Mabon et al. 2019). These strategies were implemented to consider wind, water runoff and various types of heat gain to the urban fabric. The good practices that Fukuoka employed were not only driven by environmental reasoning but also considered human wellbeing and place making that provides recreational activities as well as spaces for retreat and respite from a modern human life (Ibid).

Criticism surrounding the effectiveness of integrating BGI into intensely densifying city centres largely questions the adequacy and performance of such interventions to contribute to environmental sustainability (Trainer 2014). This does not take into account sustainability of livelihoods and only considers the environmental aspect of sustainability. Effectiveness should also take into consideration the social and emotional wellbeing of a cities inhabitants of which natural elements contribute to.

8.5 Policy and Planning

8.5.1 Global Frameworks

Global frameworks for sustainable development have developed extensively over the last three decades. More recently the strategic goals of these frameworks have placed emphasis on disaster preparedness and mitigation implementation from all levels of governance as well as the inclusion of community and private sector actors (Cradock-Henry et al. 2018). These strategies align with transformational goals and focus attention towards transformational systemic change (Pahl-Wostl et al. 2013). For example, Sustainable Development Goals (SDGs) one, 11 and 13 are directly related to disaster risk reduction; while goals 11a and 11a.1 specifically support socio-economic and environmental links between urban, PU and rural areas by strengthening national and regional development planning alongside urban development plans that include an assessment of resource needs based on size (Gupta et al. 2017; UNDP 2019).

The vulnerabilities associated to climate change appear both locally and globally. As the world becomes increasingly more connected, policymakers need to factor globalisation into the vulnerability equation with indirect and incremental stresses acknowledged (Tyler and Moench 2012). Strengthening the urban–rural link facilitates a whole system approach that places value on the co-benefits of natural and human systems. EbA at the local level can contribute not only to the sustainability of livelihoods but make bigger contributions and impressions to the sustainability of modern human life through progressive implementation of green and blue infrastructure.

8.5.2 Political Will, Policy Making and Planning

Community leaders in government and private sectors need to be engaged to get the right levels of implementation. The private sector is important to the continuity of livelihoods as a major employer; however, it is often underrepresented (Le Masson 2015). The move towards resilient and adaptive programmes lends itself to decentralisation of responsibility from governments and onto private and civil actors emphasising co-production and consistent representation that does not depend on election time cycles (Archer et al. 2017).

Policy making does not give an accurate representation of how the world works nor how it should work (Handmer and Dovers 2016). As such, policy should be created to allow community actors to implement flexible strategies providing resilience based on local capacities and perceptions of vulnerability (Le Masson 2015). This form of policy making is akin to the uncertainty and long-time frames associated with sustainability (Dovers and Hussey 2013).

Environmental policy is differentiated from planning regulations and general compliance policies due to its cross-sectoral nature. The use of multiple policy instruments allows for various sectors of society to engage with concepts of sustainability and enrich the knowledge base in which policies are developed (Dovers and Hussey 2013; Giebels and de Jonge 2014). Without inclusive environmental frameworks, it is unlikely that the root causes embedded within human systems will be addressed (Dovers and Hussey 2013). Adaptive and multi-level governance provides communities with the ability to participate and remain engaged with the process of DRR and CCA. This style of governance aligns with the natural adaptive and multifunctional tendencies of PUA. Urban agriculture practices have often been implemented from the bottom up and spontaneously as the need for survival arises (La Rosa et al. 2014). Political will needs to not only incorporate the driving forces that stimulate policy change but also the inclusion of adaptability at a local community level to implement global sustainability targets and induce transformation (Bahadur and Tanner 2014; Cradock-Henry et al. 2018).

To achieve this more emphasis on network governance and processes of social and societal learning is required. Governance considers different models of governing where state and non-state actors are required for transformative change (Pahl-Wostl 2009). Adaptive and multi-governance systems are often self-organising networks of teams and actor groups that can draw on multiple streams of knowledge and experiences to develop a common understanding (Pahl-Wostl et al. 2013). This style of governance utilises a wide spectrum of actors and is better equipped to deal with issues including gender sensitive policies, finance, cooperation on technology development as well as enabling participation, transparency, capacity building and learning among different actors at different levels (IPCC 2019).

The introduction of urban green growth is relatively new in developing nations and relies on the integration of social and economic objectives with environmental goals (Gupta et al. 2017). Developing urban BGI is therefore going to require unique and novel ways of finding solutions conducive with inter and transdisciplinary approaches

(Pauleit et al. 2019). Multi-level or adaptive governance can bring together objectives to strengthen the urban–rural link, thus enhancing functional inter-relationships to secure livelihoods (Zasada 2001). Developed nations such as the United Kingdom (UK) and Denmark have utilised zoning to designate areas of ecological value close to urban centres. In the UK a 'Green Belt' strategy has been actioned, which restricts or limits urban development in certain natural environments. However with the pressures of development the 'Green Belt' strategy is often is often compromised and boundaries adjusted to suit the needs of growing populations, reducing the overall effect of the 'Green Belt' to provide a safe and diverse environment to flora and fauna (James and O'Neill 2016). Preservation and protection of natural environments is important, but it also needs to be strategic. The sanctioning of land, whilst effective to the boundaries of that area, also relocates pressures to neighbouring areas that are 'unprotected'. City master planning needs to take into consideration the pressures it places on PUA and rural environments. Utilising precautionary principles builds in tolerances that allow policy to be flexible and actions adapted (Dovers and Hussey 2013). The permanency of development as well as the complexity of land titles further highlights the need for multi-level governance with long-term goals focused on the integration of connected ecosystem services that drive urban development sustainably, in contrast to the current reactionary approach to population growth.

8.6 Conclusion

Urbanisation is one of the biggest drivers of climate change. The density of these environments has the potential to create real change if transformation can occur. Transformation implies a change in paradigm and relies on informal settings and institutions being able to collaborate with formal processes of governance (Pahl-Wostl 2009; Pahl-Wostl et al. 2013). All over the world urbanisation is creating a physical transformation of open space into built-up areas and with this physical transformation follows socio-cultural transitions. These are often related to the adoption of urban lifestyles by rural migrants. The concentration of populations and process of transformation of natural areas to accommodate urban growth needs to be reassessed and a paradigm shift needs to occur. With continuing trends of population in-migration, national and local governments should be focusing on creating sustainable city growth plans to ensure the substantiality and longevity of livelihoods.

Poverty, gender, ethnicity, and age define marginalised groups and contribute to the differential vulnerability of social groups which typically reside in the city fringe (Tyler and Moench 2012). The inclusion of allotment gardens, urban farming and hobby farms into PUA enables these groups to influence their own livelihoods contributing to the regions overall capacity whilst providing service and amenity to the urban community. More emphasis needs to be placed on flexible planning that establishes co-benefits using ecosystem-based approaches.

Through effective and adaptive multi-level governance environmental issues which are closely related to social issues such as hunger, consumption patterns, health, education, inequality, waste and sanitation, refugees, migration and conflicts can begin to be addressed. Ecosystem focused planning enables DRR and CCA practices to be embedded into urban centres. Maintaining and establishing open areas both in and around urban centres mitigates the effects of natural hazards. Although the effectiveness of retrofitting green and blue infrastructure into existing cities is questioned in terms of the scale of its contribution in reducing GHG emissions it does draw attention to the availability of PUAs to provide this service. *Implementation of ecosystem-based approaches within PUA strengthens the urban–rural link preserving natural environments and the effectiveness of natural environments to reduce the effects of anthropogenic climate change. BGI creates continuity in planning approaches linking cities with productive environments to influence systemic change for sustainable modern human life.*

References

Alves A, Gersonius B, Kapelan Z, Vojinovic Z, Sanchez TA (2019) Assessing the co-benefits of green-blue-grey infrastructure for sustainable urban flood risk management. J Environ Manage 239:244–254

Archer D, Monteith W, Scottand H, Ga S (2017) Developing city resilience strategies: lessons from the ICLEI–ACCCRN process. In: IIED Asian cities climate resilience WP series, No 41/2017, pp 1–38

Bahadur A, Tanner T (2014) Transformational resilience thinking: putting people, power and politics at the heart of urban climate resilience. Environ Urban 26(1):200–214

Baxter B (1996) Ecocentrism and persons. Environ Values 5(3):205–219

Butler JRA, Bohensky EL, Suadnya W, Yanuartati Y, Handayani T, Habibi P, Puspadi K, Skewes TD, Wise RM, Suharto I, Park SE, Sutaryono Y (2016) Scenario planning to leap-frog the sustainable development goals: an adaption pathways approach. Clim Risk Manag 12:83–99

Cork S, Stoneham G, Lowe K (2007) Ecosystem services and Australian natural resource management (NRM) futures. Paper to the Natural Resource Policies and Programs Committee and the Natural Resource Management Standing Committee. Australian Government Department of Environment, Water, Heritage and the Arts

Cradock-Henry NA, Fountain J, Buelow F (2018) Transformations for resilient rural futures: the case of Kaikoura, Aotearoa-New Zealand. Sustainability 10:1–19

Davis I, Alexander D (2015) Recovery from disaster. Taylor and Francis, ProQuest Ebook Central. Retrieved from https://ebookcentral-proquest-com.ezproxy.cdu.edu.au/lib/cdu/detail.action?docID=3570429

Dovers S, Hussey K (2013) Environment and sustainability: a policy handbook, 2nd edn. The Federation Press

Filippini R, Lardon S, Bonari E, Marraccini E (2018) Unraveling the contribution of periurban farming systems to urban food security in developed countries. Agron Sustain Dev 38(2):1–15

Giebels D, de Jonge VN (2014) Making ecosystem-based management effective: identifying and evaluating empirical approaches to the governance of knowledge. Emergence Complex Organ 16(1):60–76

Gupta A, Singh S, Shiraz W, Nivedita M, Singh AK (2017) Urban resilience and sustainability through peri-urban ecosystems: integrating CCA and DRR

Handmer J, Dovers S (2016) Handbook of disaster and emergency policies and institutions. The nature of policy and institutions, chap 2. Oxfordshire Routledge, Abingdon, pp 29–46

IPCC (2019) Global warming of 1.5 degrees. Retrieved from https://report.ipcc.ch/sr15/pdf/sr15_spm_final.pdf
IRP (2018). The weight of cities: resource requirements of future urbanization. In: Swilling M, Hajer M, Baynes T, Bergesen J, Labbé F, Musango JK, Ramaswami A, Robinson B, Salat S, Suh S, Currie P, Fang A, Hanson AK, Reiner M, Smit S, Tabory S (eds) A report by the International Resource Panel. United Nations Environment Programme, Nairobi, Kenya
James S, O'Neill P (2016) Planning for peri-urban agriculture: a geographically specific, evidence-based approach from Sydney. Aust Geogr 47(2):179–194
Jayakody RRJC, Amaratunga D, Haigh RP (2018) Plan and design public open spaces incorporating disaster management strategies with sustainable development strategies: a literature synthesis. In: MATEC web of conferences, vol 229, p 04001
La Rosa D, Barbarossa L, Privitera R, Martinico F (2014) Agriculture and the city: a method for sustainable planning of new forms of agriculture in urban contexts. Land Use Policy 41:290–303
Le Masson V (2015) Considering vulnerability in disaster risk reduction plans: from policy to practice in Ladakh, India. Mountain Res Dev 35(2):104–114
Llausàs A, Buxton M, Beilin R (2016) Spatial planning and changing landscapes: a failure of policy in peri-urban Victoria, Australia. J Environ Plan Manag 59(7):1304–1322
Mabon L, Kondo K, Kanekiyo H, Hayabuchi K, Yamaguchi A (2019) Fukuoka: adapting to climate change through urban green space and the built environment? Cities 93:273–285
Mackey BG, Watson JEM, Hope G, Gilmore S (2008) Climate change, biodiversity conservation, and the role of protected areas: an Australian perspective. Biodiversity 9(3–4):11–18
Millennium Ecosystem Assessment (2005) Living beyond our means, natural assets and human well-being. Millennium Ecosystem Assessment
Morandin Ahuerma I, Contreras Hernández A, Ayala-Ortiz D, Pérez-Maqueo O (2019) Socio-ecosystemic sustainability. Sustainability 11:3354. https://doi.org/10.3390/su11123354
Pahl-Wostl C (2009) A conceptual framework for analysing adaptive capacity and multi-level learning processes in resource governance regimes. Glob Environ Chang 19:354–365
Pahl-Wostl C, Becker G, Knieper C, Sendzimir J (2013) How multilevel societal learn processes facilitate transformative change: a comparative case study analysis on flood management
Pauleit S, Andersson E, Anton B, Buijs A, Haase D, Hansen R, Kowarik I, Olafsson AS, Van der Jagt S (2019) Urban green infrastructure—connecting people and nature for sustainable cities. Urban For Urban Greening 40:1–3
Prabhakar S, Srinivasan A, Shaw R (2009) Climate change and local level disaster risk reduction planning: need, opportunities and challenges. Mitig Adapt Strat Glob Change 14:7–33
Renn O (2008) Concepts of risk: an interdisciplinary review part 1: disciplinary risk concepts. GAIA—ecological perspectives for science and society, vol 17, pp 50–65. http://doi.org/10.14512/gaia.17.1.13
SAGE: Sustainable Agriculture Education (2005) Urban edge agricultural parks toolkit. Sustainable Agriculture Education, Berkeley, California
Schipper L (2009) Meeting at the crossroads? Exploring the linkages between climate change adaptation and disaster risk reduction. Clim Dev 1(1):16–30
Shih M (2017) Rethinking displacement in peri-urban transformation in China. Environ Plan A 49(2):389–406
Stagoll K, Manning AD, Knight E, Fischer J, Lindenmayer DB (2010) Using bird-habitat relationships to inform urban planning. Landsc Urban Plan 98:13–25
Teng P, Sombilla M, Ewing A, Escaler J (2012) Feeding Asia in the 21st century: building urban-rural alliances: summary of the main findings of the international conference on Asian food security held in Singapore on 10–12 August 2011. Food Secur 4(1):141–146
Trainer T (2014) Some inconvenient theses. Energy Policy 64(C):168–174
Tyler S, Moench M (2012) A framework for urban climate resilience. Clim Dev 4(4):311–326
UN (2017) New urban agenda. Habitat III retrieved from http://habitat3.org/the-new-urban-agenda/

UN Environment (2019) Global environmental outlook 6. Healthy Panel Healthy People. Retrieved from https://content.yudu.com/web/2y3n2/0A2y3n3/GEO6/html/index.html?page=38&origin=reader

UN Habitat (2016) Urbanization and development: emerging futures world cities report 2016. Retrieved from https://unhabitat.org/sites/default/files/download-manager-files/WCR-2016-WEB.pdf

UNDP (2019) Sustainable development goals. Retrieved from https://www.undp.org/content/undp/en/home/sustainable-development-goals.html

UNDRR (2019) Global assessment report on disaster risk reduction. Retrieved from https://gar.unisdr.org/

United Nations, Department of Economic and Social Affairs, Population Division (2019) World urbanization prospects: the 2018 revision (ST/ESA/SER.A/420). United Nations, New York

Wisner B (2016) Vulnerability as concept, model, metric, and tool. Oxford Res Encycl Nat Hazard Sci. https://doi.org/10.1093/acrefore/9780199389407.013.25

World Bank Group (2015) Investing in urban resilience, protecting and promoting development in a changing world. Retrieved from https://openknowledge.worldbank.org/bitstream/handle/10986/25219/109431-WP-P158937-PUBLIC-ABSTRACT-SENT-INVESTINGINURBANRESILIENCEProtectingandPromotingDevelopmentinaChangingWorld.pdf?sequence=1&isAllowed=y

Zasada I (2001) Multifunctional peri-urban agriculture—a review of societal demands and the provision of goods and services by farming. Land Use Policy 28:636–648

Ms. Celeste Norman is an architect working on projects focused on sustainability and resilience building. Celeste completed postgraduate study in Emergency and Disaster Management at Charles Darwin University where she pursued research related to urban resilience, exploring the opportunities and role of ecosystem-based services in urban contexts.

Associate Professor Akhilesh Surjan is Research and Theme Leader of the Humanitarian, Emergency and Disaster Management Studies programme at Charles Darwin University. He has a background in disaster and climate risk reduction in the context of human settlements. He previously held positions at the Kyoto University and United Nations University.

Ms. Miranda Booth is a Lecturer and Ph.D. candidate in the Humanitarian, Emergency and Disaster Management Studies programme at Charles Darwin University. She has a background in international relations and previously held positions at the University of Melbourne and University of New South Wales, Canberra. Her current research focuses on the politics of humanitarian aid new humanitarian actors, measuring effective humanitarian aid and global and regional humanitarian ecosystems.

Chapter 9
Innovations to Reduce Disaster Risks of Water Challenges

Piyalee Biswas, Neelima Alam, and Sanjay Bajpai

Abstract There is increasing realisation over the world about the necessity to manage the ecosystems responsibly in a sustainable manner to reduce both risk and impact of disasters. Sustainable developments demand that various ecosystems such as wetlands, coastal zones, river valleys etc. are managed effectively and efficiently to achieve sustainable and resilient development. India is one of the most disaster-prone countries in the world due to its distinctive geo-climatic and socio-economic conditions making it vulnerable to floods, droughts, cyclones, earthquakes, urban flooding, landslides, avalanches and forest fire. The disaster risk is further compounded as India is also one of the world's fastest-growing economies. India is facing several water challenges related to water availability, quality and management. There have been several initiatives to integrate sustainable solutions into disaster risk reduction approaches, which include natural water management and treatment approaches and have resulted in provision of water of desired quality in required quantity. The chapter will present the framework of the innovation system, key policy objectives and methodologies for fostering innovations on reducing disaster risks. The approach to funnel innovations into implementation is also being discussed citing few case studies.

Keywords Disaster risk reduction · Technology solutions · Innovations · Water quality · Water scarcity · Holistic approach

---

Neelima Alam is the Co-Corresponding author in the Chapter

P. Biswas · N. Alam · S. Bajpai (✉)
Technology Mission Division (Energy, Water and Others), Department of Science and Technology (DST), Technology Bhawan, New Mehrauli Road, New Delhi 110016, India
e-mail: sbajpai@nic.in

N. Alam
e-mail: neelima.alam@nic.in

© The Author(s), under exclusive license to Springer Nature Singapore Pte Ltd. 2021
M. Mukherjee and R. Shaw (eds.), *Ecosystem-Based Disaster and Climate Resilience*, Disaster and Risk Research: GADRI Book Series,
https://doi.org/10.1007/978-981-16-4815-1_9

9.1 Introduction

A disaster is quite pervasive and is not bound by political, geological or geographical boundaries. By definition, a disaster means not only the natural ones but it can also be man-made. Disaster can be caused by the natural factors pre-existing in the ecosystem as well as anthropogenic factors. At the same time, disasters directly affect the ecosystem of any area.

There has been an increase in disaster cases in India over the century. Rapid and unrestricted human expansion led to higher damage potential of ecosystems due to disasters. There can be no two opinions that the fundamental elements of reducing risk and impact of disasters need to be included in development agenda. Development policies, approaches and programmes must try to find ways to avoid the undesirable effect of disasters by successfully integrating elements of disaster risk reduction.

Water, in spite of being the most essential resource for human and economic development, is one of the most inefficiently used resources. By influencing the output of agriculture, industry and other sectors, it not only affects the Gross Domestic Product (GDP) of a country but also affects the basic human traits including the potential to cause conflicts (NITI Aayog 2019).

Water sector in India has the need to focus more on disaster risk management. Most policies, plans and programmes found are related to augmentation of service levels, which, though improve quality of life, does not necessarily lead to risk reduction. In many cases, it was the indigenous knowledge and systems that have come to the rescue while finding sustainable solutions for water problems, thus reducing the risk of disasters. These interventions could mitigate the challenges in water-scarce areas, which have resulted in provision of water of desired quality in required quantity (Chattopadhyay and Prasad 2006). These interventions have also prevented intrusion of seawater in coastal areas. Similar efforts focussing on utilisation of less water-intensive crops coupled with massive conservation effort has provided a means to people in water-deprived locations for sustainable livelihood. These initiatives were triggered and sustained through appropriate technical, social and economic innovations.

9.1.1 Water Scarcity

Although India occupies only 3.29 million km^2 geographical area, which forms 2.4% of the world's land area, it supports over 15% of the world's population. India also has a livestock population of 500 million, which is about 20% of the world's total livestock population. The total utilisable water resources of the country are assessed as 1086 km^3. The utilisable annual surface water of the country is 690 km^3.[1] However, temporal and spatial variability of availability of this resource is a major challenge

[1] National Institute of Disaster Management, Mainstreaming Disaster Risk Reduction in Environment Sector, Guidelines and Tools.

leading to scarcity in few regions while excess in many areas during monsoons. The availability of water is highly uneven in India. Rainfall is restricted to only about three to four months in a year and varies widely geographically.

With the population already crossing the 1.3 billion mark, the annual per capita water accessibility of the country is just enough to satisfy its current needs. One of the goals of India's Sustainable Development, related to drinking water, is to improve the quality of water by 2030 (Dobhal et al. 2011). Water unavailability and poor water quality are the two main critical challenges the nation is facing. Domestic and industrial water needs have largely been concentrated in or near the major cities, but the claim for water from rural hinterland is expected to increase sharply as the development programmes improve socio-economic situations in the rural areas.[2]

Water is crucial for domestic, industrial and agriculture needs and is becoming limited day by day due to ever-increasing demand of these sectors. Therefore, water use efficiency and sensible use of available water is of utmost importance. In order to provide water that is safe for all and always, it is imperative that water is available. As water is a resource that has to meet competing requirements, availability needs to be seen holistically and demand-supply mismatch should be addressed through appropriate prioritisation. The shortage of water is affecting the lives of people in the country. Due to the lack of water, farmers find it difficult to grow crops. The lack of water makes it arduous to maintain livestock, which makes life harder for farmers as they find it hard to generate income and sustenance.

As per the NITI Aayog report, India will be water-stressed country from now onwards, which implies < 1000 m^3 of water availability per person per annum. The scarcity of water can have many catastrophic effects including desertification, posing disaster risks to ecosystem by affecting biodiversity and livelihood (NITI Aayog 2018).

9.1.2 *Water Quality*

Besides availability issue, the quality of the water may not be appropriate for the intended use. Water-borne diseases are the principal causes of death in the world. Unsafe water can lead to serious health effects and pose health risksto vast population.

Water in its pure form occurs rarely in nature. When water in its precipitate form reaches earth, it has already collected a number of substances and properties that characterise natural water. Around 70% of surface water in India is unfit for consumption, primarily due to bacterial contamination and untreated wastewater streams. Besides natural geogenic causes, anthropogenic influence on the quality of groundwater is quite apparent and is a major concern.[3]

---

[2] See Footnote 1.

[3] Ministry of Jal Shakti, Water Quality Activities in Central Water Commission.

To address water-related environmental problems, it is a must to have accurate information and to know precisely what the problem is, where it is occurring, how serious it is and what is causing it. Water quality is influenced by several factors such as hydrology, land use, source, pollution etc. Water quality of a water body can be monitored by development of sensors and remote sensing-based technologies (e.g. hyper-spectral imaging for water quality monitoring and biological indicators). This real-time monitoring and modelling of water quality in catchments (including assessment of existing systems and appropriate calibration) can help us understand water quality problems of water bodies and rivers. Quality of water can also be estimated by understanding the sources, transport or pathways, transformation, interactions and fate of pollutants and emerging contaminants in surface and groundwater including organics, plastics, pharmaceutical and personal care products (PPCPs) (Sankar et al. 2013). There is a need to strengthen monitoring techniques to include emerging contaminants. New techniques should be integrated and calibrated with existing systems. This will require new empirical approaches for studying water quality and the development of water quality models.

Highly water-intensive crops, grown using sub-optimal water irrigation methods lead to excessive water use which causes high runoffs. Runoffs from farms are a leading cause of water quality impairment through nutrient and organic pollutants. Modern agricultural practices emphasise the application of inorganic chemicals for increased yields. This has produced adverse impacts on the agroecology, including widely prevalent nitrate toxicity and contamination in aquifers, surface water bodies and soils impacting micro-flora and decreased resilience of agroecosystems. The use of low-quality irrigation water impacts agricultural production and food safety with health implications across the value chain [Quality Unknown, World Bank 2019].

The negative impacts above are reinforced by the effect of climate change. Climate change is projected to have and already has significant impact on the water cycle, by altering rainfall patterns and affecting the availability and quality of both surface and groundwater.

As water in many of India's rivers is unfit for drinking, 85% of the rural population of the country uses groundwater for drinking and domestic purposes. This overdependence on groundwater is a major problem as fluoride contamination in groundwater is prevalent throughout the country. Nearly three-fourths of Indian provinces are affected by fluoride contamination. Arsenic contamination is another major problem. The problem is severe in parts of West Bengal and Uttar Pradesh. Salinity of water is another well-known problem. Bacteriological contamination of water, which is a major problem with surface water, can lead to diarrhoea and cholera.

Besides health issues, unpotable water is a huge economic liability on state, resulting in heavy investments by state to provide clean and safe drinking water to its people. The federal and state governments have prioritised these initiatives and made significant investments. Notwithstanding these commendable efforts, technically feasible, socially acceptable, environmentally benign and economically viable solutions are still needed which requires custom-designed approach. Besides ongoing initiatives, there is need to look for 'out of box' thinking and look for alternatives such as seawater desalination, wastewater treatment etc. to provide lasting solutions.

The socio-economic conditions of the community can considerably improve if the beneficiaries are provided good drinking water. This would also reduce the disaster risk to the ecosystems.

9.2 Innovations as Policy Instruments to Reduce Water Challenge Risks

The countries world over have realised enormous potential of innovations which are essentially technology-based solutions, critical for addressing the challenges. Innovations are now at centre-stage of policy spectrum of the country to find solutions to various issues in an effective and efficient manner. Innovation can take place both in formal system as well as informal system. As India has a large population which does not move through formal echelons of education and research system, capturing of ideas of these people and providing them avenues for their innovative endeavour is important.

The national policy on Science and Technology has evolved over the last 60 years with varying but connected focus. India's Scientific Policy Resolution (SPR) of 1958 had pronounced to foster scientific temper and encourage cultivation of scientific research. The scientific temper created through this policy instrument later became important tool to encourage innovative spirit. In pursuance of the aspiration of SPR, 1958, scientific infrastructures were set up which was expected to lead to development of technologies. This was made explicit in the Technology Policy Statement of 1983 which emphasised on technology and self-reliance (Ministry of Science and Technology 2013).

Subsequently, science and technology were brought together through a single policy instrument in Science and Technology Policy of 2003. This policy emphasises application of science and technology for socio-economic sector for addressing national problems. The deployment of technology in the field brought home the importance of understanding the needs and requirements of society. As the potential of successful deployment of technology-based solutions was recognised, the importance of innovation was highlighted as an instrument of policy. It was recognised that sustainable innovation needs to be country, location and context-specific.

The Science, Technology and Innovation Policy of 2013 brought innovation to the forefront and stressed the need for strengthening the innovation ecosystem in the country. Besides strengthening of existing structures and programmes, the policy emphasised on creating necessary framework to address the pressing challenges of energy and environment, water and sanitation etc. The policy explicitly recognised the challenges emanating from climate variability and changes. Thus the generation of strategic knowledge to cope with the challenges which include disaster as well was also prioritised (Ministry of Science and Technology 2013). The policy aimed at creating ecosystem in which innovations would flourish. While the policy made a reference to address the socio-economic challenges through innovations, specific mention of water challenges was not evident, perhaps due to presumption that innovations will percolate every societal domain if the enabling environment was created. However, the broader societal objective translated into a few specific programmes focussing on innovations in water sector. While Department of Science and Technology (DST) through its water programmes emphasised research-led innovations and technology solutions along with ground-truthing of their application potential, Ministry of Drinking Water and Sanitation launched a specific drive to scout innovative technologies for wider implementation.

Parallel to evolution of innovation ecosystem, successive water policies also identified a number of issues and challenges that do exist for disaster risk reduction in various ecosystems of the country. The first water policy was formulated in 1987, which was subsequently revised and updated in 2002 and 2012 based on learning and experience. Water scarcity has been identified as one of the many prominent threats to ecosystem resulting in higher desertification and loss of various plant and animal species. The water required for various usages such as agriculture, industry, power and municipal use requires delicate balance. Distortion in this balance results in severe impacts on the ecosystem and makes it more prone to disaster and associated risk.

It was recognised that floods and droughts affect many parts of the country and go beyond state boundaries. Out of 40 million ha of the flood-prone area in the country, on an average, floods affect an area of around 7.5 million ha per year. Approach to management of droughts and flood has to be coordinated and guided at the national level. It has been emphasised that while every effort should be made to avert water-related disasters like floods and droughts, through structural and non-structural measures, emphasis should be equally on preparedness for flood/drought with coping mechanisms as an option. Greater emphasis should be placed on rehabilitation of natural drainage systems (Central Water Commission 2002).

In order to mitigate the risk associated with indiscriminate water use, integrated and coordinated development of surface and groundwater resources and their consumption and use is of paramount importance. Integrated and coordinated development of surface water and groundwater resources and their conjunctive use should be envisaged right from the project planning stage and should form an integral part of the project implementation. This would reduce disaster proneness of the development project during its lifetime and reduce the risk to habitants and biodiversity (Central Water Commission 2012). The importance of groundwater resources in terms of quantity and potential cannot be overemphasised. Some of the sustainable measure for maintaining ecology of overexploited areas includes utilising efficient technologies of water use, promoting community-based aquifer management, encouraging artificial aquifer recharge and promoting less water-intensive crops. Industries in water-scarce regions need to be allowed to either withdraw only the makeup water and should have an obligation to return treated effluent to a specified standard back to the hydrologic system. Concepts like Zero/near Zero Liquid Discharge (ZLD) need to be encouraged.

The water policies have also identified growing pollution of water sources as a major factor affecting the ecosystems. Growing pollution of water sources, especially through industrial effluents, is affecting the availability of safe water besides causing environmental and health hazards. In many parts of the country, large stretches of rivers are both heavily polluted and devoid of flow to support aquatic ecology.

Effective flood control and management for each flood-prone basin require that the flood forecasting is accurate, timely and extended to other uncovered areas. Inflow forecasting to reservoirs should be instituted for their effective regulation. Flood protection needs to be inbuilt for settlements and economic activity in the flood plain zones to minimise the loss of life and property on account of floods (Dobhal et al. 2011).

While emphasising need for capacity building of water managers, successive water policies emphasised management approach to regulation of water resources. Implementation of such policies and associated regulatory framework would need scientific tools and techniques and also skilled manpower, aware of holistic management of water resources. The innovations are need-based disruptive tools, techniques and analytical capabilities that modern technologies offer to facilitate this implementation. This connection between S&T policies and water policies were not vividly evident.

Science, Technology and Innovation have important roles to play in water resources development in general. Water disaster risk is cross-cutting issue that should be systematically evaluated in a holistic way through the identification of actions that will increase the adaptability of the community. These activities comprise of preventing disasters (prevention), decreasing the effect of disasters (mitigation) and readiness of the society to deal with the after-effects of a disaster when it has occurred (preparedness). There is a conscious effort to emphasise connectivity of innovations ecosystem with water sector in the formulation of New Water Policy and New Science, Technology and Innovation Policy (STIP). While STIP 2020 is likely to have sectoral approach (includes Water) with implementation strategy, Water Policy is expected to foster connection of R&D and innovations to address field-level challenges.

9.3 Innovative Approaches to Address Water Unavailability

Unavailability of water is the major challenge, especially in arid areas, primarily due to scanty rainfall. With larger part of the population relying on agriculture directly or indirectly, the impact of rainfall on human life and other living beings is mammoth. Optimum quantity of right quality of water for irrigation is the panacea for challenges faced by Indian agriculture. Due to water shortage, the problems of desertification of landmass and expanse of arid land has increased (Bajpai et al. 2019).

In dry season, low rainfall and substantial depressed agricultural production and erosion of productive resources in a region can result in malnutrition, immigration and shifts in work-related patterns. This emphasises the need to develop innovative water conservation strategies in water-scarce areas. Water shortage can be addressed by conserving rain-water, water-efficient irrigation, reducing evaporation losses, avoiding water-intensive crops and reducing runoffs, which also causes water quality impairment through nutrient and organic pollutants.

Due to shortage of water, the recurrence of drought or land becoming water-stressed has increased in many parts of India. It has thus become important that the focus should be shifted from crisis management to risk management. Drought has emphasised the significance of innovations for societies and communities to survive water shortages. This could be done by strengthening the community-based water shortage preparedness planning and holistic water risk management plans. Any holistic intervention should focus on information, communication and awareness activities, ensuring community participation, assessing current and projected demand–supply scenarios and designing the solution to meet criterion of technical, social, environmental and economic sustainability.

An innovative project in Rajasthan has successfully demonstrated the scientific methodology to address the challenges of low per capita availability and quality deficit of available water. The innovation followed two-pronged approaches viz. water conservation techniques to recharge groundwater and tapping of rainwater as substitute safe source for drinkable water. This intervention was implemented in Chirawa block, which lies in almost central part of Jhunjhunu District of Rajasthan. It comes under semi-arid zone of Rajasthan. Average rainfall of Chirawa block is 430 mm. 21% population of Chirawa block is residing in areas where groundwater quality is not potable and is affected by fluoride and nitrate. Most of the inhabitants are farmers and they used to cultivate crops like wheat and cotton.

Before the implementation of this project, Chirawa had almost nil surface water availability and there was total dependency on groundwater. Gross groundwater draft for domestic and other use was 17.70% and irrigation used 82.30% of total groundwater. The water problem was so acute that some villages of Rajasthan's Jhunjhunu district had just five feet of groundwater left. Chirawa was a dark zone and the withdrawal rate of groundwater was 328% of the replacement rate which was alarming.

This innovative intervention has resulted in the creation of a whopping 1597 rainwater harvesting tanks, 3362 soak pits and 66 Recharge wells This has helped 19,164 direct beneficiaries in 35 villages covering a population of 57,954. The intervention has renovated existing water harvesting structures (abandoned wells, Ponds) to recharge groundwater. The intervention instead of developing new solution used the traditional Rain Water Harvesting (RWH) technique using modern techniques for reducing bacterial contamination and first flush filter developed indigenously as a source for safe drinking water. This locally developed technology of rainwater harvesting is helping families save 20,000 L of water (Fig. 9.1a). 793 lakh L of water is recharged annually and has triggered a green revolution in what was once dry arid expanse (Fig. 9.1b).

Fig. 9.1 **a** Water harvesting tanks at Chirawa. **b** Complete pond site at Mahrampur. **c** Fruit Orchards at Ismailpur. **d** Awareness meetings at Jhunjhunu

There is a major reduction in crop water requirement as farmers were encouraged to grow low water intake crops like gram and mustard which takes 26% less water than wheat. There has been an increase in crop production by 22–50% which led to increase in income of farmers' upto 33.08%. At the same time, fruit orchards consuming less water have also been created to increase the income of farmers (Fig. 9.1c). The intervention has created water bank of 92.8 lakh L. The intervention has also provided equity and sustainability in water resource management in the water-scarce area of Chirawa block. The intervention has resulted in mitigating the challenges in this semi-arid area, which have resulted in delivery of water of desired quality in required quantity.

The scientific design of this intervention is evident from the holistic approach adopted to address the challenge of water scarcity, which have significantly decreased disaster proneness of this region. Scientific rainwater harvesting coupling traditional wisdom and modern device provided safe drinking water while recharge wells improved groundwater quality and availability. Soak pits and toilets improved sanitation and water availability for plants. The use of water-saving devices, seed treatment, deep ploughing and low water intake crops reduced expanse of desertification. Afforestation activity on community land decreased disaster risk through supporting climate-resilient agriculture. Capacity building of community and awareness exercises further built resilience of community and promoted collective action for disaster risk mitigation (Fig. 9.1d).

The success of this project is evidence of the transformation potential of innovations to reduce water scarcity and related disaster risks.

9.4 Innovative Approaches to Address Water Quality

While geogenic conditions affect groundwater quality, anthropogenic influence on the quality of water is also a major issue. Quality of water is affected when water is mixed with municipal and industrial wastewaters which results in severe changes in the quality of water. Advancement in existing strategies and the enhancement of new techniques based on a well-built science and technology base is needed to eradicate the pollution of surface and groundwater resources and to improve water quality and step up the recycling and reuse of water.

9.4.1 Holistic Intervention for Inland Salinity

Inland salinity is one of the major causes for degradation of soil and reducing vegetation cover in an ecosystem. It is widely prevalent in various parts of the country. It has been also observed that inland water salinity is more prevalent in water-scarce areas where evaporation losses are high. It affects food security and livelihood of the people and livestock alike (Mohan and Routray 2015). These problems are further compounded by cultivation of water-intensive crops which lower the water table further.

A novel programme undertaken at Ausa, Latur in Maharashtra province has demonstrated the potential of innovative approaches to address these critical water challenges. Ausa had a population of 30,876. Males constitute 51% of the population and females 49%. Average rainfall, during the monsoon, is between 600 and 800 mm. Summers, which begins from March to July, are dry and hot. The temperature may vary from 25 to 39.6 °C and can reach 45 °C during peak summer season and from 12 to 20 °C during the winter.

People were dependent on groundwater as the only lake in the region is saline and brackish. Groundwater which is the real lifeline of the region has fallen by more than 3 m in last 5 years. The problem was aggravated due to cultivation of water-intensive crops such as sugarcane which further lowered the water table. The demand scenario of the water indicated acute water shortage for drinking, municipal and agricultural use.

As the overall availability of water was quite less and quality was highly saline, provision of drinking water was possible only through reverse osmosis (RO) technology, which involved significant water loss. In order to reduce the wastage of water, high recovery of water was required. The RO system was integrated with forward osmosis (FO) unit to maximise the recovery of water (about 92%) and simultaneously produce concentrated sludge utilising drying bed. The concentrated sludge, which was around 8%, was converted into salts which would find utility in the vicinity and it did not adversely affect the ecosystem also (Fig. 9.2a, b).

In order to augment the water availability further, an innovative low cost, low maintenance, less energy-intensive, decentralised wastewater management system was also set up. Treated wastewater from the treatment plant is being used for agricultural and horticultural irrigation. As per the water scarcity scenario of Ausa city, the treated wastewater will also be used for general washing, to maintain greenery of Ausa in summer season and also can be a revenue generation model for Ausa Municipal Council by selling this treated water by tankers for roadside plants and construction works.

Fig. 9.2 **a** 4000 LPH RO plant. **b** Integration with FO unit to extract additional water from RO reject. **c** Fountain for lake rejuvenation to aeration of lake water. **d** Training to students and residents

Another important objective of the project is the rejuvenation of the Ausa Lake. As the lake is stagnant and deprived of oxygen, fountains were installed to oxygenate the stagnant water (Fig. 9.2c). Beautification of the lake and setting up of a walking trail are also being pursued simultaneously.

The project is also creating awareness among people regarding the best sanitary practices, methods of reuse and recycling of water (Fig. 9.2d). Similar efforts focussing on production of water together with vast conservation efforts can provide a means to people in other water-deprived locations of India for sustainable livelihood.

In another innovative initiative, community managed sustainable solutions for water-scarce villages of Mewat, established self-sustainable community-managed model to fulfill the domestic water demand through creating freshwater pockets within saline aquifer, for drinking and cooking, developing ponds to collect surface runoff and exploiting collected runoff for other domestic use and creating village institutional mechanism for management, water sharing and operation and maintenance of infrastructure developed under this project. The above examples vividly illustrate that holistic approach for mitigation of water challenges reduces the disaster risk significantly.

9.4.2 Integrated Intervention for Coastal Salinity

India has a fairly long coastline. Many of the towns and habitations falling on this coastline are water-scarce. Some of these areas fall in the rain shadow area and get very scanty rainfall. Several options have been tried to provide sustained supply of drinking water to these areas. However, cost-effective desalination appears to be the real solution to address this scarcity. The challenges are to provide safe drinking water without creating carbon footprint and it's in this area, where renewable energy becomes of paramount importance (Mohan and Routray 2015).

Ramanathapuram district situated in the South-East corner of Tamil Nadu, India, is highly drought-prone. Ramanathapuram District is 4089.57 km^2 and has 3.14% of the geographical area of Tamil Nadu State with a total population of 1,183,321. The district has a rural population of 883,508 as against urban 299,813. Ramanathapuram has a long coastline of approximately 260 km. Majority of the population are farmer and 45% of land is used for farming to produce Rice, Chilli, Groundnut, Coconut etc. The main food crop of the area is Rice.

The rainfall is the main source of water for surface and undergroundwater. The rainfall in this district is 827 mm which is much below the average of the whole state, which is 925 mm, with the rainfall maximum around the coast and decreases towards inland. Due to rainfall being restricted within a short duration, recharge of surface and undergroundwater is less and runoff is more. However, the withdrawal of surface and groundwater is growing enormously every year. This has increased the disaster risk on the water sources like deteriorating water level and worsening of the quality of water. The availability of groundwater is less due to over-extraction than recharge. The area is suffering from scarcity of potable water, due to salinity, brackishness and also poor sources, for many years and the supply of even the minimum 40 lpcd of potable water is a challenge and is severely affecting economic and industrial development of the district.

As the surface water was not adequate and groundwater was not of desired quality, the only option for meeting the drinking water needs is desalination. The innovation was devised to understand the locally available resources. There was sunshine available almost throughout the year and the location was rich in Jolly flora which was a biomass not used for other applications. The technology is developed to produce both energy and water simultaneously to meet the drinking water needs as well as energy needs. The reject brine of the desalination plant was disposed of responsibly in the sea without damaging the aquatic life. Further integration in the project was to develop bio-crude oil for energy needs utilising algal as well as waste (Fig. 9.3). These innovations not only met the energy needs of the local people but also sustained the environment by utilising locally available resources. Since the plant is renewable

Fig. 9.3 Solar-biomass based multi-effect desalination (MED) system for coastal salinity in Narippaiyur, Ramanathapuram, Tamil Nadu

energy-based, it is carbon neutral and does not affect the environment. This intervention is an example of how integrated solutions can be developed based on locally available resources to address sectoral requirements.

Development also increases the energy needs of the people which put stress on locally available resources, making them more vulnerable to disasters. The innovative application of advanced technologies in sectors of water and energy has the potential to reduce the risk of disasters in various ecosystems. Innovations in application of technologies for real-time sensing, communication and data analysis of critical environmental parameters affecting disaster risk of ecosystems are the innovations needed for taking appropriate measures for mitigating these risks. Adoption of holistic and multi-sectoral approaches for innovations based on thorough understanding of disaster risks has facilitated evolution of sustainable solutions.

9.4.3 Integrated Intervention for Sea Water Intrusion

Seawater intrusion is a major challenge in hinterland of coastlines. One of the major reasons for these inclusions is falling water table. Generally, seawater intrusion is coupled with other major challenges such as low per capita availability. The solution to reduce seawater intrusions is raising the water table by improving recharge (Oudejans 2017). However, the immediate requirement needs to be met through appropriate treatment which is environmentally benign.

BhujaBhuja, a village in Nellore District of coastal Andhra Pradesh has a typical tropical maritime climate, with hot, humid summers and mild winters. The hottest months are May and June and coolest months are January and February. The maximum temperature is 36–46 °C in summer and the lowest temperature is 23–25 °C during winter. The Bay of Bengal is at a distance of 24 km. The area receives rainfall during the months of October and December. Cyclones are common in the area during this time causing floods and havoc.

Water in the BhujaBhuja village was contaminated with salinity due to seawater intrusion. It was also affected by hardness, iron and silica. The bad quality of water was causing many health issues like increase in occurrence of gastrointestinal illness, jaundice, dysentery, kidney stones, viral infections and dental fluorosis.

It could be seen that the challenge of the location was the water salinity coupled with water scarcity. This was affecting not only human life but also aquatic life and animal husbandry. The innovation for addressing this challenge understood the demand for drinking water treatment, which was treated to the required quality. The wastewater was treated for irrigation as per agriculture water standards and rejects were utilised for production of manure. The water treatment plant provides water to a population of around 12,000 people for their drinking water needs and also caters to the other domestic needs of 3000 more beneficiaries (Fig. 9.4a).

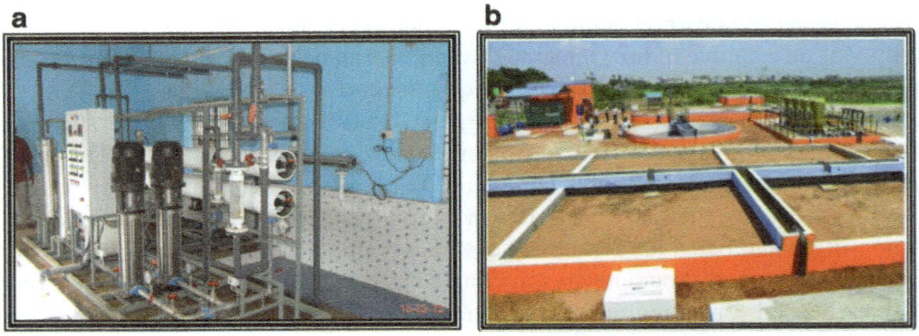

Fig. 9.4 **a** Reverse osmosis plant installed at BhujaBhuja village, Nellore Andhra Pradesh. **b** Wastewater treatment plant installed at BhujaBhuja, Nellore Andhra Pradesh

The reject water after proper treatment was used for irrigation and groundwater recharge. The wastewater ponds were utilised for aquaculture. The biological slurry and manure were for development of village green belt (Fig. 9.4b). This integrated intervention for providing potable drinking water and treating wastewater has improved local ecology and contributed to the economic upliftment of local population.

9.5 Addressing Health-Related Disasters Due to Poor Water Quality

9.5.1 Innovations to Address Arsenic Menace

Besides the contamination of surface water due to pathogenic contamination, contamination due to geogenic contaminants such as fluoride, iron, arsenic is widely prevalent. Emerging contaminants such as pharmaceutical and personal care products (PPCP) and endocrine disruptor compounds are another category of contaminants polluting the water stream. Amongst geogenic contaminants, while fluoride is most prevalent, arsenic is most toxic.

In India, arsenic contamination was first reported in West Bengal in 1983. Now, Assam, Bihar, Jharkhand, Manipur, Uttar Pradesh, West Bengal, Punjab and some parts of Chhattisgarh and Karnataka are reported to have arsenic contamination. In spite of a number of measures, the spread of arsenic contamination in groundwater continues to increase with each new survey and new arsenic affected areas and more people suffering from arsenic-related diseases are being reported.

Drinking water is the main way to intake of arsenic in human system as well as eating food that has been adulterated with arsenic. Symptoms of arsenic poisoning are headaches, severe diarrhoea, confusion, drowsiness, spasm and pigmentation in nails of fingers.

Arsenic groundwater contamination has wide-ranging implications including its consumption through food chain. Persistent presence of arsenic in food chain can lead to food chain transfer. Food chain transfer of contaminants takes place when an animal or plant absorbs contaminants at a rate that is faster than the removal of the contaminants from the body. Thus, even if the level of toxin is not high in environment, the longer the contaminants stay in the body, greater is the risk of chronic poisoning. The food crops grown using arsenic-contaminated water are traded to other non-contaminated areas, where the populaces may ingest arsenic from the contaminated food.

If the surface water could be treated, the problem would have become much simpler. In the long run, this is the sustainable solution. However considering the habitations location and investment needed, this solution is still far off. Further, contamination due to arsenic has also been spreading to various areas which do not have adequate surface water and are dependent on groundwater supply. The challenge for removal of arsenic starts with the detection process which has to be quite sensitive. The treatment process has to use benign materials and disposal of toxic sludge, particularly in decent light systems is a big challenge. Several technologies developed indigenously qualify the criteria of sensing and treatment to desired level. Development of technologies by embedding sludge into immobilised matrix makes these removal technologies environment friendly. Many of these technologies use simple materials such as iron nails and specially treated laterite soil. Nanotechnology has further improved efficiency of materials with higher adsorption capacity and longer life. The disposal of sludge has also become more responsible. The technologies based on these materials are being successfully used in various parts of the country. These innovative technologies are evidence of the potential of applications of science and technology to address challenging tasks successfully.

AMRIT Filter Developed by IIT Madras

AMRIT-Arsenic and Metal Removal by Indian Technology is an affordable, compact, gravity-fed water purification unit which is providing clean drinking water in Yadgiri district of Karnataka, Murshidabad district in West Bengal and in Bihar (Fig. 9.5a). Community filters are also in use in villages across Murshidabad and Nadia districts of West Bengal and Punjab. Cost of treated water is 3 paise/L (Fig. 9.5b). Media can be easily disposed of in the environment or can also be used for brick making as it is prepared with iron oxides and eco-friendly materials.

Fig. 9.5 **a** AMRIT domestic arsenic filter. **b** AMRIT community arsenic filter

Laterite Based Arsenic Filter Developed by IIT Kharagpur

The filter uses naturally plentiful raw laterite customised for higher adsorption capacity by appropriate chemical treatment (acid–alkali treatment). The low cost of the filter media and no power requirement for functioning of the filters makes them ideal for underprivileged people. Arsenic being adsorbed chemically does not leach out and sludge can be used for road laying purposes after its adsorption capacity is exhausted. The cost of the treated water is 3 paise/L. The filter has been approved by Arsenic Task Force, Government of West Bengal. Technology has been transferred to Vas Bros Enterprises Private

Limited and demonstrated as affordable solutions for domestic and community filters in the Arsenic pockets of West Bengal (Fig. 9.6a, b).

Fig. 9.6 **a** Laterite based arsenic domestic filters. **b** Community-scale filter of 500 L/day installed in a primary school in Kashinathpur in Rajarhat, near Kolkata

Arsenic Filter Developed by Indian Institute of Technology—Bombay

These community-scale hand pump attached arsenic removal filters uses Zero-Valent Iron (ZVI). The method is based on corrosion of ZVI. The uniqueness of the project is that it is based on indigenous technology using iron nails and locally available aggregates and can be fabricated by local plumbers and masons. The filters function in the absence of electricity and with direct inlet from hand pump. The cost for producing 1 cubic metre of water for a family of 5persons varies from Rs. 0.10 to Rs. 1. Media can be used for brick making as it is prepared with iron oxides and eco-friendly materials. These units are installed in various parts of Uttar Pradesh, Bihar, West Bengal and Assam (Fig. 9.7).

Fig. 9.7 IIT Bombay developed arsenic removal units

9.5.2 Reducing Disaster Risk in Wetlands

India has 26 wetlands of international significance, with a surface area of above 6 lakh ha, 4.7% of the total geographical area of the nation. Wetlands are important as they carry out crucial ecological functions like moisture maintenance and flood mitigation. Wetland ecosystems function as natural sponges that trap and slowly release surface water, rainwater, groundwater and floodwaters. Hence conserving wetland is very important to diminish the impacts of climate change and also for the livelihood of the poor and marginalised communities.

While wetlands are very important for ecosystem, anthropogenic pollution has contributed to spreading chronic diseases in some of the wetlands. Indicators such as early detection of water clarity, colour and assessment of algal pollution show the proneness of the wetlands to infestation which can cause the epidemic. One simple device for monitoring water quality in lakes has shown the capability to measure these parameters and report them well in time. This and similar other innovations can preserve the health of people and mitigate risks to ecosystems.

Popularly known as the Backwaters, Vembanad Lake is the largest water body in the state of Kerala, a Ramsar site, an Indian cultural icon and has been identified as the Vembanad–Kol Wetland system by India's National Wetlands Conservation Programme. Around 1.6 million people are directly or indirectly depend on lake for agriculture, fishing, tourism, navigation and household activities along its 100 km long stretch.

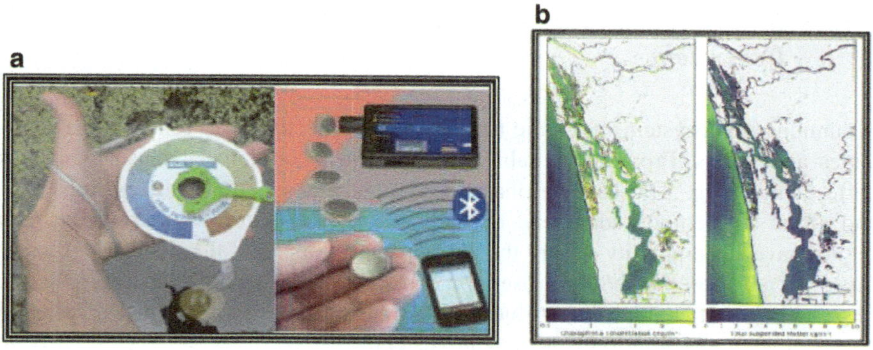

Fig. 9.8 **a** A secchi disk with temperature sensors for distribution to public as part of citizen science. **b** Remote sensing image showing the distribution of chlorophyl and suspended matter in Vembanad lake

This innovative programme assessed the disaster risk and suggested mitigation measures for the reservoirs of Vibrio cholera in the Vembanad lake, the largest estuary which runs parallel to the major axis of Kerala. The lake receives freshwater from 10 rivers of the state. Vibrio cholera, present throughout the lake and its reservoirs are phytoplankton, zooplankton and macro benthos, with the bacteria showing highest association with the phytoplankton in the lake. They are also found in suspension in the water. The management measures include establishing water clinics for continuous monitoring of water quality of the lake, rivers and wells in the area. These water clinics successfully provide information about the condition of wells (turbid or clear, non-usable or usable, contaminated with faecal coliforms or not). A citizen science network gathers information on turbidity and colour of Vembanad lake using a low-cost innovation of 3D printed mini Secchi disc (Fig. 9.8a) and a mobile application (TurbAqua). Turbidity and water colour are two easily identifiable properties of some types of pollution.

It has also been noted that alterations in phytoplankton composition and chlorophyll concentration in response to monsoons and associated seasonal variations in the environmental factors help in identifying vibrio reservoirs and can be used for risk mapping. Such risk maps generated using a combination of in-situ, satellite remote sensing and modelling results will be useful in generating operational advisories to alert the authorities and stakeholders who depend directly or indirectly on water in the lake for routine commerce, agriculture, fisheries, tourism and other livelihood activities.

Parallel to these technical innovations, organising awareness programmes and establishing WASH (Water, Sanitation and Hygiene) facilities in rural areas was also taken up. As WASH facilities tend to break down under floods and other forms of natural disasters, the satellite-based methods for mapping the extent of open water in the Vembanad Lake region are useful in these situations. Such tools help regional authorities to implement spatial planning for evacuation, rescue and relief missions and prioritise regions for immediate and urgent action (Fig. 9.8b).

9.5.3 Innovations to Address Disaster Risk in Sensitive Ecosystem

Mountainous ecosystems are quite fragile and are affected by torrential rains, landslides and floods. Providing potable water to the inhabitants of these regions is quite challenging. The conventional water treatment scheme involves huge cost and affect the ecology adversely. Further, the surface water abstraction schemes do not have sustainability during the extraordinary rains and floods. Natural water treatment on the other hand utilises the property of the aquifers and are capable of providing sustainable water solutions. An innovative experiment in Uttarakhand provided technical solutions for several clusters.

This intervention was based on recognition of the fact that suitable hydrogeological conditions exist in Uttarakhand for River Bank Filtration (RBF). Riverbank filtration is a process of obtaining naturally filtrated groundwater from aquifers that are hydraulically connected to river or lake. During riverbank filtration, surface water is subjected to a combination of physical, chemical and biological processes such as filtration, dilution, adsorption and biodegradation that significantly improves the raw water quality (Dobhal et al. 2011). It was also realised that conventional surface water abstraction techniques while being cost intensive failed to provide sustainable solutions to water challenges prevalent in the area.

Four sites namely Satpuli, Augustmuni, Karanprayag and Srinagar were identified where the demand was not sustainably met with the prevailing arrangements of obsolete river water surface abstraction methodology (Fig. 9.9a, b). The RBF interventions have resulted in final water content of drinking quality at these sites which was impure prior to the process. Dependability on abstraction of water from surface level of sources becomes minimal confirming RBF as a tested process of water treatment. RBF has thus resulted in providing technical solutions at all four locations. The intervention has resulted in increasing per capita water availability from 44 to 103 lpcd. The quality of water has also improved and only disinfection is required as against coagulation/ filtration along with disinfection, required previously.

Fig. 9.9 **a** RBF well at Alaknanda in Srinagar in Uttarakhand. **b** RBF well at Augustmuni in Uttarakhand

9.6 Conclusion

Due to fast population growth, industrialization, climate change, improvements in people's living standards and changing lifestyles there is a rise in water demand and reduction in per capita water availability in India. Due to this increase in demand for water, using water efficiently, developing methods to improve water quality, affordable tools and techniques for water reuse and recycling have become important for augmenting the per capita availability of water.

Understanding and awareness of challenges is the first step to come up with innovative solutions to the problem. Besides data on demand, supply and source sustainability, online remote water quality or quantity monitoring technologies facilitated by existing available technologies of sensors and Internet of Things (IoT) can help in identifying status of water scarcity and quality dynamically. These can help in generation and updation of water quality and quantity maps at relatively smaller intervals. Online water quality monitoring techniques can help in pointing out pollution sources and take remedial action in real time. Empowering the communities to assess water quality at affordable cost can reduce the risk of health-related disasters.

Innovations are needed to develop irrigation practices which help not only in saving huge quantity of water but also helps in increasing crop intensity, irrigated areas and higher crop yields. Huge scope exists in improving the water productivity in agriculture, i.e. "more crop per drop of water" can translate into huge water savings. As agriculture is a major water consumer, bringing irrigation efficiency in this sector has potential to result in major savings. Cost-effective conversion of saline to fresh is the key to water security of coastal areas.

Improved rainwater harvesting practices coupled with technical innovations such as first flush filters and disinfection technologies can be a sustainable approach for augmenting domestic water in water scanty areas. The treatment technologies for potable and wastewater technologies need to not only treat water to levels required as per standards but also meet the criterion of public acceptance.

Innovations are also needed for excess water situations like floods. Potable water is first casualty in these situations. Several innovations such as mobile water purification units have been deployed. However, affordable in-situ purification of floodwater to potable water still eludes us. While there are several such systems commercially available, mass deployment at affordable cost is a challenge. This would significantly bring down water-borne diseases and reduce the health risks.

Technology solutions to water-related problems should meet several criteria specific to the location and community behaviour and are required to be tailored to the socio-economic context. It is, therefore, imperative that solutions are designed for ease of access from the users' perspective, economic viability from solution providers' as well as users' perspective, technical viability, environmental sustainability and social viability.

The innovations in water sector can significantly reduce disaster risk and also increase resilience of community to face disaster-like situations resulting in reduced losses. These innovations will also empower community with better understanding of the challenge and consequences. This, in turn, could bring disaster risk reduction as a major component in projects aimed at addressing water challenges.

Acknowledgements The authors would like to thank Professor Ashutosh Sharma, Secretary, Department of Science and Technology, Government of India for his constant encouragement and support to the innovations in water sector. The authors would also like to acknowledge the support received from principal investigators (PIs) of various projects supported by Technology Mission Division of Department of Science and Technology, Government of India.

References

Bajpai S, Alam N, Biswas P (2019) Evaluating water quality to prevent future disaster. In: Present and potential water quality challenges in India, vol 1, pp 85–112

Central Water Commission (2002) National water policy

Central Water Commission (2012) National water policy

Chattopadhyay S, Prasad, N (2006) Essays on water, 1st edn. IRIS Publication Pvt. Ltd, New Delhi, pp 92–94

Dobhal R, Grischek T, Uniyal HP, Uniyal DP, Sandhu C (eds) Drinking water source, treatment and distribution. Bishen Singh MahendraPal Singh, Dehradun (UK), pp 89–90

Ministry of Science and Technology (2013) Science technology innovation policy

Mohan NS, Routray S (2015) Sharing blue gold: locating water conflicts in India; SP6–2015. National Institute of Advanced Studies, Bangalore (KA), pp 10–11

National Institute of Disaster Management, Mainstreaming Disaster Risk Reduction in Environment Sector, Guidelines and Tools

NITI Aayog (2018) Water index report

NITI Aayog (2019) Composite water management index

Oudejans L (2017) Report on the 2016 U.S. environmental protection agency (EPA) international decontamination research and development conference. U.S. Environmental Protection Agency, Washington, DC, EPA/600/R-17/174

Sankar MU, Aigal S, Chaudhary A, Anshup, Maliyekkal SM, Kumar AA, Chaudhari K, Pradeep T (2013) Biopolymer reinforced synthetic granular nano composites for affordable point-of-use water purification. Proc Natl Acad Sci USA 110:8459–8464

Part II
Science Investigation, Technology and Planning Intervention

Chapter 10
Future Heat Risk in South Asia and the Need for Ecosystem Mitigation

Peter J. Marcotullio and Michael T. Schmeltz

Abstract Cities in South Asia commonly experience high heat events. These so-called heat waves, however, are increasing in intensity and are projected to increase in frequency, as the climate continues to change. Given the large and growing urban population in the region, urban planners need information on the state and trends of urban heat, the risk of this heat to human wellbeing, and ways to modify heat within the city. This chapter attempts to broadly address these issues by providing an overview of the state and trends of urban heat, urban heat-island formation, urbanization, climate change-related future heat, human risk to heat waves, and the potential for ecosystem mitigation in South Asia. The chapter complements other chapters in this volume by providing the background to the importance of future heat shock events and the potential for blue-green infrastructure to address these hazards.

Keywords South Asia · Heat waves · Blue-green infrastructure · Scenarios

10.1 Introduction

Climate scientists have argued that extreme heat events will be increasing in frequency and intensity in locations around the world. Recent studies suggest the number of high-temperature events has already increased over the past decades (Christidis et al. 2015; Rahmstorf and Coumou 2011; Sun et al. 2014), including in urban areas (Matthews et al. 2017; Mishra et al. 2015). A number of heat events have made headlines for their destructive impacts including during 1987 in Athens,

P. J. Marcotullio (✉)
Department of Geography, Hunter College, City University of New York, 695 Park Ave, New York, NY 10065, USA
e-mail: peter.marcotullio@hunter.cuny.edu; pmarcotu@hunter.cuny.edu

M. T. Schmeltz
Department of Public Health, California State University, East Bay, 25800 Carlos Bee Boulevard, CA 94542 Hayward, USA
e-mail: michael.schmeltz@csueastbay.edu

P. J. Marcotullio
Institute for Sustainable Cities At Hunter College, 695 Park Ave, New York, NY 10065, USA

© The Author(s), under exclusive license to Springer Nature Singapore Pte Ltd. 2021
M. Mukherjee and R. Shaw (eds.), *Ecosystem-Based Disaster and Climate Resilience*, Disaster and Risk Research: GADRI Book Series,
https://doi.org/10.1007/978-981-16-4815-1_10

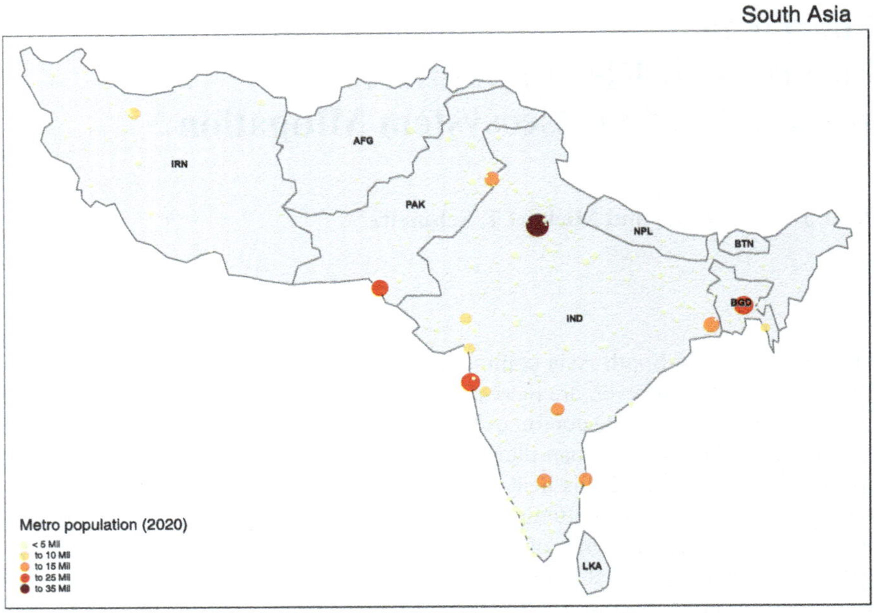

Fig. 10.1 South Asia with large urban centers. Data source: UN (2018), figure created by authors

1995 in Chicago, 2003 in Europe, during 2010 in Russian, and in 2015 in Southern Asia. Epidemiologists have pointed out that much of this devastation occurred in cities where people are concentrated and heat waves are amplified through the urban heat-island (UHI) effect.

Asia, and specifically South Asia, is already an area of high heat and increasingly high-intensity heat events We define the South Asia region as per the UN (2018), which includes Iran, Afghanistan, Pakistan, India, Maldives, Sri Lanka, Nepal, Bhutan, and Bangladesh nations (Fig. 10.1). This chapter provides a review of extreme heat events, UHI, and how these might develop in the South Asia region under climate change. It also reviews the epidemiological literature on the impacts of heat with specific focus on South Asia. The chapter ends with recommendations for future urban development that include blue-green infrastructure.

10.2 Extreme Heat Events and the Urban Heat-Island Effect

Elevated temperatures in our environment are often described in various ways, especially when discussing how temperatures affect human health. Studies often characterize heat-health relationships with high ambient temperatures, extreme heat, or most often, as heat waves. Despite common usage, however, scientists have no common empirical definition or standard measure of a heat wave (Horton et al. 2016, Perkins 2015, Dosio et al. 2018). Qualitatively, heat waves are observed high-temperature extremes over consecutive days at the global, regional, or local scales (Alexander et al. 2006; Della-Marta et al. 2007). Heat wave characteristics (e.g., intensity, frequency ,and duration) and their impacts vary by geographic scale and region (Perkins et al. 2012). As a result, there is no universal agreement on heat wave definitions; they vary substantially across countries and localities, as each country determines thresholds and indices in accordance with the local climate and population vulnerability (Nissan et al. 2017). For example, the India Meteorological Department (IMD) declares a heat wave when weather stations in the plains region reach maximum temperatures of 40 and 30 °C or more for the hilly regions. Severe heat waves are defined when actual maximum temperature ≥ 47 °C.[1] In the USA, a heat wave is defined as "A period of abnormally and uncomfortably hot and unusually humid weather. Typically a heat wave lasts two or more days".[2] In Adelaide, Australia, a heat wave is defined as "five consecutive days at or above 35 °C, or three consecutive days at or over 40 °C.[3]" In Denmark, a national heat wave is defined as a period of at least 3 consecutive days of which period the average maximum temperature across more than fifty percent of the country exceeds 28 °C.[4]

Climate scholars attempt to provide quantitative definitions of heat waves, but they also differ. One definition of a heat wave is at least three days of heat where the first daily maximum temperature is above the 97.5th percentile of temperature distribution and the remaining days are above the 81st percentile of temperature distribution (Meehl and Tebaldi 2004). Taking human vulnerability into consideration, others have defined heat waves using a day-and-night index, which focuses on elevated day- and nighttime temperatures above the 95th percentile for 3 consecutive days (Nissan et al. 2017).

High temperatures, however, alone make up only a portion of the impact of extreme heat. Among a variety of weather conditions that affect human comfort (radiation, wind, etc.,) a critical component of heat waves is humidity. Together these weather aspects define how ambient conditions "feel", sometimes called "apparent temperature", and arguably provide a better indicator of human comfort than either temperature or humidity alone (Epstein and Moran 2006). Combining these two factors creates a heat index. The most common technique used in heat index research today is a variant originally proposed by Steadman (1979a, b) with subsequent refinement (Anderson et al. 2013). The US National Oceanographic and Atmospheric Administration heat index is based on this model.

---

[1] See http://internal.imd.gov.in/pages/heatwave_mausam.php.

[2] See https://www.weather.gov/glossary/index.php?word=heat+wave.

[3] See http://www.bom.gov.au/announcements/media_releases/sa/20100115_First_Heatwave_SA_Jan.shtml.

[4] See https://web.archive.org/web/20080723170544/dmi.dk/dmi/danmark_faar_varme-_og_hedeboelge.

For cities, extreme heat events concern is exacerbated by the higher temperatures in the urban core as compared to surrounding areas. This is called the urban heat island (UHI) which provides a clear example of anthropogenic impacts on climate. It has been known for centuries, as it was identified 200 years ago by Howard (1818) and has subsequently been found in cities around the world (Kataoka et al. 2009). Urban climate analysts consider UHI one of the major environmental problems of the Twenty-first Century (McKendry 2003; Rizwan et al. 2008; Arnfield 2003).

Over recent years, four different types of UHI have been defined, including subsurface UHI, surface UHI, canopy layer UHI, and boundary layer UHI. These different UHIs relate to differences between the urban and rural temperatures, but each is measured at different altitudes and with different techniques. The subsurface UHI is under the city and measured in the soil and subterranean built fabric. The surface UHI is measured at the interface of the outdoor atmosphere with the solid materials of the city or rural areas and often through satellite or other remote sensors and thermal cameras (Small et al. 2018). The canopy UHI is measured between the surface and the tree canopy or below the average building height of the city with stationary sensors or those mounted on vehicles. The boundary UHI is measured above the canopy layer, which is similar to the atmospheric boundary layer (Oke et al. 2017), using sensors on balloons, helicopters, or other airborne devices.

Research suggests that these UHIs also differ in intensity. For example, surface UHI can be an order of magnitude higher than canopy UHI (U.S. EPA 2008). These temperatures relate to the energy budgets defined by surface properties and structure (roof, canyon, walls, and floor). For example, roof surface temperatures can reach or even exceed 60 °C, but these are not the temperatures experienced by residents. For studies interested in human vulnerability, however, canopy UHI, which is typically measured at 1.5–2 m above the ground, maybe the most appropriate indicator among the different UHI types for identifying heat risk. This UHI is taken at the most appropriate geographic scale (i.e., the immediate surrounding area of an instrument) and is the most relevant temperature experienced by people (i.e., the temperature that people feel). We use canopy UHI (combined with a humidity value when appropriate) and henceforth-referenced UHI refers to canopy UHI.

Abundant research on UHI suggests that the details of intensities differ among cities although commonalities have emerged. UHI varies with time of day, month and year. While ideal conditions are clear (cloudless) skies and weak winds over the course of the day, UHI emerges under other conditions. In many cases, the highest UHI occurs during the summer or warmer seasons (Erell and Williamson 2007; Makrogiannis et al. 1998; Wang and Hu 2006), while other studies find it is highest during the winter or cooler months (Hinkel et al. 2003; Salvati et al. 2017). In tropical areas, the UHI can be experienced during the monsoon season (Chow and Roth 2006) or during the dryer months (Khan et al. 2019). In most cases, maximum UHI intensities are experienced during the evenings (Arnfield 2003), as urban areas fail to cool as rapidly as rural areas. Changes in daily UHI are driven by differential rates of urban warming and cooling in urban and non-urban areas. On the other hand, there are studies that record maximum UHI during the day (Oguntoyinbo 1984). There are even examples, during certain times of the day when UHI is reversed and urban areas are cooler than surrounding rural and suburban areas, sometimes called urban cool island effects (Yang et al. 2017). In general, UHIs are typically 3–4 °C (Oke 1997; Voogt 2002), but can vary between 0.4 and 12 °C (Santamouris 2015). Some have argued that the 12 °C maximum maybe a self-limiting characteristic of cities (Oke et al. 2017).

Most UHI studies have been undertaken in cities in the West (North America and Europe) (Arnfield 2003), but increasingly UHI has been studied in sub-tropically and tropically located cities (Roth 2007). Recent reviews have identified a significant number of studies in Asia (Roth 2007; Santamouris 2015; Kotharkar et al. 2018; Tzavali et al. 2015).

Factors driving UHI are complex. UHI is controlled by air temperature modifiers at the local level including, inter alia, street geometry, building fabric, and anthropogenic activity (i.e., vehicle traffic and space heating/cooling), as well as the soil moisture and plants in the rural sites. Urban climatologists characterize the physical aspects of UHI in two important indices: the canyon aspect ratio (the ratio of the height to width of buildings) and the sky view factor (a dimensionless value defined by the ratio of the radiation received by a surface to the radiation emitted by the surface to the atmosphere affected by the proportion of the visible sky or open canyon space in urban streets). Typically, lower sky view values and higher canyon aspect ratios are associated with higher UHI intensities. The unique nature of built forms and surrounding areas and the variety of urban functions may help to explain the variety of UHI intensity results.

Importantly, all urban settlements demonstrate "cliff, plateau, hill, valley, and peak" (Oke et al. 2017) temperature features in cross-section with rural areas. Smaller cities resemble miniature versions of large cities. Mega cities demonstrate more extreme distinctions. This characteristic is found across the world despite differences in macroclimate and culture (Oke et al. 2017). Furthermore, scholars have demonstrated that the absolute values of the UHI in a city relates to the size of a city, measured in population, city area, or diameter (Oke 1973). Several studies have noted the significant increase of UHI with urban size (Oke 1973; Roth 2007; Santamouris 2015).

10.3 Extreme Heat and UHI in South Asia

Extreme heat events are important in Asia. Observers have singled out the region as currently experiencing increasing high ambient temperatures for extended periods of time (Dash and Mamgain 2011; Panda et al. 2017; Rohini et al. 2016). Since 1980, there have been over 16,800 deaths attributed to extreme heat in Asia. Approximately 80% of these deaths occurred in South Asia. As Fig. 10.2 demonstrates, since 1980, extreme heats impacts in South Asia are common and increasing in intensity. These devasting events include the so-called, 1998, 2002–2003, and 2015 heat waves affecting India, Pakistan, and Bangladesh.

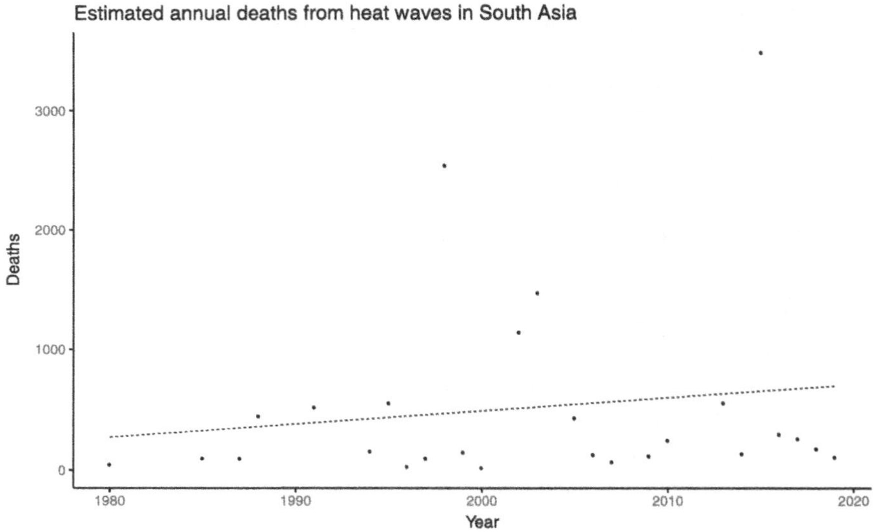

Fig. 10.2 Heat wave impact in South Asia since 1980. Data source: EM-DAT, CRED / UCLouvain, Brussels, Belgium, www.emdat.be (D. Guha-Sapir)

Moreover, UHI is evident in Asia's cities (Santamouris 2015). We examined over 36 UHI studies performed in Asia cities that identified the phenomena during the evening. We limited the study to those performed with stationary instruments and examined maximum differences in rural and urban UHI intensities. There are uncertainties in this analysis. For example Stewart (2011) has pointed out that UHI studies are often performed differently. Our findings confirm that UHI rises with the population size of the Asian cities (Fig. 10.3). Among the 36 cities examined in our database, the average UHI for megacities was approximately 6.6 °C. These values dropped for smaller cities to an average of 3.3 °C (Table 10.1).

The UHI values are significant in that they increase the nighttime temperatures, which can increase the daily minimum (and mean) temperature. The increase in heat during the nighttime hours also can reduce the ability for people to recover from extreme heat of the day. As population in the region is expected to increase so is the size of urban agglomerations, hence affecting UHI. For a sense of what cities in the future might look like, we examine projections of total and urban populations in the region in the next section.

10.4 Future Urbanization in South Asia

There have been several attempts to model future urbanization (Angel et al. 2010; Güneralp et al. 2017; Hoornweg and Pope 2014, 2016; Seto et al. 2012). These studies

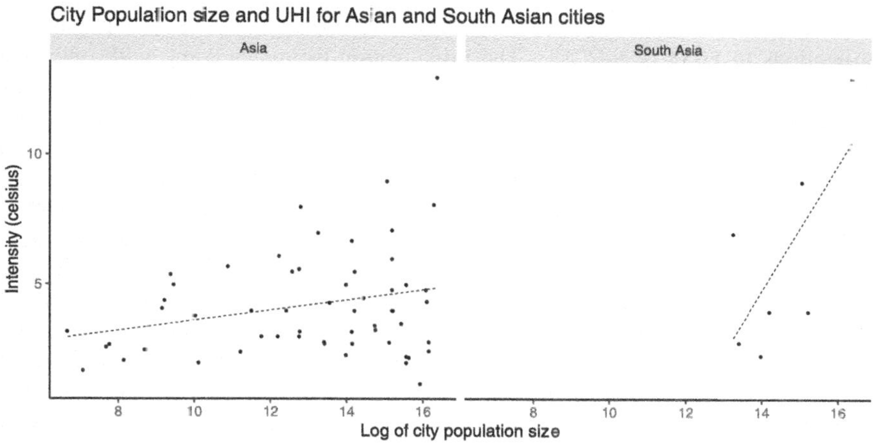

Fig. 10.3 UHI intensity (degrees celsius) in Asian and South Asian cities by population size. Data source: literature review, figure by the authors

Table 10.1 Mean UHI intensities by city size for urban areas in Asia

City size	Intensity	Mean	Count
Category	(°C)	Population	(*n*)
Very small	3.29	9000	12
Small	4.03	145,000	7
Medium	5.06	340,000	5
Large	4.21	675,000	4
Very Large	4.14	4,000,000	25
Mega city	6.59	11,500,000	4

Source Literature review

are often based upon urban growth models and one of two datasets of population projections; the UN urbanization and total population projections (UN 2017, 2018) and the new Shared Socioeconomic Pathways (SSPs) (Jiang and O'Neill 2017; Samir and Lutz 2017; O'Neill et al. 2014). In this section, we examine and compare selected modeling attempts in terms of both urban population and growth of megacities in South Asia.

The United Nations Department of Economic and Social Affairs has engaged in projecting urban populations since the late 1980s and the total population projections have been published since the early 1950s. These datasets are currently one of the most authoritative sources of international projections, are updated every two years, and include a comprehensive set of demographic data and indicators for population and urbanization monitoring. Among the UN total population projections are the low, medium, and high variants, which project national population annually to 2100 (UN 2017). The urbanization projections include urban population, urban growth rates, and urbanization levels to 2050 and selected urban agglomeration population projections to 2035 (UN 2018). The urbanization estimates come with only one version.

According to UN data, total world population projections for 2100 range from approximately 7.2–16.4 billion with a medium estimate of 11.1 billion. These data suggest that the global urban population will reach 6.68 billion by 2050. To obtain urban population figures for 2050 and 2100, we project the UN urban population using low, medium, and high variants of the total population and urbanization projections based upon the average increase in urbanization level found for each nation by urbanization decile of cities within their borders. We then apply the percent change of urbanization level by countries based upon their projected population projections.

A new set of projections attempt to assess future mitigation and adaptation, called the Shared Socioeconomic Pathways (SSPs). These projections present plausible alternative trends in the evolution of society and natural systems over the twenty-first century at the regional and global levels through narrative descriptions and quantifications of socioeconomic variables (such as population growth, economic development, and the rate of technology change) (Ebi et al. 2014). The SSPs were created to address questions of climate change, but are "reference" pathways and assume no climate change or climate impacts, and no new climate policies (Riahia et al. 2017). The differences between SSPs are development distinctions. Each pathway poses different global development trajectories ending up in differential mitigation and adaptation challenges. The five SSPs are named, from 1 to 5: "Taking the green road", "Middle of the road", "A rocky road", "A road divided" and "Taking the highway," respectively. SSP 1 is the sustainable development pathway and has the lowest mitigation and adaptation challenges. SSP 5 has fewer challenges to adaptation to climate change, although poses significant challenges for climate mitigation (higher energy use and subsequent fossil fuel combustion). For SSP 4, the world has lower mitigation challenges although there are great challenges to adaptation. For SSP 3 the world experiences both adaptation and mitigation challenges. SSP 2 is in the middle pathway, between SSP1 and SSP3. From these SSPs, we use total population and urbanization projections to identify total and urban population for 10-year periods from 2010 to 2100 (Jiang and O'Neill 2017; Samir and Lutz 2017).[5]

A comparison of South Asian total and urban population projections between those of the modified UN levels and the SSPs demonstrates different possible futures (Fig. 10.4). For example, by 2100, the lowest projection suggests the total population in South Asia is approximately 1.3 billion (UN low variant) and the highest projection exceeds 3.5 billion (SSP 3, or the worst-case scenario from the Shared Socioeconomic Pathways analysis). All other projections, including the high variant from the UN, are all within this range (approximately 2.2 billion from lowest to highest). By 2100, for the urban population, the lowest projection is again the low variant from the UN at slightly more than 1 billion. The highest projection is the high variant from the UN, which suggests the urban population in South Asia will reach approximately

[5] Data was obtained from SSP Database (Shared Socioeconomic Pathways)—Version 1.1 from: https://tntcat.iiasa.ac.at/SspDb/dsd?Action=htmlpage&page=about.

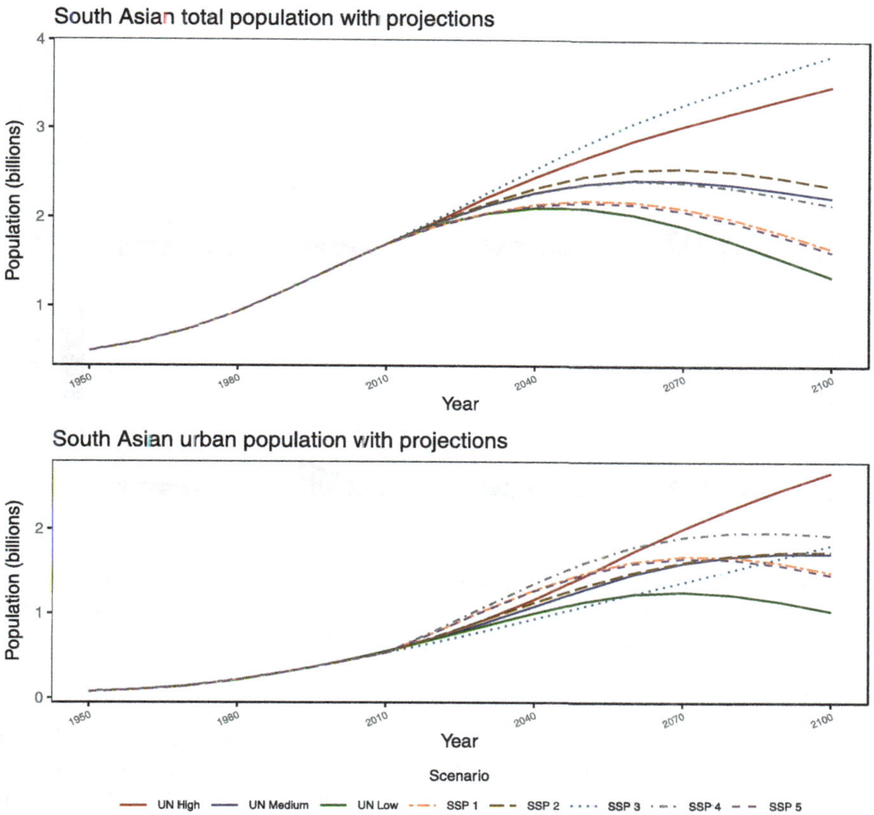

Fig. 10.4 Historical and future projected urban and total populations for South Asia. Data sources: UN (2017, 2018). The SSP database is hosted by the IIASA Energy Program at https://tntcat.iiasa.ac.at/SspDb. The underlying scientific data was published in Riahi et al. (2017). Figure created by authors

2.7 billion, or 57% higher than the current total population in the region. The range in the 2100 values among all projections is over 1.6 billion from lowest to highest values. Much of the range in values for all scenarios depends upon the population growth and urbanization rates in India and Pakistan (Fig. 10.5).

Certainly, this set of projections describe a large range of potential population futures for the region. However, what is interesting is that all projections suggests increases in the urban population from current conditions. These increases vary between 90 million and 2.1 billion, with a mean of over 1 billion. Arguably, the best estimate between these projections suggest that a billion people will be added to the region's cities. What might that mean for urban development and urban heat (UHI)?

Among a variety of potential outcomes is the increasing size of cities in the region. Southern Asia is already the location of a significant share of the world's megacities. According to the UN (2018) most of the world's largest cities and most

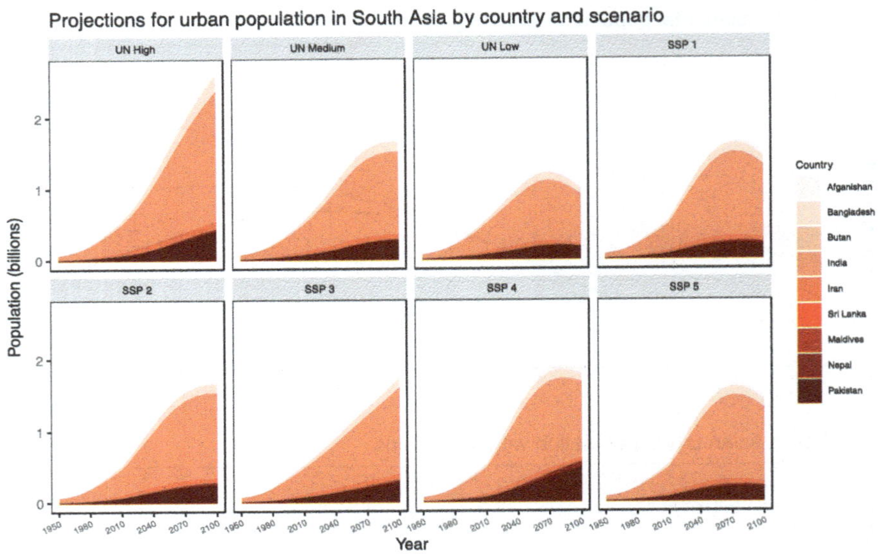

Fig. 10.5 Urban population and projections for South Asian nations by scenario. Data sources: UN (2017, 2018). The SSP database is hosted by the IIASA Energy Program at https://tntcat.iiasa.ac.at/SspDb. The underlying scientific data was published in Riahi et al. (2017). Figure created by the authors

of the world's population living in large cities is located in Asia (Fig. 10.6). Within Asia, the Southern sub-region has a rapidly increasing number of megacities with a large and growing population (Fig. 10.7). By 2035, the UN projects that 244 million people will be living in megacities (those greater than 10 million) in Southern Asia. If these 12 cities grow at the same rate as the region, using the best estimate projection, by 2100, they will hold over 385 million people. The average population size between them is 32 million.

Other analysts have also identified the emergence of these extremely large cities in the future. For example, Hoornweg and Pope (2014, 2016), using both the UN urban projections and the SSPs (1–3), estimate that between 45 (SSP 3) and 60 (SSP 1) cities will exceed 10 million by 2050 with the largest city growing to between 37 and 54 million (SSP 3) (Hoornweg and Pope 2014). These analysts project that by 2100 the largest city in the world might exceed 100 million (SSP 3) and estimate that by that time, between 2 and 10 cities will exceed 50 million. Other observers have also prognosticated that many cities around the world could reach very large sizes, perhaps 100 million or more, given the potential increase in the total population and the tendency towards urbanization (Ridgwell 2018; Vidal 2018).

Among the cities that Hoornweg and Pope (2014, 2016) project by 2100 to reach enormous sizes (between 25 and 58 million) are Karachi, Mumbai, Dhaka, Delhi, Kolkata, and Lahore. If these cities develop in ways typical of contemporary urban development, which includes large expanses of built up area without adequate green

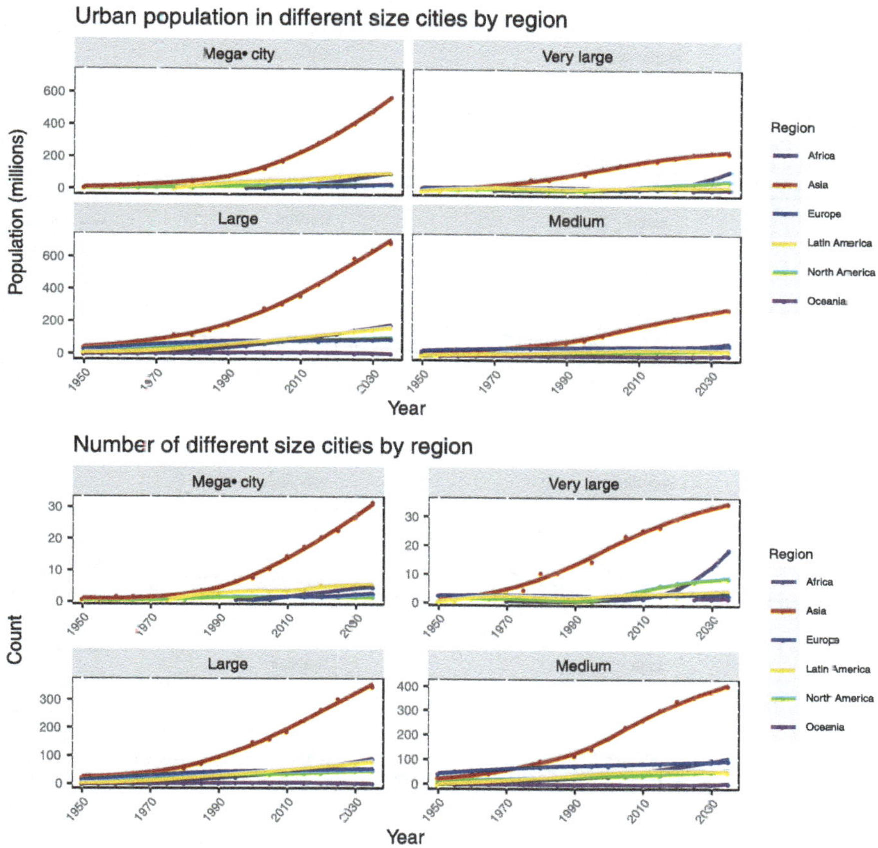

Fig. 10.6 Regional comparison of growth of urban population and cities by size category. Data Source: UN (2018), figures by authors

space and natural watercourses, residents will suffer from increasing UHI, perhaps up to 12 °C. With increasing climate change and without mitigation, people living in these cities could be highly affected by extreme heat events (see below). In the next section, we discuss potential scenarios for climate change impacts in South Asia focusing on increasing heat.

10.5 Climate Change and Extreme Heat in South Asia

Under climate change, the Asian region is projected to experience significant increases in future extreme heat events (Mishra et al. 2015, 2017, Mora et al. 2017; Pal and Eltahir 2015; Panda et al. 2017). Of particular concern is South Asian arid

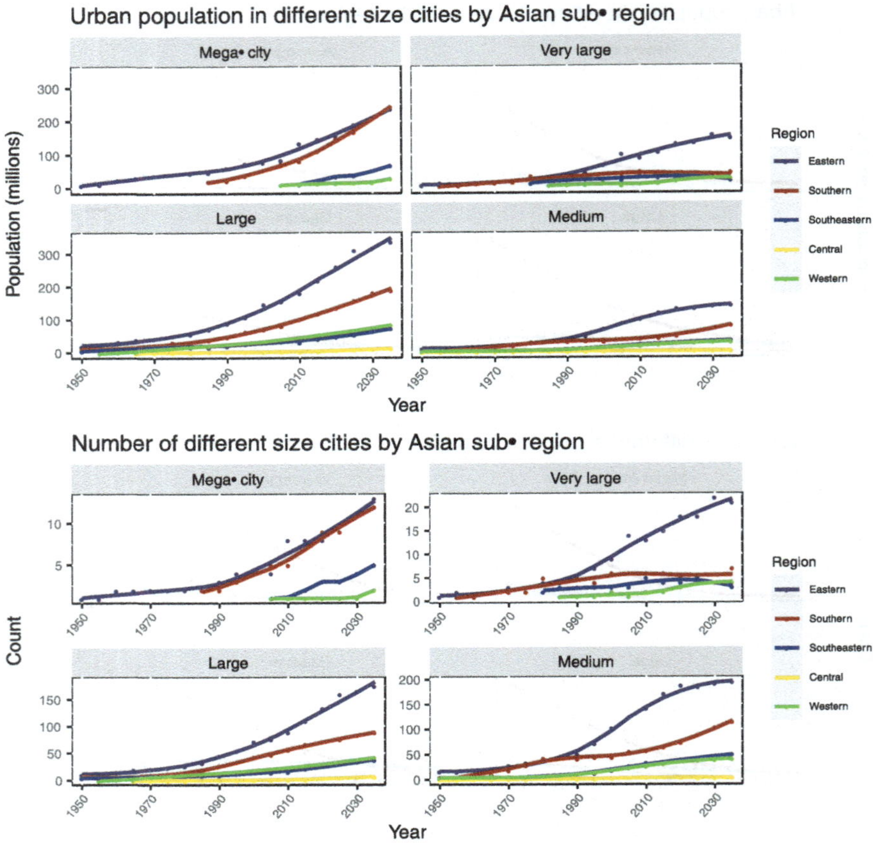

Fig. 10.7 Comparison of urban population and city growth by Asian sub-region and city size category. Data Source: UN (2018), figures by the authors

and semi-arid areas, which are frequently exposed to strong heat (Masood et al. 2015; UN ESCAP 2015). We provide a glimpse of the potential extent and intensity of future heat.

This example uses the HadGEM2, a coupled Earth System Model produced by the Met Office Hadley Center and used in the Inter-sectoral Impact Model Intercomparison Project (ISIMIP) ensemble (Warszawski et al. 2014). From daily temperature and humidity data, we estimate the strongest 5-day heat events for the period of 2070–2099 for RCPs 2.6, 4.5, 6.0, and 8.5. The results demonstrate that with increasing climate change, South Asia becomes covered by plausible extreme heat events of very high intensity (> 45 °C and even > 50 °C) (Fig. 10.8). Even under RCP 2.6 heat intensity of large magnitude is expected to cover many of the large cities in the region.

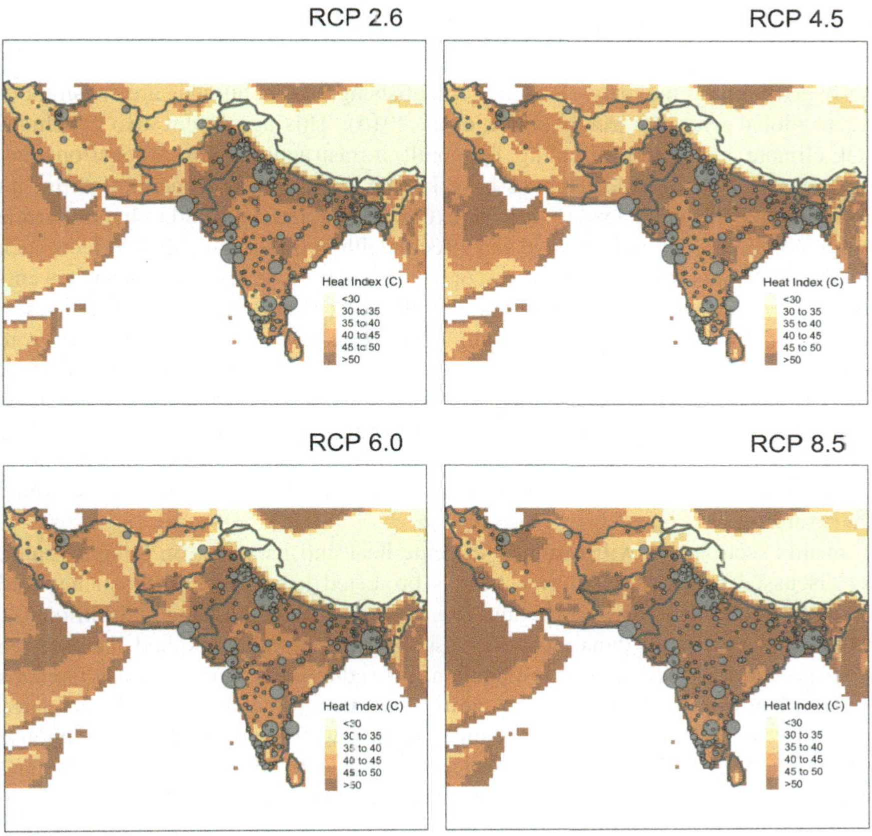

Fig. 10.8 Projected 5-day extreme heat event intensities for South Asia for the period 2070–2099 by RCP. Data sources: UN (2018) and the underlying RCP data from ISIMIP-6 (Warszawaki, et al. 2014). Figure by the authors

This analysis complements that of (Mora et al. 2017) where a "deadly threshold" of heat was identified through the examination of over 700 studies of heat mortality. Their analysis suggests that much of land between 10° N and 10° S will experience daily heat exceeding the deadly threshold into the future, under RCP 8.5 conditions.

Moreover, these data do not include UHI. Exactly how global climate change will affect UHI and the impacts of UHI is an emerging topic of research, but much remains unknown (Huebler et al. 2007; Roy et al. 2011). Consensus among researchers suggest that climate change will affect UHI (Shastri et al. 2017; Tran et al. 2006; US EPA 2018). If we add this increase in temperature to the results of this study, the intensities will increase dramatically. How this might affect the urban population's health is discussed in the next section.

10.6 Climate Change and Human Health

An overwhelming majority of climate scientists agree that humans are causing our recent global warming trends (Cook et al. 2016). This agreement also concludes that climate change is affecting our built infrastructure, natural environment, economy, and our health. Humans have known weather and climate affect health for millennia. Droughts lead to famine, extreme heat, and cold lead to hyperthermia and hypothermia. Climate change can affect health in direct immediate effects such as extreme temperatures and wildfires, or have more long-term effects through events such as drought and flooding (Frumkin et al. 2008; Haines et al. 2006). To understand how climate change affects health researchers rely on empirical studies from past and current observations to examine associations between climate and disease. This evidence is used to help project the potential future impacts climate will have on health. As simple as this may sound, climate impacts on health are modified by social, economic, geographic, and other factors. For example, the built and natural environments can enhance, or protect us against the negative effects of extreme temperatures.

In this section, we will examine extreme heat and its effect on human health. As discussed previously, extreme heat is projected to increase in frequency and intensity due to global warming and the effects of climate change. Additionally, changes in population dynamics, decreases in rural populations, and increases in urban populations, will present new challenges to addressing extreme heat as it relates to the urban heat-island effect. Our ability to reduce the incidence of heat-related morbidity and mortality will be determined by how well we can mitigate greenhouse gas emissions and adapt to rising global temperatures. In response, cities, states, and other municipalities are exploring investments in blue-green infrastructure to lessen the heat-related risks to urban populations.

10.7 Extreme Heat and Human Health

Heat-related morbidity and mortality include direct heat illnesses and illnesses or conditions that maybe exacerbated by exposure to extreme heat, such as cardiovascular disease. Physiologically heat morbidity is attributed to the elevation of core body temperature due to high ambient temperatures in the environment and the inability of the body to dissipate heat and cool down.

Heat illnesses are a spectrum of disorders due to exposure to heat, usually high ambient temperatures in the environment. These illnesses range in severity and can include minor conditions such as heat cramps or heat rash. Heat illnesses can advance in severity from these minor conditions to serious conditions, like heat exhaustion, or the most severe condition, heat stroke. Heat stroke, and other heat illnesses, are often clinically described as exertional, affecting individuals involved in strenuous physical activity in a hot environment, or non-exertional (classical) which affect children, elderly, or individuals with underlying medical conditions by just being in a hot environment. Heat stroke is defined as having a core body temperature greater than 40.6 °C with central nervous system dysfunction that can quickly lead to multi-organ failure with permanent disability or death (Bouchama and Knochel 2002; Epstein and Yanovich 2019).

In addition to direct heat illnesses, researchers and clinicians include underlying medical conditions, which are exacerbated by exposure to extreme heat in their assessment of heat-related illnesses. As temperatures increase, our bodies are stressed as it tries to cool itself, increasing cardiac output and vasoconstriction. This may increase the incidence of severity among individuals with kidney disorders, respiratory illnesses, cardiovascular diseases, diabetes, and other conditions, requiring medical intervention during extreme heat events (Hausfater et al. 2010).

These heat-related illnesses occur when our body's cooling mechanisms are unable to dissipate heat. These mechanisms include conduction, through direct contact with cooler objects, convection, when cooler air passes over exposed skin, radiation, which is the release of heat directly into the environment, and evaporation, through sweating (Miners 2010). With the future climate projections of higher temperatures and higher humidity in urban environments, particularly tropical and sub-tropic climates in South Asia, it becomes more difficult for the body to cool itself resulting in heat-related illnesses.

In the United States, extreme heat events during the 1990's lead to increased research and examination of the risk factors associated with heat-related morbidity and mortality. After an excess of 700 deaths in Chicago in 1995, public health officials initially identified risk factors such as cardiovascular disease, individuals unable to care for themselves, living alone, and lack of air conditioning as being associated with heat-related mortality. In the years after the 1995 Chicago heat event, researchers examining extreme heat events came to identify additional risk factors such as age (elderly and children), individuals living in poverty, those with mental illnesses, and race/ethnicity, among others (McGeehin and Mirabelli 2001; Semenza et al. 1996; Naughton et al. 1999; Bouchama et al. 2007). The research and evidence on risk factors associated with heat-related morbidity and mortality continue to grow and help to informed public health and other government officials on how to protect vulnerable populations and reduce the risk of negative health outcomes.

10.8 Heat-Related Morbidity and Mortality in South Asia

There have been a number of notable extreme heat events, for example, the European heat wave of 2003 captured global attention with an initial estimation of 35,000 heat-related deaths. Investigations later estimated heat-related mortality as high as 70,000 (Robine et al. 2008). There have also been notable extreme heat events in South Asia, with India experiencing heat waves in 1998, 2002, and 2015, and in Pakistan in 2015 (Salve et al. 2018). More recently, in 2019 a heat wave in Pakistan and Northern India brought temperatures above 50 °C. While heat waves are notable events due to their length and high maximum temperatures, epidemiological evidence has shown that high ambient temperatures that occur through the summer months play a significant role in heat-related mortality and morbidity.

In the review of the evidence of heat-related morbidity and mortality in South Asia, it has been noted that the majority of research on the temperature effects on human health have come from high-income countries in the mid-latitudes (Europe and North America), while fewer research studies on heat-related morbidity and mortality have been conducted in low- to middle-income tropical and sub-tropic regions. While studies from the West have identified certain risk factors, vulnerability is determined by the exposure, sensitivity, and adaptive capacity of specific populations. South Asia is the most densely populated region in the world and home to over one-fifth of the global population. A significant portion of the economy is based on agriculture, though urbanization is increasing rapidly. South Asia is also home to a majority of the world's population that are considered poor (Sivakumar 2010). All of this is to say that vulnerability is expressed differently in different regions. Given South Asia's unique demographics, geography, and climate, it is important to understand how these populations maybe vulnerable to extreme heat.

Table 10.2 presents an overview of epidemiological studies that have identified temperature-related morbidity or mortality relationships in South Asian populations. Of the fourteen studies reviewed, nine were from India, four from Bangladesh, and one from Nepal. As many of the authors in these studies, and other studies on this topic have mentioned, there is a significant gap in epidemiological evidence examining the effect of high ambient temperatures and heat waves on human health in South Asia. Much of the evidence that we obtain on temperature relationships to health come from reports and counts of deaths during extreme heat events. There is no universal definition of what constitutes a heat wave or extreme heat event and thus mortality attributed to temperature will differ by country and region. Some of the first evidence of risk factors associated with heat morbidity come from case reports of hospitalizations and deaths during extreme heat events. For example, Saleem et al. (2017) reviewed case reports of deaths during the 2015 heat wave in Karachi, Pakistan, and found that the majority of deaths were among males, ages greater than 50 years, and among homeless individuals and those living with limited electricity and water (Saleem et al. 2017). These types of initial descriptive reviews can identify specific risk factors associated with temperature and mortality, which gives public health officials insight into identifying vulnerable populations who will be at risk for future extreme heat events.

Initially, data from South Asia regarding population risk to extreme temperatures came from these sources, mortality counts, and case reviews. However, a more comprehensive examination of risk factors and vulnerability is needed to assess heat-related morbidity and mortality. Studies from high-income countries use meteorological data and human health data to assess this relationship. These key factors relate to access of data on temperature and health records. The dearth in epidemiological evidence in South Asia maybe related to the inability to collect complete and accurate health records. This was evident in a number of the studies reviewed with authors citing limited information and availability of health records in many study

Table 10.2 Review of studies that show epidemiological evidence of temperature-mortality and morbidity relationships in South Asia

Authors	Year	Location	Temperature threshold	Outcome	Findings
Hajat et al. (2005)	2005	Delhi, India	20 °C	All-cause mortality	For every 1 °C increase in temperature above threshold, there was approximately a 5% increase in all-cause mortality Mortality in Delhi identified 'other causes' (includes infectious disease) as the main cause of death Children under the age of 15 accounted for 48% of all deaths
McMichael et al. (2008)	2008	Delhi, India	29 °C	All-cause mortality, cardiorespiratory, and non-cardiorespiratory mortality	For every 1 °C increase in temperature above threshold, there was approximately a 4% increase in all-cause mortality The threshold for cardiorespiratory-related deaths was 17 °C while for non-cardiorespiratory deaths it was 30 °C
Hashizume et al. (2009)	2009	Matlab, Bangladesh	30 °C	All-cause mortality, and various cause-specific mortality	While there was no clear high-temperature effect on all-cause mortality, heat effects were seen among cardiovascular mortality and deaths among the elderly (65 + years) at temperature thresholds of 30 °C. Additionally, there was an increase in infectious disease deaths as average temperature increased, though estimates were uncertain

(continued)

Table 10.2 (continued)

Authors	Year	Location	Temperature threshold	Outcome	Findings
Burkart et al. (2011)	2011	Bangladesh	Various (28–33 °C)	All-cause mortality	Mortality in rural areas increased up to 5%, and up to 15% in urban areas, for every 1 °C increase in temperature above 29 °C. Cardiovascular heat-related mortality was more pronounced in urban areas, compared to rural areas
Azhar et al. (2014)	2014	Ahmedabad, India	N/A	All-cause mortality	Authors compared all-cause mortality from one year before and one year after the 2010 heat wave to estimate excess mortality. There was a 43.1% increase in all-cause mortality above the reference period for the 2010 heat wave
Burkart et al. (2014)	2014	Bangladesh	34–35 °C (based on UTCI)	All-cause mortality (excluding accidental and maternity-related deaths)	Study found that all-cause mortality increased by 31.3% per 1 °C increase above threshold. In the study, heat effects were strongest in the elderly, males, and those living in urban and high SES areas
Dholakia et al. (2015)	2015	Ahmedabad, Mumbai, Hyderabad, Bangalore, and Shimla—India	Various (20–36.5 °C)	All-cause mortality	Using temperature-related mortality relationships from 2000 to 2009, the study estimated future temperature-related mortality to increase by an average of 140% compared to reference period

(continued)

Table 10.2 (continued)

Authors	Year	Location	Temperature threshold	Outcome	Findings
Fu et al. (2018)	2018	India	34.2 °C (moderately hot) 39.7 °C (extremely hot)	All-cause mortality	The study identified a 9% increase in mortality for every 1 °C temperature increase from 35 to 40 °C Additionally, deaths attributed to extremely hot temperature threshold occurred in all ages 30 + years and among cause-specific mortality for stroke (in ages 30–69 years)
Ingole et al. (2017)	2017	Vadu, India	39 °C	All-cause mortality	Study indicated that a 33% increase in heat-related mortality was seen above the defined temperature threshold. Heat-related deaths were specifically increased in non-infectious diseases, men, and among 12–59 year olds
Rathi et al. (2017)	2017	Surat, India	40 °C	All-cause mortality	Study results state that a 20% increase in all-cause mortality was seen at and above threshold temperatures. Portions of the city that experienced more temperature-related deaths included high-density zones; being further distant from sea; and low SES communities

(continued)

Table 10.2 (continued)

Authors	Year	Location	Temperature threshold	Outcome	Findings
Shrestha et al. (2016)	2016	Nepal	20 °C	Hospitalizations (waterborne diseases, vector-borne diseases, urinary disease, heart disease) and all-cause mortality	Based on different climate regions (Mountain, Hill, and Terari (Plains)); hospitalizations for heart diseases and all-cause mortality ranged from 0.9 to 8.2% per 1 °C increase in average temperature above the threshold
Ingole et al. (2017)	2017	Vadu, India	31 °C	All-cause mortality	Results indicate that there was a 48% increased risk of dying for each 1 °C increase in temperature above the threshold Sociodemographic factors that increased the risk of dying, at or above 31 °C, included women, farmers, and those with low education attainment
Babalola et al. (2018)	2018	Matlab, Bangladesh	25.71 °C (mean) and 33.15 °C (max)	All-cause infant mortality	There was a protective association between monthly temperature and mortality For every 1 °C increase in monthly mean temperature, there was a reduction of 3.7 under 5 deaths per 1000. Similar results were found for the monthly maximum temperature threshold
Dutta et al. (2020)	2020	Bhubaneswar, India	36.2 and 40.5 °C	All-cause mortality	Every 1 °C increase in temperature above the lower threshold mortality risk increased by 2%; for the higher threshold mortality increased by 6%

locations. While there are still limitations in obtaining data, this did improve allowing researchers and public health officials to increase the use of epidemiological studies to assess risk factors associated with heat-related morbidity and mortality.

By 2014, a handful of studies from Bangladesh and India started to quantify the relationship between mortality and temperature. From Table 10.2, we see that temperature thresholds examining heat-related mortality vary significantly by region. As stated previously, temperature-related vulnerability will be expressed differently in different populations and geographic regions, allowing researchers and public health officials to identify the specific temperature-health relationships among local populations. From these studies, we start to get a sense of what risk factors are associated with heat-related morbidity and mortality in South Asia. Some of the risk factors are the same, like those from Western high-income countries, including advanced age and underlying cardiorespiratory issues. However, some specific regional risk factors show up such as infectious disease (including waterborne and vector-borne diseases), urban living conditions, and certain occupations. For example, Rathi et al. (2017), in their examination of temperature-related health effects in Surat, India, identify urban zones further away from the sea and recent migrants to the city as key risk factors related to heat mortality. It is clear from the literature that epidemiological evidence on the heat morbidity and mortality relationship is lacking in South Asia and more information is needed to address localized vulnerabilities within these populations (Rathi et al. 2017). In general, compared to high-income countries in the West, low- to middle-income countries in South Asia have less access to health care, inadequate housing and infrastructure, and low economic advancement. By examining the specific localized heat-related human health relationships in these areas, we have the opportunity to derive specific actions and adaptation plans to reduce the incidence of heat-related morbidity and mortality in the future (Green et al. 2019). While changing individual knowledge and behavior is one approach, the increased use of blue-green infrastructure can help to improve the built environment to reduce negative health outcomes associated with extreme heat.

10.9 Reducing the Impacts of Heat-Related Morbidity and Mortality in South Asia Through the Use of Blue-Green Infrastructure

The review of the epidemiological evidence on risk factors associated with heat morbidity and mortality in South Asia identified urban infrastructure as a key factor that potentially increased risk of negative heat-health outcomes. As discussed above, the urban heat-island effect will increase high ambient temperatures in cities and cities in the region are expected to grow in size. Although the UHI occurs in cities around the world, it will have a greater impact on urban locations in warmer climates like South Asia. Heat-related risk of illness and death will remain one of the main climate-related hazards in cities due to UHI. Urbanization can have both negative and positive effects to human health, allowing populations increased access to health care or through increasing their exposure to air pollutants and extreme heat. Municipalities and urban planners must account for the potential negative impacts to health and design urban areas to reduce surface temperatures that contribute to the UHI.

Administrative changes and public health interventions can help to some extent. After the 2010 heat wave in Ahmedabad, India, researchers and city officials developed South Asia's first Heat-Health Action Plan. Extensive planning with government agencies, healthcare professionals, and high-risk populations the heat preparedness plan included a weather forecast alert that was triggered when temperatures were expected to hit dangerous levels (Knowlton et al. 2014). This type of plan requires significant coordination among officials and citizens, though addresses adaptation to extreme heat. A different approach, through mitigation, is the use of blue-green infrastructure changes to urban environments, potentially reducing the overall ambient temperature or preventing temperatures in urban areas from reaching extreme highs.

Blue-green infrastructure changes to the urban landscapes increase natural environments (e.g., vegetation and water) or integrate them into the man-made built environment (e.g., buildings and roads). A study in Colombo, Sri Lanka assessed 'green walls', vegetation alongside buildings, versus bare walls, man-made structures. The researchers found that there was a 1.6–1.7 °C reduction in temperature around buildings with green walls. A reduction in temperature outside a building will also greatly reduce the temperature inside a building as well (Galagoda et al. 2018). While 1.7 °C may not seem like a significant reduction in temperature, our review of the epidemiological evidence showed that even a 1 °C change in temperature above a certain temperature threshold increased heat-related mortality significantly. Studies from North America and Europe have also shown that by increasing vegetative cover and reducing impervious paved surfaces can reduce surface temperatures by as much as 10 °C (Venter et al. 2020; Hamstead et al. 2016). In New York City, researchers showed that neighborhood-level mortality rates of older adults (65 + years) on days exceeding 38 °C were positively associated with increased impervious surface cover, indicating that unvegetated surfaces likely increase the risk of heat-related mortality for populations in these areas (Rosenthal et al. 2014). In addition to reducing the risk of heat-related morbidity and mortality, the increased access to green space and bodies of water can improve physical activity and healthy lifestyles, including positive impacts on mental health and wellbeing. While less studied than green infrastructure, blue infrastructure, including decorative and recreational waters, have mitigating effects in reducing the UHI and increase evapotranspiration in urban areas. For both green and blue infrastructure, heating effects are influenced by the scale and spread of these mitigating interventions, as well as the climate (Gunawardena et al. 2017). Policy-makers, engineers, and urban planners will need to assess which green–blue infrastructure will be best for their city, though the benefits should be towards reducing the UHI. By reducing temperatures in urban areas through these structural interventions, there will be a co-benefit to health in reducing the potential impact of extreme heat events (Heaviside et al. 2017).

10.10 Future Implications of Extreme Heat and Blue-Green Infrastructure in South Asia

This chapter identified the potential future exposure of urban populations to heat in South Asia. Projections explored in this chapter forecast increases in total population in South Asia, and in particular, urban areas. Given these growing population trends, urban areas in South Asia will grow from medium to large, but more likely to very large or megacities, with millions living in urban areas. It is also important to note that given current trends, we can expect climate change to continue and potentially increase. Unmitigated warming of the globe will result in the potential for increased risk of extreme heat events in the region. The chapter also reviewed the epidemiological evidence, which indicates that large segments of the population in South Asia are experiencing extreme heat and there have been increases in heat-related morbidity and mortality associated with these events. Specific risk factors, such as housing, poverty, age, and underlying medical conditions add to the vulnerability of certain populations. In particular, the urban heat-island effect increases the risk of urban population to extreme heat. The chapter also identified the importance of mitigation, through blue-green infrastructure, which is needed to reduce the urban heat-island effect in metropolitan areas. A focus on urban infrastructure, including 'cool roofs', increasing vegetation, and urban water features will potentially reduce ambient temperatures in urban areas. By reducing urban temperatures, municipalities will potentially reduce heat-related morbidity and mortality in South Asia. In order to prepare for impacts decisions on how to adapt and mitigate extreme heat in urban areas should be made now, particularly in regards to infrastructure, as infrastructure is long-lived. As such, decisions made currently will have impacts on the mid and even long-term future. Appropriate ecosystem-based mitigation solutions, as described above, can help to reduce heat now and into the future, reducing the vulnerability of urban populations to extreme heat. In this volume, other chapters will demonstrate how this can be accomplished in more detail.

References

Alexander LV, Zhang X, Peterson TC, Caesar J, Gleason B, Klein Tank AMG, Haylock M, Collins D, Trewin B, Rahimzadeh F, Tagipour A, Kumar KR, Revadekar J, Griffiths G, Vincent L, Stephenson DB, Burn J, Aguilar E, Brunet M, Taylor M, New M, Zhau P, Rusticcucci M, Vazquez-Aguirre JL (2006) Global observed changes in daily climate extremes of temperature and precipitation. J Geophys Res Atmos 111:D05109. https://doi.org/10.1029/2005JD006290

Anderson GB, Bell ML, Peng RD (2013) Methods to calculate the Heat Index as an exposure metric in environmental health research. Environ Health Perspect 121:1111–1119

Angel S, Parent J, Civco D, Blei AM (2010) The presistent decline in urban densities: global and historical evidence of "sprawl". Lincoln Institute of Land Policy. https://www.researchgate.net/profile/Jason_Parent/publication/265570541_The_Persistent_Decline_in_Urban_Densities_Global_and_Historical_Evidence_of_'Sprawl'/links/5510448f0cf2a8dd79bc490a/The-Persistent-Decline-in-Urban-Densities-Global-and-Historical-Evidence-of-Sprawl.pdf

Arnfield AJ (2003) Two decades of urban climate research: a review of turbulence, exchanges of energy and water, and the urban heat island. Int J Climatol 23:1–26
Azhar G, Mavalankar D, Nori-Sarma A, Rajiva A, Dutta P, Jaiswal A, Sheffield P, Knowlton K, Hess J (2014) Ahmedabad Heatclimate study group. Heat-related mortality in India: excess all-cause mortality associated with the 2010 Ahmedabad heat wave. PLoS One 9
Babalola O, Razzaque A, Bishai D (2018) Hotter months, lower mortality. PloS one 13
Bouchama A, Knochel J (2002) Heat stroke. N Engl J Med 346:1978–1988
Bouchama A, Dehbi M, Mohamed G, Matthies F, Shoukri M, Menne B (2007) Prognostic factors in heat wave–related deaths: a meta-analysis. Arch Internal Med 167:2170–2176
Burkart K, Schneider A, Breitner S, Khan M, Krämer A, Endlicher W (2011) The effect of atmospheric thermal conditions and urban thermal pollution on all-cause and cardiovascular mortality in Bangladesh. Environ Pollut 159:2035–2043
Burkart K, Breitner S, Schneider A, Khan M, Krämer A, Endlicher W (2014) An analysis of heat effects in different subpopulations of Bangladesh. Int J Biometeorol 58:227–237
Chow WTL, Roth M (2006) Temporal dynamics of the urban heat island of Singapore. Int J Climatol 26:2243–2260
Christidis N, Jones GS, Stott PA (2015) Dramatically increasing chance of extremely hot summers since the 2003 European heatwave. Nat Clim Chang 5:46–50
Cook J, Oreskes N, Doran PT, Anderegg WR, Verheggen B, Maibach EW, Carlton J, Lewandowsky S, Skuce A, Green S, Nuccitelli D (2016) Consensus on consensus: a synthesis of consensus estimates on human-caused global warming. Environ Res Lett 11:048002
Dash SK, Mamgain A (2011) Changes in the frequency of different categories of temperature extremes in India. J Appl Meterol Climatol 50:1842–1858
Della-Marta PM, Haylock MR, Luterbacher J, Wanner H (2007) Doubled length of western European summer heat waves since 1880. J Geophys Res Atmos 112
Dholakia H, Mishra V, Garg A (2015) Predicted increse in heat mortality under climate change in urban India. Working Paper 2015-5-2. Indian Institute of Management, Ahmedabad
Dosio A, Mentaschi L, Fischer EM, Wyser K (2018) Extreme heat waves under 1.5 °C and 2 °C global warming. Environ Res Lett 13:054006
Dutta A, Bhattacharya S, Ak K, Pati S, Swain S, Nanda L (2020) At which temperature do the deleterious effects of ambient heat "kick-in" to affect all-cause mortality? An exploration of this threshold from an eastern Indian city. Int J Environ Health Res 30:187–197
Ebi KL, Hallegatte S, Kram T, Arnell NW, Carter TR, Edmonds J, Kriegler E, Mathur R, O'Neill BC, Riahi K, Winkler H, van Vuuren DP, Zwickel T (2014) A new scenario framework for climate change research: background, process, and future directions. Clim Change 122:363–372
Epstein Y, Yanovich R (2019) Heatstroke. N Engl J Med 380:2449–2459
Epstein Y, Moran DS (2006) Thermal comfort and the heat stress indices. Ind Health 44:388–398
Erell E, Williamson T (2007) Intra-urban differences in canopy layer air temperatures at a mid-latitude city. Int J Climatol 27:1243–1255
Frumkin H, Hess J, Luber G, Malilay J, Mcgeehin M (2008) Climate change: the public health response. Am J Public Heal 3
Fu S, Gasparrini A, Rodriguez P, Jha P (2018) Mortality attributable to hot and cold ambient temperatures in India: a nationally representative case-crossover study. PLoS Med 17
Galagoda RU, Jayasinghe GY, Halwatura RU, Rupasinghe HT (2018) The impact of urban green infrastructure as a sustainable approach towards tropical micro-climatic changes and human thermal comfort. Urban Forest Urban Green 34:1–9
Green H, Bailey J, Schwarz L, Vanos J, Ebi K, Benmarhnia T (2019) Impact of heat on mortality and morbidity in low and middle income countries: a review of the epidemiological evidence and considerations for future research. Environ Res 171:80–91
Gunawardena KR, Wells M, Kershaw T (2017) Utilising green and bluespace to mitigate urban heat island intensity. Sci Total Environ 584:1040–1055

Güneralp B, Zhou Y, Ürge-Vorsatz D, Gupta M, Yu S, Patel PL, Fragkias M, Li X, Seto KC (2017) Global scenarios of urban density and its impacts on building energy use through 2050. Proc Nat Acad Sci USA 114:8945–8950

Haines A, Kovats RS, Campbell-Lendrum D, Corvalan C (2006) Climate change and human health: impacts, vulnerability and public health. Public Heal 120:585–596

Hajat S, Armstrong B, Gouveia N, Wilkinson P (2005) Mortality displacement of heat-related deaths: a comparison of Delhi, Sao Paulo, and London. Epidemiology 1:613–620

Hamstead ZA, Kremer P, Larondelle N, McPhearson T, Haase D (2016) Classification of the heterogeneous structure of urban landscapes (STURLA) as an indicator of landscape function applied to surface temperature in New York City. Ecol Indicators 70:574–585

Hashizume M, Wagatsuma Y, Hayashi T, Saha S, Streatfield K, Yunus M (2009) The effect of temperature on mortality in rural Bangladesh—a population-based time-series study. Int J Epidemiol 38:1689–1697

Hausfater P, Megarbane B, Dautheville S, Patzak A, Andronikof M, Santin A, André S, Korchia L, Terbaoui N, Kierzek G, Doumenc B (2010) Prognostic factors in non-exertional heatstroke. Intensive Care Med 36:272–280

Heaviside C, Macintyre H, Vardoulakis S (2017) The urban heat island: implications for health in a changing environment. Curr Environ Health Rep 4:296–305

Hinkel KM, Nelson FE, Klene AE, Bell JH (2003) The urban heat island in winter at Barrow, Alaska. Int J Climatol 23:1889–1905

Hoornweg D, Pope K (2014) Socioeconomic pathways and regional distribution of the world's 101 largest cities. Global Cities Institute, Toronto

Hoornweg D, Pope K (2016) Population predictions for the world's largest cities in the 21st century. Environ Urban 29:195–216

Horton RM, Mankin JS, Lesk C, Coffel E, Raymond C (2016) A review of recent advances in research on extreme heat events. Curr Clim Change Rep 2:242–259

Howard L (1818) The climate of London, deduced from meteorological observations made at different places in the neighborhood of the metropolis, in two volumes London. W. Phillips, George Yard

Huebler M, Klepper G, Peterson S (2007) Costs of climate change. The effects of rising temperatures on health and productivity in Germany. Kiel Working Paper No. 1321, Kiel: Kiel Institute for the World Economy

Ingole V, Kovats S, Schumann B, Hajat S, Rocklöv J, Juvekar S, Armstrong B (2017) Socioenvironmental factors associated with heat and cold-related mortality in Vadu HDSS, western India: a population-based case-crossover study. Int J Biometeorol 61:1797–1804

Jiang L, O'Neill BC (2017) Global urbanization projections for the shared socioeconomic pathways. Glob Environ Chang 42:193–199

Kataoka K, Matsumoto F, Ichinose T, Taniguchi M (2009) Urban warming trends in several large Asian cities over the last 100 years. Sci Total Environ 407:3112–3119

Khan N, Shahid S, Ismail T, Ahmed K, Nawaz N (2019) Trends in heat wave related indices in Pakistan. Stoch Env Res Risk Assess 33:287–302

Knowlton K, Kulkarni S, Azhar GS, Mavalankar D, Jaiswal A, Connolly M, Nori-Sarma A, Rajiva A, Dutta P, Deol B, Sanchez L (2014) Development and implementation of South Asia's first heat-health action plan in Ahmedabad (Gujarat, India). Int J Environ Res Public Health 11:3473–3492

Kotharkar R, Ramesh A, Bagade A (2018) Urban heat island studies in South Asia: a critical review. Urban Clim 24:1011–1026

Makrogiannis T, Santamouris M, Papanikolaou N, Koronaki I, Tselepidaki I, Assimakopoulos D (1998) The Athens urban climate experiment—temperature distribution. ACTA Universitatis Lodziensis. Folia Geogr Phys 3:33–44

Matthews TKR, Wilby RL, Murphy C (2017) Communicating the deadly consequences of global warming for human heat stress. Proc Natl Acad Sci 114:3861–3866

McGeehin M, Mirabelli M (2001) The potential impacts of climate variability and change on temperature-related morbidity and mortality in the United States. Environ Health Perspect 109:185–189

Mckendry IG (2003) Applied climatology. Prog Phys Geogr 27:597–606

McMichael A, Wilkinson P, Kovats R, Pattenden S, Hajat S, Armstrong B, Vajanapoom N, Niciu E, Mahomed H, Kingkeow C, Kosnik M (2008) International study of temperature, heat and urban mortality: the 'ISOTHURM' project. Int J Epidemiol 37:1121–1131

Meehl GA, Tebaldi C (2004) More intense, more frequent, and longer lasting heat waves in the 21st century. Science 305:994–997

Miners A (2010) The diagnosis and emergency care of heat related illness and sunburn in athletes: a retrospective case series. J Can Chiropr Assoc 54:107

Mishra V, Ganguly AR, Nijssen B, Lettenmaier DP (2015) Changes in observed climate extremes in global urban areas. Environ Res Lett 10. https://doi.org/10.1088/1748-9326/10/2/024005

Mishra V, Mukherjee S, Kumar R, Stone DA (2017) Heat wave exposure in India in current, 1.5 °C, and 2.0 °C worlds. Environ Res Lett 12:124012

Masood I, Majid Z, Sohail S, Zia A, Raza S (2015). The deadly heat wave of Pakistan, June 2015. Int J Occup Environ Med 6(247)

Mora C, Dousset B, Caldwell IR, Powell FE, Geronimo RC, Bielecki CR, Counsell CWW, Dietrich BS, Johnston ET, Louis LV, Lucas MP, McKenzie MM, Shea AG, Tseng H, Giambelluca TW, Leon LR, Hawkins E, Trauernicht C (2017) Global risk of deadly heat. Nat Clim Chang 7:501–506

Naughton MP, Henderson A, Mirabelli MC, Kaiser R, Wilhelm JL, Kieszak SM, Rubin CH, Mcgeehin MA (1999) Heat-related mortality during a 1999 heat wave in Chicago. Am J Prev Med 22:221–217

Nissan H, Burkart K, Coughlan De Perez E, Van Aalst M, Mason S (2017) Defining and predicting heat waves in Bangladesh. J Appl Meteorol Climatol 56:2653–2670

O'Neill BC, Kriegler E, Riahi K, Ebi KL, Hallegatte S, Carter TR, Mathur R, van Vuuren DP (2014) A new scenario framework for climate change research: the concept of shared socioeconomic pathways. Clim Change 122:387–400

Oguntoyinbo JS (1984) Urban climates of tropical Africa. In: Oke TR (ed) Urban climatology and it applications with special regard to tropical areas. Proceeding of the technical conference Mexico, 26–30 Nov. World Meteorologial Organization, Geneva, Switzerland

Oke TR, Mills G, Christen A, Voogt JA (2017) Urban climates. UK, Cambridge University Press, Cambridge

Oke TR (1973) City size and the urban heat island. Atmos Environ 7:769–779

Oke TR (1997) Urban climate and global change. In: Perry A, Thompson R (eds) Theoretical and applied climatology. Routledge, London

Pal JS, Eltahir EAB (2015) Future temperature in southwest Asia projected to exceed a threshold for human adaptability. Nat Clim Change 6:197–200

Panda DK, Kouchak AA, Ambast SK (2017) Increasing heat waves and warm spells in India, observed from a multiaspect framework. J Geophys Res Atmos 122:3837–3858

Perkins SE, Alexander LV, Nairn JR (2012) Increasing frequency, intensity and duration of observed global heatwaves and warm spells. Geophys Res Lett 39. https://doi.org/10.1029/2012GL053361

Perkins SE (2015) A review on the scientific understanding of heatwaves—their measurement, driving mechanisms, and changes at the global scale. Atmos Res 164:242–267

Rahmstorf S, Coumou D (2011) Increase of extreme events in a warming world. Proc Natl Acad Sci 108:17905–17909

Rathi S, Desai V, Jariwala P, Desai H, Naik A, Joseph A (2017) Summer temperature and spatial variability of all-cause mortality in Surat city, India. Indian J Commun Med Off Publ Indian Assoc Prevent Soc Med 42:111

Riahi K, Van Vuuren DP, Kriegler E, Edmonds J, O'neill BC, Fujimori S, Bauer N, Calvin K, Dellink R, Fricko O, Lutz W, Popp A, Cuaresma JC, Samir KC, Leimbach M, Jiang L, Kram T, Rao S, Emmerling J, Ebi K, Hasegawa T, Havlík P, Humpenöder F, Da Silva LA, Smith S, Stehfest E, Bosetti V, Eom J, Gernaat D, Masui T, Rogelj J, Strefler J, Drouet L, Krey V, Luderer

G, Harmsen M, Takahashi K, Baumstark L, Doelman JC, Kainuma M, Klimont Z, Marangoni G, Lotze-Campen H, Obersteiner M, Tabeau A, Tavoni M (2017) The shared socioeconomic pathways and their energy, land use, and greenhouse gas emissions implications: an overview. Glob Environ Change 42:153–168. ISSN 0959-3780. https://doi.org/10.1016/j.gloenvcha.2016.05.009

Riahia K, Van Vuuren DP, Kriegler E, Edmonds J, O'Neill BC, Fujimori S, Bauer N, Calvin K, Dellink R, Fricko O, Lutz W, Popp A, Cuaresma JC, Leimbach M, Jiang L, Kram T, Rao S, Emmerling J, Ebi K, Hasegawa T, Havlik P, Humpenöder F, Aleluia Da Silva L, Smith S, Stehfest E, Bosetti V, Eom J, Gernaat D, Masui T, Rogelj J, Strefler J, Drouet L, Kreya V, Luderer G, Harmsen M, Takahashi K, Baumstark L, Doelman JC, Kainuma M, Klimont Z, Marangoni G, Lotze-Campen H, Obersteiner M, Tabeaun A, Tavoni M (2017) The shared socioeconomic pathways and their energy, land use, and greenhouse gas emissions implications: an overview. Glob Environ Change, 42:153–168

Ridgwell H (2018) Africa could see world's first 100-million-person city by century's end. VOA

Rizwan AM, Leung DYC, Liu C (2008) A review on the generation, determination and mitigation of urban heat island. J Environ Sci 20:120–128

Robine JM, Le Roy S, Van Oyen H, Griffiths C, Michel JP, Herrmann FR (2008) Death toll exceeded 70,000 in Europe during the summer of 2003. Comptes Rendus Biol 331:171–178

Rohini P, Rajeevan M, Srivastava AK (2016) On the variability and increasing trends of heat waves over India. Sci Rep 6. https://doi.org/10.1038/srep26153

Rosenthal JK, Kinney P, Metzger KB (2014) Intra-urban vulnerability to heat-related mortality in New York City, 1997–2006. Health Place 30:45–60

Roth M (2007) Review of urban climate research in (sub)tropical regions. Int J Climatol 27:1859–1873

Roy J, Chakrabarti A, Mukhopadhyay K (2011) Climate change, heat stress and loss of labor productivity: a method for estimation. Global Change Programme, Jadavpur University, Kolkata

Saleem SG, Ansari T, Ali AS, Fatima S, Rizvi MH, Samad MA (2017) Risk factors for heat related deaths during the June 2015 heat wave in Karachi, Pakistan. J Ayub Med Col Abbottabad 29:320–324

Salvati A, Roura HC, Cecere C (2017) Assessing the urban heat island and its energy impact on residential buildings in mediterranean climate: Barcelona case study. Energy Build 146:38–54

Salve HR, Parthasarathy R, Krishnan A, Pattanaik DR (2018) Impact of ambient air temperature on human health in India. Rev Environ Health 33:433–439

Samir K, Lutz W (2017) The human core of the shared socioeconomic pathways: population scenarios by age, sex and level of education for all countries to 2100. Glob Environ Chang 42:181–192

Santamouris M (2015) Analyzing the heat island magnitude and characteristics in one hundred Asian and Australian cities and regions. Sci Total Environ 512–513, 582–598

Semenza J, Rubin C, Falter K, Selanikio J, Flanders W, Howe H, Wilhelm J (1996) Heat-related deaths during the July 1995 heat wave in Chicago. N Engl J Med

Seto KC, Güneralp B, Hutyra L (2012) Global forecasts of urban expansion to 2030 and direct impacts on biodiversity and carbon pools. Proc Natl Acad Sci 109:16083–16088

Shastri H, Barik B, Ghosh S, Venkataraman C, Sadavarte P (2017) Flip flop of day-night and summer-winter surface urban heat island intensity in India. Nat Sci Rep 7:40178

Shrestha S, Shrestha I, Shrestha N (2016) Region-wise effects of climate sensitive variables on some specific disease burdens in Nepal. Open Atmos Sci J 10

Sivakumar MV, Stefanski R (2010) Climate change in South Asia. In: Climate change and food security in South Asia. Springer, Dordrecht

Small C, Okujeni A, Van Der Linden S, Waske B (2018) Remote sensing of urban environments. In: Loboda TV (ed) Comprehensive remote sensing, vol 6. Mapping Land Surface Types and Changes. Elsevier

Steadman RG (1979a) The assessment of sultriness. Part I: A temperature-humidity index based on human physiology and clothing science. J Appl Meteorol 18:861–873

Steadman RG (1979b) The assessment of sultriness. Part II: effects of wind, extra radiation and barometric pressure on apparent temperature. J Appl Meteorol 18:874–885
Stewart ID (2011) A systematic review and scientific critique of methodology in modern urban heat island literature. Int J Climatol 31:200–217
Sun Y, Zhang X, Zwiers FW, Song L, Wan H, Hu T, Yin H, Ren G (2014) Rapid increase in the risk to extreme summer heat in Eastern China. Nat Clim Chang 4:1082–1085
Tran H, Uchihama D, Ochi S, Yasuoka Y (2006) Assessment with satellite data of the urban heat island effects in Asian mega cities. Int J Appl Earth Obs Geoinf 8:34–48
Tzavali A, Paravantis JP, Mihalakakou G, Fotiadi A, Stigka E (2015) Urban heat island intensity: a literature review. Fresenius Environ Bull 24
UN (2017) World population prospects, 2017 revisions. UN Department of Economic and Social Affairs, New York
UN (2018) World urbanization prospects, 2018 revisions. United Nations, Department of Economic and Social Affairs, New York. https://esa.un.org/unpd/wup/
U.S. EPA (2008) Reducing urban heat islands: compendium of strategies. EPA, Washington, DC
US EPA (2018) Measuring heat islands [Online]. Environmental Protection Agency. [Accessed]. https://www.epa.gov/heat-islands/measuring-heat-islands
UN ESCAP (2015) Asia-pacific disaster report 2015, disasters without borders: regional resilience for sustainable development. Bangkok, Thailand: United Nations Publication
Venter ZS, Krog N, Barton DN (2020) Linking green infrastructure to urban heat and human health risk mitigation in Oslo, Norway. Sci Total Environ 709:136–193
Vidal J (2018) The 100 million city; Is 21st century urbanization out of control? The Guardian
Voogt J (2002) Urban heat island. In: Douglas I, Munn T (eds) Encyclopedia of global environmental change, Volume III causes and consequences of global environmental change. Wiley Ltd., Chichester
Wang Y, Hu F (2006) Variations of the urban heat island in summer of the recent 10 years over Beijing and its environment effects. Chin J Geophys 49:59–67
Warszawski L, Frieler K, Huber V, Piontek F, Serdeczny O, Schewe J (2014) The inter-sectoral impact model intercomparison project (ISI-MIP): project framework. Proc Natl Acad Sci 111:3228–3232
Yang X, Li Y, Luo Z, Chan PW (2017) The urban cool island phenomenon in a high-rise high-density city and its mechanisms. Int J Climatol 37:890–904

Chapter 11
Urban Risk Assessment Tools and Techniques for Ecosystem-Based Solutions

Aditya Rahul, Siva Ram Edupuganti, Vickyson Naorem, Mahua Mukherjee, and Talbot Brooks

Abstract Urbanization has picked up rapid pace since the dawn of the industrial revolution. This unprecedented and unplanned urbanization has brought about an integral change in natural fabric influencing climate at the global and the local level and has led the urban areas to become more prone to climate risk. The chapter is subdivided into three parts. The first part introduces and discusses various facets of heat and flood risks in urban areas and explores the causes and impacts of various parameters associated with them. The second part discusses the concept of indices and explores the tools for quantitative estimation of the multiple risks associated to the urban area. The third and final section introduces various tools and techniques that can be used for risk assessment and investigation. It discusses various datasets (Remote sensed data, In Situ measurement, Numerical Modelling (CFD)) and methodologies to identify hot spots and specific problems associated with them. This chapter introduces various tools and techniques to identify the susceptible hot spots, investigate the intensity and possible reason and check the potential of ecosystem-based solutions for risk attenuation.

Keywords Climate risk · Urban heat · Urban flood · Remote sensing · Microclimatic model

11.1 Introduction

Unprecedented urbanization has brought about a fundamental change in the natural fabric that influences climate at the global and local levels. This has led the urban areas to become more prone to heat risk and instances of flooding. Urban areas house 50% of the world's population (Liu et al. 2014). Ecosystem-based solutions are most suited to tackle the climate risk posed by unplanned urbanization. This

A. Rahul (✉) · S. R. Edupuganti · V. Naorem · M. Mukherjee
IIT Roorkee, Roorkee, India
e-mail: arahul@ar.iitr.ac.in

T. Brooks
Delta State University, Cleveland, USA

© The Author(s), under exclusive license to Springer Nature Singapore Pte Ltd. 2021

M. Mukherjee and R. Shaw (eds.), *Ecosystem-Based Disaster and Climate Resilience*, Disaster and Risk Research: GADRI Book Series,
https://doi.org/10.1007/978-981-16-4815-1_11

chapter provides insight into various investigation methods for climate risk (heat stress and floods). These investigative methods help identify the problematic areas and pave the way for the provision of ecosystem-based solutions.

11.1.1 Understanding and Assessing Risk

Risk is a combination of the probability that an event will occur and the subsequent consequences associated with the event. Consequences may be primary (raging floodwaters topple a building); secondary (commodity prices rise after cropland floods); or even tertiary (the combination primary and secondary consequences cause a third-order effect such as a nation-wide loss of credit standing and external investment). Disaster management is a continuous cycle of managing the elements of risk: event planning, event mitigation actions, response and recovery. Planners and decision-makers must make every effort to use a high-quality data-driven approach to understand and pair risk reduction activities with coping strategies that yield the highest possible returns.

A quantitative approach, guided by qualitative information and community input, provides a solid, standardized approach for understanding and communicating risk. A basic quantitative approach is mathematically written as:

$$\text{Risk} = \text{Disaster Threat} \times \text{Consequence} \times \text{Vulnerability} \tag{11.1}$$

The score assigned to each risk factor (disaster threat, asset value and vulnerability) has a quantitative basis adjusted for local considerations using and publishing community input. Establishing and publishing community-based definitions is especially important because it accounts for cultural specifics that eventually link to a willingness to bear the cost of risk reduction activities. Table 11.1 is an example which illustrates how risk factors may be quantified. Note that each risk factor may contain several sub-factors. Community input may be used to either modify weights when calculating factor scores from sub-factors or may be qualitatively expressed in a plain language such as "High," "Medium," or "Low."

Vulnerability is the degree to which a population can cope with any particular threat. While related, vulnerability has less to do with physical factors such as housing type—though a relationship exists—and is far more related to socio-economic factors and the availability of support systems. Examples of extremely vulnerable populations include the elderly, poor, incarcerated and similar. The United States of America has used census demographic data to create a Social Vulnerability Index score. The score is amalgamated and published using the US Bureau of the Census tabulation areas (tract level) and published as a geographic layer.

Resiliency is the ability to resist a disaster across all risk factors such that threats are reduced, values protected and vulnerabilities minimized. The outcome of managing the disaster management lifecycle is to increase community resiliency through improved risk management. The urban planning process must embrace the

Table 11.1 An example matrix for calculating the consequence of loss

Consequences: structures					
Building characteristics	Lightweight/wood frame	Moderate Streel truss/ordinary masonry	Heavy steel truss/reinforced concrete	Multi-story steel and reinforced concrete	Extreme multi-story steel and reinforced concrete
Locality type	Rural/sub-urban	Semi-urban/light industrial	Industrial	Urban	Dense urban
Number of occupants					
Non-dense urban	< 100	100–500	500–2000	2000–5000	> 5000
Score	*1*	*3*	*5*	*8*	*10*
Dense urban	< 1000	1000–3000	3000–7000	7000–10,000	> 10,000
Score	*1*	*5*	*8*	*10*	*10*
Replacement value					
Non-dense urban	< $1M	$1M–$10M	$10M–$100M	$100M–$500M	> $500 M
Score	1	3	5	7	9
Dense urban	< $20 M	$20M–$50M	$50M–$250M	$250M–$1B	> $1B
Score	*4*	*5*	*6*	*8*	*10*
Historic value	Low	Medium	Low	Medium–High	High
Score	*1*	*3*	*5*	*7*	*10*
Ease of replacement					
Score	*1*	*3*	*5*	*7*	*10*
Impact of loss	Local	Regional	Province	National	International
Score	*1*	*3*	*5*	*7*	*10*

In this instance, structural types by density type are illustrated. Rather than tabulating across rows or down columns, this table is meant to serve as a reference that allows users to mix and match sub-factors as needed use of a cost–benefit analysis approach. A matrix of potential actions across the risk equation is translated into an improvement in resiliency such that maximum return on investment is realized. The use of geospatial technologies as an intelligence gathering, management and analysis system provides a tremendous advantage for the management of risk. The latter set of processes are often referred to as GEOINT.

11.1.2 The GEOINT Advantage

Tobler's First Law of Geography simply states, "everything is related to everything else, but near things are more related than distant things" (Tobler 1970). Geospatial technologies, which include geographic information systems, remote sensing, global positioning satellite systems and spatial analysis methods, provide a means to contextualize all aspects of risk to geography (space/place). Individual factors inherent in risk may be quantified using the definitions and examples provided above and then linked to geographic layers that depict their location. These layers may be viewed and assessed individually or used in combination to develop sophisticated risk analysis models.

The integration of positioning information is what powers any geospatial analysis. There are two fundamental data types commonly in use: (1) raster data, which are pixel-based representations of a surface and (2) vector data, which uses geometric shapes described by coordinate pairs and topology to associate information about a location with the place. Spatial data result from either direct measurement, such as the use of a global positioning satellite system-based descriptor added to a point or polygon or remote sensing. Remote sensing is the measurement of reflected, transmitted, or absorbed energy using a sensor system that does not physically interact with the object or scene of interest. Aerial photography and satellite-based images that are referenced to the surface of the earth are the most frequently found measurement types and represent passive data collection where the sensor is simply an observer. Advanced techniques use "active remote sensing" where the sensor platform emits specific, known forms of energy and measures a return. RADAR, LIDAR and side-aperture radar systems use such active means.

Vector and raster data sources are often intertwined. The location of a single weather observation system observing a continuous data type (temperature) may be observed in a raster-based satellite image. Its position is known because the image is georeferenced and is thus easily converted to a point (vector) having the attribute (information about the feature) recorded as the temperature at a particular time. The location of several weather stations may be combined into a single geographic layer in vector format and statistical interpolation routines (e.g. inverse distance weighted, kriging) run to produce a raster (pixel) surface representative of temperature across a region. Thus, numerous models and techniques are possible because location serves as the anchor for comparing raster images across time (temporal change detection), interpolation from point data versus time to create raster images for temporal change detection, direct observation and visualization and much, much more. The US Federal Emergency Management Agency (FEMA) has developed and/or recommend the use of several such GEOINT-based models, most notably HAZUS and the CAMEO/MARPLOT analysis suites (US Dept. of Environmental Protection).

The HAZUS modelling suite is integrated with Esri's ArcGIS software product and available at no cost. The model includes national coverage for base geographic layers such as social vulnerability index, housing stock, vegetative cover, economic activity, critical infrastructure and much more. Similarly, the CAMEO/MARPLOT suite uses environmental and hazardous materials to analyze the environmental impact of spills and plumes. In both cases, each layer of data may be modified by users and integrates into several routines, which allow the user to model the effects of flood, earthquake and cyclones of varying intensities and magnitudes. Output products are typically presented in map format with supporting tables. The modelling and presentation of risk in such a visual format are more impactful and often easier to understand than other means; it provides the opportunity to identify areas of particularly high risk and allows users to model the potential efficacy of mitigation activities.

The core of all risk modelling efforts rests firmly upon the availability, completeness and quality of spatially-based data. The most critical factor in understanding risk thus rests in understanding the nature of the threat on a geographic basis. The collection and sharing of high-quality geospatial data are the utmost priority for gaining the return on investment with respect to all risk reduction efforts. Two specific threats are explored hereafter: urban heat and flooding.

Both urban heat and flooding are spatially-based phenomena and geospatial analysis provides an excellent pathway for understanding them within the context of risk and the disaster management lifecycle.

11.2 Urban Heat

The urban environment influences the movement of wind, heat absorption and dissipation and, hence, takes longer to dissipate heat than a rural environment resulting in higher temperatures in urban areas. This temperature differential is attributed to Urban Heat Island. A city of 1 million people is expected to see a temperature differential of 1–3 °C. This can reach as high as 12 °C (US Environmental Protection Agency 2008). Further, temperature differentials are exacerbated by climate change (Fig. 11.1). Existing research has established that urban morphology and surface properties are among the primary reasons for the increasing urban ambient temperatures and creating the phenomena of urban heat islands (Pandey et al. 2014; Thomas et al. 2014; Mathew et al. 2016). City centre areas store twice as much heat when compared to rural areas (Christen and Vogt 2004).

The phenomenon of urban heat islands can have an undesirable impact on people's health and comfort and the city's energy efficiency. Sensitive populations like senior adults, children are more at risk in high air temperature conditions (Kalkstein 1991).

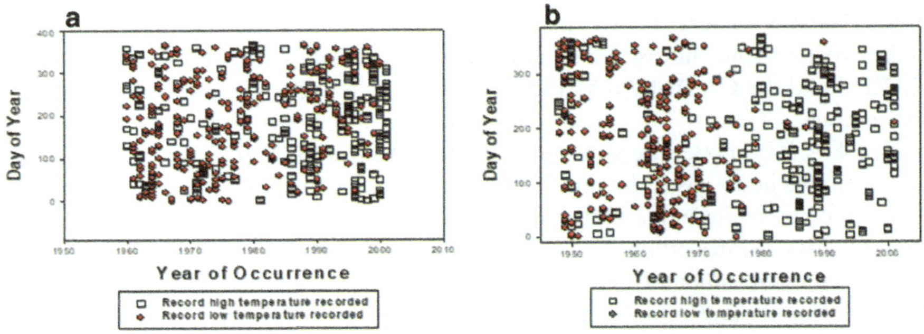

Fig. 11.1 Year of occurrence for daily record high and low temperatures at a rural community located 45 km south of Phoenix, AZ (Panel A) and in and at Phoenix Sky Harbor International Airport (Panel B). While record high temperatures are clustered to the right in panel A, record low temperatures were occasionally set after 1975. Conversely, very few record low temperatures were set after 1975 in the urban core, strongly suggesting the collapse of the diurnal temperature range and formation of an urban heat island (Brooks, unpublished data from the Arizona Meteorological Network (AZMET))

Heat islands result in reduced energy efficiency, air pollution, heat-related illness and mortality and deterioration in water quality Celsius (US Environmental Protection Agency 2008). Higher electricity consumption of 5–10% has been observed in urban city centres due to the urban heat island effect (Akbari and Konopacki 2005). Understanding the root causes of urban heat and analyzing opens potential routes for reducing vulnerability to urban heat. Urban surfaces have a high heat capacity and low thermal conductivity and store twice as much heat than rural areas (Christen and Vogt 2004). Altering the urban landscape through the use of reflective materials and vegetation is one such method of reducing vulnerability (Jenerette et al. 2015). Hence, it is imperative to understand the risks of urban heat. Further, these risks have to be analyzed and assessed to quantify these risks. This process is critical to identify ecosystem solutions and inform the decision making for the policymakers.

11.3 Urban Flood

Urban areas all over the world are facing various climate-related risks. Urban floods are one of the more predominant risks in urban areas. The level of vulnerability also depends on the size of the floods. Understanding the various factors that lead to this risk will help develop better contextual solutions for resilience building.

Flooding in an area refers to when the amount of water present in the region is larger than its water accommodating capacity. It occurs when the inundation rate of water exceeds the outflow capacity (drainage). Precipitation is the most dominant source of water, but other factors such as inflow from another water body, infrastructure failure, etc., can also act as contributors.

Urban areas are much more susceptible to flooding events. Lack of natural vegetation cover and many impervious areas significantly reduces the water infiltration capacity of the land surface and causes an increase in surface runoff. The speed of surface runoff is influenced by surface roughness and dimensions. Urbanized areas tend to form highly restrictive flow environments with low surface roughness, enabling a higher water flow rate. The amount of kinetic energy associated with these events can cause severe damage to public and private assets and negatively impact the quality of life and human safety.

A paradigm shift is needed to replace the outdated resistance-based approach (e.g. requiring the construction of new hard infrastructure) with an ecosystem-based approach that seeks to restore, enhance, or create ecosystem services within the urban matrix. The latter would promote the conservation and restoration of natural systems, specifically for the benefits they provide to humans.

11.4 Assessment Indices

Quantitative estimation of risk provides valuable insight into a thorough understanding of the study area's behaviour under stress. This estimation helps identify hotspots in the area and highlights the most critical parameter that needs support through intervention for risk attenuation. Different indices are available to measure urban heat and urban floods. Some of the critical heat specific and urban flood-related indices are highlighted in the following sections.

11.4.1 Heat Specific Indices

Heat stress in urban areas is one of the major climate-related risks. Urban heat creates an uncomfortable living environment which further increases morbidity and mortality among its residents. Thus a human-centric approach is the need of the hour for heat risk investigation. The impact of heat on human body can be measured through various thermal comfort indices (Rahul et al. 2020).

These indices measure the human thermal stress taking into account various factors of the human body such as height, weight etc. and multiple climate variables such as air temperature, humidity, wind etc. A total of about 162 different thermal comfort indices have been developed until now by various previous studies (de Freitas and Grigorieva 2015). Some of them are listed in Table 11.2. These thermal comfort indices can be classified into the following three categories based on their evolution approach:

1. Indices based on linear equation
   Indices based on linear equation refer to human thermal comfort as a function of the thermal environment. These indices do not take into account the

Table 11.2 Heat specific assessment indices

Index	Source
Wet bulb globe temperature (WBGT)	Yaglou and Minard (1957)
Discomfort index (DI)	Thom (1959)
Index of thermal stress (ITS)	Givoni (1963)
Heat index (HI)	Steadman (1984)
Thermal strain index (TSI)	Lee (1958)
Predicted mean vote (PMV)	Fanger (1970)
Physiologically equivalent temperature (PET)	Mayer and Höppe (1987)
Perceived temperature (PT)	Staiger et al. (2012)
Outdoor variant, OUT–SET*	Pickup and Dear (2000)
Universal thermal climate index (UTCI)	Jendritzky et al. (2012)
Modified physiologically equivalent temperature (mPET)	Chen and Matzarakis (2014)

microclimate and human factors. Some of the prominent indices based on the linear equation approach are Effective Temperature (ET), Corrected Effective Temperature (CET) and Wet Bulb Globe Temperature (WBGT).

2. Empirical indices
   Empirical indices refer to human thermal comfort specific to a particular climate. These are based on linear regressions of data collected through field studies. Thermal Sensation (TS) is one of the examples of empirical indices. It is generated through multiple regressions of field data obtained by onsite measurement and questionnaire surveys.
3. Indices based on Human Thermal Models (HTM)
   These indices are based on the human body's energy balance. It explores the interrelationship between human factors such as metabolic activity, clothing parameters etc. and various environmental factors such as wind, relative humidity etc. Indices based on HTMs have also evolved over the last few years. The HTMs can be classified based on dividing the human body (segments) and calculating the temperature of the body parts (nodes or elements) (Katić et al. 2016). The human body can be considered as a single segment or as multiple segments which are further divided into various layers, also known as 'nodes'. Various indices based on HTMs are either one node model or two node model or a multi node model. Predicted Mean Vote (PMV) is one of the examples of an index based on one node model. Physiologically Equivalent Temperature (PET) index, which employs Munich Energy balance Model for Individuals (MEMI), is one of the prominent examples of two node model whereas Universal Thermal Climate Index (UTCI) employs a multi node model.

11.4.2 Flood Specific Indices

Some suggested indices to measure flood risk assessments include Urban Flood Risk (Bansal et al. 2017). The components of Urban Flood Risk (UFR) are Urban Flood Hazard (UFH), Urban Flood Social Vulnerability (UFSV) and Urban Flood Physical Vulnerability (UFPV) by using weightage comprehensive analysis and z-score standardization. Sharma et al. (2012) calculated flood risk index using infrastructure, population and land use in the event of frequent river-flood disasters. Hazarika et al. (2018) assessed and mapped flood hazard, vulnerability and risk in Upper Brahmaputra River valley using stakeholders' knowledge and multicriteria evaluation (MCE).

Batica et al. (2013) evaluated the flood resilience and examined existing floods management strategies and their effectiveness in decreasing flood damage; the Flood Resilience Index (FRI) can be used to compare different cities. Limitations of the FRI evaluation were that the index still depended on some assumptions. For example, addressing the flooding process in urban systems and the measurement of indicators relied on weights to assign for each indicator. Three directions were listed to prevent the instability of the urban system and identify the resilience of urban systems.

- Adjusting the thresholds of a system with respect to changes in response to flood waves.
- Defining the level to which the system was capable of self-organizing.
- Defining the level to which the system was able to build and increase learning and adaptation capacity.

11.5 Risk Investigation

Investigating the risk of urban heat and floods involves collecting specific data sets that enable researchers to adopt assessment tools and techniques to identify these specific risks. These investigations become the basis for identifying risks but also the implementation of mitigation strategies. Further, these investigations generate a body of knowledge for the local development bodies to apply for planning and policymaking. The authenticity of these investigations depends upon the norms and processes followed in the tools and techniques adopted. The tools and techniques vary from observational approaches like statistical analysis and numerical simulation approaches. Investigation of risk associated with an urban area can be broadly divided into the following stages:

a. Data Collection
b. Data Analysis and Assessment techniques.

11.5.1 Data Collection

This stage of investigation involves the collection of relevant data corresponding to the risk in urban areas. Various possibilities are available to facilitate this stage. These can range from In situ measurements to collect data or remote sensed data like satellite, LiDAR or UAV data. One of the contemporary approaches is to take advantage of crowdsourced data sources. Nowadays, even in developing countries, local government agencies are setting up extensive data collection networks in terms of sensors and weather stations to collect relevant data for present and future purposes. The datasets collected are crucial for analysis and subsequently informing the policymaking.

11.5.1.1 In Situ Measurement

In Situ Measurements is a vital tool at the researcher's disposal. Various Meteorological Parameters can be measured onsite at various scales to give a very accurate insight into urban heat behaviour or urban flood detection. Any measurement which involves direct contact with the medium is called In Situ Measurement. This can be both static and a mobile-based approach depending on the context and requirements.

a. **Fixed-point measurement**

   i. **Urban Heat specific data collection**
   Several metrological parameters like air temperature, relative humidity, wind velocity and direction, surface temperature are essential for urban heat assessment. These parameters can be measured onsite by the researcher using various instruments like automatic weather stations (AWS) (Fig. 11.2). Data can also be accessed from existing meteorological parameter databases such as National Oceanic and Atmospheric Administration (NOAA) or India Meteorological Department (IMD). Urban heat is very sensitive to variations in the environment, both spatially and temporally (Roth and Chow 2012). Hence, the location of the sensor is very vital to collect data. The location should consider the external factors that impact data quality and the location should be representative of the research question. For example, if the measurement is specific to a Land Use and Land Cover (LULC) type or a Local climate zone (LCZ); then the location and placement have to match with LCZ or LULC type. The meteorological instrument has to be placed in a way that prevents any ambiguity. This is essential to get representative measurements of the LULC or LCZ (Stewart and Oke 2012).

The instrument placement has to take cognizance of the external impacts on the sensor. For example, air temperature measurements, direct radiation on the sensor has to be avoided. For this purpose, the sensor has to be equipped with a radiation shield, preferably an aspirated shield which encourages convection in

Fig. 11.2 Automatic weather station (AWS) installed onsite for fixed-point measurements. A typical weather station has various sensors which capture different metrological parameters and send them to the data storage unit of AWS

the proximity of the sensor. Improper considerations of the placement and inadequate shielding of the sensor can lead to errors in the measured data. Infrared or thermal imagers can be used to measure surface temperature. For Infrared based instruments, the ratio of the distance between the measurement surface and instrument with respect to the area of the measurements is vital. For, thermal imagers' sensitivity and resolution of the imager are paramount. For wind-based parameters, the morphology of buildings and surrounding landscape can create locational turbulence that is not representative of the study area. According to WHO guidelines, the instruments should be deployed at the height of 1.5–2.1 m above any horizontal surface (Oke 2004a, b). For microscale measurements, the instrument is deployed at the ground level. For a mesoscale measurement, the instrument is deployed at a height representative of a larger study area that is not affected by micro-level factors (Oke 2006). Further, while selecting the sensor, consideration should be given to the sensor's response time and accuracy.

ii. **Urban flood specific data collection**
   A rain gauge traditionally measures precipitation measurements. Modern precipitation sensors have an advantage over the traditional rain gauge as it can also measure intensity. The main impediment for deploying these instruments is the wind, which can significantly impact data's accuracy.

Hence, the windshield is essential to protect the gauge. Further, the instrument's placement has to be in the open yet protected area from wind elements; the deployed area should uniform height obstacles so that turbulence is not generated. Due to advances in technologies, urban flooding can be tracked in real-time through sensors installed at crucial areas like pump stations and drainages, sumps which may overflow. Further, Inundation sensors can also be installed in hotspot areas like traffic junctions to monitor flooding in real-time. These sensors can be linked to a real-time management system to send immediate updates and warnings.

b. **Mobile measurement**
A mobile approach can be adapted to measure meteorological parameters based on the study's context and feasibility. The instrument can be deployed on a motor vehicle or a bicycle, depending on the context. This helps in studying urban heat phenomena over a larger area. Adopting a fixed instrument approach requires a large number of instruments compared to a mobile-based study. UHI is very sensitive to variations in the environment, both spatially and temporally and selecting the time and day of the experiments is vital (Roth and Chow 2012). Stable atmospheric conditions are ideal, while windy and rainy days should be avoided. The ideal time for UHI measurements is 4 h after sunset (Oke 2004a, b; Kotharkar and Bagade 2018). Mobile measurements are not restricted to study UHI and can also be adopted for thermal comfort research. Strict guidelines have to be followed to get reliable measurements as the possibility of errors in data collected is high. The localized impact of anthropogenic heat has to be avoided; for example, emissions from vehicles on roads can significantly impact the measurements. Hence, the mobile routes have to be designed in a way suited to minimize anthropogenic heat and document their occurrences using GPS loggers in sync with the mobile instruments. Further, daytime measurements should use aspirated sensors with a radiation shield. This will help in getting ambient measurements and nullify the impact of direct solar radiation (Nichol et al. 2009). The impact of natural and manmade elements such as flyovers, water bodies and parks needs to be isolated depending on the study's context. For mobile measurements specific to air temperature, vehicle's speed is ideally maintained between 25 and 30 km/h (Kotharkar and Bagade 2018). For certain metrological parameters like relative humidity, mobile measurements are not ideal.

11.5.1.2 Remote Sense Data

Remote sensing technique refers to the collection of field data remotely from sensors fixed in satellite or any other airborne vehicle (drones, aircraft). With the recent advances in remote sensing, it has become one of the predominantly used methods for urban risk investigation due to its spatial and temporal resolution.

Table 11.3 Selected operational sensors capable of retrieving LST from TIR bands, adapted from Tomlinson et al. (2011)

Sensor	Satellite	Spatial resolution	Orbital frequency
SLSTR	Sentinel-3	1 km	Twice daily
VIIRS	Suomi NPP	750 m	16 days
TIRS	Landsat-8	30 m	16 days
ETM+	Landsat-7	30 m	16 days
MODIS	Aqua	1 km	Twice daily
MODIS	Terra	1 km	Twice daily
ASTER	Terra	90 m	Twice daily
AVHRR	NOAA platforms	1.1 km	Twice daily
AVHRR	MetOp-A	1 km	29 days
AVHRR	MetOp-B	1 km	29 days
SEVIRI	Meteosat-8	3 km	Geostationary
GOES Imager	GOES 8–15	4 km	Geostationary

Satellite Data

i. **Urban Heat Investigation**
Recent developments in remote sensed imagery obtained from various satellites provide a viable alternative to studying an urban area's thermal environment. Remotely sensed data from satellites overcomes various drawbacks of in situ measurement. Better temporal and spatial resolutions are the major advantages of remotely sensed data. Table 11.3 lists the available operational sensors and their basic information.

While satellite remote sensing offers a seamless spatial and temporal resolution, it has its own set of limitations. The presence of clouds in the atmosphere is one of the major obstacles, as it can easily contaminate the data. Thus the accuracy of retrieved LST is dependent upon the cloud-screening algorithm (Wan 2008).

ii. **Urban flood investigation**
Urban flood investigation is carried out through an additive approach with consideration to flood hazard, vulnerability and exposure. Firstly, flood hazard exploration is done using the source of water masses. Secondly, the investigation of vulnerability is carried out through the lens of five dimensions: social, economic, physical, institutional and attitudinal. We must investigate urban floods with the high-resolution modelling of dynamic exposure to flood disasters and disaster-hit bodies during an event to satisfy the increasingly elaborate modelling and management of urban floods (Zhu et al. 2019). So, we are discussing the different data sources, which helps in investigating the urban flood.

LiDAR Data

Light Detection and Ranging (Lidar) is a remotely sensed method that uses light in the form of a pulsed laser to measure ranges (variable distances) to any surfaces beneficially. The data measured from this method is used for the elevation of any objects on the earth surface to generate the bathymetric maps for urban flooding. These light pulses generate precise, three-dimensional information about the shape of the earth and its surface characteristics. Lidar systems allow users to examine both natural and manmade environments with accuracy, precision and flexibility.

UAV Data

Unmanned Aerial Vehicles (UAVs) are small airborne platforms equipped with several onboard sensors to acquire remotely sensed data at the desired time and spatial resolution. UAVs can acquire more accurate and precise data than satellite images because they fly at low altitudes of around 100 m and thus mitigate the influence of weather conditions.

11.5.1.3 Crowdsourced Data

One of the major drawbacks in urban climate research is the lack of high resolution of data. This limits a lot of urban climate research. Crowdsourcing is a development that enables in tackling this lack of data resolution (Meier et al. 2017). Crowdsourcing takes advantage of technologies available for the general public throughout the day (GPS, sensors on smartphones). A work done in Brazil monitored the battery temperature of smartphones to derive the urban air temperature (Droste et al. 2017). Similarly, mobile applications can also be developed specifically for data collection. While measuring the flood risk to understand the vulnerability and exposure, social, economic and environmental aspects are essential. By taking advantage of the accessibility of smartphones and the internet, traditional household surveys can be replaced with a mobile-based approach. This can reduce manpower and time and also eliminate human and instrumental errors. In the mobile user interface, questionnaires can be used to assemble different attributes accordingly.

11.5.2 Assessment Tools and Techniques

Data collected from all the sources mentioned above gives the quantitative estimate of parameters affecting urban heat and urban flood phenomenon. However, to quantify these risks, the data needs to be analyzed. There are various tools and techniques to facilitate the investigation of urban heat and flood risk. Though there is a conformity of data collection processes, the analysis techniques vary significantly. Adherence to scientific processes will result in a better generalization of results. Contemporary urban heat and flood investigation methods can be classified broadly into two approaches: (a) observational approach and (b) numerical simulation approach. The following section discusses these two approaches in detail with their investigation methods:

11.5.2.1 Observational Approach

Investigation through observational approach refers to methods that employ observing the risk through its various parameters. The following sections discuss in detail the assessment of urban heat and urban floods through an observational approach.

i. **Urban Heat Assessment**
   Assessment of urban heat in urban areas through observational approach can be classified based on the data collection method:

   a. **In situ measurement-based assessment**
      All the field measurements collected either through static or mobile methodologies have to be checked for data noise. Any anomalies of data have to be filtered statistically. In mobile measurements, data filtering is essential; to remove sections of data under the influence of external factors and filter it based on the ideal speed required for the study; as it is unrealistic to maintain the same speed throughout the study. Hence, the measurements taken outside the desired velocities have to be purged. Further, data collected in coordination with LCZ and LULC classes have to be filtered to remove buffer areas. For LCZ specific study, a 200–400 m buffer is considered and accordingly, the collected data has to be filtered (Stewart and Oke 2012). Further, any anthropogenic influences have to be filtered out. This is an essential part of the analysis, a lack of which will result in unreliable data. This filtered data can be plotted and analyzed using a statistical or suitable Geographic Information System (GIS) program. Any programming language such as R or Python can be used to plot and analyze the data. Spatial data can be interpolated spatially and temporally through GIS or programming languages using appropriate interpolation algorithms (Fig. 11.3).
   b. **Thermal remote sensing**
      The recent developments in remote sensing have brought to the forefront another method for urban heat analysis. Thermal remote sensing provides a viable alternative for urban heat stress investigation due to its better spatial and temporal resolution than the traditional field measurement techniques.

The extraction of the skin temperature of an object from its emitted thermal radiance is based on Planck's law. Planck's law states that total radiative energy emitted by any object is directly proportional to its temperature.

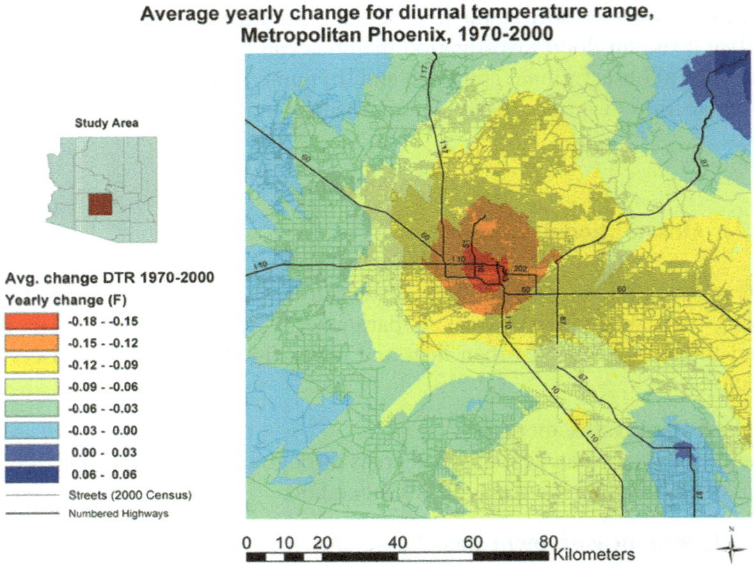

Fig. 11.3 Methods based on Tobler's First Law use spatial autocorrelation to interpolate continuous data types from point observations. The average daily diurnal temperature range was averaged by year and interpolated across the Phoenix metropolitan area and its surrounding rural landscape for a 30-year period. The average yearly collapse of diurnal temperature range across the interpolated surface was calculated and determined to be a change of 0.6 °F (0.5 °C) per year. Overlaying the street network highlights the urban heat island effect and assumes that the area developed outwards from the urban core, illustrates the expansion of the effect over time (Brooks, unpublished analysis)

Estimating of earth's surface temperature is carried out by investigating radiometric emission found in the spectral region of the sensors fitted on the satellite observe the thermal radiance emitted by any object in the thermal infrared region. The radiance observed at the sensor is then atmospherically corrected to calculate the radiance emitted by the surface.

Various particles present in the atmosphere contaminate the reflectance through absorption and scattering. Atmospheric correction refers to the process of removing the effects of the atmosphere from the reflectance observed by the sensors. There are various atmospheric correction algorithms available to aid this process. These algorithms are made up of two steps. The first step involves the estimation of optical characteristics of the atmosphere and then radiative transfer algorithms can calculate the various quantities required for atmospheric correction by using the atmospheric optical property. The second step involves the inversion procedure to derive the corrected surface reflectance (Fig. 11.4).

The surface emitted radiation obtained after atmospheric correction is a function of temperature and emissivity. Emissivity plays a critical role in LST retrieval as a 0.015 (1.5%) error in emissivity is equivalent to approximately a 1 K error in the LST for a material at 300 K (Hulley and Hook 2009). Thus to accurately estimate

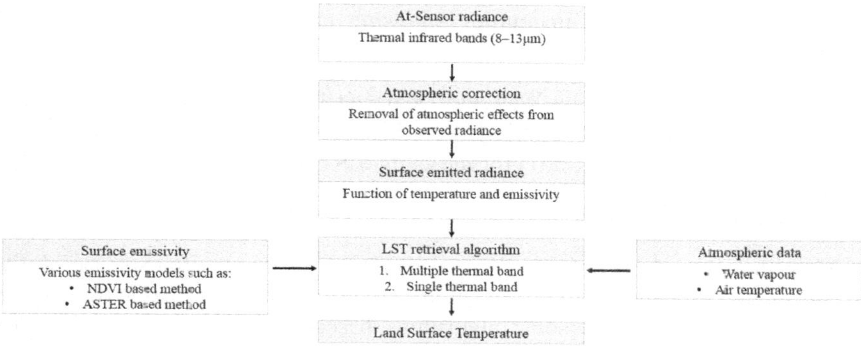

Fig. 11.4 Land surface temperature calculation process

the temperature, accurate estimation of the emissivity of the earth's surface should be ensured. Earth's surface is a conglomeration of different surfaces having different emissivity. This, in turn, makes the process of calculation of the overall emissivity of any area extremely difficult. Irrespective of these challenges, various models have been developed for emissivity estimation, but all of these models have their own set of limitations. Following are some of the commonly used methods to accurately estimate the land surface emissivity:

- Land cover classification methods
  This method advocates the surface area be classified based on its physical cover and assigning each land cover class its emissivity. Land cover classification can be carried out through two different methods (i) by conducting a physical survey of the site and (ii) using remote sensed data. Both of these methods have their own set of shortcomings. The physical survey is very resource intensive and takes a lot of time during which the land cover is susceptible to change. On the other hand classification through remote sensed data is plagued by problems inherent to remote sensed data such as cloud cover and pixel heterogeneity. Thus the emissivity too derived from this method is susceptible to error.
- ASTER Global Emissivity Dataset (GED)
  The ASTER GED product provides land surface emissivity maps of the Earth's surface derived from ASTER data in 1 km spatial resolution with an average accuracy of 0.01 (Hulley et al. 2012; Hulley and Hook 2009).
  After calculating the surface emissivity the thermal infrared bands are further analyzed through LST retrieval algorithms to estimate the skin temperature of the object. These algorithms can be briefly divided into following categories:

1. **Multiple thermal band approach**
   Algorithms employing more than one thermal band use the band's absorption difference to compensate for the atmosphere. This approach does not employ the radiative transfer model; thus it does not need an atmospheric profile to

retrieve land surface temperature. Following are some of the commonly used algorithms based on multiple thermal band approach:

- Split-window algorithm by Wan and Dozier (1996) for obtaining land surface temperatures from AVHRR and MODIS data.
- Algorithm by Qin et al. (2001) for retrieving LST from AVHRR images.

2. **Single thermal band approach**

- Algorithms based on a single thermal band need accurate atmospheric profile information as the temperature and emissivity cannot be separated in the emitted radiance. Following are some algorithms based on the single thermal band approach:

- Mono window algorithm developed by Qin et al. (2001).
- Single channel algorithms developed by Jiménez-Muñoz and Sobrino (2004).

Irrespective of all the precautions, LST products are largely based on approximations. Following are some of the factors which can lead to error in LST retrieval:

- Intra pixel heterogeneity

LST data represents the average temperature of a pixel. Since any urban area is heterogeneous in nature and consists of different surfaces with varied thermodynamic and radiative properties, the pixel-based LST estimate is not a true representative.

- Surface spectral emissivity

As discussed in the previous section, accurate assessment of surface emissivity of an urban surface is very difficult due to its heterogeneous nature. Any error in surface emissivity estimation will reflect in LST calculation.

- Atmospheric temperature and humidity variations
- Clouds and large aerosol particles

Cloud cover and suspended aerosol particles can lead to significant thermal anisotropies in urban areas. These, in turn, can significantly affect the accuracy of LST products.

11.5.2.2 Numerical Modelling Approach

a. **Urban heat assessment**
   In urban climate research, Numerical modelling is preferred due to flexibility and versatility for the investigation. In the past, computational power requirements were a major impediment for numerical modelling. Due to easy access to higher computational resources, in recent times, Numerical modelling has

become the predominant tool for urban climate research (Salata 2017). Numerical modelling typically can be classified under two different types, Energy Balance Models (EBM) and Computational Fluid Dynamics (CFD) (Toparlar et al. 2017). When compared to EBM models, the CFD approach has few inherent advantages.CFD enables higher spatial resolution and coupling with other aspects such as pollution. As a result, CFD is more taxing computationally than EBM models.

EBM's predict the surface energy balance fluxes (Oke 1982). Some of the examples of EBM's are the Town Energy Balance Model (TEB) (Masson et al. 2002), Temperatures of Urban Facets in 2D (TUF2D) (Krayenhoff and Voogt 2007), Met Office Surface Exchange Scheme (MOSES1T) (Best 2005). The representation of the environment varies between different EBM's. The main points of contention are the depiction of the built environment and vegetation. Based on the type of built environment modelling the EBM's are classified as slab models (MOSESIT), single layer (TEB) and multi-layer models (TUF2D). Single and multi later models are sometimes also referred to as Urban Canopy Models (UCB). Vegetation can be represented in a couple of different ways when it comes to EBM's. It can either be defined as a separate surface or tile (TEB, MOSES) or embedded into the urban area (MOSESIT).

CFD can be done at multiple scales, mesoscale and microscale. CFD analysis in mesoscale is called Mesoscale Meteorological Models (MMM). Historically, this was done two-dimensionally. In recent times, three-dimensional MMM is also possible. Also, another approach is to couple mesoscale models with models at a lower scale. The mesoscale models serve as input to the coupled model. CFD analysis done at less than 2 km^2 is considered microscale (Toparlar et al. 2017). At this scale, various parameters like thermal comfort, wind flow, pollutant dispersion can be studied. Typical input parameters for CFD simulation include the computational domain and the boundary conditions along with meteorological parameters.

Envi-met is CFD bases simulation software and is widely accepted for studying outdoor microclimate scenarios in an urban setting at a neighbourhood scale (De and Mukherjee 2017; Middel et al. 2014; Johansson et al. 2014). Boundary conditions need to be defined based on the scale and context of the site; the model's grid size dictates the simulation process's detail. Increasing the number of grids in the three dimensions dramatically increases the simulation time. The following parameters are also needed to perform the numerical simulation.

- Meteorological Parameters: Parameters generally used but not restricted to air temperature, surface and ground temperature (vegetation modelling), relative humidity, wind direction and velocity, globe temperature (essential for thermal comfort calculations) etc.
- Material Parameters: Generalized predominant construction materials and surfaces like roads and pavements should be included in the modelling process
- Vegetation parameters: Vegetation has a significant impact on urban heat. Vegetation type, the root and canopy structure have to be modelled accurately.

- Morphological Parameters: Morphology of the built forms has to be accurately modelled using site photographs and satellite imagery (Edupuganti and Mukherjee 2017)

ENVI-met numerical modelling process can give more accurate results if forced with fixed instrument measurements onsite. Multiple instruments can be strategically placed and the collected data is used in forcing the simulation process. These instruments are placed at the ground level (sensor at the height of 2.1 m) to collect mot of the parameters like air temperature, relative humidity and globe temperature. For wind parameters, a separate instrument is installed at a height of 10 m. A minimum logging interval of 30 min is ideal for the simulation process. Using the simulated results, various heat specific indices like PET, Mean radiant Temperature, ET can be output spatially and temporally (Fig. 11.5). Other Simulation softwares also have a very similar structure in terms of boundary conditions and various input parameters.

b. **Urban flood assessment**

Numerical modelling for urban flood assessment aims to characterize real hydrologic features and systems through computer simulation and mathematical analogues. Each individual flow of a hydrological system and the various related parameter (climate, soil and land) are modelled for analysis. These models can

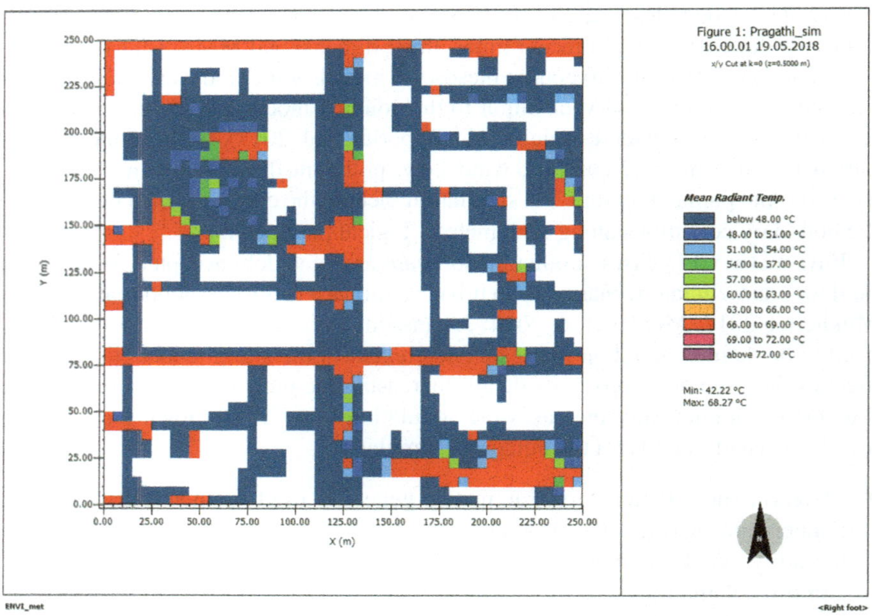

Fig. 11.5 Numerical approach using Envi-Met simulation modelling. Mean radiant temperature is the output of the simulation, which has strong correlation to heat stress and thermal comfort (Edupuganti and Mukherjee 2017)

be further classified into three categories 1D, 2D and 3D models depending upon the resolution of the flow in the floodplain.

1. **1D model**
   1D models are the set of models which represent the flow as one dimensional along the centre line in a floodplain (Brunner 2016). These models employ the idea of mass and momentum conservation between two cross-sections and solve the equations derived from them. HEC-RAS and MIKE 11 are some of the most commonly employed 1D models for flood simulation.
   Hydrologic Engineering Centre's River Analysis System (HEC-RAS) is a free 1D software that can perform steady and unsteady flow hydraulics, sediment transport/mobile bed computations and water temperature modelling.
   MIKE 11 can perform flood analysis, dam break analysis, water quality analysis, sediment transport analysis, optimization of reservoir and river structure operations, river salinity intrusion analysis, integrated flood and catchment modelling and wetland restoration studies. Some of the studies using MIKE 11 are river stage simulation (Panda et al. 2010), groundwater response investigation to overland flow and topography (Liu et al. 2007), lowland wet grassland modelling (Thompson et al. 2004) and water resource management (Doulgeris et al. 2012).
2. **2D model**
   The 2D set of models resolve the floodplain flow in two dimensions. These models are the most predominantly used models for urban flood investigation. These models offer a comprehensive representation of flow hydrodynamics and small scale topographic features, which in turn is a huge advantage in flood risk estimation and flood extent mapping. FLO 2D, TUTFLOW, SOBEK and MIKE 21 are some of the examples of 2D models.
3. **3D model**
   While 1D and 2D models are easier to run, they are plagued by various limitations such as assumption of hydrostatic pressure and viscous shear stresses and bed friction on fluid components. Models based on three-dimensional flow of water employ Navier-Stoke equations. Thus overcoming the shortcomings of 1D and 2D models. TELEMAC is one of the 3D models available for flood analysis.
4. **Coupled models**
   Recent advances in numerical modelling have given rise to a family of coupled models. These models combine the approach discussed above to minimize the limitations. Many techniques have been developed to link 1D and 2D models. These include a lateral link, longitudinal link and vertical link. They model the exchange flows between 1 and 2D models differently and are bound by limitations such as: omitting momentum exchange between channel and floodplain; neglecting return flow from floodplain

back to channel, or assuming the flow is purely one dimensional until the river reaches bankfull stage and then switching to 2D modelling.

11.5.2.3 Contemporary Technologies

The advancement of computational resources and technologies has bought various novel techniques to the fore. These technologies have opened up new avenues of research and methodologies. Some of these approaches are big data, machine learning, artificial intelligence, IoT (Rathore et al. 2018). Advances in sensing have made the Internet of Things (IoT) technologies possible. These enable real-time observations of environmental parameters. By installing a network of sensors, real-time patterns in the data can also be established. Also, there has been a proliferation of geolocated smartphone usage, social media, cloud-based remote sensing data and various other public databases. There is also a vast network of CCTV networks available in urban areas. This enables a heightened level of spatial and temporal information flows. This massive input of data can be analyzed through machine learning, artificial intelligence and used in various domains of energy efficiency, smart urban planning, weather forecast, natural disaster management, etc. Further, big data enables observing and monitoring, understanding, predicting and optimizing in real-time (Hassani et al. 2019). These various technological advances can be combined to take advantage of the large datasets available. These advances will enable cities to be smart and sentient.

11.6 Conclusion

This chapter introduces and discusses a framework for heat and flood risk analysis and checks the suitability of resilience-building solutions in urban areas. Various tools and techniques employed for climate risk assessment and their contextual suitability and limitations are discussed in detail. In ground reality, data collection networks have become more proliferated and are lesser of an issue. But, more needs to be done in analyzing the data while adhering to accepted scientific processes. A thorough analysis will result in more precise risk identification for the present and the future. Urban heat and urban floods are two of the most predominant risks associated with an urban area due to the integral change in natural fabric brought about by unsustainable development in the region. The specific factors which lead and contribute to these phenomena are inherent to the contemporary urbanization parameters. Understanding each parameter specific to the two risks identified is the first step in proposing contextual ecosystem-based solutions for resilience building against these risks.

Once the specific parameters are identified and their behaviour analyzed, a quantitative estimation is necessary to establish the problem's extent. The various specific risk-related indices help in scoping the issue and establishing the need for intervention. The human-centric approach of these indices integrates the risk to its impact on human beings. After scoping the risks, identifying the areas most susceptible to them is of utmost importance as it helps in specifying the areas in need of intervention. This framework is essential for proper identification of risks and eventual implementation of mitigative solutions. Further, this process will also inform policy and decision making. The tools and techniques discussed above provide a basket of analysis approaches, which can be used with respect to the problem's context.

An ecosystem-based solution is the way forward for resilience building in urban areas and the above-discussed framework helps in employing them.

Bibliography

Akbari H, Konopacki S (2005) Calculating energy-saving potentials of heat-island reduction strategies. Energy Policy 33(6):721–756. https://doi.org/10.1016/j.enpol.2003.10.001

Bansal N, Mukherjee M, Gairola A (2017) Smart cities and disaster resilience, pp 109–122. https://doi.org/10.1007/978-981-10-2141-1_8

Batica J, Gourbesville P, Hu F-Y (2013) Methodology for flood resilience index. In: International conference on flood resilience: experiences in Asia and Europe

Best MJ (2005) Representing urban areas within operational numerical weather prediction models. Bound -Lay Meteorol 114:91–109. https://doi.org/10.1007/s10546-004-4834-5

Brunner GW (2015) HEC-RAS river analysis system hydraulic reference manual version 5.0. In: Hydrologic engineering center, pp 547

Chen YC, Matzarakis A (2014) Modification of physiologically equivalent temperature. J Heat Island Inst Int 9 (2)

Christen A, Vogt R (2004) Energy and radiation balance of a central European city. Int J Climatol 24(11):1395–1421. https://doi.org/10.1002/joc.1074

Coutts AM, Tapper NJ, Beringer J, Loughnan M, Demuzere M (2012) Watering our cities: the capacity for Water Sensitive Urban Design to support urban cooling and improve human thermal comfort in the Australian context. Prog Phys Geogr 37(1):2–28. https://doi.org/10.1177/0309133312461032

De B, Mukherjee M (2017) Optimisation of canyon orientation and aspect ratio in warm-humid climate: case of Rajarhat Newtown, India. Urban Clim 0–1. https://doi.org/10.1016/j.uclim.2017.11.003

De Freitas CR, Grigorieva EA (2015) A comprehensive catalogue and classification of human thermal climate indices. Int J Biometeorol 59(1):109–120. https://doi.org/10.1007/s00484-014-0819-3

Doulgeris C, Georgiou P, Papadimos D, Papamichail D (2012) Ecosystem approach to water resources management using the MIKE 11 modeling system in the Strymonas River and Lake Kerkini. J Environ Manage 94(1):132–143. https://doi.org/10.1016/j.jenvman.2011.06.023

Droste AM, Pape JJ, Overeem A, Leijnse H, Steeneveld GJ, Van Delden AJ, Uijlenhoet R (2017) Crowdsourcing urban air temperatures through smartphone battery temperatures in São Paulo, Brazil. J Atmos Oceanic Technol 34:1853–1866. https://doi.org/10.1175/JTECH-D-16-0150.1

Edupuganti SR, Mukherjee M (2017) Impact of unregulated developments in urban villages of Hyderabad city on outdoor thermal comfort. Cities People Places 2019. Int J Urban Environ. ISSN: 2345-9530

Fanger PO (1970) Thermal comfort. Analysis and applications in environmental engineering. In: Thermal comfort. Analysis and applications in environmental engineering

Givoni, B. (1963). *Estimation of the effect of climate on man : development of a new thermal index.* Hebrew University.

Hassani H, Huang X, Silva E (2019) Big data and climate change. Big Data Cogn Comput 3(1):12. https://doi.org/10.3390/bdcc3010012

Hazarika N, et al (2018) Assessing and mapping flood hazard, vulnerability and risk in the Upper Brahmaputra River valley using stakeholders knowledge and multicriteria evaluation (MCE). J Flood Risk Manag 11:S700–S716. https://doi.org/10.1111/jfr3.12237

Hulley GC, Hook SJ (2009) Intercomparison of versions 4, 4.1 and 5 of the MODIS land surface temperature and emissivity products and validation with laboratory measurements of sand samples from the Namib desert, Namibia. Remote Sens Environ 113(6):1313–1318. https://doi.org/10.1016/j.rse.2009.02.018

Hulley GC, Hughes CG, Hook SJ (2012) Quantifying uncertainties in land surface temperature and emissivity retrievals from ASTER and MODIS thermal infrared data. J Geophys Res Atmos 117(23). https://doi.org/10.1029/2012JD018506

Jendritzky G, de Dear R, Havenith G (2012) UTCI—why another thermal index? Int J Biometeorol 56(3):421–428. https://doi.org/10.1007/s00484-011-0513-7

Jenerette GD, Harlan SL, Buyantuev A, Stefanov WL, Declet-Barreto J, Ruddell BL, et al (2015) Micro-scale urban surface temperatures are related to land-cover features and residential heat related health impacts in Phoenix, AZ USA. Landscape Ecol 31(4):745–760. https://doi.org/10.1007/s10980-015-0284-3

Jiménez-Muñoz JC, Sobrino JA (2004) A generalized single-channel method for retrieving land surface temperature from remote sensing data. J Geophys Res 108. https://doi.org/10.1029/2003JD003480

Johansson E, Thorsson S, Emmanuel R, Krüger E (2014) Instruments and methods in outdoor thermal comfort studies—the need for standardization. Urban Clim 10(P2):346–366

Kalkstein LS (1991) A new approach to evaluate the impact of climate on human mortality. Environ Health Perspect 96:145–150. https://doi.org/10.1289/ehp.9196145

Katić K, Li R, Zeiler W (2016) Thermophysiological models and their applications: a review. In: Building and environment, vol 106. Elsevier Ltd. , pp 286–300. https://doi.org/10.1016/j.buildenv.2016.06.031

Kotharkar R, Bagade A (2018) Evaluating urban heat island in the critical local climate zones of an Indian city. Landscape Urban Plan 169:92–104. https://doi.org/10.1016/j.landurbplan.2017.08.009

Krayenhoff ES, Voogt JA (2007) A microscale three-dimensional urban energy balance model for studying surface temperatures. Bound-Layer Meteorol 123(3):433–461. https://doi.org/10.1007/s10546-006-9153-6

Lee DHK (1958) Proprioclimates of man and domestic animals. Climatol Arid Zone Res X 102–125

Liu HL, Chen X, Bao AM, Wang L (2007) Investigation of groundwater response to overland flow and topography using a coupled MIKE SHE/MIKE 11 modeling system for an arid watershed. J Hydrol 347(3–4):448–459. https://doi.org/10.1016/j.jhydrol.2007.09.053

Liu Z, He C, Zhou Y, Wu J (2014) How much of the world's land has been urbanized, really? A hierarchical framework for avoiding confusion. Landscape Ecol 29(5):763–771. https://doi.org/10.1007/s10980-014-0034-y

Manteghi G (2015) Influence of street orientation and distance to water body on microclimate temperature distribution in tropical coastal city of Malacca. Int J Appl Environ Sci 10(2):973–6077. http://www.ripublication.com

Masson V, Grimmond CSB, Oke TR (2002) Evaluation of the town energy balance (TEB) scheme with direct measurements from dry districts in two cities. J Appl Meteor 41:1011–1026. https://doi.org/10.1175/1520-0450(2002)041%3c1011:EOTTEB%3e2.0.CO;2

Mathew A, Sreekumar S, Khandelwal S, Kaul N, Kumar R (2016) Prediction of land-surface temperatures of jaipur city using linear time series model. IEEE J Select Top Appl Earth Observ Remote Sens 9(8):3546–3552. https://doi.org/10.1109/JSTARS.2016.2523552

Mayer H, Höppe P (1987) Thermal comfort of man in different urban environments. Theoret Appl Climatol 38(1):43–49. https://doi.org/10.1007/BF00866252

Meier F, Fenner D, Grassmann T, Otto M. Scherer D (2017) Crowdsourcing air temperature from citizen weather stations for urban climate research. Urban Clim 19:170–191. https://doi.org/10.1016/j.uclim.2017.01.006

Middel A, Häb K, Brazel A, Martin C, Guhathakurta S (2014) Impact of urban form and design on mid-afternoon microclimate in phoenix local climate zones. Landsc Urban Plan 122:16–28

Nichol JE, Fung WY, Lam K, se, & Wong, M. S. (2009) Urban heat island diagnosis using ASTER satellite images and "in situ" air temperature. Atmos Res 94(2):276–284. https://doi.org/10.1016/j.atmosres.2009.06.011

Oke T (1982) The energetic basis of urban heat island. Q J R Meteorol Soc 108:1–24. https://doi.org/10.1002/qj.49710845502

Oke T (2004b) Initial guidance to obtain representative meteorological observations at urban sites. World Meteorol Organ 81:51

Oke TR (2004a) Siting and exposure of meteorological instrument at urban sites. In: 27th NATO/CCMS international technical meeting on air pollution modelling and its application, vol 1, pp 25–29. https://doi.org/10.1007/978-0-387-68854-1_66

Oke T, Canada (2006) Initial guidance to obtain representative meteorological observations at urban sites

Panda RK, Pramanik N, Bala B (2010) Simulation of river stage using artificial neural network and MIKE 11 hydrodynamic model. Comput Geosci 36(6):735–745. https://doi.org/10.1016/j.cageo.2009.07.012

Pandey AK, Singh S, Berwal S, Kumar D, Pandey P, Prakash A et al (2014) Spatio–temporal variations of urban heat island over Delhi. Urban Clim 10(P1):119–133. https://doi.org/10.1016/j.uclim.2014.10.005

Pickup J, Dear RD (2000) An outdoor thermal comfort index (Out-Set*)-Part I –the model and its assumptions. In: de Dear R, Kalma J, Oke T, Auliciems A (eds) Biometeorology and urban climatology at the turn of the millennium. World Meteorological Organization, pp. 1–7. https://www.researchgate.net/publication/268983313

Qin Z, Karnieli A, Berliner P (2001) A mono-window algorithm for retrieving land surface temperature from Landsat TM data and its application to the Israel-Egypt border region. Int J Remote Sens 22(18):3719–3746. https://doi.org/10.1080/01431160010006971

Rahul A, Mukherjee M, Sood A (2020) Impact of ganga canal on thermal comfort in the city of Roorkee, India. Int J Biometeorol 1–13. https://doi.org/10.1007/s00484-020-01981-2

Rathore A, Rao S, Rajasegarar S, Vanz E, Gubbi J, Palaniswami M (2018) Real-time urban microclimate analysis using internet of things. IEEE Internet Things J 5(2):500–511. https://doi.org/10.1109/JIOT.2017.2731875

Roth M, Chow WT (2012) A historical review and assessment of urban heat island research in Singapore. Singap J Trop Geogr 33(3):381–397. https://doi.org/10.1111/sjtg.12003

Salata F, et al (2016) Urban microclimate and outdoor thermal comfort. A proper procedure to fit ENVI-met simulation outputs to experimental data. Sustain Cities Soc 26:318–343. Elsevier B.V. https://doi.org/10.1016/j.scs.2016.07.005

Sharma S, Rao G, Bhanumurthy V (2012) Development of village-wise flood risk index map using multi-temporal satellite data: a study of Nagaon district, Assam, India. Curr Sci 103(6):705–711. Retrieved 26 Aug 2020, from http://www.jstor.org/stable/24088804

Staiger H, Laschewski G, Grätz A (2012) The perceived temperature—a versatile index for the assessment of the human thermal environment. Part A: scientific basics. Int J Biometeorol 56(1):165–175. https://doi.org/10.1007/s00484-011-0409-6

Steadman RG (1984) A universal scale of apparent temperature. J Climate Appl Meteorol 23(12):1674–1687. https://doi.org/10.1175/1520-0450(1984)023%3c1674:AUSCAT%3e2.0.CO;2

Stewart ID, Oke TR (2012) Local climate zones for urban temperature studies. Bull Am Meteor Soc 93(12):1879–1900. https://doi.org/10.1175/BAMS-D-11-00019.1

Thom EC (1959) The discomfort index. Weatherwise 12(2):57–61. https://doi.org/10.1080/00431672.1959.9926960

Thomas G, Sherin AP, Ansar S, Zachariah EJ (2014) Analysis of urban heat island in Kochi, India, using a modified local climate zone classification. Procedia Environ Sci 21:3–13. https://doi.org/10.1016/j.proenv.2014.09.002

Thompson JR, Sørenson HR, Gavin H, Refsgaard A (2004) Application of the coupled MIKE SHE/MIKE 11 modelling system to a lowland wet grassland in southeast England. J Hydrol 293(1–4):151–179. https://doi.org/10.1016/j.jhydrol.2004.01.017

Tobler W (1970) A computer movie simulating urban growth in the Detroit region. Econ Geogr 46(Supplement):234–240

Tomlinson CJ, Chapman L, Thornes JE, Baker C (2011) Remote sensing land surface temperature for meteorology and climatology: a review. In: Meteorological applications, vol 18, Issue 3. Wiley Ltd., pp 296–306. https://doi.org/10.1002/met.287

Toparlar Y, Blocken B, Maiheu B, van Heijst G (2017) A review on the CFD analysis of urban microclimate. Renew Sustain Energy Rev 80(January):1613–1640

U.S. Environmental Protection Agency (2008) Reducing urban heat islands: compendium of strategies. Draft. https://www.epa.gov/heat-islands/heat-island-compendium

Wan Z (2008) New refinements and validation of the MODIS land-surface temperature/emissivity products. Remote Sens Environ 112(1):59–74. https://doi.org/10.1016/j.rse.2006.06.026

Wan Z, Dozier J (1996) A generalized split-window algorithm for retrieving land-surface temperature from space. IEEE Trans Geosci Remote Sens 34(4):892–905. https://doi.org/10.1109/36.508406

Yaglou CP, Minard D (1957) Control of heat casualties at military training centers. A.M.A. Arch Ind Health 16(4):302–316. http://europepmc.org/abstract/MED/13457450

Zhu X, Dai Q, Han D, Zhuo L, Zhu S, Zhang S (2019) Modeling the high-resolution dynamic exposure to flooding in a city region. Hydrol Earth Syst Sci 23(8):3353–3372. https://doi.org/10.5194/hess-23-3353-2019

Chapter 12
Scaling-up Nature-Based Solutions for Mainstreaming Resilience in Indian Cities

Shalini Dhyani, Rudrodip Majumdar, and Harini Santhanam

Abstract Rapid growth of urban areas has attracted foremost global attention in the last few decades. By 2050, India is expected to be the center of urbanization with 68 Indian cities having population of around a million. Rapid urbanization has resulted in huge loss of urban blue-green infrastructure (BGI) resulting in loss of biodiversity and ecosystem services. There is growing need of disaster proofing fast expanding Indian cities. It is fundamental to explore, customize and integrate Nature-based Solutions (NbS) based resilience planning by restoring existing BGI in Tier I Indian megacities to curb the growing risks within acceptable limits. There is huge scope for exploration and mainstreaming of NbS in Tier II and Tier III cities that have space to plan BGI as per the growing population requirements and anticipated future risks. Scientifically planned NbS can help towards building resilience by maintaining the socio-cultural, economic, and ecosystem sustainability. The present chapter provides an overview of BGI as an efficient NbS and mainstreaming of nature-based interventions to ensure sustainable habitat in Indian megacities and growing small cities. The chapter highlights the factors that determine the use of urban NbS and underlines the problems associated with the implementation of NbS.

Keywords Megacities · Tier II cities · Nature-based solutions · Blue-green infrastructure · Urban resilience

12.1 Introduction

Despite their insignificant land occupancy, urban areas have revealed their intense, powerful, and unanticipated impacts on global environments (Chen et al. 2020). Global urbanization has resulted in high concentrations of urban population in hazard-prone, sensitive and vulnerable regions (Andersson et al. 2017). Although,

S. Dhyani (✉)
CSIR-National Environmental Engineering Research Institute, Nagpur, Maharashtra 440020, India

R. Majumdar · H. Santhanam
National Institute of Advanced Studies, Bangalore, Karnataka 5560012, India

© The Author(s), under exclusive license to Springer Nature Singapore Pte Ltd. 2021

M. Mukherjee and R. Shaw (eds.), *Ecosystem-Based Disaster and Climate Resilience*, Disaster and Risk Research: GADRI Book Series,
https://doi.org/10.1007/978-981-16-4815-1_12

urbanization depreciates environment, due to substantial loss of biodiversity, degraded ecosystems, fragmented landscapes, and altered climate systems (Haase et al. 2013) however, urban areas have been considered as a fundamental link to attain regional and global sustainability (Lovell and Taylor 2013; Wu et al. 2015). The enhanced focus on sustainability, that has been well proposed in Urban Sustainable Development Goal (SDG) 11 as one of the pertinent SDG by the UN in 2015 to develop sustainable cities by 2030, has brought growing urbanization under focus (Bhattacharya 2018). Projections by the UN have predicted nearly 70% of the global population to live in urban areas by 2050 (UN 2005). This increasing urban population is expected to exhibit a higher congregation in India and China, the predominant Asian countries. India has witnessed an unprecedented growth in urbanization over the past few decades and is steadily moving towards housing no less than 1/8th portion of the urban population of the world by 2050. By 2030, India is expected to have around 68 cities with more than 1 million populations (McKinsey 2010). By 2050, urban India is expected to be having more than 14% of the global urban population that will lead to remarkable alteration of landscapes, financial arrangements, and also the societal setups (Swerts et al. 2013). The concentration of urbanization has never been limited to the megacities in the country and has significantly influenced rapid expansion and growth in Tier I and Tier II cities1 that will lead to major urban transformations in coming decades. The cities that have undergone rapid urbanization in last few decades in the country are mostly of Tier I and Tier II cities (McKinsey 2010). Urban areas in developing and underdeveloped countries are continuously facing challenges because of the rapid change in land use land cover, old drainage systems, enhanced frequency of flash flood incidences, fatal heat stress, and droughts (Liu et al. 2014; Majidi et al. 2019). Decreasing green cover in urban areas and increasing infrastructural development indicated by enhanced levels of concretized surfaces have resulted in higher average city temperatures, heat islands, and diminished rate of aquifer recharge (Lahoti et al. 2019a, b). Rapidly growing urban sprawls in the country have augmented human dominated landscapes and resulted in enhanced dependency of growing urban population on existing formal and informal blue-green infrastructure in the city. The crucial challenge of developing the sustainable urban growth plan for the country that can be effectively implemented in megacities and have similar success rates for growing smaller urban sprawls has equally attracted urban planners and policy-makers (Bhattacharya 2018). So far, the governing authorities in India have mostly taken up initiatives by focusing on basic infrastructure amenities and benefits without much needed attention to achieve sustainability in their approaches (Randhawa and Kumar 2017).

With a view towards solving some of the key challenges of growing urbanization across the world, i.e., environmental vulnerability, human wellbeing, prevailing policy documents are shifting their focus from ecosystem-based approaches to Nature-based solutions (NbS) (Dhyani et al. 2020; Dhyani 2019; Raymond et al. 2017). In order to ensure urban resilience, globally NbS and their importance have continuously been highlighted to mainstream them in urban planning for addressing growing environmental vulnerabilities (Suárez et al. 2018). There is growing recognition of NbS to address growing disaster risks in fast expanding urban sprawls (Dhyani et al. 2018). The potential of an array of NbS options viz. green infrastructure, natural infrastructure, urban green spaces, urban forests, ecological engineering, urban agriculture, Eco-DRR (ecosystem-based disaster risk reduction), EbA (ecosystem-based adaptations), has been approved by multiple stakeholder groups to deliver cost efficient and sustainable outcomes for climate adaptation, mitigation and disaster risk reduction (Armson et al. 2013; Dhyani and Thummarukuddy 2016; Raymond et al. 2017; Dhyani et al. 2018, 2019, 2020). Interface of plausible influences of climate variability and unrestrained urban developments is expected to have severe strain on urban planning for small cities (Reckien et al. 2015). NbS and resilience are complementary approaches to counter the problems linked with environmental change in urban planning (Suárez et al. 2018) are considered to be fundamental for achieving urban sustainability (Wendling et al. 2018; Acharya et al. 2020). Urban blue-green infrastructure (BGI) can bridge the gap between nature and urban dwellers by developing resilience and reducing environmental vulnerability to emerging and present stresses and threats. There are evidences to endorse benefits of NbS mainstreaming in urban areas that include but are not limited to improving air quality (Calfapietra et al. 2015), urban ecosystems (Oliveira and Mell 2019), heat islands (Makido et al. 2019), flash floods (Majidi et al. 2019), public health (Bennett et al. 2015) and overall urban sustainability (Perez and Perini 2018). The strategic approach for exploration and identification of an appropriate NbS assessment scheme that cautiously supports one or more sub-goals of urban SDG 11 can support to exploit focused ability by developing collaborations in various urban schemes (Wendling et al. 2018; Acharya et al. 2020). Due to limited municipal autonomy, there is growing consideration on lack of sufficient financial support for implementation of NbS and that is where public expenses on NbS will be important to understand (Droste et al. 2017). Mainstreaming NbS will need larger partnerships amongst diverse policy areas, sectors, and stakeholders (Bhardwaj et al. 2020). It will require broader collaboration and support that includes financial support from private sector, much needed local participation, for realizing and localizing the NbS applications and benefits in urban areas (van Ham and Klimmek 2017). In the context of NbS, urban green spaces (UGS) and blue-green infrastructure (BGI) are relevant state-of-the-art solutions to address emerging environmental concerns (Lovell and Taylor 2013). Although the BGIs often do not span over large areas, they are frequently characterized by high levels of species richness and different microhabitats within the city limits (Ghofrani et al. 2017). There is growing need for smart use of available urban open spaces and integration of urban BGI into the built-up infrastructure of the growing urban sprawls. BGI, being a network of natural and near-natural areas, has a positive effect on the quality of urban environment (Well and Ludwig 2020). In order to understand the growing urban systems broadly, the rewards provided by urban green and blue spaces/infrastructure are important to be understood. The ecosystem services (ES) from the BGI in the urban areas encompass microclimate regulation (Stewart and Oke 2012), flood regulation (Sörensen and Emilsson 2019), carbon sequestration (Chen 2015), noise pollution (Dzhambov and Dimitrova 2015), air pollution control (Hewitt et al. 2020) and many others. Owing to the growing urban sprawls, most of the cities in India have undergone haphazard growth with just a few outliers. There has been no specific consideration of including urban blue-green spaces to enhance resilience against the warming world. Most of the major megacities are on the threshold of getting toppled in terms of the fine balance that they struggle to maintain for supporting basic livable services to growing urban population; whereas, the Tier II cities still have scope to localize and mainstream appropriate and customized NbS with a view to reducing the climate-induced impacts on the modern urban habitat (Aithal and Ramachandra 2012). Green infrastructure (GI) planning in India has the potential to streamline current urban development constraints relating to loss of green infrastructure and rapid urban expansion (Dhyani et al. 2019). Potential funding to protect, manage and restore green infrastructure through state interactions can provide opportunity to plan economic growth *vis a vis* ecological sustainably effectively (Mell 2015).

This chapter attempts to address the larger sensitivity of growing urban sprawls of India following NbS that has multiple co-benefits as compared to the short term grey engineered solutions. The present work provides an overview of key NbS that can be included and implemented based on the site-specific requirements of different cities in different agro-climatic zones of the country. The chapter provides an overview about mainstreaming NbS in cities to target long-term climate resilience and disaster risk reduction following Sendai Framework of Disaster Risk Reduction, 2015. With a focus on understanding the role and importance of NbS in urban planning, the chapter provides an overview of implementing NbS in two important cities that are present in different stages of urbanization. The garden city of India-Bengaluru, which used to have the most diverse urban green spaces previously, has lost a significant area under green cover in last few decades, has been taken as an example of an overgrown megacity. Nagpur's case emerges from the pursuit of studying the fate of BGI in a fast-growing Tier II city of India that was second greenest city at one point of time and has significantly lost green cover over the past few decades. While discussing the benefits of quintessential BGI, this work also tries to understand the key challenges in mainstreaming NbS by underlining the factors that determine the use of NbS and problems in implementation of NbS.

12.2 Identification of Appropriate NbS for Sustainability Concerns of Growing Urban Sprawls

The capacity of adaptation is an essential element towards urban resilience. For the development, as well as, the implementation of the resilience framework, the focus should be on creating more open green spaces to facilitate the sustainability of socio-cultural, commercial, and ecological attributes of the growing urban centers (Sharifi and Yamagata 2014). Investment in urban areas through the provision of creating efficiently planned biodiversity hotspots is believed to be a form of mitigation by natural space for anticipated environmental change phenomenon. Facing the onslaught of accelerated urban sprawl, BGI such the gardens, parks, playgrounds, lakes, ponds, and wetlands are some of the key indicators of quality of life. The infrastructure development, land-use changes, and competing land use pose major threats to the urban BGI. It has been time and again affirmed and reaffirmed that the foremost importance should be assigned to the natural resource management (NRM) strategies while forming the core of compulsory binding guidelines during the course of developmental planning towards urban expansion. The NRM strategy includes but is not solely limited to plantation of trees, because several densely packed and heavily concretized city centers lack enough open space required for large-scale plantation. From a sustainability point of view, it is important to assess the co-benefits and trade-offs associated with each NbS in an urban environment before the design and establishment of the same with respect to the urban scales and boundaries. For example, the mutual association of small-scale NbS such as rain gardens, splash blocks, rain barrels, rain saucers, rain chain collectors, recharge pits, or bores installed in residential complexes can provide holistic opportunities to harvest stormwater and to remove pollutants if integrated with large-scale rainwater harvesting systems and underground cisterns to store the filtered water.

The co-benefits of employing such NbS options outweigh the tedious efforts required to install these systems at the levels of individual houses to a neighborhood. Compared to the implementation of other larger-scale NbS such as constructed wetlands or bio-retention swales, the investment on the holistic placement of these NbS would be comparatively less expensive. Constructed wetlands and bio-retention swales, with comparative co-benefits, can offer further the functions of secondary and or tertiary treatment of stormwater. However, the trade-off for choosing an NbS option for stormwater management is reliant on the fiscal and cultural factors. The choice and conceptualization of city-level planning of NbS requires inclusion of the spatial, economic, and policy-led considerations, as depicted schematically in Fig. 12.1.

Selection of a suitable NbS will depend on the nature and scale of interactions between an urban center's internal (I_s) and external (E_s) sustainability components, as well as, in the buffer spaces of their nexus. Let us take the examples of lake ecosystems, collectively serving as NbS themselves for urban environments as a part of the BGI, with strong I_s and E_s components. I_s encompasses the key factors, viz. surface area, depth, and water quality characteristics; while E_s includes factors such as the level of shore concretization, urban landscape development, presence of heat islands, etc. which function across multiple spatio-temporal scales. In the context of NbS, I_s will determine the frameworks suitable for enhancing the ecosystem services at a system level, whereas E_s will determine the selection of appropriate scales of operation as well as the economic feasibility of using the NbS. The presence or absence of buffer spaces (i.e., nexus) where I_s and E_s functions merge to provide the total ecosystem service of the BGI (i.e., lakes) are subjected to the constraints placed on policy frameworks and the economic features associated with a likely NbS option.

A typical example of the relationships between I_s and E_s can be illustrated by the use of permeable pavements as spaces of water-land exchanges (Santhanam and Majumdar 2020). In this case, the Is solutions require the construction of buffer spaces which will block and regulate excessive flow of solids as well as non-point

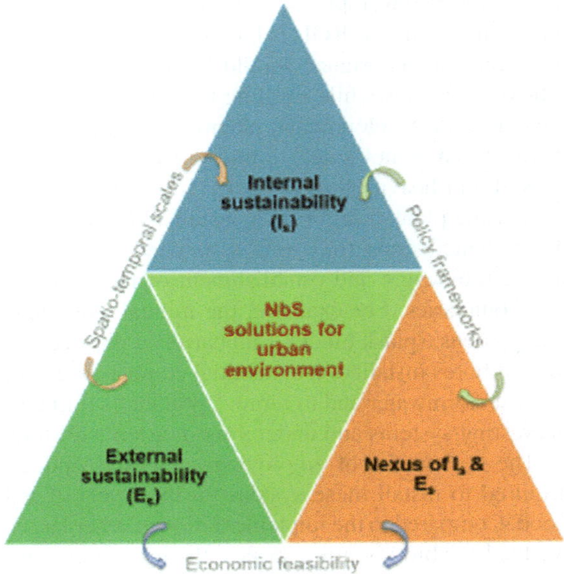

Fig. 12.1 Relationships between the sustainability components of an urban center and the choice of appropriate NbS

source streams, which can be satisfied by the use of permeable pavements. However, the scale of operation i.e., the area of laying the buffer area by the use of permeable pavements may be restricted to the shore area of the lake alone (in this case, the optimum E_s option); in contrast to the option of the existence of a natural wetland which is lost to dense urbanization. Nevertheless, the ecosystem function of the lake can be restored somewhat close to normal, when the NbS achieves a balance of I_s and E_s. Nevertheless, backing of the NbS options by strong policy frameworks, as well as, the illustration of their value-addition in terms of lesser environmental costs are very important for implementation of permeable pavements as NbS, which cannot be satisfactorily achieved without considering both I_s and E_s options.

Some of the pertinent NbS interventions that should be adopted and scaled up from Tier 1 to later Tier II and Tier III cities (Dhyani et al. 2020) are listed below:

1. The large-scale implementation of cost-effective NbS as BGI or GI, such as permeable pavement systems with integrated sub-surface recharge cisterns.
2. Effective use of brownfield spaces and growth of fringe green spaces and buffer zones.
3. Multi-scalable utilization of NbS with support from local designers. This integrating more natural products in building material that reduces energy, water, and carbon footprint.
4. Engagement of proper spatial planning and design to distribute NbS throughout urban areas.
5. Inclusion of NbS concepts in urban planning and design, policy formulation, and city-level decision-making processes

6. Encouragement of dialogues on the co-benefits, as well as, disservices and/or limitations associated with the NbS already implemented or planned.
7. Evolving flexible institutional frameworks, as well as, incentives to promote NbS at corporate and individual levels to realize the transformative potentials of NbS over long terms.
8. Mainstreaming NbS in decision-making, policy, and urban planning.

12.3 NBS Options for Urban Environment

Integration and conservation of the remaining green islands by linking them through green corridors, conservation of natural water bodies, restoration of the ecosystems, ecosystem services have not been envisaged during the planning of urban expansion and the developmental stages of infrastructure projects. Subsequently, the insufficient availability of open spaces in the urban settings has emerged as a major bottleneck in augmenting and effective planning of the GI. The standards for access to public green spaces vary from one country to another, the best available practices of per-capita provisions recommend green space of 20 m^2 which is "minimum of 1.25 ha open space per 1000 residents (Govindarajulu 2014). The Urban and Regional Development Plan Formulation Implementation (URDPFI) guidelines, 2014 by Ministry of Housing and Urban Affairs are used by local bodies, to determine the facility requirements (Udas-Mankikar 2020). URDPFI guideline recommends per-capita green space of 10–12 m^2 and 1.2–1.4 ha of green expanse per 1000 population[2]. There remains huge demand supply gap in accessible and available public green spaces to urban dwellers.

12.4 Case Study Approach to Understand the Context of BGI in Indian Cities

12.4.1 *Tier I City Context: Bengaluru*

The metropolitan area known as 'Greater Bangalore' (77° 37′ 19.54″ E and 12° 59′ 09.76″ N), comprising of the districts Bangalore Urban and Bangalore Rural, is the principal hub of the administrative, industrial, commercial, cultural and academic activities in the state of Karnataka (Sudhira and Nagendra 2013). This area has experienced an unprecedented pace of urban sprawl in recent times, primarily due to the dense industrial development in the fringe areas of the city, aimed at ensuring regional economic growth (Online Resources: Geo-visualization of Urbanization in Greater Bangalore). Based on the real estate growth, as well as, the cost of living, Bangalore is classified as a metropolitan (or Tier 1) city in India (Online Resources: The 31 Cities in India). To accommodate ever-growing population in the developing urban setting, large-scale concrete infrastructure has been developed all over the city without considering the long-term environmental impacts. Analyzes of land-use patterns exhibit a 584% growth in built-up area during the past four decades, whereas, the water bodies witnessed a shrinkage of about 74%. Historical records suggest that Bengaluru city had about 262 water bodies till 1960. Over the period of past five decades, the number has been reduced to 81. Out of the remaining water bodies, only 34 are currently recognized as live lakes.

Although, this city has been known as the 'Garden city', the green cover of the city has diminished alarmingly from 68% in 1973 to 6% in 2017 (Online Resources: Balram December 2018). Unplanned activities are slowly transforming the city into a concrete jungle, and in many localities, the city dwellers are facing a series of serious challenges, such as lack of basic amenities (e.g., electricity, water, and sanitation), enhanced level of GHG emissions and warmer urban microclimate, known as the Urban Heat Island effect (UHI). Studies indicate that urban ecosystems are rich in biodiversity and provide critical natural capital for climate change adaptation and mitigation. Hence, uncurbed urbanization and induced changes in the microclimate are likely to impart detrimental effects on biodiversity hotspots, urban species, and critical ecosystem services. Therefore, focus should be on maintaining the quality and quantity of green infrastructure to ensure better quality of life, improved human health, and social wellbeing (McPhearson et al. 2018). It is expected that by 2030, the number of houses will increase by about 70% in the key urban circles in India (Online Resources: Balram December 2018). Therefore, systematic long-term nature-based approaches are needed for Indian cities, so that critical urban and peri-urban ecosystem services (UES) are given due importance while preparing a broader strategy towards climate adaptation and mitigation. Strategies need to be prepared to ensure sustainable harnessing of the nature-based products, as well as, the services over the different time scales, ranging from medium to long-term, viz. 20, 50, and 100 years, as applicable (McPhearson et al. 2018). Ecosystem-based planning can prove to be instrumental in strengthening the linkages between urban, peri-urban, and rural ecosystems through participatory planning for management of NBS at both the cities, as well as, regional levels. Above all, an implementation-oriented, policy-based strong framework is needed to ensure energy-efficient, resilient, sustainable human habitat.

For ensuring effective urban–rural linkage, the multifunctionality of NbS needs to be evaluated as it can potentially result in high social return on investments by simultaneously addressing several societal goals both in the urban, as well as, the rural set-up (Droste et al. 2017). For example, in the urban and peri-urban zones, the green buildings can be made using natural and locally sourced materials, such as the straw bales, which possess the potential to replace the use of concrete, plaster, gypsum, and other building materials in walls (Online Resources: 7 Sustainable Construction Materials). Recently, amenities such as furniture and utensils made of bamboo have gained enough popularity owing to the light weight, durability, and aesthetic finish (Online Resources: The Benefits of Organic and Why Your Utensils Should be Organic too). Bamboo can be used for construction purposes as well, owing to its light weight, good tensile strength, and renewability, and subsequently, the need for the expensive imported materials can be significantly reduced (Online Resources:

Bamboo as a Building Material—its Uses and Advantages in Construction Works). The above-mentioned NbS options for the urban environment can be acquired from a rural setting on a regular basis, provided the rural people are incentivized and encouraged on a long-term basis towards growing and maintaining such natural resources. In this way, the urban and rural settings can forge a lasting bond through creation of 'green jobs'.

In terms of the internal and external sustainability components (Fig. 12.1), the trading of NbS options through the urban–rural connect would provide a sound transfer of knowledge for adopting specific NbS practices that would enhance the functionality of I_s and E_s components using city as a unit of development. For example, the use of 'buffer spaces' around urban centers could be considered an essential component to plan the local or regional-scale E_s options as mentioned above. The buffer space provides for nexus of the operability of the NbS operations, i.e., flow of options to maintain I_s that can strengthen the urban–rural linkage over longer periods of time.

12.4.1.1 Factors that Determine Mainstreaming of Urban NbS for Bengaluru—Tier I City

Four factors can affect the selection, as well as, the use of NbS options for a densely populated and economically important urban center such as Bengaluru city (Fig. 12.2). The need for coarser or finer spatial resolution of some selected GI or BGI options, as well as, their placements in the city itself can influence the level of their usage.

For example, the proximity of lake ecosystems within the Bengaluru city to the urban water bodies is known to influence the urban heat island characteristics to a large extent (e.g., Ramachandra and Kumar 2010; Ambinakudige 2011). However, closely located lake systems can add to overland flooding in times of excessive

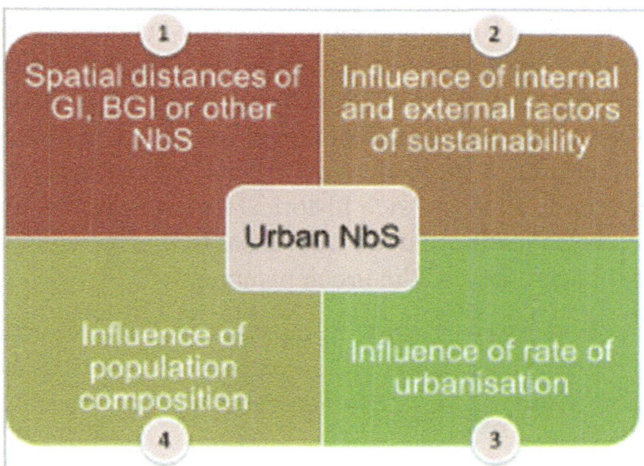

Fig. 12.2 Factors affecting the selection and performance of NBS options for an urban center to improve resilience

rainfall in the absence of interconnecting canal systems or *rajakaluves* which were historically in existence (Rajeshuni et al. 2017). Table 12.1 shows the list of NbS planned and/or implemented at various locations in Bengaluru urban area. Some of these options are illustrated in Fig. 12.3.

Each type of NbS planned or in existence has a few advantages and limitations, which are highlighted in Table 12.1. In the case of Bengaluru city (both urban and rural divisions), the above scenario is quite relevant as development around the city's several lakes progresses rapidly, converting the permeable buffer spaces around these lakes into impermeable concretized walkways which interrupt the normal exchanges of green and blue water with the actual lake ecosystems (Fig. 12.3). These factors in turn determine the success or the failure of the NbS adopted for the Tier 1 city as Bengaluru. However, some examples of NbS implementation in Bengaluru are largely attributable to the fiscal provisions and renewed active support from the government and non-government agencies alike, that have been instrumental in promoting sustainable development practices starting from the grassroots levels over the past 25 years. The positive outcomes of the public–private partnerships are evident at several locations in Bengaluru where NbS options have been adopted (Rajeshuni et al. 2017; Table 12.1). These NbS can serve as examples for planning the urban management of Tier 1 cities.

In order to ensure a sustainable built environment within the Bengaluru city premises, the concept of 'Green Building' is being promoted recently. As per the recent reports, the city of Bengaluru has 332 active green building projects registered with Indian Green Building Council (IGBC) (Online Resources: Salvekar March 2019). A green building consumes less water, conserves the natural resources around it, enhances energy efficiency, generates less waste, and provides healthier indoor environment for the occupants, as compared to a conventional building (Online Resources: Diwaker November 2018). Experts suggest that the construction cost is a mere 1% higher for green buildings; whereas the use of recycled water for toilet flushing, as well as, the reduction in usage of air-conditioning and cooling appliances can potentially lead to about 20–30% savings in the energy bills and around 30–50% savings in domestic water consumption (Online Resources: IGBC Green New Buildings Rating System version 3.0; Diwaker November 2018). The 'Green Buildings' and their adherence to the suggested norms are monitored by IGBC formed in the year 2001, which is a part of the Confederation of Indian Industry (CII) (Online Resources: Indian Green Building Council). Bangalore as Tier I megacity will require more proactive efficient, effective urban planning approaches for integrating BGI. A few rich historical experiences can also help in this direction. Here we have mentioned a few traditional BGI approaches that can help in mainstreaming and integrating BGI in urban planning:

- The tree groves in the city outskirts need to be monitored and preserved on a regular basis. The green coverage areas need to be expanded wherever possible. Fruit-bearing trees (e.g., *Psidium guajava*, *Ficus carica*, *Artocarpus heterophyllus* etc.)

Table 12.1 List of NbS options, their contributions to enhancement of ecosystem services and building urban resilience, their advantages and limitations as well as corresponding locations in the Bengaluru and Nagpur urban area

Type of NbS	Enhancement of ecological services	Contribution towards enhancing resilience	Advantage	Disadvantage	Some prominent locations in Bengaluru urban area	Some prominent locations in Nagpur urban area
Vertical greenery systems including green roofs and moss walls	Improve temperature regulation and sound isolation, reduce buildings' energy use	Mitigate urban heat island effect, mitigate excess energy consumption and provide means to manage local peaks in energy demand	Low cost of implementation, improving aesthetics as well as to regulate radiative properties, promote circular economy at community levels	Require monitoring, as well as maintenance, access to gardeners need supportive infrastructures like high-rise ladders etc.	Lalbagh, Hosur road electronic city flyover pillars, CSG international office spaces	Sitabuldi, Dharmpeth, Mahal, Gandibagh, Jaripatka, Sadar, Ajni, Ramdaspeth, Reshmibagh, Irwari
Integrated lake-park complexes	Improve the blue-green infrastructure, provide lung spaces for clean air, interactive socio-ecological centres, provide thermal regulation	Mitigate urban flooding, reduce urban heat island effect, filter air due to vehicular pollution	Add to effective socio-ecological management of urban infrastructure	Need periodic monitoring, support of active public-private participatory frameworks over long-term agreements; need to provide strong fiscal frameworks for management	Lalbagh. Sankey, Nagavara. Ulsoor, Rachenahalli, Jakkur, Hebbala lakes and many other lakes in Bengaluru	Futala, Ambazari, Gandhisagar, Sakkardara, Pandrabodhi Gorewada and many other lakes in Nagpur

(continued)

Table 12.1 (continued)

Type of NbS	Enhancement of ecological services	Contribution towards enhancing resilience	Advantage	Disadvantage	Some prominent locations in Bengaluru urban area	Some prominent locations in Nagpur urban area
Multifunctional, multi-modal public green spaces	Improve biodiversity, improve air quality	Mitigate urban heat island effect	Less space requirement	Reduction in number of by-lanes etc.	Cubbon park	127 small parks, gardens and playgrounds managed by NMC, NIT, Nagpur
Urban forestry and green belts	Decrease inner-city ambient air temperature	Sequester carbon, regulate the micro-climate, purify the air and reduce urban noise	Provide long-term green, belt area development in urban environments	Need for effective administrative involvement and financial investments for maintenance	Cubbon Park, UAS campuses, Lalbagh, Jakarabanda forest, Sankey forest belt	Ambazari, Seminary hills, Telankhedi garden, Gorewada reserve forests and other small *zudpi* jungles
Private community gardens and backyard forests	Enhance the air quality of neighbourhoods	Provide in-situ flood control and control of excess run-offs	Several possible designs and optimisations possible	Expensive option, may not be available to middle or lower-income households due to high maintenance needs	Individual private residences housing Miyawaki forests and backward forests, Indian Institute of Science Campus	Empress Van, Bharat van etc. and Campus forests of CSIR-NEERI, ICAR, VNIT etc.

(continued)

Table 12.1 (continued)

Type of NbS	Enhancement of ecological services	Contribution towards enhancing resilience	Advantage	Disadvantage	Some prominent locations in Bengaluru urban area	Some prominent locations in Nagpur urban area
Permeable pavement systems	Capturing over 60% of annual rainfall	Provide in-situ flood control and percolation of excess run offs into ground, promoting recharge of green water	Low cost of implementation and maintenance	Requires local administrative and civil bodies' involvement, requires good maintenance to preserved permeability characteristics	Several locations within Bengaluru city	Several densely populated and built up areas of Nagpur city
Converted brownfield sites into eco-friendly landscapes	Revival of ecosystem services	Avoid contamination of groundwater and as stagnant water-filled pools breeding disease vectors	Promote circular economy as well as promoting new investments on land-use	Require high-level administrative and judicial involvement and huge capital investments	HAL Airport (converted into cargo, defence and VIP airport; terminal converted into lounges and air traffic management buildings); proposal for extension of ex-Binny mills area in Malleshwaram into railway platforms for South Western Railways	Industrial park (MIHAN), 11 Maharashtra Industrial Corporation (MIDC) in Butibori as Asia's largest industrial suburb, Multi modal International Cargo Hub and Airport (MIHAN), International Airport and Special Economic Zone (SEZ)

(continued)

Table 12.1 (continued)

Type of NbS	Enhancement of ecological services	Contribution towards enhancing resilience	Advantage	Disadvantage	Some prominent locations in Bengaluru urban area	Some prominent locations in Nagpur urban area
Green ventilation or aeration spaces, open air-nooks, sky lighting including intuitive HVAC systems and natural lighting	Reduce buildings' energy use, promote easy wind and light circulation	Mitigate excess energy consumption and provide means to manage energy demand	Offer nature-based ventilation	Costs are dependent on the quality of construction, optimal design and maintenance criteria	Adobe office spaces, several hotels and restaurants	Research institutes, university, government and private offices, schools, colleges, hospitals
Mainstreaming sponge city concept by developing rain gardens, splash blocks, rain barrels, rain saucers, rain chain collectors, recharge pits or bores	Collect and remove pollutants from the stormwater	Improve local flood control, blue and green water harvesting	Use of native plants, local soil, and mulch to remove pollutants from water and to recharge groundwater, easy to construct and use	Need for constant monitoring and maintenance	Private residential complexes and house	All wards and residential complexes
Bio retention swales and constructed wetlands with or without floating islands	Collect rainwater and excess run-off	Offer adequate flood control, biological secondary and tertiary treatment options simultaneously	Easy to construct, long-term impacts	Need for constant monitoring and maintenance, initial investment may be high	Hebbagodi lake, Rachenahalli lake fringe	Naik Talav, Ambazari, Futala and Gandhi sagar lakes and also different *nala* in the city

(continued)

Table 12.1 (continued)

Type of NbS	Enhancement of ecological services	Contribution towards enhancing resilience	Advantage	Disadvantage	Some prominent locations in Bengaluru urban area	Some prominent locations in Nagpur urban area
Combined NbS with underground cisterns	Purified runoff can be stored for re-use	Periods of drought to prevent water scarcity	Easy to construct, long-term impacts	Need for constant monitoring and maintenance, initial investment may be high	Several residential apartments and townships	All wards and residential complexes
Urban farms and agriculture; Peripheral green fringe landscapes	Locally based urban and peri-urban food production; sequester carbon	Mitigate urban blight and urban food deserts, balance the loss of arable land to urbanisation	Promote social cohesion; provide alternative food systems	Can be commercialised leading to loss of intended NbS services	Bannerghatta, Krishnarajapuram areas	MIHAN and nearby areas
Rainwater, grey water and stormwater harvesting systems	Improve drainage connectivity	Recharge of groundwater to avoid droughts, blue and green water harvesting	Increase local to regional water availability	Need regular maintenance and monitoring	Most residential apartments, houses and townships, St. John's hospital campus, BMS college campus etc.	Most residential apartments, houses and townships, schools, colleges, government and private offices and institutes etc.

(continued)

Table 12.1 (continued)

Type of NbS	Enhancement of ecological services	Contribution towards enhancing resilience	Advantage	Disadvantage	Some prominent locations in Bengaluru urban area	Some prominent locations in Nagpur urban area
Curb side or avenue plantations, Optimized street trees networks	Optimised leaf area index, air pollution tolerance index, and water demand on local scales	Improve the thermal regulation within urban spaces, efficient corridors to link gardens and help mitigate urban heat island effect	Promote greenery without huge demands on space and water	Need for long-term planning and protection from tree thefts or vandalism	Small tree lanes are located within big residential communities, malls and workspaces at several sites Bengaluru	Strengthened network in entire Nagpur city
Biophilic buildings	Multicompartmental nature-based living/working spaces	Improve the biological integration of urban spaces with environment and mitigate heat island effects, regulate energy consumption, and add to urban flood-resilient landscapes	Integrated solutions offer numerous options for implementing NbS at different spatial scales	Expensive and not affordable for small institutions or residences	Indian Institute of Science campus, TITAN 'Integrity' workspace in Electronics City, CSG International offices	Research institutes, schools, colleges

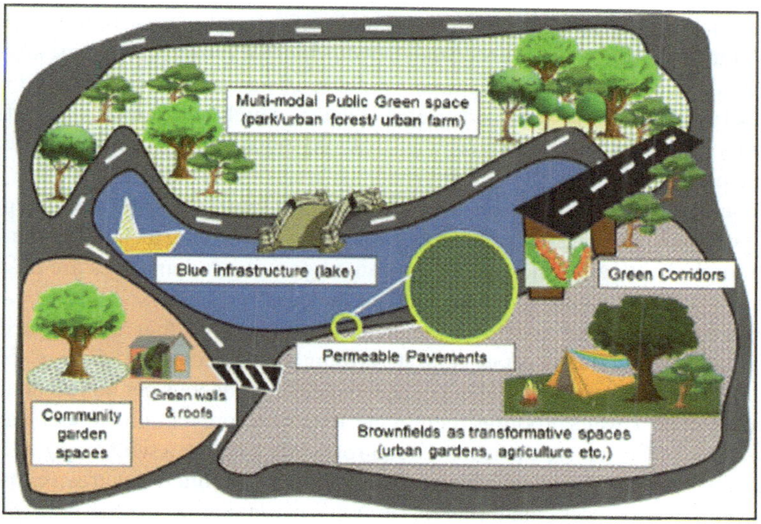

Fig. 12.3 Illustrations of different nature-based solutions (NbS) observed at different localities in fast-growing urban sprawls of India

need to be planted in higher numbers in such groves, as they would provide food sources to the frugivorous urban biodiversity, viz. a few species of fruit-bats, green pigeons and hornbills, palm civet cats, squirrels and tree-shrews, langurs and the bonnet macaques. Ample availability of food and presence of green canopy will reduce their interactions with the urban people, leading to a more harmonious, sustainable co-existence. Furthermore, the fruit-bats and the hornbills serve as active vectors for seed dispersal, and hence, they can help in regeneration of the deforested lands.

- Peri-urban groves have helped in curbing the rise in ambient temperature, aid in recharging the ground water table, and also meet the needs of local people (Online Resources: Khanna April 2016). The peri-urban areas of Bangalore used to have several such tree groves, popularly called 'Gundu Thopes', comprising of large trees, e.g., tamarind and mango trees; however, most of them have vanished due to the pressure of residential and commercial constructions. Remnant populations need immediate attention and appropriate protection, which must be accompanied by enforcement of stringent laws. Trees like Sal (*Shorea robusta*), Mahogany (*Swietenia mahagoni*) etc. should be planted and nurtured for the services like leaves and hard-wood timber.
- Perennial trees with dense foliage, viz. *Mimusops elengi* (Bakul), *Albizzia lebbek* (Siris), *Azadirachta indica* (Neem) help in enhancing the air quality of the urban settings. Pollution-free atmosphere is known to attract pollinating insects, which helps in further regeneration of urban green cover, leading to long-term better health of the city.

- The smaller gardens within residential and commercial places should be planted with medicinal herbs, such as Kalmegh (*Andrographis paniculata*), Vasaka (*Adhatoda vasica*), Fenugreek (*Trigonella foenum-graecum*) etc., in order to ensure conservation of these rare and useful species, which substantially elevates the quality of the urban biodiversity, as well as, enhances the services offered by these plants.
  New emerging BGI approaches that can help in mainstreaming and integrating BGI in urban planning include few proven green landscaping techniques that could be found effective for the space-constrained growing Indian cities are described below (Online Resources: Urban Green Landscaping, IGIN):
- Green roof: The green roofs refer to the roofs of buildings that are designed to be covered with suitable vegetation. Mat-forming plant species, i.e., *Sedum* sp. (e.g., *S. acre, S. rupestre*, and *S. album*), *Sempervivum*, herbaceous perennials and moss, planted over a waterproofing membrane to ensure prevention of seepages and erosion is one effective measure. Green roofs can assume a wide array of forms, ranging from roof gardens with raised beds and pots, to rolled-out green carpets, as well as, the shed-roofs containing low growing perennials (e.g., *Rudbeckia fulgida*, *Rudbeckia hirta* etc.), and dense, tufted, evergreen grasses (e.g., *Helictotrichon sempervirens*).

  Some of the key benefits of using green roofs in the urban areas are as below:

1. Reduces temperature of roof surface by up to 31 °C and the ambient temperature by about 4 °C, thereby aiding in the mitigation of UHI effect. It also reduces the cooling cost of the building by 25% by lowering the peak cooling load by 80%.
2. The planted vegetation filters airborne particles and is capable of removing about 80% of the suspended dirt particulates.
3. Green roofs absorb CO_2 and facilitate oxygen exchange in the atmosphere, thereby augmenting the flow of fresh air.
4. Planned rooftop farming can provide fresh vegetables that are grown on toxin-free soil (Dhyani et al. 2019). Hydroponic technology has the potential to take the rooftop farming to the next level, where soil-less farming (known as sky-farming) would be able to partially meet the fresh food requirements (Online Resources: Gardens grow up: nature reclaims cities with green walls and rooftop gardens; Farming in the Sky; Crumpacker February 2019).

- **Green Wall**: Green walls are important BGI in urban areas arranged in vertical structures with vegetative cover that are either incorporated inside a building or installed as free-standing units. These are popular among the prevalent vertical greenery systems (VGSs) (Online Resources: Urban Green Landscaping, IGIN). Generally, very low growing plants (less than a foot in height), viz. groundcovers (e.g., *Muhlenbergia capillaris*, *Bouteloua dactyloides*), ferns (e.g., *Dryopteris erythrosora*), perennial flowers (e.g., *Salvia coccinea*, *Sisyrinchium atlanticum*), small shrubs, are frequently used in the green wall panels (Online Resources: Plants for green walls) (Online Resources: Plants for green walls). Some of the key benefits of using green walls in the urban areas include:

1. Protection of the sides of a building from direct sunlight during the scorching summer sun, thereby reducing the need for indoor cooling appliances, leading to a considerable reduction (~74%) in energy consumption (Wong et al. 2009; Jaafar et al. 2013).
2. Reduction in the noise transmission via the acoustic insertion loss provided by the green insulation layer (Wong et al. 2010).
3. Improvement in the air quality by enhanced absorption of CO_2 and dust particulates.
4. Substantial reduction in the mean radiant temperature of the living rooms (by ~10 °C) (Wong et al. 2009); accompanied by moderate increment in the humidity ensuring comfort, particularly in dry conditions (Jaafar et al. 2013).

- **Vertical Garden**: The primary difference between vertical gardens and green walls from the design point of view lies in the use of felt fabric for the plants instead of soil media. The plants that are suitable for such a VGS configuration include ferns, mosses, sedums, and several species of epiphytes such as orchids and bromeliads (Online Resources: Plants for green walls). The benefits of the vertical gardens include beautification of the residential and commercial environments within the cities, improvement in the air quality, and efficient thermoregulation of the urban microclimate.

12.4.2 Tier II City Context: Nagpur

Nagpur is the 13th largest and fast-expanding urban sprawl of India after Mumbai and Thane, and it has been placed as fourth highly urbanized center of the state (Dhyani et al. 2018). Nagpur urban is located in the central part of the Indian Subcontinent in the north-eastern Vidarbha region of the Maharashtra state and the city is being developed under the Atal Mission for Rejuvenation and Urban Transformation, 2015 (AMRUT). Nagpur is a Tier II city with an area of 217.56 km^2 (latitude 21° 8′ 47.8788″ N, longitude 79° 5′ 19.8960″ E), geographically situated on the edge of the central Deccan plateau with an elevation of 310 m amsl (above mean sea level). Nagpur is among the largest and rapidly expanding urban sprawls of India with the largest municipal area in south-central Asia region (Kotharkar et al. 2014). City has 18% of its area under vegetative cover (forests and plantations), 17% of land area is used for cultivation and the water bodies amount to about 2% of the area (Chaturvedi et al. 2013). The land-use arrangements of the Nagpur urban show that 69% of the city area has already been exhausted for infrastructure development and it has been estimated that 73% of the area will be under development by year 2031 (Dhyani et al. 2018). Of the total built-up area, 45% accounts for the housing, 6% is either used for commercial or industrial purposes, 41% is used for various public uses, and only 8% area in Nagpur urban accounts for public green as parks or gardens. Few decades ago, Nagpur urban was the second greenest city in the state (Chaturvedi et al. 2013). Presently, the city is flecked with many public green (managed by municipal corporations or forest departments), private green and blue spaces (lakes, ponds etc.). Presence of monuments and heritage buildings in old Nagpur makes it more congested; whereas, the urban forests and the beautiful lakes are soothing sites that enhance the quality of life for the city dwellers. Existing blue-green infrastructure (BGI) in the city encompasses the urban forests, campus forests, public gardens, parks, playgrounds, fallow lands, wastelands (zudpi forests), lakes, wetlands, ponds, etc. The city has 77 parks and gardens managed by Nagpur Municipal Corporation (NMC), 50 under Nagpur Improvement Trust (NIT) and 13 are proposed to be developed in the upcoming years. Coverage area-wise a total of 49 ha space of public green spaces is managed by NMC, 39.45 ha by NIT, and 5.09 ha is proposed in different zones of Nagpur Urban. The number of parks, gardens, and playgrounds managed by the NMC in the western part of the city (the greenest area of Nagpur) is 18. NMC manages 16 gardens in the eastern and 11 gardens in the northern part of the city. NIT manages 15 parks, playgrounds, and gardens in the eastern, 12 in the western, 8 in the southern and the north eastern and 5 in the central part of the city. Among the upcoming proposed parks, playgrounds, and garden areas 5 belong to the eastern, 3 to the north eastern and the western, and 1 each to the central and the northern part of the city. Forest area apportioned to Nagpur Urban Forest Working Circle amounts to 1036.85 ha, which mostly comprises of reserve, unclassed forest, and zudpi jungle. Ambazari, Seminary hills reserve forests, and well-managed campus forests, viz. those found in the premises of National Environmental Engineering Research Institute (CSIR-NEERI), Visvesvaraya National Institute of Technology (VNIT), and PDKV College of Agriculture are some of the major remaining protected and healthy green spaces within the city premises. Nagpur city was developed over more than three centuries hence, different zones of Nagpur urban have a mix of low and mid-rise buildings distributed in different densities. Mahal, the oldest area of Nagpur, has 76% built-up area with fully paved land cover. It exhibits mixed residential and commercial land use having dense low and mid-rise constructions and narrow lanes, resulting in few open spaces and scarce trees; whereas, Shankar Nagar developed in 1960s has planned constructions with 23% built-up area, having detached low rise buildings and sufficient number of interspersed public green spaces (Godbole and Bhargava 2017). Rapid expansion of urban sprawl has resulted in shrinking sizes, increased number of fragmentation, as well as, fast degradation of the blue-green infrastructure in the Nagpur city (Dhyani et al. 2018). Owing to the recent population growth (Fig. 12.4) and the socio-economic changes, Nagpur is witnessing spatial expansion in its administrative boundaries. Rise in population and urban expansion has been considered as the important influences towards the increased vulnerability of the habitat and residents from the exposure to changing climate.

In rapid urban transition scenario faced by the city of Nagpur, the natural and managed public blue-green spaces, drainage basins of polluted Nag and Pili River, are under tremendous pressure of destruction and degradation in the upcoming years. Air pollution (Maji et al. 2016), increasing frequency and intensity of heat islands (Jain et al. 2020) and flash floods (Dhyani et al. 2018) are important concerns for the growing Tier II city of Nagpur. Less green spaces in central and north zone of

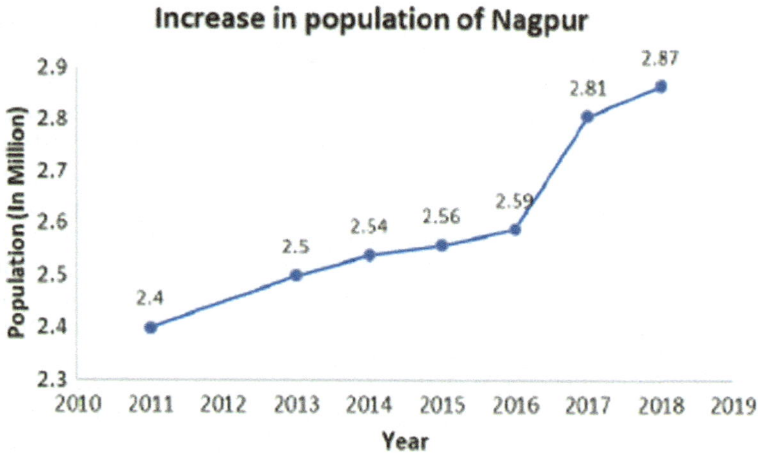

Fig. 12.4 Showing predicted rapid rise in population of Nagpur

Nagpur is due to dense build-up area that leaves no space for developing more green spaces. The space available for avenue plantation is also not substantial and hence, the corridors linking the green spaces are not significant in terms of impact. This aspect particularly enhances the vulnerability of urban dwellers during the summers and the monsoons.

12.4.2.1 Factors that Determine the Mainstreaming Urban NbS for Nagpur—Tier II City

Investment in the urban areas through creating the provision for efficiently planned biodiversity hotspots is believed to be a major strategy for mitigation of anticipated environmental change phenomenon using natural space (Govindarajulu 2014). The provision for using BGI as a public asset can potentially minimize the negative impacts of environmental change in the fast-growing Tier II cities like Nagpur, which falls in high-temperature belt. Nagpur regularly faces water shortage during the summers, flash floods during monsoons, and also the urban heat island effects are observed year-round. All of these signature events need to be considered for adaptation planning (Pickering and Hill 2007; Dhyani et al. 2018; Jain et al. 2020). Although, the city is witnessing rapid expansion and over densification of developmental activities, the conservative statistics-based land-use planning lacks availability of local-level data that may hamper the long-term vision towards the conservation of fragmented vegetative cover, as well as, the provisioning and planning for GI. Understanding of localized scenarios and response of locals would be helpful in effective planning of GI, as the locals often come up with intricate information about their neighborhood. Representatives of different urban stakeholders should be involved in planning to ensure effective and sustainable GI development. The massive infrastructural growth in Nagpur and rapid increase in the built-up surfaces without avenue plantations is likely to result in reduced level of percolation of rainwater and inadequate recharging of groundwater aquifers. In April 2019, NMC became possibly the first civic body in the country to make it obligatory to carry out compensatory afforestation before felling of the trees having a height of at least 6 ft. Rainwater harvesting and green roofs are being put into practice and a sizable fraction of area for every upcoming built-up surface is now having green patches as a part of urban municipal regulation. Curbside plantation and vertical wall and/or gardens are efficient NbS options towards creating GI in the densely populated zones of the city like Gandhibagh, Mahal, Bardi, etc. where open space for plantation is not available (Table 12.1). Curbside plantation requires selection of trees and plants in accordance with the width of road available. Such interventions can help in lowering the effect of vehicular pollution on nearby residential and commercial establishments, as well as, in reducing the intensity of heat islands. The plantation below the bridges and the vertical gardens can collectively act as an urban green belt. Vertical gardens on the sides of pillars of metro corridors, as well as, the flyover pillars supporting the girder provide the dwellers with pleasant sight. It also has the potential to reduce air pollution effects up to a certain extent leading to enhanced quality of ecosystem services. Table 12.1 shows the list of NbS planned and/or implemented at various locations in Nagpur urban area.

12.4.2.2 Key Recommendations for Realizing the Benefits of NbS in Tier II Cities

- Perception-based participatory surveys can help in better understanding and planning BGI and it will also facilitate a focused implementation. A step in this direction has already been initiated by the first author.
- Availability of ample spatial data can help in maximizing the BGI benefits to a larger population. Identifying the healthy and appropriate regions with green, the regions that demand management and interventions with orange, and the vulnerable zones with red codes can help the city planners to prioritize their efforts and allocate requisite funds in improving BGI in respective wards and zones.
- Participatory survey results should be given due importance to understand and address demand–supply gap and the actual scenario is necessary to understand the zones in different color grades of green, orange and red (representing the available open spaces, built-up area density, and population, respectively) for better planning of BGI following recommendations of URDPFI guidelines.
- The avenue plantations should be assigned the foremost priority in the cities, especially in the orange and red zones, to enhance green patch connectivity and to develop corridors for creating a larger BGI network. Avenue plantations can help in addressing, as well as, in reducing impacts of air pollution and heat islands.
- There are many wards in an urban center that are densely constructed and are devoid of sufficient space for avenue plantations or creating green parks, and

hence, the idea of plantation will not effective there. In those places, alternate measures like developing vertical gardens on metro pillars, flyover pillars, and greenery under the bridges should be planned.

- Road dividers can be efficiently used for raising healthy plants, with specific focus on the hardy and tolerant fast-growing native species, to reduce air pollution and heat island impacts. This can also be followed at ward level roads by making it compulsory for each road in the city premises.
- NMC needs to make a 10-year monitoring and review plan for the vulnerable zones for understanding and identifying locations for immediate plantations and for developing vertical GI along with Natural Infrastructure (NI). If no due effort and care is ensured these sensitive wards will be more vulnerable in coming years to increasing environmental change impacts in the city.
- In the orange zones, suitable locations should be identified for avenue plantation and installation of vertical gardens, as appropriate. Monitoring of remaining BGI for effective conservation and restoration would also prove to be instrumental in such zones. Special effort should go into identifying the reasons behind deterioration of quality and quantity of green spaces and a 5-year plan should be prepared for avoiding the degradation, with a provision to monitor and review it annually, so that the orange zones do not add to the list of red zones. The primary goal of such an initiative should be aimed at the reversal of trend.
- In the green zones, the prime focus should be towards reducing the anthropogenic interferences within the BGI. Strong legal frameworks are to be created and implemented to stop any kind of encroachment or felling, and on the other hand, additional plantations should be heavily promoted and encouraged by the local administrative bodies, as well as, the non-profit organizations. Most of the parks lack sufficient, and therefore, regular monitoring of trees should become a compulsory activity for the garden department of NMC with involvement of the district forest department, in order to facilitate understanding of the dynamic changes in the tree density. Scientific organizations should also be involved as partners to provide advanced scientific advice as per the requirement, as well as, to disseminate technical support as a part of citizen science awareness program.

As urbanization is getting intensified across India, existing BGI is having enhanced pressure from diverse threats with fast land-use change and infrastructure build up being the main ones. Given the rapid-paced technological advancement and the ever-increasing demands of growing urban populations, there is a need for frequent and routine horizon scanning for growing urban sprawls and ecosystem degradation. The case of a rapidly growing megacity and tier II city both demands integration of BGI, as potential NbS for reducing the threats and disaster risks. Integration of BGI is going to have substantial implications for developing resilience that needs to be well integrated with the urban policy and planning in future city's context. Overall, the key requirement of integrating BGI was found strongly relevant for both the expanded megacities and fast-growing urban sprawls. However, there are some major differences that will ensure success of BGI in these two contexts. While it looks tough and will need specific proactive efforts to mainstream and integrate BGI in already expanded megacities like Bangalore, tier II cities are seen as big hopes and have huge scope to integrate BGI following the lessons learned from megacities. We suggest that future initiatives aimed at identifying priority areas of BGI integration in both the cities and horizon scanning will allow simultaneous identification of both current and future BGI integration implications and priority issues.

12.5 Conclusion

The rate of growth of NbS in urban constructed environments has not yet matched the accelerated pace of urban sprawl in India. Regardless of various regulations and conservation bylaws, the planning, management, and decision-making aspects pose immense challenges for the stakeholders towards creating effective blue-green infrastructure governance framework, which becomes even steeper due to lack of reliable and well-maintained information repository. This is partially owed to the lack of effective and systematic practices in urban planning, and more because of the rapid expansion which is an important element that adds complexity in the context of monitoring urban resources. Thus, revisiting planning process is suggested as an effective way forward over the next two decades, in order to save potentially arable lands. Factors such as commitment of individuals and the random needs of the administrators to reflect upon a few eco-friendly approaches towards the urban planning have produced a few good examples of green NbS designs across the growing urban sprawls in India. However, at a policy level, NbS such as green corridors, vertical green spaces, multi-modal public green spaces, etc. need to be strengthened with the initiatives led by local administrators. More numbers of memorandum of understanding (MOUs) need to be established to enhance the efforts in realizing the NbS planning at city levels, learning from the successful cases of lake management initiatives through public–private partnerships (PPPs) with lake fringe communities (e.g., PLNIT, 2011). There is an urgent need to assess and evaluate the institutional and social governance capabilities to identify, evolve and evaluate the indicators of urban resilience that should be based on the performances of NbS at different scales in the densely populated urban centers. Despite the high level of awareness and positive public perception towards NbS among the city dwellers, intuitive NbS options incorporating the eco-friendly designs into urban building and construction have a long way to go and substantial research needs to be directed towards bringing down the huge transformational cost involved in establishing and maintaining NbS-aided engineering solutions. Further, upon engaging the corporate partnership along with government initiatives, BGI in the cities can be managed and novel NbS options can be customized to develop city resilience in the warming world.

Acknowledgements First author is grateful to KRC, CSIR-NEERI for plagiarism check by i-thenticate software under the number CSIR-NEERI/KRC/2020/MAY/WTMD/2.

References

Acharya P, Gupta AK, Dhyani S, Karki M (2020) New pathways for NbS to realise and achieve SDGs and post 2015 targets: transformative approaches in resilience building. In: Dhyani S, Gupta A, Karki M (eds) Nature-based solutions for resilient ecosystems and societies. Disaster resilience and green growth. Springer, Singapore

Aithal B, Ramachandra TV (2012) Spatial pattern analysis of two urbanising Tier II cities in Karnataka using open source GIS-GRASS

Ambinakudige S (2011) Remote sensing of land cover's effect on surface temperatures: a case study of the urban heat island in Bangalore, India. Appl GIS 7(1)

Andersson E, Borgström S, McPhearson T (2017) Double insurance in dealing with extremes: ecological and social factors for making nature-based solutions last. In: Kabisch N, Korn H, Stadler J, Bonn A (eds) Nature-based solutions to climate change adaptation in urban areas: linkages between science, policy and practice. Springer International Publishing, Cham, pp 51–64

Armson D, Stringer P, Ennos AR (2013) The effect of street trees and amenity grass on urban surface water runoff in Manchester, UK. Urban For Urban Greening 12:282–286. https://doi.org/10.1016/j.ufug.2013.04.001

Bennett EM, Cramer W, Begossi A et al (2015) Linking biodiversity, ecosystem services, and human well-being: three challenges for designing research for sustainability. Curr Opin Environ Sustain 14:76–85. https://doi.org/10.1016/j.cosust.2015.03.007

Bhardwaj S, Gupta AK, Dhyani S, Thummarukudy M (2020) Nature-based solution entry points through sectoral policies, strategic instruments and business continuity. In: Dhyani S, Gupta A, Karki M (eds) Nature-based solutions for resilient ecosystems and societies. Disaster resilience and green growth. Springer, Singapore

Bhattacharya S (2018) Urban sustainability in India: evolution, challenges and opportunities. In: The Palgrave handbook of sustainability: case studies and practical solutions, pp 673–698

Calfapietra C, Barbati A, Perugini L et al (2015) Carbon mitigation potential of different forest ecosystems under climate change and various managements in Italy. Ecosyst Health Sustain 1:1–9. https://doi.org/10.1890/EHS15-0023

Chaturvedi A, Kamble R, Patil NG, Chaturvedi A (2013) City–forest relationship in Nagpur: one of the greenest cities of India. Urban For Urban Greening 12:79–87. https://doi.org/10.1016/j.ufug.2012.09.003

Chen W (2015) The role of urban green infrastructure in offsetting carbon emissions in 35 major Chinese cities: a nationwide estimate. Cities 44. http://doi.org/10.1016/j.cities.2015.01.005

Chen G, Li X, Liu X et al (2020) Global projections of future urban land expansion under shared socioeconomic pathways. Nat Commun 11:1–12. https://doi.org/10.1038/s41467-020-14386-x

de Oliveira FL, Mell I (2019) Planning cities with nature: theories, strategies and methods. Springer, Berlin

Dhyani S (2019) Mainstreaming nature based solutions for ecosystem and community resilience. Clim Change Environ Sustain 7(2):123–124. Online ISSN: 2320-642X

Dhyani S, Thummarukuddy M (2016) Ecological engineering for disaster risk reduction and climate change adaptation. Environ Sci Pollut Res 23:20049–20052. https://doi.org/10.1007/s11356-016-7517-0

Dhyani S, Lahoti S, Khare S et al (2018) Ecosystem based disaster risk reduction approaches (EbDRR) as a prerequisite for inclusive urban transformation of Nagpur City, India. Int J Disaster Risk Reduction 32:95–105. https://doi.org/10.1016/j.ijdrr.2018.01.018

Dhyani S, Dasgupta R, Kumar P et al (2019) Exploring the current and future potential of urban agriculture in growing urban sprawls of India: strengths and challenges. Clim Change Environ Sustain 7:97. https://doi.org/10.5958/2320-642X.2019.00012.7

Dhyani S, Karki M, Gupta AK (2020) Opportunities and advances to mainstream nature-based solutions in disaster risk management and climate strategy. In: Dhyani S, Gupta A, Karki M (eds) Nature-based solutions for resilient ecosystems and societies. Disaster resilience and green growth. Springer, Singapore

Droste N, Schröter-Schlaack C, Hansjürgens B, Zimmermann H (2017) Implementing nature-based solutions in urban areas: financing and governance aspects. In: Kabisch N, Korn H, Stadler J, Bonn A (eds) Nature-based solutions to climate change adaptation in urban areas: linkages between science, policy and practice. Springer International Publishing, Cham, pp 307–321

Dzhambov A, Dimitrova D (2015) Green spaces and environmental noise perception. Urban For Urban Greening 14:1000–1008. https://doi.org/10.1016/j.ufug.2015.09.006

Ghofrani Z, Sposito V, Faggian R (2017) A comprehensive review of blue-green infrastructure concepts. Int J Environ Sustain 6:15–36

Godbole SP, Bhargava R (2017) Study of summer-time intra urban heat island intensity in residential areas of Nagpur. https://www.semanticscholar.org/paper/Study-of-Summer-Time-Intra-Urban-Heat-Island-in-of-GodboleBhargava/64ac34171ed6f598ac1b1cde059faa95e9658001. Accessed 2 May 2020

Govindarajulu D (2014) Urban green space planning for climate adaptation in Indian cities. Urban Clim 10. http://doi.org/10.1016/j.uclim.2014.09.006

Haase D, Frantzeskaki N, Elmqvist T (2013) Ecosystem services in urban landscapes: practical applications and governance implications. https://doi.org/10.1007/s13280-014-0503-1

Hewitt CN, Ashworth K, MacKenzie AR (2020) Using green infrastructure to improve urban air quality (GI4AQ). Ambio 49:62–73. https://doi.org/10.1007/s13280-019-01164-3

Jaafar B et al (2013) Impact of vertical greenery system on internal building corridors in the tropic. Procedia Soc Behav Sci 105:558–568

Jain S, Sannigrahi S, Sen S et al (2020) Urban heat island intensity and its mitigation strategies in the fast-growing urban area. J Urban Manag 9:54–66. https://doi.org/10.1016/j.jum.2019.09.004

Kotharkar R, Bahadure P, Sarda N (2014) Measuring compact urban form: a case of Nagpur city, India. Sustainability 6:4246–4272. https://doi.org/10.3390/su6074246

Lahoti S, Kefi M, Lahoti A, Saito O (2019a) Mapping methodology of public urban green spaces using GIS: an example of Nagpur City, India. Sustainability 11:2166. https://doi.org/10.3390/su11072166

Lahoti S, Lahoti A, Saito O (2019b) Benchmark assessment of recreational public urban green space provisions: a case of typical urbanizing Indian City, Nagpur. ScienceDirect

Liu W, Chen W, Peng C (2014) Assessing the effectiveness of green infrastructures on urban flooding reduction: a community scale study. Ecol Model 291:6–14. https://doi.org/10.1016/j.ecolmodel.2014.07.012

Lovell ST, Taylor JR (2013) Supplying urban ecosystem services through multifunctional green infrastructure in the United States. Landscape Ecol 28:1447–1463. https://doi.org/10.1007/s10980-013-9912-y

Maji KJ, Dikshit AK, Deshpande A (2016) Human health risk assessment due to air pollution in 10 urban cities in Maharashtra, India. Cogent Environ Sci 2:1193110. https://doi.org/10.1080/23311843.2016.1193110

Majidi AN, Vojinovic Z, Alves A et al (2019) Planning nature-based solutions for urban flood reduction and thermal comfort enhancement. Sustainability 11:6361. https://doi.org/10.3390/su11226361

Makido Y, Hellman D, Shandas V (2019) Nature-based designs to mitigate urban heat: the efficacy of green infrastructure treatments in Portland, Oregon. Atmosphere 10:282. https://doi.org/10.3390/atmos10050282

McKinsey (2010) India's urban awakening: building inclusive cities, sustaining economic growth

McPhearson T et al (2018) Urban ecosystems and biodiversity. In: Rosenzweig C et al (eds) Climate change and cities: second assessment report of the urban climate change research network. Cambridge University Press, New York, pp 257–318

Mell I (2015) Establishing the rationale for green infrastructure investment in Indian cities: is the mainstreaming of urban greening an expanding or diminishing reality? AIMS Environ Sci 2:134–153. https://doi.org/10.3934/environsci.2015.2.134

Perez G, Perini K (2018) Nature based strategies for urban and building sustainability. Butterworth-Heinemann, Oxford

Pickering CM, Hill W (2007) Impacts of recreation and tourism on plant biodiversity and vegetation in protected areas in Australia. J Environ Manage 85:791–800. https://doi.org/10.1016/j.jenvman.2006.11.021

Rajeshuni U, Dongare P, Sheriff VA (2017) New perspectives for rainwater conservation and water-sensitive urban design for the Bangalore metropolitan area. WIT Trans Ecol Environ 210:701–711

Ramachandra TV, Kumar U (2010) Greater Bangalore: emerging urban heat island. GIS Dev 14(1):86–104

Randhawa A, Kumar A (2017) Exploring sustainability of smart development initiatives in India. Int J Sustain Built Environ 6:701–710. https://doi.org/10.1016/j.ijsbe.2017.08.002

Raymond CM, Frantzeskaki N, Kabisch N et al (2017) A framework for assessing and implementing the co-benefits of nature-based solutions in urban areas. Environ Sci Policy 77:15–24. https://doi.org/10.1016/j.envsci.2017.07.008

Reckien D, Flacke J, Olazabal M, Heidrich O (2015) The influence of drivers and barriers on urban adaptation and mitigation plans—an empirical analysis of European cities. PLOS ONE 10:e0135597. http://doi.org/10.1371/journal.pone.0135597

Santhanam H, Majumdar R (2020) Permeable pavements as sustainable nature-based solutions for the management of urban lake ecosystems. In: Dhyani S, Gupta A, Karki M (eds) Nature-based solutions for resilient ecosystems and societies. Disaster resilience and green growth. Springer, Singapore

Sharifi A, Yamagata Y (2014) Resilient urban planning: major principles and criteria. Energy Procedia 61:1491–1495. https://doi.org/10.1016/j.egypro.2014.12.154

Sörensen J, Emilsson T (2019) Evaluating flood risk reduction by urban blue-green infrastructure using insurance data. J Water Resour Plan Manag 145:04018099. https://doi.org/10.1061/(ASCE)WR.1943-5452.0001037

Stewart ID, Oke TR (2012) Local climate zones for urban temperature studies. Bull Am Meteorol Soc 93:1879–1900. https://doi.org/10.1175/BAMS-D-11-00019.1

Suárez M, Manuel BFD, Méndez-Fernández L et al (2018) Nature-based solutions and resilience as complementary strategies for urban governance and planning: a review of assessment methodologies. 1. http://doi.org/10.3390/ifou2018-05959

Sudhira HS, Nagendra H (2013) Local assessment of Bangalore: graying and greening in Bangalore—impacts of urbanization on ecosystems, ecosystem services and biodiversity. In: Elmqvist T et al (eds) Urbanization, biodiversity and ecosystem services: challenges and opportunities. Springer, Dordrecht, pp 75–91

Swerts E, Pumain D, Denis E (2013) The future of India's urbanization. Futures 54. http://doi.org/10.1016/j.futures.2013.10.008

Udas-Mankikar S. Formulating open-space policies for India's cities: the case of Mumbai. In: ORF. https://www.orfonline.org/research/formulating-open-space-policies-for-indias-cities-the-case-of-mumbai-65007/. Accessed 4 May 2020

United Nations (2005) Human development report 2005

van Ham C, Klimmek H (2017) Partnerships for nature-based solutions in urban areas—showcasing successful examples. In: Kabisch N, Korn H, Stadler J, Bonn A (eds) Nature-based solutions to climate change adaptation in urban areas: linkages between science, policy and practice. Springer International Publishing, Cham, pp 275–289

Well F, Ludwig F (2020) Blue–green architecture: a case study analysis considering the synergetic effects of water and vegetation. Front Architectural Res 9:191–202. https://doi.org/10.1016/j.foar.2019.11.001

Wendling LA, Huovila A, zu Castell-Rüdenhausen M et al (2018) Benchmarking nature-based solution and smart city assessment schemes against the sustainable development goal indicator framework. Front Environ Sci 6. http://doi.org/10.3389/fenvs.2018.00069

Wong NH et al (2009) Energy simulation of vertical greenery systems. Energy Build 41(12):1401–1408

Wong NH et al (2010) Acoustics evaluation of vertical greenery systems for building walls. Build Environ 45(2):411–420

Wu J, Xie W, Li W, Li J (2015) Effects of urban landscape pattern on PM2.5 pollution—a Beijing case study. PLOS ONE 10:e0142449. http://doi.org/10.1371/journal.pone.0142449

Online Resources

7 Sustainable construction materials. https://c-r-l.com/content-hub/article/sustainable-construction-materials/. Accessed 22 June 2020

Air purifying plants: 20 best air cleaning plants. https://www.bhg.com.au/best-air-cleaning-plants. Accessed 29 Mar 2020

Balram S (December 2018) Bengaluru's green cover is on the rise, square foot by square foot. https://economictimes.indiatimes.com/news/politics-and-nation/bengalurus-green-cover-is-on-the-rise-square-foot-by-square-foot/articleshow/67071316.cms. Accessed 29 Mar 2020

Bamboo as a building material—its uses and advantages in construction works. https://theconstructor.org/building/bamboo-as-a-building-material-uses-advantages/14838/. Accessed 22 June 2020

Crumpacker M (February 2019) Farming in the sky—a look at 6 innovative rooftop farms. https://medium.com/@MarkCrumpacker/farming-in-the-sky-a-look-at-6-innovative-rooftop-farms-348812a8e132. Accessed 29 Mar 2020

Diwaker RK (November 2018) Almost 14 lakh houses in India are now 'green'. https://realty.economictimes.indiatimes.com/news/industry/almost-14-lakh-houses-in-india-are-now-green/66828963. Accessed 22 June 2020

Farming in the sky. https://www.popsci.com/cliff-kuang/article/2008-09/farming-sky/. Accessed 29 Mar 2020

Gardens grow up: nature reclaims cities with green walls and rooftop gardens. https://urban-hub.com/energy_efficiency/gardens-grow-up-nature-reclaims-cities-with-green-walls-and-rooftop-gardens/. Accessed 29 Mar 2020

Geo-visualisation of urbanisation in Greater Bangalore. https://www.geospatialworld.net/article/geo-visualisation-of-urbanisation-in-greater-bangalore. Accessed 29 Mar 2020

Green roofs. https://www.rhs.org.uk/advice/profile?pid=289. Accessed 29 Mar 2020

IGBC green new buildings rating system version 3.0. https://igbc.in/igbc/html_pdfs/abridged/IGBC_Green_New_Buildings_Rating_System_(Version_3.0_with_Fifth_Addendum).pdf. Accessed 22 June 2020

Indian Green Building Council. https://igbc.in/igbc/redirectHtml.htm. Accessed 22 June 2020

Khanna B (April 2016) Groves on city outskirts make way for 'development'. https://www.deccanherald.com/content/542001/groves-city-outskirts-make-way.html. Accessed 29 Mar 2020

Plants for green walls. http://extension.msstate.edu/plants-for-green-walls. Accessed 29 Mar 2020

Salvekar V (March 2019) Green powered buildings in Bengaluru—the need of the hour. https://www.cleanmax.com/blog/green-buildings-in-bangalore.php. Accessed 22 June 2020

The 31 cities in India. A guideline report by Promar Consulting, Sponsored by USDA, Foreign Agricultural Service, Emerging Economies Program. http://www.promarconsulting.com/site/wp-content/uploads/files/STC%20India%20Report%20FINAL.pdf. Accessed 29 Mar 2020

The benefits or organic and why your utensils should be organic too. https://www.bambuhome.com/blogs/bambuliving/the-benefits-of-organic-and-why-your-utensils-should-be-organic-too. Accessed 22 June 2020

Urban Green Landscaping, India Green Infrastructure Network (IGIN). http://www.iginasia.org/green_landscaping.php. Accessed 29 Mar 2020

Chapter 13
Incorporation of BIM Based Modeling in Sustainable Development of Green Building from Stakeholders Perspective

Raju Sarkar, Karan Narang, Abhinav Daalia, Vidushi Gautam, Ujjawal Nathani, and Rajib Shaw

Abstract Building information modeling (BIM) is a set of tools that represent the physical and functional properties of a building digitally. BIM is used to document building designs or simulate construction and operation of new facilities. The emerging model is information-rich and object-based, where data can be inserted, extracted, updated, modified, and analyzed to improve the design of facility. Owners initiate and finance building projects. By selecting service providers and deciding on the type of delivery process to be used, owners make very strategic decisions in the facility delivery process. The effectiveness of BIM on an integrated project depends on these conclusions and decisions. Nowadays, construction projects are also moving towards sustainability through owners being committed to having green-rated buildings and infrastructure. This paper focuses on building information modeling with an owner's perspective since owners being committed to having green-rated buildings and how this approach and interpretation of BIM may differ from that of an engineer, or contractor. This paper involves a study of the performance of a construction project in which BIM was implemented by the owner in the project from a later stage and a comparative earned value analysis between the two stages.

Keywords Building information model · Virtual design construction · Earned value Analysis · Green Building industry

Project Outline

R. Sarkar (✉) · V. Gautam · U. Nathani
Department of Civil Engineering, Delhi Technological University, New Delhi, India
e-mail: rajusarkar@dce.ac.in

K. Narang
Project Controls, Bechtel Oil, Gas&Chemicals, Houston, TX, USA

A. Daalia
Management Consultant Mckinsey & Company, Mumbai, India

R. Shaw
Graduate School of Media and Governance, Keio University, Fujisawa, Japan

© The Author(s), under exclusive license to Springer Nature Singapore Pte Ltd. 2021
M. Mukherjee and R. Shaw (eds.), *Ecosystem-Based Disaster and Climate Resilience*, Disaster and Risk Research: GADRI Book Series,
https://doi.org/10.1007/978-981-16-4815-1_13

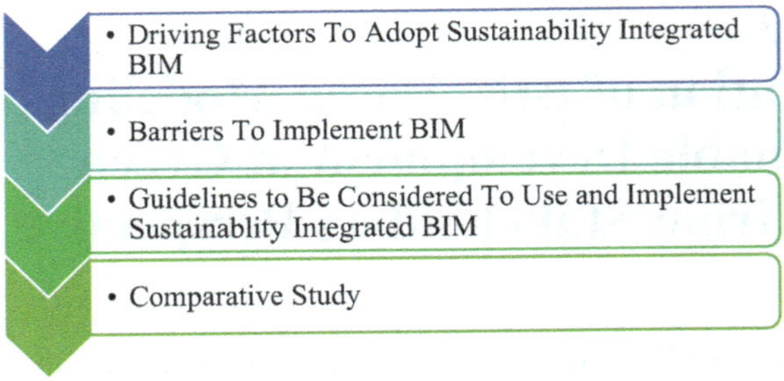

13.1 Introduction

> Living services should be driven by citizen experience, enabled by smart technologies, and based on circular economy.
>
> ~ Pekka Tuominem

Day-to-day services to be used in the structure can be provided by building a living and sustainable environment. Smart built environment technologies and materials could be used rather than using natural building materials to make the building eco-friendlier and sustainable.

Building Information modeling can be termed as a multilayered operating system data model which could be used in the process of creating, documenting, and enhancing the building design, providing momentum for construction and working of new facilities or refurbishing modern and enhanced facilities (Zeng and Qiang 2012). The BIM model created using BIM software programs is an intelligent 3D model which provides informative data and a digital representation facility, which can be used for the analysis and enhancement of the design and also in facilitating the generation of user's feedback. BIM, which is also known as VDC, i.e., virtual design and construction uses the intellectual and calculatable multifaceted model of the project which is created to upgrade the design, functioning, maintenance, and construction of the project (Sawhney 2014). The main motive of the architecture framework designing, engineering analysis, and construction (AEC) industry has always been the use of improvised techniques to increase productivity, quality, decrease cost, and reduce project delivery time. All of the above purposes can be achieved by the potential and technology that BIM offers (Azharet al. 2008). The data generated during designing process using BIM can be used to build over the different phases of a project during its lifecycle which will enable safer, cost-effective, faster, less wasteful, sustainable functioning, maintenance, and scraping of the structure. (Shaikh et al. 2017). The concept of integrated project delivery (IPD) is also supported by BIM. The IPD is a novel project delivery approach to integrate system, people, and business structures and practices it into a cooperative process to reduce waste and optimize efficiency through all phases of the project life cycle (Glick and

Guggemos 2009). Some of the sustainable building materials which can be used in the construction industry are Bamboo in place of timber, shape memory alloys, hydrogels, etc. Bamboo serves as a perfect sustainable alternative to timber. Since it's grass it regenerates very quickly as compared to trees and resembles wood aesthetically. Among different shape memory alloys, the most commonly available is Nitinol. These materials could be used in high strain energy-absorbing applications and can be used alternative to steel. Hydrogel is another smart building material that could be used on rooftops or some externally exposed surfaces. Many other construction practices provide single document on its application and use, but it's not the case with BIM (Associated General Contractors of America 2005). There has been very little progress in drafting model BIM contract documents (Post 2009). About two to three decades back, once a construction project was completed, there wasn't basically much anything which the firm could do afterward to improve the basic features or in the continuing working of the project. The task was left to the owner or facility manager who used to get it done only when the repairs and retrofitting work became necessary.

But, nowadays due to the precise detailing which is available and easily shareable 3D models which are designed using BIM, much of the designed data and overviews can be handed over to the owner of the building or the manager managing it, and hence it would help in the improving their ability to have the building working at its utmost potential. The very essence of BIM, i.e., time-saver, efficiency improvement, enhanced design, and reduced errors provides a computable opportunity to reduce the environmental impact that would be caused due to the building. Thus BIM fully supports the whole theory behind sustainable construction. BIM processes and applications can provide the owners to obtain remarkable gains in streamlining the delivery of better quality and high performing buildings (Zhang 2009). The effect of BIM is such that, in today's time the owners are rephrasing contract language, criteria, and specifications of the project, just to include the use of BIM-based process and technology and its application into their project to the maximum possible extent. Those owners who had participated in this effort are reaping the benefits of it in the market by delivering higher value facilities and minimized operational costs. (Eastman et al. 2008). All the BIM tools and applications which have currently made their presence in the market revolve around service providers as contractors, engineers, architects, and the tools which should be specifically targeted for sustainability motivated facility managers and owners are scarcely available.

The main objective of this paper is to focus on the less explored concept of sustainable design philosophy, whose extent is not only limited to the basic concept of construction and design but extends to the concept of sustainable use of the constructed structure and its preservation.

13.2 Driving Factors to Adopt Sustainability Integrated BIM

The benefits and driving factors for an owner to implement BIM are listed as in Fig. 13.1.

13.2.1 Design Assessment

Matching with their requirements, owners can handle and evaluate the scope of design.

Due to the transparent feature of the BIM, an early on knowledge of building in the context of integration, construction, and working of the building is available. This feature provides the owner with much more needed specifications before any money is spent on supplies and materials and thus makes the project more sustainable in the design phase.

This involves:

- Integrating development of programmatic requirements
- Using BIM spatial analyzes to improve program compliance
- Using visual simulation to receive more valuable input from project stakeholders
- Rapidly reconfiguring and exploring design scenarios
- Simulating facility operations.

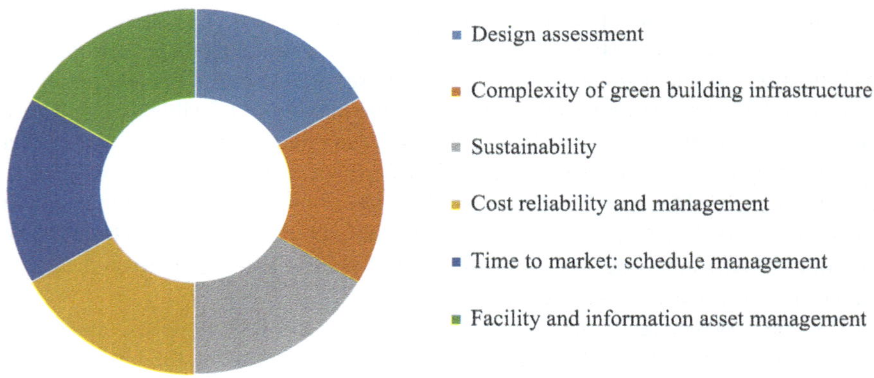

Fig. 13.1 Driving factors to adopt sustainability integrated BIM

13.2.2 Complexity of Green Building Infrastructure

With respect to the physical infrastructure and the financial, legal, and organizational structures associated with it, modern facilities have become more complicated. Sometimes due to the incoordination, inconsistency, and non-completion of construction documents, it may lead to poor instruction, low standard of work, and hence repairs works leading to rebuilding efforts.

All of the above-stated problems can cause extra cost to the builder or owner and much more loss in terms of materials and energy. The increasingly complicated and tortuous green building infrastructure and regulatory process can be managed by the owner using sustainability integrated BIM processes like:

- Integration of 3D models of MEP, architectural, and structural systems for better coordination of infrastructure
- An interactive evaluation and inspection of well-coordinated models to produce high quality and maintainable infrastructure
- Collaborative modeling and sign-off of building information models to prevent litigative liabilities.

The methods stated above not only save time, money, or delays in the project but also ensure that more environment-friendly methods and materials are used in the construction.

13.2.3 Sustainability

The energy efficiency of facilities and overall comprehensive environmental impact of the projects is being considered by many owners owing to the recent trend of green buildings. BIM can help in improving design and construction efficiency. Engineers can make an assessment about the environmental impacts and materials so as to support the decision needed to produce green sustainable buildings. From an owner's viewpoint, BIM processes can help by:

- Using energy analysis for reducing energy consumption
- Using model creation and simulation tools for improving operational productivity
- Designing and implementing effective day lighting practices
- Optimizing performance of equipment
- Evaluating sources of renewable energy as an alternative form of energy
- Enhancing natural ventilation of the structure
- Providing infrastructure to help reusing and recycling wastewater
- CO_2 emission analysis and evaluation help in achieving some degree of carbon neutrality in the project.
- Helps in minimizing construction waste, thereby making it more sustainable and economic.

13.2.4 Cost Reliability and Management

Contingencies are generally added by owners and service providers for mitigating risk related to cost overruns, unreliable estimates, and unpredictability during construction. Cost can be integrated with BIM applications by owners to provide:

- conceptual BIM estimating to obtain more dependable estimates at the start of the process
- Reduce crew redundancy to minimize cost as well as carbon footprint
- BIM quantity takeoff tools for faster, superior-detailed, and more precise estimates.

13.2.5 Time to Market: Schedule Management

Firms having well-defined time to market requirements empower them to deliver facilities quicker, cheaper, and of enhanced quality. The following time to market needs can be tackled by using BIM tools:

- Integrated use of parametric models to reduce time to market
- 3D coordination and prefabrication to reduce schedule duration
- Reducing schedule-related risk by use of BIM-based planning
- Quickly responding to unforeseen field conditions by use of 4D-coordinated BIM models.

13.2.6 Facility and Information Asset Management

A building information model can be used productively and successfully by owners who view the total lifecycle ownership of their projects for:

- Efficient Commissioning of a building and its equipment
- Quickly populating a facility management database
- Using BIM asset management tools to manage facility assets and prevent wastage of material
- Rapid evaluation of the impact of retrofitting or maintenance work on the facility, thereby increasing efficiency of operational phase
- providing a quantifiable opportunity for the building's lifelong environmental impact to be reduced.

13.3 Barriers to Implementing BIM

Alterations in work processes are associated with high risks. There are process barriers which include legal and organizational hurdles that prevent BIM implementation; and obstacles in the form of technical barriers related to preparedness and execution. The market is still in an early phase. Many owners presume that they will not receive competitive bids if the contracts are changed to create a requirement for new types of deliverables, specifically 3D or building information models, which would hamper their potential business and eventually increase the price of the project. The barriers involved in BIM implementation are mentioned in the flowchart (Fig. 13.2).

13.3.1 High Cost of Training

Implementation of new technologies is costly with respect to training and bringing changes to workflows and work processes. The training costs and losses in initial productivity add to the investment in software and hardware. More often than not almost all service providers do not intend to make an investment towards imparting training to users unless they comprehend a long-term benefit.

13.3.2 Implementing BIM After Completion of Finance and Design

As a project approaches construction phase, the owners and the design team would not be able to take advantage of BIM tools for conceptual estimating and program compliance. There are apprehensions whether implementation of BIM is still profitable in the latter phase of design and whether adequate time and opportunity is still available in early stages of construction.

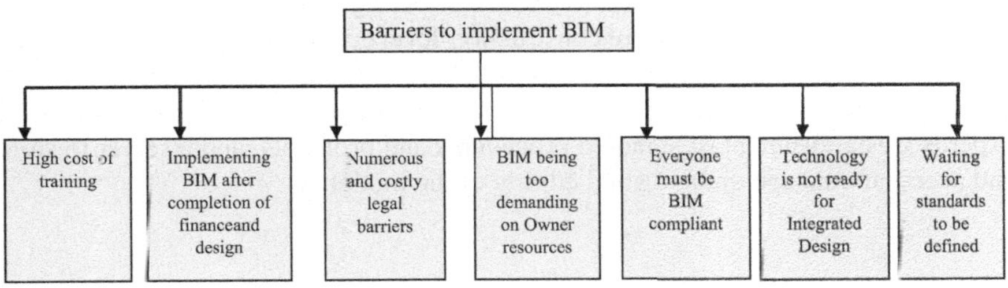

Fig. 13.2 Flowchart showing the barriers involved in BIM implementation

13.3.3 Numerous and Costly Legal Barriers

For use and interoperability of BIM on more collaborative project teams, major contractual and legal modifications are required on several fronts. Also, there is a risk associated with digital exchange of project data and teams are often forced to issue only paper drawings and rely specifically on contracts that are old-fashioned.

13.3.4 Building Information Model Being Too Demanding on Owner Resources

Conventionally, it is a construction manager who administers communication and oversees project documentation. Alignment of the process with particular deliverables and milestones is also checked by the facility manager. By using BIM tools, the current lax in the delivery is appreciably reduced, resulting in an increase of direct owner involvement. Discovery of issues and identification of problems occur early when BIM is implemented but this requires more frequent and extra inputs from owner.

13.3.5 Everyone Must Be BIM Compliant

It has been somewhat tough to ensure that all project participants are willing and technically capable enough to participate in the use of the building information model. It is also challenging to recreate information from organizations in which are not a part of the modeling effort.

13.3.6 Technology is not Ready for Integrated Design

As of now, BIM is typically good for works involving collaboration of only one or two disciplines. The integration of construction-level detail, work process, and model management is a more tedious task. As collaborating multiple disciplines requires multiuser access to BIM, there is an additional requirement in the form of technical expertise, establishment of standard procedures and protocols, and a server to store and access to manage updating and editing of the model.

13.3.7 *Waiting for Standards to Be Defined or Widely Adopted*

Although software companies have improved their IFC import and export functions, designers have not yet learned to make optimal use of the exchange standards, and proprietary formats are used by many organizations for model exchange. The short, as well as long-term investments by owners stand at risk in any building information modeling endeavor. Owner-specific standardization attempts can be used to mitigate the risk.

13.4 Guidelines Thatowners Can Consider to Use and Implement Sustainability IntegraTED BIM

13.4.1 *Scope and Level of Detail*

Owners need to assess the types of BIM tools and applications available and also be conversant with the level of detail they want to be implemented for the building model of their project. To take benefit of post-construction BIM tools, owners need to integrate with the service providers to ensure sufficient scope, level of detail, and data in the building model for the purposes preconceived. Generally, owners mandate the scope of work defined by the service providers for post-construction use of the model.

13.4.2 *Leading Implementation of BIM on a Project*

The preference of design service providers, final specifications, type of delivery processes, procurement is managed by the owner. Lamentably, quite a few owners do not perceive their ability to change or control the way in which a project is delivered. Owners can provide paramount value to their corporation by developing and implementing BIM guidelines, building internal knowledge and leadership, appointing service providers having BIM tools experience, and changing contractual obligations and requirements.

13.4.3 *Develop Guidelines for BIM on Projects*

Many organizations, particularly owners that build and manage multiple facilities, have developed guidelines for BIM. Some can be the identification of goals for BIM use and its alignment with organizational goals, scope and use of BIM across phases of the project (for example, a checklist of BIM applications, scope of standards or formats related to BIM, and the exchange of BIM and roles of participants in the BIM process and handovers between all participants).

13.4.4 Build Internal Leadership and Knowledge

Owners identify hindrances relating to the management of 2D information and the broad diversity of project data. Owners also develop knowledge by reviewing and evaluating the internal business models and work processes related to operation and delivery of facilities. Inadequacies associated with their current work processes are well understood by them.

13.4.5 Service Provider Selection

With no market leaders, owners usually take reference from their competitors in the industry's latest practices and technology trends. Furthermore, numerous owners are lacking expertise to take up a leadership position as they initiate only one project. But one common aspect, though, is the control over selection of service providers and the formatting of deliverables. Job skill requirements can be altered by owners to incorporate certain specific BIM-related skills and proficiency, including BIM-specific prequalification criteria and interview prospective vendors.

13.4.6 Change Deliverable Requirements

Owners can choose the type of project delivery process and thereby select the BIM applications to be implemented on the project than altering the requirements is preferred over changing the delivery process as it is usually more arduous. Owners can begin with three sectors: scope of the model information, uses of model information, and arrangement of model details.

13.4.7 Establishing Metrics to Analyze Progress

To assess the incorporation of new tools, technologies, and processes, metrics are used. Project metrics can include rework, and reduction in standard square footage cost and variance from baseline schedule or baseline cost.

13.4.8 *Doing a Prototype Run or Running a Pilot Project*

Owner or the Facility Manager can initiate a small pilot project with a clear path forward or does a prototype run on a small-scale model. This ensures sufficient coordination and collaboration of various disciplines involved.

13.5 Comparative Study

In the present study, an earned value analysis (EVA) of a multistory RCC building structure project having 10 towers supported on pile foundation has been carried out.

Details of the project are as follows:
Name of the project: 88A (Oasis and Icon)
Name of the builder: Godrej Properties Limited
Location: Gurgaon, Haryana, India
Towers No.: 10
Tower structure: 1B + G + 12F to 3B + G + 31F
Where

B = Basement
G = Ground floor
F = Floors over ground floor

Construction stage: All towers in progress
Shuttering used: Conventional formwork

Conventions used

$$\text{Cost variance} = \text{EV} - \text{AC}$$

$$\text{Schedule variance} = \text{EV} - \text{PV}$$

$$\text{Cost performance index} = \text{EV}/\text{AC}$$

$$\text{Schedule performance index} = \text{EV}/\text{PV}$$

where

EV = the value of project estimated on the basis of physical work completed.
AC = the amount of money consumed of funds burned at a particular time of construction cycle.
PV = the value of the project at that time which was scheduled well before in advance.

Cumulative planned, earned, and actual values of the building project were plotted against each quarter. As shown in Fig. 13.3, at the end of the eighth quarter (for simplicity it is written as Q8 in the graph), cumulative planned, earned and actual

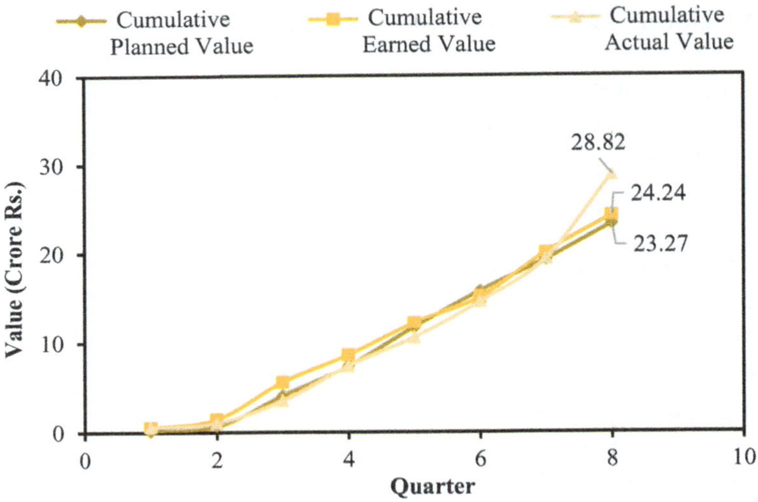

Fig. 13.3 Variation of value with respect to quarter (Till Q8)

values calculated were 23.27, 24.24, and 28.82 crore (10 million = one crore) rupees respectively. Again, BIM was implemented by the owner after the 10th quarter (for simplicity it is written as Q10 in the graph). Figure 13.4 shows that upto Q10, the

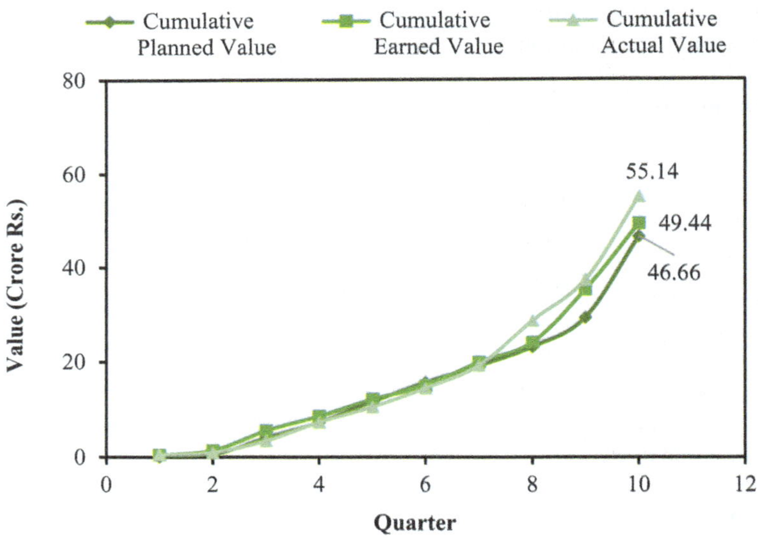

Fig. 13.4 Variation of value with respect to quarter (Till Q10)

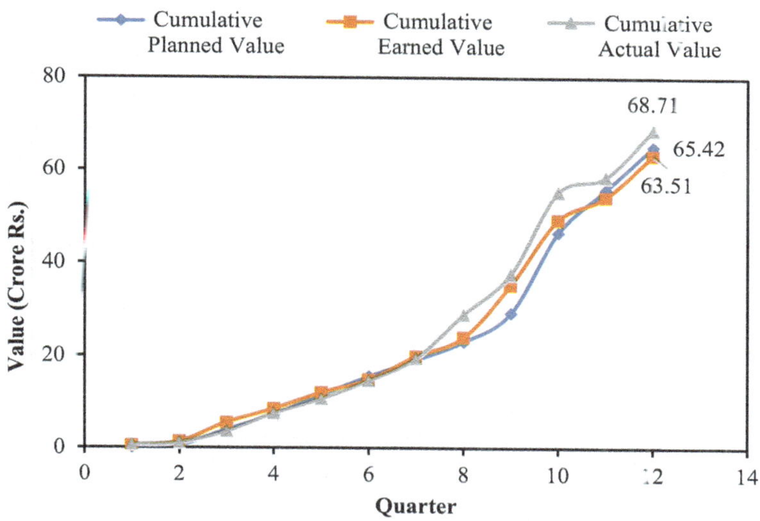

Fig. 13.5 Variation of value with respect to quarter (Till Q12)

cumulative planned, earned and actual values calculated were 46.66,49.44 and 55.14 crore rupees, respectively.

After the implementation of integrated BIM sustainability tool on the project by the owner, cumulative planned, earned and actual values calculated till the 12th quarter were 63.51, 65.42 and 68.71 crore rupees, respectively (Fig. 13.5). As shown in Fig. 13.6, the cumulative planned, earned and actual values calculated till the 13th quarter were 75.27, 75.05, and 75.62 crore rupees, respectively, and later increased to 85.40, 85.76, and 85.81 crore rupees, respectively, till the 14th quarter (Fig. 13.7). The calculated value of cost performance index is 0.84, 0.90, 0.92, 1.00, and 1.00 at the end of Q8, Q10, Q12, Q13, and Q14, respectively (Fig. 13.8). Figure 13.9 shows the variation of schedule performance index. From Fig. 13.9, the calculated value of schedule performance index is 1.04, 1.06, 0.97, 1.00, and 1.00 at the end of Q8, Q10, Q12, Q13, and Q14, respectively. Table 13.1 shows the variation of EVA parameters over different quarters.

13.6 Conclusion

Owners can use BIM for managing risk, improvement of project schedules, and delivering value to their businesses while enhancing sustainable practices for construction and operation. Although most of the BIM applications target towards service providers like engineers, designers, architects, developers, checkers, fabricators, contractors, owners, and facility managers will see many benefits on implementing

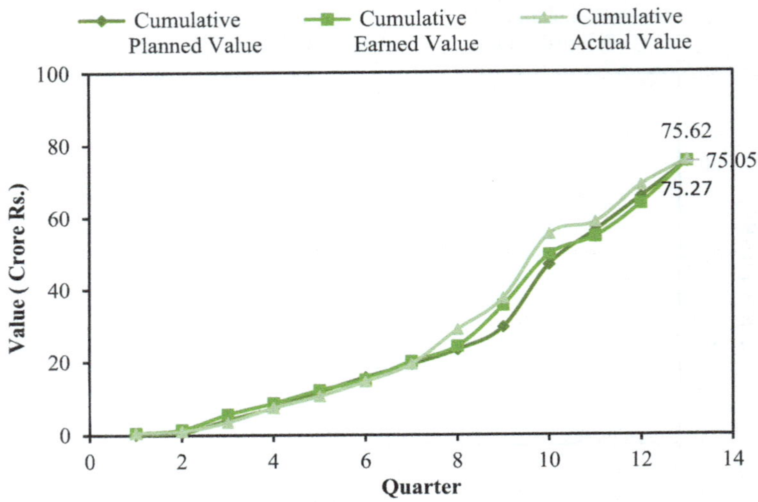

Fig. 13.6 Variation of value with respect to quarter (Till Q13)

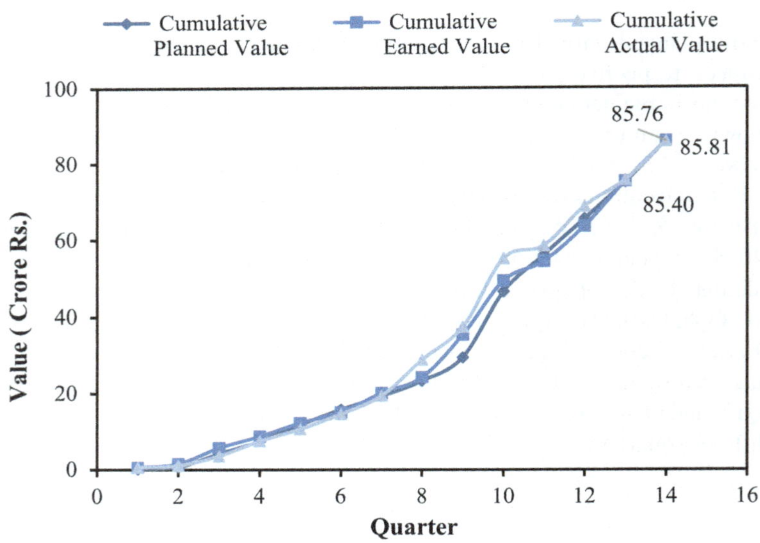

Fig. 13.7 Variation of value with respect to quarter (Till Q14)

BIM. The concept of integrated design, multiple stakeholder coordination, and contribution, common goal, efficient concept presentation, fast decision-making, and dialogue between stakeholders are required for BIM process as well as in the process of sustainable design (Dowsett and Hardy 2014).

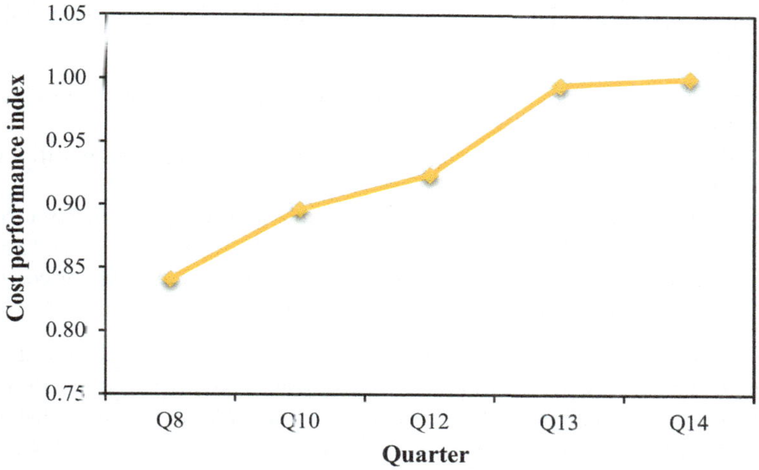

Fig. 13.8 Variation of cost performance index (EV/AC)

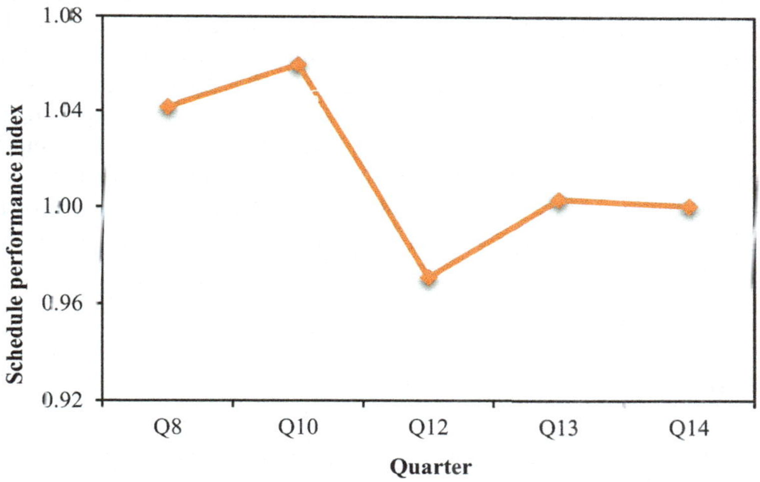

Fig. 13.9 Variation of schedule performance index (EV/PV)

Table 13.1 Variation of EVA parameters over different quarters

Quarter number	Cost variance	Schedule variance	CPI	SPI
Q8	−45.8	9.7	0.84	1.04
Q10	−57	27.8	0.90	1.06
Q12	−52	19.1	0.92	0.97
Q13	−3.5	2.2	1.00	1.00
Q14	0.8	0.2	1.00	1.00

In the comparative study between the first and second stages of the project, following factors were observed

(a) There was a decrease in cost and schedule variance in the second stage when BIM was incorporated.
(b) CPI shows a slower growth rate till Q10 as compared to the quarters after which BIM was implemented.
(c) SPI shows a decrease when BIM was implemented. This can be attributed to the fact that project took some time initially to get accustomed to BIM applications and sustainability practices. Subsequently, SPI factor comes out to be 1.00 which shows that project is on schedule.

The above results strengthen the observations of economic as well as sustainable advantages that BIM can offer on a sustainable construction project. A significant leadership role is played by owners and clients in the building industry. Sustainable building materials could be used instead of natural building materials for a better and sustainable building. All the materials such as Bamboo, Hydrogel, and Shape Memory Alloys could be used by the owner in the construction for a better and efficient building. Since green building is not only a wise choice but also a necessary choice, the construction industry should adopt eco-friendly practices and materials to reduce the increasing adverse impact on the environment. BIM helps to employ the concept of sustainability at three main stages, i.e., at the design phase by providing greater transparency and efficiency, and at the construction phase and the operation phase by providing greater control. There are different ways for owners to implement BIM applications can be implemented by owners on their projects in different ways such as prequalification of service providers, conducting training seminars, developing guidelines with respect to contractual requirements, and changing internal processes of their projects.

References

Associated General Contractors of America (2005) The contractor's guide to BIM (1st edn). AGC Research Foundation, Las Vegas, NV

Azhar S, Nadeem A, Mok JYN, Leung BHY (2008) Building information modeling (BIM): a new paradigm for visual interactive modeling and simulation for construction projects. In: Proceedings of the first international conference on construction in developing countries, Karachi, Pakistan, pp 435–446

Dowsett RM, Harty CF (2014) Evaluating the benefits of BIM for sustainable design—a review. In: Proceedings 29th annual association of researchers in construction management conference, ARCOM 2013, pp 13–23

Eastman CM, Teicholz P, Sacks R, Liston K (2008) BIM Handbook: a guide to building information modelling for owners, managers, designers, engineers and contractors. Wiley, Hoboken, NJ

Glick S, Guggemos A (2009) IPD and BIM: benefits and opportunities for regulatory agencies. In: Proceedings of the 45th Associated Schools of Construction National Conference, Gainesville, FL

Post N (2009) Building team members see progress and problems. Eng News-Rec 262(12):28

Sawhney A (2014) State of BIM adoption and outlook in India. RICS school of built environment, Amity University
Shaikh M, Shah DA, Shelke K, Giniwale A, Chheda S (2017) Sustainable development with BIM. Int Res J Eng Technol (IRJET), 1784–1788
Zhang Y (2009)The ways of building information modeling to improve the value chain of building process. In: International conference on management and service science, 2009. MASS '09, Wuhan, pp 1–4 (2009)
Zeng XD, Qiang WZ (2012) Research into the building information model during the whole building life-cycle. Advanced Materials Research 368:3797–3800

Chapter 14
Road to Ecosystem-Based Disaster Risk Reduction: Comprehensive Approach for Smart Urban Areas Management

Norio Maki

Abstract This article starts from the history of flood management policy in Japan. Then the present trend of Eco-DRR activities in Japan would be discussed. Finally, the land use planning scheme to adapt to the future climate change will be discussed. It should start from understanding the future impact of disaster and society. The techniques used for pre-disaster recovery planning can be used. It also starts from understanding the future impact of disaster and society. The possibility of using pre-disaster recovery planning techniques for the land use planning of climate change adaptation and Eco-DRR would be discussed.

Keywords Land use regulation · Indigenous knowledge · Pre-disaster recovery planning · Urban green farm · Rainwater tank

14.1 Introduction

Over act in the modern age was established in 1896 and the main target was flood control by the levee system. The 1964 amendment added irrigation responding to rapid economic growth. Priority of modern flood management is "flood control" and "Irrigation". And a river was managed as a cannel and the bottom and side walls are covered by concrete. However, those river management policies are dramatically changing in 1990s. The 1990 notice about "natural-oriented river work" became a turning point of river management policy in Japan (Soda and Kazuhiro 2012). And the eco-friendly or "Green" became the important concept in the river management policy. Nice Landscape with eco-friendly became the new evaluation criteria in the new policy.

Be natural became the hot point in the river management policy, but the area of implementation is limited to a river. In 1977, the river council made a report titled "Promoting comprehensive flood control policy" and the flood management expanding to urban areas was kicked off. But the trials were done only at the rivers

N. Maki (✉)
Kyoto University, Kyoto, Japan
e-mail: maki.norio.8v@kyoto-u.ac.jp

© The Author(s), under exclusive license to Springer Nature Singapore Pte Ltd. 2021
M. Mukherjee and R. Shaw (eds.), *Ecosystem-Based Disaster and Climate Resilience*, Disaster and Risk Research: GADRI Book Series,
https://doi.org/10.1007/978-981-16-4815-1_14

running urbanized areas. And finally, the discussion about land use regulation in high flood risk areas or comprehensive flood risk management was started. This discussion comes from the urban planning side thinking about a compact city reflecting depopulation in Japan. In addition to a compact city, Green infrastructure was also spotlighted and Ministry of land, infrastructure, transportation and tourism, MLIT made the report on green infrastructure in 2019 (MLIT 2019). This report discusses about the combination of eco-friendly technology, land use and social infrastructure development. It means that two post-modern river management trials such as "natural-oriented river work" and "comprehensive flood management" could be merged into one through "green infrastructure concept.

Japan has a long history of ecosystem-based disaster risk reduction or green infrastructure. Indigenous way of flood control was "Green". "Seigu" which means the sacred caw in Japanese, is the name of a wood structure to reduce the speed of water flow at the time of flooding. "Kasumi-tei" which means the haze shaped bank in Japanese, is the traditional open levee system. People did not live in flood prone areas and those who live in flood prone areas owned small boats to accommodate flooding and constructed their homes on the high mound with an evacuation hut at the time of severe flooding. People prepared for flooding in a more comprehensive manner.

The concept of green infrastructure or comprehensive approach is good. However, there are many barriers to implement those concepts. For example, there are policies of comprehensive flood management in urban areas having detention basins, having rainwater tanks in each house, keeping green areas in the city. But the implementation of those policy was limited. Then the depopulation makes comprehensive discussion of DRR, including eco-DRR possible. This article discusses about the eco-DRR trials in Japan especially about flood management from the historical perspective from following perspectives such as indigenous techniques, river channel, land use and green infrastructure concept (Fig. 14.1). The case in Kyoto would be spotlighted in this article. And the tool to accomplish ecosystem-based disaster risk reduction will be introduced. By merging all the eco-DRR trials, land use planning is the key concept. There are land use planning trials preparing for future disasters. The techniques for pre-disaster recovery planning for the future earthquake and tsunami in the western part of Japan would be introduced. Those techniques could be used for planning to adapt to future climate change in an ecosystem-based manner.

14.2 A River Became Beautiful and Eco-friendly

The 1990 note titled "Natural-Oriented Style River Work" was the turning point toward ecosystem-based of river management in Japan. "Natural-Oriented Style River Work" is defined as follows in the manual (MLIT 1990).

> Promoting river works which consider a natural growth environment of living things in the river, so that it preserves and creates beautiful land scape.

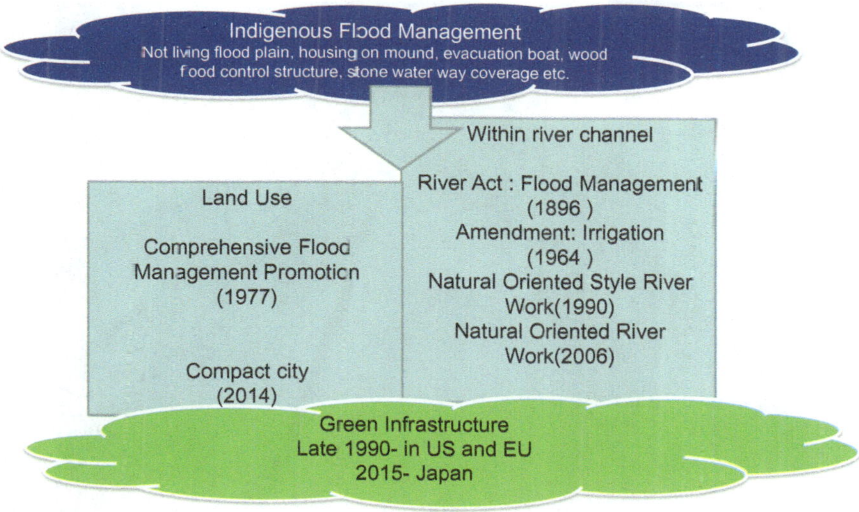

Fig. 14.1 Road to comprehensive approach

As defined in the manual, Landscape and ecology were major concerns of this policy. By the effort starting in 1990s, landscape of rivers became beautiful with many "natural things" (Photo 14.1).

Photo 14.1 Landscape of Kamo River, Kyoto in 2020

Photo 14.2 River as a function of water drainages. River was covered by concrete

Instead of concrete, natural stones and woods were used as a material for river maintenance work. Old technology using natural material such as fascine mattresses made from woods and natural stones were revived in the context of using "natural material". And riverbank was covered by natural stone instead of concrete mortal. It was the counterculture to the concrete covered river as drainage (Photo 14.2). Water park or front (Shinsui in Japanese) can be the keyword to understand those natural materials using. Architectural Institute of Japan, AIJ published a series of books guiding the design with water, such as "Layout between Architecture and Water" in 1984 (AIJ 1984) and "Planning of water environment for Architecture and City" in 1991 (AIJ 1991). The title of the 1991 book expands the target to City. It means a designer was also interested in waterfront with the context of city design from 1990s. For the design of a park or a beautiful landscape, natural materials were used. In the case of Kamo river, Kyoto, a beautiful walking path along the river was constructed (Photo 14.3). Design to reach river or water was promoted. Soda and Kazuhiro (2012) points that "Beautiful landscape or esthetics of river was one of major concern in this policy change".

However, the evaluation of those new approaches was negative. In 2005, MLIT set up the committee evaluating the 1990 policy toward "Natural-Oriented Style River Work" and the committee published the report telling negative aspects of those trials such as standardized methods of river management not considering to ecosystem of each location. Reflecting those reviews, the name of the policy was amended to "Natural-oriented river work" from "Natural-oriented *style* river work" in 2006.

Photo 14.3 Plate explain about history of a traditional festival in Kyoto

Photo 14.4 Urban reservoir in Kyoto; it is usually used as a park

By removing the word "*Style*", Japanese river management policy really try to be ecosystem-based or "natural". The definition of the 2006 policy is as follows.

> Nature-Oriented River Management is method of river restoration from the view of natural dynamism. 1) From a site-by-site nature-oriented approach to an integrated approach taking into consideration the workings of nature in the entire river, 2) Nature-oriented river management that take river management in general into consideration and 3) River management closely connected to local life, history and culture. (MLIT 2020)

History and culture are the new components of the 2006 policy. The plate explains about the history of culture added to the river park (Photo 14.4).

14.3 Story at Outside of Rivers

Though History of Ecosystem-based flood management history was discussed in the previous chapter, the scope of discussion was limited to the river itself or "river area" which is a technical term of land use and river relating departments is in charge of management. And this chapter discusses about the story outside of "river area".

In 1977, the river council of MLIT published the mid-term report titled "Promotion policy for comprehensive flood management". This report was the kick-off for holistic flood management. However, the target of the comprehensive flood management was limited to rivers in urban areas. In urban areas, there is no space for levee construction, so comprehensive flood management reducing rainwater inflow into rivers became the strategy in urban river flood management. Urban reservoir using school campus or parks (Photo 14.4) and water tank collecting rain falls at each home was used for flood management. Water tank activity is very famous in Japan (Photo 14.5). That water is also used for community firefighting. However, those activities were limited to the urban river and did not expand to smart land use for example not living in the flood prone area or regulating development at farmland being reservoir at the time of flooding.

In 2000, the river council of MLIT made a new report titled "Effective flood management including consideration all the river basin". Though the targeting area or tools for disaster reduction became wider, there exist difficulties to implement a new policy. Isaka (2010) points out the difficulties coordinating various stakeholders

Photo 14.5 Rain water tank

such as forest management, agriculture, land use, sewage and education etc. included in comprehensive food management.

For comprehensive flood management, the role of land use management specialists is important. But a government is the stove-pipe organization. Land use planners are not interested in flood management issues because the effort of flood management was done in the river management section. However, in 2014, a land use planner was forced to think about a FLOOD RISK for the amendment of the Act on Special Measures concerning Urban Reconstruction. This amendment set the new scheme for a compact city corresponding to depopulation in Japan, it says that we had better live safe places from natural hazards such as flooding, landslide. In a compact city planning, we can set an urban function promotion zone and a living promotion zone and the manual of the planning tells that those promotion zones should not be located in HIGH FLOOD RISK AREA.

Land use planners joined the effort for food management. However, there are many issues to deal with flood disasters from Land use side. Until this effort, the flood risk analysis was done for the design of levee construction and evacuation and there is no risk data fitting to land use regulation. Shiga prefecture makes a unique effort, it publishes flood maps of various risk levels such as 10-year, 100-year and 200-year interval flood risk. Showing various levels of risk is essential for comprehensive flood management involving land use planning. Not only the risk information but there are also many issues to be discussed for the comprehensive flood management, such as land taking, land use regulation and consensus building of stakeholders. However, it is important the holistic flood management by the collaboration manner was kicked off in Japan.

In 2019, National government in Japan published the report "Promoting Strategy of Green Infrastructure" (MLIT 2019). This is the first national government report on Green Infrastructure having a close relation with ecosystem based DRR. This report points out eight things for the green infrastructure such as (1) Adaptation to climate change, (2) Creative city, (3) Biophilic Design, (4) Sustainable Land use and management, (5) Compact city, (6) Smart urban management, (7) Ecology, Natural-Oriented River management and (8) Living Space with natural environment. Though nothing new about the contents of the report, this official report says that Japanese government promote green infrastructure or "Ecosystem-Based Disaster and Climate Resilience". This report compiles both natural-oriented river management and compact city, which are the Japanese trials for ecosystem base DRR. Then the question is how we can implement those concepts to accomplish Ecosystem-Based Disaster and Climate Resilience.

14.4 To Make a City Safe and Beautiful by Eco-friendly Way

As mentioned before, land use planning or management can be the tool for implementing holistic food management merging all the existing eco-DRR efforts. Many people tell the importance of land use management for adapting to climate change. But the trials for land use planning considering future climate change or flooding are very limited.

In the field of recovery research, several efforts to develop the techniques for making land use plans considering the damage of future disasters were conducted. Those efforts are called "Pre-disaster Recovery Planning" or "Post-disaster Redevelopment Planning". Effort for flood management considering to the future climate change has similarity, such as (1) targeting near future event, (2) compiling all the activities for DRR and (3) having difficulty sharing the future risk with stakeholders.

In case of Japan, the targeting disaster of pre-disaster recovery plan is the Nankai trough earthquake which hits western part of Japan near future, possibly 80% within 30 years. Preparing for this event, the planning tools for holistic recovery plan and land use plan has been developed. Figure 14.2 shows the process of pre-disaster recovery planning (Kim et al. 2017). The first step of the planning is "Setting future vision of community". In this step, community member makes the asset map of the community by using the site model of the community. People tell what is good and share their memory about their community. Those asset maps compiling what the community member thinks about could be the future vision of the community.

The second step is "Impact assessments". People learn about what would happen to their community both from disaster and societal conditions. To share the risk of

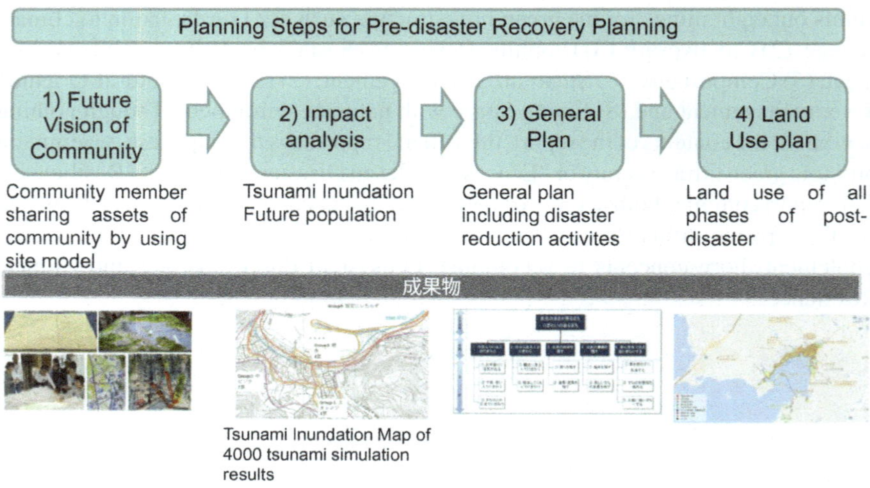

Fig. 14.2 Planning procedure of pre-disaster recovery planning

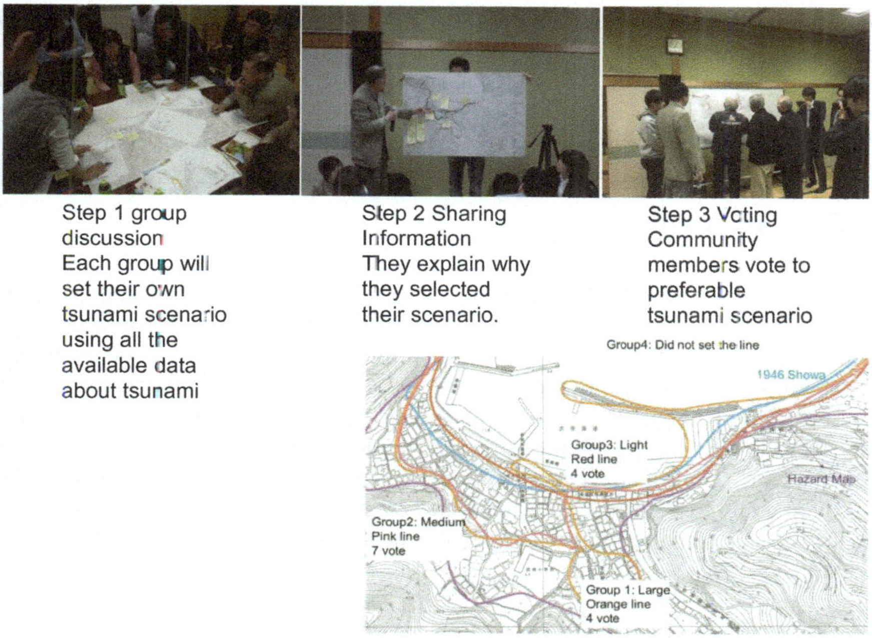

Fig. 14.3 People setting the risk of their community by themselves

tsunami, the democratic way of risk setting would be used. The important thing is the expecting risk is different among targets such as human life and community. For human life, the worst-case scenario shall be used for risk setting, though the risk for community or pre-disaster recovery planning could be set by the decision of community members. Figure 14.3 shows how the community people set their risk for recovery planning. Thousands of tsunami inundation simulation results were used for the risk scenario setting. This way of risk setting could be useful for a community risk setting for a climate change adaptation.

The third step is "Making general plan". The general plan after the disaster would be developed in this step. Disaster management can be one objective in the plan. To make the adaptation strategies for climate change or managing floods in a more comprehensive way, it is very important to organize the adaptation strategies with their future development plan. This is the step to coordinate future development, general plan and eco-DRR systems.

And the final step is "Setting land use plan". This is the step of implementing the vision set in the general plan. In the case of the DRR, catalogues of eco-DRR trials and efforts can be used to design the community and smart land use preparing for the future climate change.

In Japan, there are many efforts or trials for an eco-based approach for DRR. However, those efforts were done in each field and there is no way to compromise all those efforts. And the strategies of green infrastructure are the good promotor to facilitate those activities into one. And the planning for the compact city can be a good opportunity to compile all those efforts. The techniques for pre-disaster reduction planning can be the tool for planning. Japanese activities hope to be good lessons for the global activates.

14.5 Conclusion

Each country has their traditional way of DRR, however, we are losing those techniques within the process of modernization, Japanese experiences of losing and revitalizing traditional DRR techniques, which are eco-based DRR, are good textbooks for countries now losing those techniques.

The effort to compiling efforts for an eco-based approach for DRR is also important. The strategies of green infrastructure are the good promotor to facilitate those activities into one. And the planning for the compact city can be a good opportunity to compile all those efforts. The techniques for pre-disaster reduction planning can be the tool for planning. Japanese activities hope to be good lessons for the global activates.

References

AIJ (1984) Kenchiku to Mizu no Layout, Maruzen. (in Japanese)

AIJ (1991) Kenchiku to Toshi no Mizu Kannkyou Keikaku, Shokokusya. (in Japanese)

Isaka N (2010) River basin management against political and institutional flood disaster risk. Public Policy Stud 10:104–115 (in Japanese)

Kamogawa River Meeting (2006) Sennenno Miyako to Kamogawa. https://www.pref.kyoto.jp/kamogawa/documents/1175491663974.pdf, referring on 25 May 2020. (in Japanese)

Kim M, Sato K, Maki N, Hirata T, Inachi S, Kishikawa H, Tanaka H (2017) Challenges in realizing sustainable community development: case study of pre-disaster recovery planning in Ena. Wakayama Pref ISSS J 30. (in Japanese)

Ministry of Land, Infrastructure, Transportation and Tourism (2020) Natural oriented river management. https://www.milt.go.jp/pdf/conf_06-0.pdf, referred on 20 May 2020

Ministry of Land, Infrastructure, Transportation and Tourism (1990) Natural oriented river work manual. (in Japanese)

Ministry of Land, Infrastructure, Transportation and Tourism (2019) Green infrastructure promoting strategy. (in Japanese)

Soda R, Kazuhiro Y (2012) What was the aim of "nature-oriented river work?" E-J Geo 7(2):147–157. (in Japanese)

Part III
Case Studies

Chapter 15
Path Towards Sustainable Water Management: A Case Study of Shimla, India

Kamakshi Thapa, Chetna Singh, Sameer Deshkar, and Rajib Shaw

Abstract About four billion people around the world experience severe water scarcity during at least one month of the year. These issues have also begun to surface in Shimla city, located in hilly terrains of India. While the city draws its water from its peripheral rural areas, the declining freshwater availability has recently caused severe unrest among the local natives. To address this issue, land use and land cover of Shimla district were analysed for the year 2005–2006 to 2015–2016. It has been found that rapid urbanisation and change in cropping pattern has largely impacted the water resources of Shimla city. Based on the study results, this chapter emphasises the nature-based solutions for integrated and comprehensive management of water resources in Shimla city.

Keywords Water stress · Urbanisation · Land use land cover change · Cropping pattern · Nature-based solution

15.1 Introduction

A popular phrase by Samuel Coleridge, "Water, water everywhere, nor any drop to drink", gives an insight into earth's freshwater availability (UNDP 2006). Freshwater accounts for 3% of the earth's total water resource, out of which only 0.5% is available

K. Thapa (✉) · S. Deshkar
Department of Architecture and Planning, Visvesvaraya National Institute of Technology, South Ambazari Road, Nagpur, Maharashtra 440010, India

S. Deshkar
e-mail: smdeshkar@arc.vnit.ac.in

C. Singh
Department of Regional Planning, School of Planning & Architecture, 4-Block-B, Indraprastha Estate, New Delhi 110002, India

R. Shaw
Graduate School of Media and Governance, Keio University, 5322 Endo, Fujisawa 252-0882, Kanagawa, Japan
e-mail: shaw@sfc.keio.ac.jp

© The Author(s), under exclusive license to Springer Nature Singapore Pte Ltd. 2021

M. Mukherjee and R. Shaw (eds.), *Ecosystem-Based Disaster and Climate Resilience*, Disaster and Risk Research: GADRI Book Series,
https://doi.org/10.1007/978-981-16-4815-1_15

due to heterogeneity in space and time (U.S. Department of the Interior 2020; Oki and Quiocho 2020). As explained by Earthwatch Institute (2020), "From a human perspective it means that water is often in the wrong place, in the wrong form or available at the wrong time". This uneven distribution of freshwater along with the constantly growing population has now raised the concern of freshwater availability and accessibility (Earthwatch Institute 2020).

As per UNESCO's 'Water for Sustainable World' report, growing population, urbanisation, migration, industrialization along with changing production and consumption pattern have intensified the freshwater demands (UNESCO 2015). Together with climate change, the water stress situation has been further aggravated in the water-stressed regions of the world (WWDR 2020). Water stress refers to a situation wherein the demand for freshwater exceeds the currently available resources or when the poor quality of water restricts its use (Veolia 2016). Cassella et al. (2019) based on World Resource Institute report highlights that, 17 countries are experiencing extremely high baseline water stress as more than 80% of supplied water are withdrawn annually for agricultural, industrial and municipal use. Of these 12 out of 17 extremely water-stressed countries are mainly located in parts of Middle East, North African and South Asian region. Herein, India ranks 13th which caters to more than three times the population of other remaining water-stressed countries (Hofste et al. 2019).

Water is a major concern in India, as it is home to 17% of world's population with only four percent of global water resources (Veolia 2016). As per Shiao et al. (2015), about 54% of the country experiences high to extremely high levels of water stress. Both the quantity and quality of water have progressively been deteriorating in the country. As per World Resource Institute report, of the 4,000 groundwater wells, around 54% of wells have deteriorated while 16% are declining by more than 1 m per year. Similarly, among 632 districts, only 59 districts have groundwater quality above the limits of Bureau of Indian Standards (BIS). Majority of the districts have high levels of fluoride, chlorine, iron and nitrate contents in the groundwater, which are unsafe for drinking (Shiao et al. 2015).

To rectify these emerging challenges of water stress, United Nations World Water Development Report (2018) highlighted the role of "Nature-based Solution" (NbS), to improve the supply and quality of water along with minimising the impact of natural disasters. The report advocates the traditional and indigenous knowledge that embraces a greener approach to manage the water resource (WWAP 2018). The concept of NbS is relatively new, which is primarily based on the relationship between biological diversity and human wellbeing (Cohen-Shacham et al. 2019). Although this relationship was well acknowledged from earlier times, its framing as 'Ecosystem Services' was initiated from the 1970s. Major scientific and political acceptance to Ecosystem Services came with the publication of Millennium ecosystem assessment in 2005, which highlighted the interdependence of people and nature (Vihervaara et al. 2010; Cohen-Shacham et al. 2019). To fully comprehend the potential of NbS, it is necessary to implant the concept of NbS in the national and state planning system, not only for water resources but also for other sectors.

15.2 Understanding Nature-Based Solution (NbS) and Water

In the late 2000s, a significant evolution was observed in the concept of NbS, where people moved from being passive beneficiaries to proactively protecting, managing or restoring the ecosystems to address societal challenges (Cohen-Shacham et al. 2019). Over the years, extensive work has been done by various international organisations to, develop the knowledge base and practically implement the concept of NbS *to rectify the problem of water security, food security, human health, disaster risk reduction and climate change*. For example, the European Commission (EC) and International Union for Conservation of Nature and Natural Resources (IUCN) attempted to define NbS in different ways but with similar goal of "safeguarding human wellbeing and enhance the resilience of ecosystems and their capacity for renewal". EC has defined NbS as "*living solutions inspired by, continuously supported by and using nature* designed to address various societal challenges in a resource-efficient and adaptable manner and to provide simultaneously economic, social and environmental benefits *(European* Commission 2017*)*. While, IUCN has defined NbS as "an action to protect, manage and restore natural or modified ecosystems that address societal challenges effectively and adaptively simultaneously providing human wellbeing and biodiversity benefits" (Cohen-Shacham et al. 2016).

In respect to water security, NbS for water resource management involves a collaboration of ecosystem services with conventional water infrastructure to improve the quantity & quality of water as well as to increase resilience against climate change (UN Environment-DHI, UN Environment and IUCN 2018). Herein, it emphasises managing the built and natural infrastructure to achieve the water management goals as shown in Fig. 15.1.

As per UN Environment-DHI, UN Environment and IUCN (2018), NbS for water management can be applied in three ways, i.e. 'Protection', 'Restoration', 'Extension'. Herein, 'Protection' means identifying, quantifying, utilising and protecting the existing ecosystem services; 'Restoration' signifies rehabilitation and restoration of degraded ecosystem services and 'Extension' implies reproducing ecosystem services for sustainable water resource management. There are ample instances around the world, where these three ways of NbS were applied to manage the water resource. For example, in South Africa, Department of Water Affairs established the intersectoral programme to increases the upstream water flow by destroying the intrusive foreign plants that consume more water. Likewise, Kongoni River farm constructed a wetland for purification of its wastewater to address the problem of eutrophication caused due to mixing of untreated wastewater released through commercial agriculture (UN Environment-DHI, UN Environment and IUCN 2018). Similar to these, there are many evidences which show NbS provides tools for attaining sustainable use of water resource.

Similarly need of managing the water resource of famous tourist destinations, i.e. Shimla city, located in northern state of Himachal Pradesh (India) has been realised.

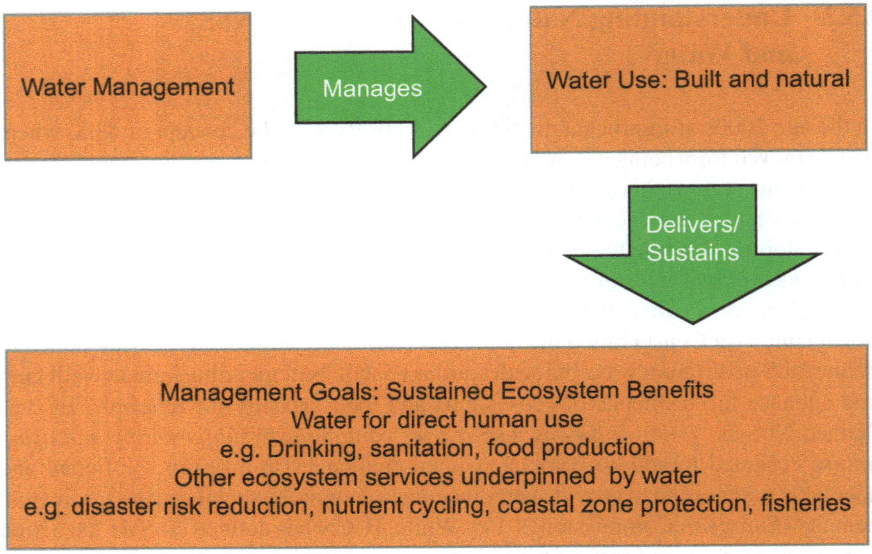

Fig. 15.1 Water ecosystem nexus. *Source* WWAP (2018)

Following sections explains the growing water concern of Shimla city and factors contributing to the water stress in the city.

15.3 Introduction of Study Area

The state of Himachal Pradesh is located at an elevation ranging from 350 m in lower Shivalik to 7000 m in Greater Himalayas (NABCONS 2015). It enjoys diverse climatic and physiographic conditions due to its locational and altitudinal variation (Chand 2013). The state is largely drained by tributaries of Satluj, Beas, Ravi and Yamuna which provides potentiality for agriculture/ horticulture. As shown in Fig. 15.2, the state is divided into 12 districts with Shimla being its administrative capital.

For administrative and development convenience, Shimla district is sub-divided into 13 tehsils and 7 sub tehsils and 10 Development Blocks (as shown in Fig. 15.2). It further comprises 363 Panchayat, 3231 villages and 11 local urban bodies (NABCONS 2015). Amongst which, Shimla Municipal Corporation is the largest urban agglomeration that serves as a major administrative, educational, heritage and tourist centre for both the state and district.

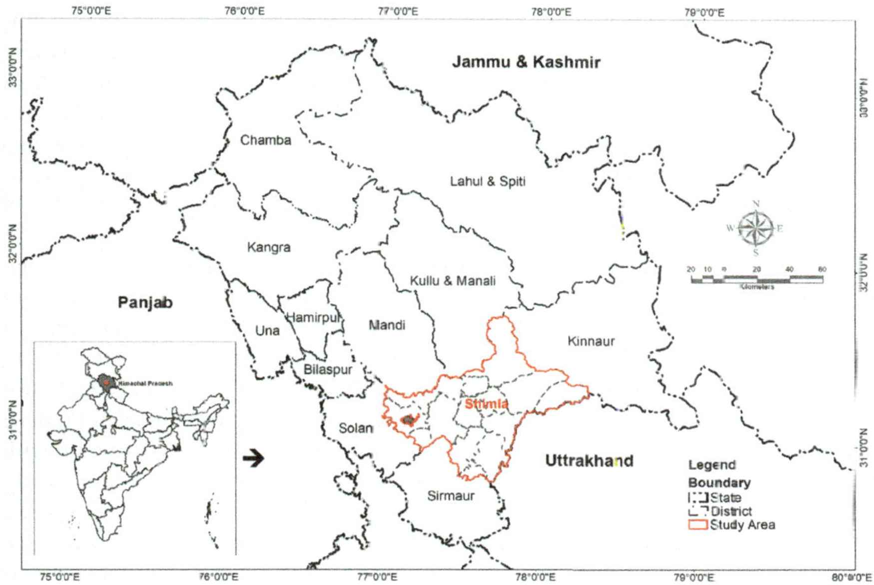

Fig. 15.2 Administrative division of Himachal Pradesh India. *Source* Census of India (2011a)

15.4 Growing Concern of Water Stress in Shimla City

As mentioned earlier, 54% of India faces high to extremely high water stress, this phenomenon is particularly noticeable in north-west part of the country (Shiao et al. 2015). One of the major tourist destinations of Himachal Pradesh, i.e. Shimla city, has been experiencing acute water stress in recent past. Generally, the state is blessed with water resources as it forms the catchment of major river systems (Indus and Ganga) of India and serves as a main source of water to its neighbouring states (TERI 2015). But, ironically in 2018, the state's own capital struggled to keep up with the unprecedented water shortage for continuously eight days (Thakur 2018).

Water supply system of Shimla city was laid down in the late 19th and early twentieth century during British time (Snyder 2014). With increasing demand, extensions were made in the water supply system and water was augmented from multiple sources namely Dhalli, Churat, Chair, Gumma, Ashwini and Giri which are located in peripheral rural areas. However, the reliability and sustainability of these sources are very low. The overall water installed capacity of the city is 61 Million Litres Per Day (MLD) out of which only 37–38 MLD water is produced (Thakur 2018). Of the many other valid reasons, water leakage due to ageing infrastructure is one of the major reasons for water shortage in the city (Bajpai, 2018).

In 1986, a study was conducted by the National Environmental Engineering Research Institute (NEERI) which has highlighted that more than 45% of water loss are due to leakages (Municipal Corporation Shimla 2009). Still, the state government failed to anticipate and take precursory measures to prevent such a situation. Later in 2015, Shimla city had also reported an outbreak of jaundice epidemic which affected many lives in and around the city. This outbreak was largely caused due to contamination of Ashwini water intake, which is located downstream of Malyana sewerage treatment plant (Chatterji 2018). The reduced efficiency of the sewage treatment plant and poor connectivity of sewerage network with habitation contributed to direct discharge of greywater in the natural streams.

Understanding the severity of these problems, ample studies have been conducted by the state government and researchers to find a new avenue for water supply in Shimla city, for instance, National Institute of Hydrology Roorkee in collaboration with Irrigation & Public Health department Shimla conducted the study on the impact of sewage effluent on drinking water sources of Shimla city and suggested ameliorative measure (NIH and I&PH 2013). Similarly, Government of Himachal Pradesh and World Bank in consultation with Deloitte and DRA consultancy came up with the situation analysis report on the drinking water & sewerage system of Shimla city (Deloitte Touche Tohmatsu India LLP/DRA Consultants Ltd 2018). Further, Sharma et al. (2015) conducted a resource assessment for suggesting short term corrective measures to improve the water supply for the city. These wide-ranging studies highlighted that water stress in Shimla city is the amalgamation of multiple factors such as weak regulation, high tourists' influx, poor operation and maintenance of water infrastructure and climate change etc.

Until now, majority of the studies related to water were confined within the Shimla Planning Area (SPA) limits with very little attention towards its catchment or watershed area which forms the major source of water to Shimla city. It has been noticed from the past few years that catchment or watershed area has been experiencing the tremendous change in land use and land cover because of urbanisation, change in cropping patterns and tourism (Vasudeva 2018). The aim of the study is to examine the change in land use and land cover of Shimla district using Natural Resource Census- land use land cover database for 2005–2006 and 2015–2016. Further, study the change in cropping pattern and rate of urbanisation in Shimla district and city.

15.5 Land Use Land Cover of Shimla District

Land is becoming a very significant resource in the wake of growing population and urbanisation. Numerous studies have accepted the fact that land use and land cover change has been one of the most prominently visible change taking place around us (Roy and Roy 2010). As per Praksham et al. (2018), Land use and land cover are two different terms that are often used together. Herein land cover means physical feature of the earth surface whereas land use implies human alteration of earth's surface for social and economic uses. Together these term, land use land cover change refers to human modification of the earth's terrestrial surface (Ellis 2007). According to Roy and Roy (2010), apart from altering the physical feature, land use land cover change also impacts the secondary processes which support the whole system of earth. For example, depletion of forest cover may lead to change in water cycle, loss of biodiversity which may directly impact the quality of air, precipitation of an area and further this loop continue impacting other relatable macro or micro phenomenon. Considering the present rate and intensity of urbanisation, it is necessary to analyse the land use land cover change to maintain the equilibrium between different uses (Kumar et al. 2018). Broadly, land use and land cover of Shimla district can be categorised into different levels as shown in Table 15.1.

Generally, Shimla district is spread over ridges, spurs and valleys, which is largely covered with shrubs or dense forests. The topographical constraints provide limited opportunity for various activities (Planning Department 2005). With time, major changes have been observed in the land use land cover pattern of Shimla district. Praksham et al. (2018) attempted to perform the change detection to understand the change in land use land cover Shimla tehsil. Wherein significant changes have been observed in the Shimla tehsil over a period of time. Figure 15.3 explains the change in the areas of different land use and land cover from 2005–2006 to 2015–2016.

Shimla district covers an area of 5131 square kilometres (km^2) which forms 9.22% of the total area of the state. As mentioned earlier, majority of the district is composed of evergreen dense and scrub forest. In 2005–2006, the evergreen forests occupied 2,267 km^2 of area which forms considerable part of Shimla district and it further experienced an increase of 1.19% in 2015–2016. While scrub forests which occupied 134 km^2. in 2005–2006 have seen a significant decline of 89.3% in 2015–2016. As agriculture is the mainstay of economy, it involves cultivation of various types of cereals, fruits and off-season vegetables (Prakasam et al. 2018). Significant changes have been observed in the agriculture of the district. In 2005–2006, agriculture cropland covered an area of 656 km^2 which has seen a significant fall of 29.8% in 2015–2016. Whereas agriculture plantation which occupied 672 km^2 of area has experienced an increase of 36.4%. As the area under agriculture, fallow land has drastically decreased to 98.4%, this depicts extensive utilisation of agricultural land. In 2005–2006, a considerable area of 732 km^2 falls under grazing land, but a notable dip of 37.19% has been observed in the grazing land. Along with agriculture and livestock rearing, tourism also contributes to the economy of the state.

Shimla is one of the major tourist and recreational centres. Corresponding to that, the urban built-up lands which include settlements, industries, government buildings and other recreational areas have experienced an increase of 34.7% from 2005–2006 to 2015–2016. However, rural built-up area has seen a decline of 9.8%. The district has witnessed a substantial increase in unproductive barren rocky, scrub and sandy land. In 2005–2006 barren scrubland accounts for 406 km^2 while barren rocky land covers only 23 km^2. But by 2015–2016 it has decreased by 39.8% and increased by 858.4% respectively. This depicts that extensive degradation of land has occurred due to human or natural activity. While glaciers and snow cover has seen drastic decrease of 38.8%, it will have a major impact on the availability of water in the longer run. Figure 15.4 spatially represent the temporal variations in the land use and land cover of the Shimla district from 2005–2006 to 2015–2016.

Table 15.1 Land use and land cover classification of Shimla district

S. No.	Level 1	Level 2	Level 3
1	Built-up	Urban	Residential, mixed, communications, public utilities, transportation, vegetated area, industrial/mine dump
		Rural	Rural
		Mining	Mine/quarry, abandoned mine pit, landfill area
2	Agriculture	Cropland	Kharif, rabi, zaid, two cropped, more than two
		Plantation	Plantation agriculture, horticulture, agro horticulture
		Fallow	Current and fallow
		Current shifting cultivation	Current shifting cultivation
3	Forest	Evergreen/semi-evergreen	Dense, closed and open
		Deciduous	Dense, closed and open
		Forest plantation	Forest plantation
		Scrub forest	Scrub forest, forest blank, current and abandoned shifting cultivation
		Swamp/mangroves	Dense, closed and open mangrove
4	Grazing	Grass/grazing	Alphine, sub-alphine, temperate/sub tropical, tropical/desertic
5	Barren land	Salt affected land	Slight, moderate and strong salt-affected land
		Gully /ravinous land	Gullied, shallow ravine and deep ravine area
		Scrub land	Dense, closed and open
		Sandy area	Desertic, coastal and riverine sandy area
		Barren rocky	Barren rocky
		Rann	Rann
6	Water bodies	Inland wetland	Inland natural land and manmade wetland
		Coastal wetland	Coastal natural and coastal manmade
		Rivers/ stream/canals	Perennial and dry river/stream and line and unlined canal/drain
		Water bodies	Perennial, dry, kharif, rabi and zaid extent of lake/pond and reservoir and tanks

(continued)

Table 15.1 (continued)

S. No.	Level 1	Level 2	Level 3
7	Snow and Glaciers		Seasonal and permanent snow

Source Prakasam et al. (2018)

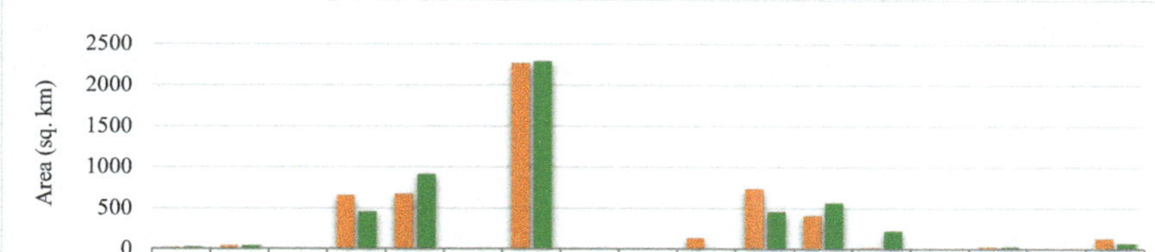

	Built up Urban	Built up Rural	Built mining	Agriculture Crop land	Agriculture Plantation	Agriculture Fallow	Forest Evergreen	Forest Deciduous	Forest Plantation	Forest scrub forest	Grass/ Grazing	Barren/ scrub land	Barren/Barren rocky	Barren/ sancy Area	Wetland/ canals	wetland/ponds	Snow/ Glacer
2005-2006	20.9	46.33	0	656.4	672.4	2.63	2267	12.48	0.38	134	732.9	406.6	23.56	0	29.06	0.4	125.8
2015-2016	28.17	41.76	0.13	460.1	917.8	0.04	2294	14.09	0	14.23	460.3	565.5	225.8	0.06	31.14	1.2	76.89

Fig. 15.3 Change in the land use land cover of Shimla district from 2005–2006 to 2015–2016, *Source* NRSC (2006); NRSC (2019)

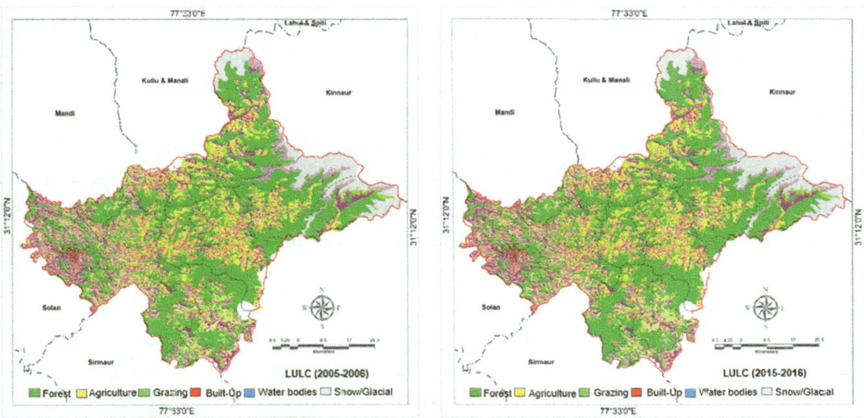

Fig. 15.4 Land use land cover of Shimla district in 2005–2006 and 2015–2016. *Source* NRSC (2019); NRSC (2006)

Further to validate the finding of land use land cover change, temporal analysis of urbanisation and change in cropping pattern of the Shimla district has been studied to build deeper understanding of the change in land use land cover.

15.6 Reasons for Increasing the Water Stress in Shimla City

This section is intended to provide support to the previous findings by discussing the factors responsible for the change in land use land cover, which directly or indirectly leads to increase in the demand for water in Shimla. This section is divided into two sub-sections. The first sub-section attempts to explain the trends of urbanisation in Shimla district and city since independence. While second sub-section provides a detailed overview of the change in the cropping pattern of Shimla district from 2005 to 2015.

15.6.1 Trend of Urbanisation in Shimla District

Urbanisation broadly refers to rural to urban transition involving population, land use, economic activity and culture (McGranahan and Satterthwaite 2014). Demographically the level of urbanisation can be measured by the percentage of the population living in urban areas (Bhagat 2011). As per Wu and Tan (2012) rapid urbanisation increases the urban water demand for different purposes and aggravates the water stress condition. Decadal population analysis of Shimla district shows a persistent and gradual rise in the level of urbanisation. At the time of attainment of Statehood, the district has only six urban centres namely Shimla Municipal Corporation, Municipal council of Rampur, Theog, Rohru and Jutogh Cantonment board. Gradually urban population grew and expanded to other parts of the district. Figure 15.5 shows, gradual rise in the urban growth rate to 15.7% in 1981 and 20.4% in 1991. Here, five settlements of Narkanda, Seoni, Chaupal, Kotkhai and Jubbal were notified as Nagar Panchayat in 1991 (Sharma 2003). The increase in urban population was largely due to natural increase, rural to urban migration and partly due to the emergence of newly notified towns by the State government. By 2011, growth rate

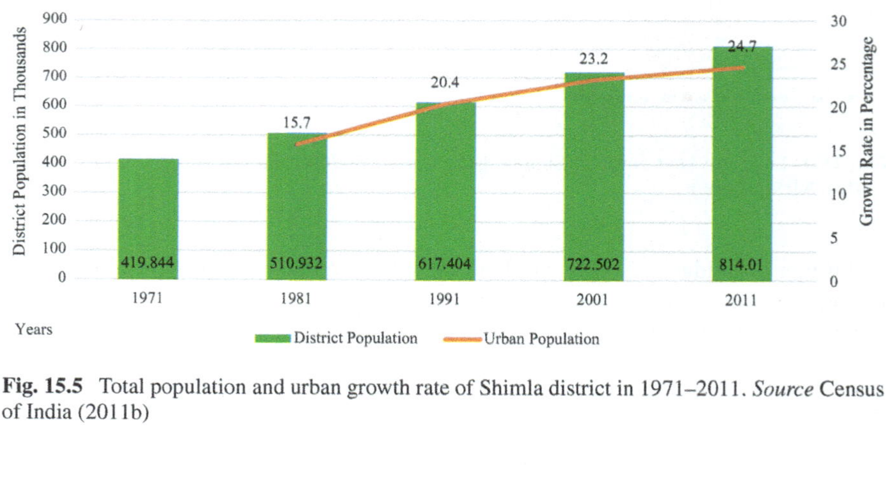

Fig. 15.5 Total population and urban growth rate of Shimla district in 1971–2011. *Source* Census of India (2011b)

reached 24.7% and Jhakhri was notified as a census town (Census of India 2011a). However, among all the urban centres, Shimla Municipal Corporation has experienced the maximum influence of urbanisation as it is the only class I city of the State which accommodates around one-fourth of the State's total urban population. This is mainly due to its early establishment, administrative capital, tourist destination and health resort (Sharma 2003). Urbanisation is largely attributed to administrative and economic primacy of Shimla city.

The historical evolution and development of Shimla city can be categorised into four stages, as explained by (Kumar and Pushplata 2015). The first stage includes the pre-independence colonial time when Shimla was originally conjointly belonged to Maharaja of Patiala and Rana of Keonthal. In 1830, the British acquired the land and established a summer resort to escape from the scorching heat of plains and to recover from homesickness. Realising the vast potentialities of Shimla, it soon became a reputed hill station and sanatorium (Jayaswal 1979). Consequently, this raised the population size of a small hamlet from less than 500 people in 1804 to 7,077 in 1860 (Jayaram 2017). The second stage of development is marked by the independence of the country, where the town continued to experience high population growth rate, largely due to partition and shifting of the East Panjab government to Shimla. However, with the relocation of the capital of East Panjab to Chandigarh, Shimla faced the worst neglect and positioned as a distant headquarter. The growth rate of population reflected the major events of the city's evolution. Figure 15.6 shows from 1941 to 1951 the population grew exceptionally high at 151.5% but suddenly dipped to −7.7% in the consecutive decade. In 1966 reorganisation of Panjab resulted in the merger of Shimla with Himachal Pradesh and it became the permanent capital, with the attainment of statehood in 1971 growth rate surged up to 71.1% which marked the third stage of development where Shimla again rejuvenated. Growth was reflected with the substantial increase in the size of the population however municipal limits increased marginally. Table 15.2 shows the population and the spatial trend of Shimla city.

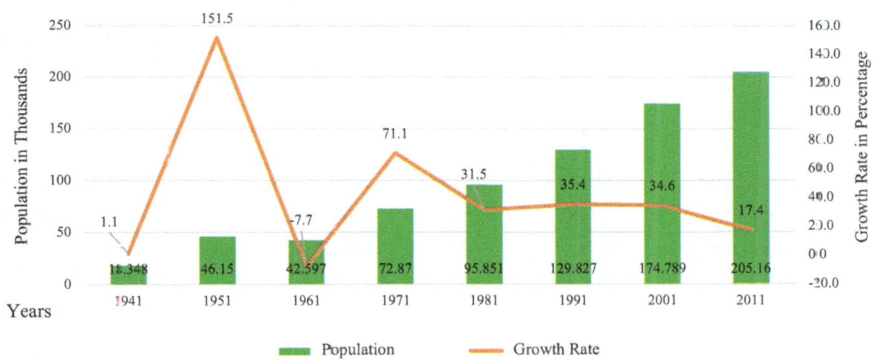

Fig. 15.6 Population and growth rate of Shimla city from 1941 to 2011. *Source* Kaistha and Sharma (1998); Deloitte Touche Tohmatsu India LLP/DRA Consultants Ltd. (2018)

Table 15.2 Total population and spatial extent of Shimla city (1961–1991)

Year	Class of Town	Population	Area in km^2	Area of agglomeration
1961	III	42,597	18.13	
1971	II	55,368	19.55	
1981	II	70,604 (73,004)	19.55 (21.66)	21.66
1991	I	82,054 (110,360)	31.60 (35.34)	35.34

Source Kaistha and Sharma (1998)

Considering the disproportionate spatial and population growth, the state government constituted Shimla Planning Area (SPA) in 1977. It was followed by the constitution of the Interim Development Plan and Shimla Development Authority to ensure planned development. SPA constituted of Shimla Municipal Corporation along with agglomeration of Dhalli, Tutu, New Shimla and Special Area of Kufri, Shoghi, Ghanahatti, which are potential growth centres for urbanisation which are experiencing rapid growth (Shekhar 2011). Major developmental changes took after independence but the issue of congestion housing, water scarcity persisted and further aggravated. This marks the last and the present stage of development, Shimla is still struggling with. Here, population continued to rise at a constant average rate of 33.8% till 2001 but suddenly dipped to 17.4% in 2011.

However, the city's available water resource is not at par with the growing urbanisation. Presently Shimla city gets its water supply from major springs/ Khads namely Dhalli, Churat, Chair, Nauti, Ashwini and Giri (Disaster Management Cell 2012). Although these sources have a total installed capacity of 62–65 MLD, the supply is limited to 33- 38 MLD. This is largely due to poor maintenance of pipelines, water reservoirs, leakage, theft and depleting resources. Considering the growing urban population and existing installed capacity the water demand of Shimla city may cater up to 2025 (GIZ ASEM 2011). The City Development Plan (2006) projected the future water demand of Shimla city as shown in Table 15.3.

In order to cater to the present and future water demand, it is important to rejuvenate the existing and augment the new water source.

Table 15.3 Water demand projection of Shimla city

	2010	2011	2021	2031	2041
Resident population	1,98,717	2,07,063	2,56,883	3,49,361	4,18,296
Floating population	70,000	76,000	1,00,000	1,25,000	1,50,000
Water demand (MLD)	36.28	38.21	48.18	64.04	76.72

Source GIZASEM (2011)

15.6.2 *Cropping Pattern of Shimla District*

Cropping patterns refers to the percentage of area under different crops in a particular period of time (Kumar et al. 2018). Analysing the change in cropping patterns helps us to understand the agricultural development of a region. As per NABCONS (2015), Shimla has 855.23 km^2 of arable land which accounts for 8% of the total district area. A wide variety of food and non-food crops are grown based on the geo-climatic conditions of a region. Here, Kharif crops of maize, rice (cereals); bajra, chulai (coarse crops); rajmah, urad (pulses) are grown in 49% of arable land and Rabi crops of Wheat, barley (Cereals), Gram and lentil (Pulses), Onion and tomatoes (Vegetables) are cultivated in 45% of land while remaining 7% includes summer crops of peas and cabbage (vegetable). Shimla also has majority of its area under production of fruits (apple, pear and cherry). Over the years there has been a prominent shift observed from agricultural cropland to agriculture plantation. Figure 15.7 also shows the prominent changes that occurred in the cropping pattern of Shimla district from 2005 to 2015. In Kharif season a substantial decrease in the area under cereal (29.1%), coarse cereals (39.4%), pulses (41.3%), and vegetable (78.6%) has been noticed except under spices. Similar decrease has been noticed in cereals (39.1%) and oilseeds (46.2%) in Rabi season, while the area under pulses, spices and vegetables has seen a substantial increase. In the same fashion area under vegetable and spices has increased prominently in the summer season. However, as land is a limited resource, the production and productivity of the crops can be increased by irrigation, use of high yielding varieties and bio fertilisers etc. So, it is important to meticulously take into consideration the available water resource and its demand for various purposes.

In Shimla, majority of arable lands (855.23 km^2) are rain fed. Here, irrigation accounts to only 6% of area, largely through Khul, community/ individual ponds and bore well and open well. Although agriculture and horticulture are largely rain fed and only 19% of arable land is irrigated, but it is imperative to consider the

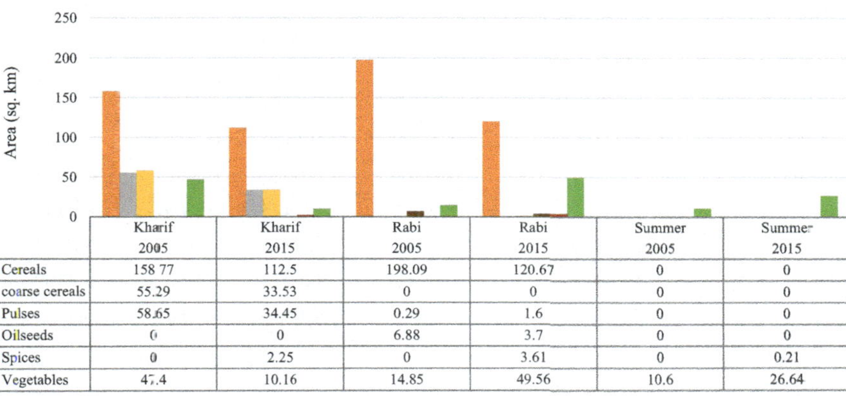

	Kharif 2005	Kharif 2015	Rabi 2005	Rabi 2015	Summer 2005	Summer 2015
Cereals	158.77	112.5	198.09	120.67	0	0
coarse cereals	55.29	33.53	0	0	0	0
Pulses	58.65	34.45	0.29	1.6	0	0
Oilseeds	0	0	6.88	3.7	0	0
Spices	0	2.25	0	3.61	0	0.21
Vegetables	47.4	10.16	14.85	49.56	10.6	26.64

Fig. 15.7 Change in cropping pattern of Shimla district from 2005 to 2015. *Source* NABCONS (2015); Planning Department (2005)

Table 15.4 Water requirement of the major crops in Shimla district

Crops	Number of watering	Total water requirement (in millimetre)
Paddy	12	600
Wheat	5	150
Maize	2	60
Vegetables	6	180
Horticulture crops	–	60

Source NABCONS (2015)

water requirement of different crops, to assess the present and future water demand. According to NABCONS (2015), as per the discussion with agronomists of Agriculture University, the water requirement of the major crops has been calculated as shown in Table 15.4, shows paddy, vegetables and wheat require more frequency and amount of water for cultivation.

Considering the (2015–2020) District Irrigation Plan of Shimla district, Table 15.5 shows the total crop water requirement of different blocks in Shimla district.

Herein, it is evident that the existing water potential of the district is 7.24 MCM against the requirement of 84.06 MCM, which means a total gap of 76.82 MCM is needed to be created to fill in the present requirement. Considering the depleting

Table 15.5 Block wise crop water demand of Shimla district

Block/Tehsils	Area sown (km^2)	Irrigated area (km^2)	Crop water demand (MCM)	Water potential required (MCM)	Existing water potential (MCM)	Water potential to be created (MCM)
Basantpur	68.61	4.21	7.3029	7.3029	0.8445	6.4584
Chauhara	102.8	6.03	12.8379	12.8379	2.3214	10.5165
Chopal	134.28	4.54	15.3276	15.3276	1.668	13.6596
Jubbal Kothkai	88.99	0.49	5.7999	5.7999	0.1722	5.6277
Mashobra	85.9	7.92	8.2785	8.2785	1.1415	7.137
Nankhari	29.77	0	2.1141	2.1141	0	2.1141
Narkanda	84.17	1.4	7.0074	7.0074	0.534	6.4734
Rampur	87.55	0.79	8.9445	8.9445	0.1986	8.7459
Rohru	73.41	0.62	5.8446	5.8446	0.1908	5.6538
Theog	99.75	0.97	10.6041	10.6041	0.1632	10.4409
Total	855.23	26.97	84.0615	84.0615	7.2342	76.8273

Source NABCONS (2015)

water resources, it is imperative to replace the water-guzzling crops with less water-intensive crops to maintain the ecological balance. It is important to revive the traditional crops with new technological intervention to maintain the food and livelihood security, natural resources and prevent the impact of climate change.

15.7 Suggestion

Change in the land use and land cover of Shimla district has become quite prevalent with increasing urbanisation and changing cropping patterns (as discussed in Sect. 6). Further, these changes have a direct impact on the water resource of the Shimla district. As Shimla is already undergoing through major water stress situation, it is necessary to undertake important measures to preserve water for present and future generations. Considering the topography and hydrogeology, it is important to apply nature-based solutions for effective management of water resources in Shimla.

1. Spring shed Development
   Most of the irrigation and domestic water supply are channelized through spring, *khads, khul, bowaries*. Proper development of springs and revival of traditional water storage (*bowaries*) can help in managing the groundwater resource of Shimla district. In recent times, a number of bore wells and hand pumps were constructed to extract the groundwater. However, considering the fragile ecosystem, drilling activities should be minimised or avoided. So, revival of traditional water storage (*bowaries*) could be a more appropriate and feasible solution to tap the groundwater resource of the district. Other than its primary use, these *bowaries* also serves as a point of cultural and social interactions.
2. Water harvesting
   As Shimla city is frequently witnessing the water shortage during lean period or summer seasons, there is an immediate need to conserve the water or augment new water resources. Based upon the climatic and topographic condition of Shimla district, rainwater harvesting and artificial groundwater recharge can be an appropriate solution to overcome the problem of water shortage during lean period. Rooftop rainwater harvesting in urban or rural areas and water harvesting structure in rural areas need to be promoted to supplement the water resource during lean period.
3. Agroforestry
   It is evident from land use land cover change, area under scrub forest has experienced a major decline over a period of time. Conversely, the area under plantation agriculture has seen a major increase. Studies have shown that forests play a very significant role in water conservation. So, to maintain the forest cover along with the agriculture, it is important to perform agroforestry in hilly areas. Agroforestry performs various other ecosystem services such as carbon sequestration, soil enrichment, biodiversity conservation and air & water quality improvement. Moreover, it also gives economic benefits to the farmers.

4. Biological and Mechanical measure for water conservation
   As the district has hilly terrain with rolling topography and sloppy lands. Heavy rainfall may cause splash erosion and low infiltration of surface water. To improve soil and water conservation, it is important to promote biological (vegetative Barriers, strip cropping, mulching) and physical (contour trenches, bunding, terracing) measures in rural and urban areas to conserve water and soil.
5. Watershed management
   Shimla district forms the watershed of major rivers system (Ashwini, Nauti, Giri, Satluj and Pabbar) of district. Change in land use and land cover has accelerated the problem of deforestation, land degradation in the watershed. This change along with other factors has collectively contributed to drying up the natural water source which has directly impacted the socio-economic life of the people in watershed. By promoting integrated watershed management programme, harmonious development can be achieved between natural and socio-economic environment of watershed.
6. Reviving Traditional Crop
   A major shift in the agriculture pattern of the Shimla district has been evident from land use land cover change. Area under agriculture cropland has seen drastic decline while area under agriculture plantation has seen tremendous increase. This is largely due to higher economic benefits from the plantation crops. It is essential to maintain the traditional food crops along with the commercial crops, as traditional crops are favourable to local climatic and physiographic condition and requires comparatively less input for production (Fig. 15.8).

15.8 Conclusion

In view of rapidly changing land use land cover of Shimla, it is important to manage the available water resource in Shimla district. By understanding the trend of urbanisation and change in the cropping pattern of Shimla district, the study attempted to understand the growing reliability on limited water resources and determine the pathway to sustainably manage the water. Based on literature review, the study highlighted the growing concern of water stress in India and Shimla and increasing recognition of nature-based solutions to tackle the problem of water insecurity.

Further, the study emphasised understanding the change in land use land cover of Shimla district from 2005–2006 to 205–2016. For analysing the change, databases on land use land cover of Shimla district from the specified years were retrieved from Natural Resource Census- land use land cover. It has been found that major changes have been observed in scrub forest cover, urban built-up area and cropland /plantation agriculture. To examine the observed change, temporal analysis of urbanisation and change in cropping pattern of the Shimla district has been studied. It has been found that Shimla has witnessed significant increase in urban population and their demand

Fig. 15.8 **a** Picture show reconstruction of water *bowaries*. **b** Picture show the use of pipe to capture the water. **c** Skyline view of forest cover in Narkanda. **d** Water collection tank in Narkanda. *Source* Author

for water. Prominent change has also been found in the cropping pattern and it has been realised that existing water potential of the district is less than the requirement of crops.

The study generated substantial evidence of growing water concern due to changes in land use land cover. Further study suggests a suitable direction of natured based solution to address the growing water issue in Shimla.

References

Bajpai I (2018) Parched Shimla struggles with leaking pipes tourist influx, Down to Earth. https://www.downtoearth.org.in/news/water/parched-shimla-struggles-with-leaking-pipes-tourist-influx-60445. Accessed 08 August 2020

Bhagat R (2011) Emerging pattern of urbanisation in India. Econ Pol Wkly 46:10–12

Cassella C (2019) Nearly 25% of the world's population faces a water crisis, and we can't ignore it, Science alert. https://www.sciencealert.com/17-countries-are-facing-extreme-water-stress-and-they-hold-a-quarter-of-the-world-s-population. Accessed 13 April 2020

Census of India (2011a) Himachal Pradesh Administrative Divisions 2011, Office of the Registrar General & Census Commissioner India. http://censusindia.gov.in/2011census/maps/administrative_maps/HIMACHAL%20PRADESH.pdf. Accessed 13 April 2020

Census of India (2011b) District census handbook Shimla village and town wise primary census abstract (PCA), Directorate of Census Operations Himachal Pradesh. http://Censusindia.Gov.In/2011census/Dchb/0211_PART_B_DCHB_SHIMLA.Pdf. Accessed 4 April 2020

Chand J (2013) Himachal Pradesh: trends of urbanization from 1971 to 2001. Asian J Multidiscip Stud 1:134–140

Chatterji R (2018) Shimla water crisis explained: sewage contaminated a stream, a jaundice outbreak followed, then a high court order left the city high and dry, Huff post. https://www.huffingtonpost.in/2018/06/05/shimla-water-crisis-explained-sewage-contaminated-a-stream-a-jaundice-outbreak-followed-then-a-high-court-order-left-the-city-high-and-dry_a_23451060/. Accessed 13 April 2020

Cohen-Shacham E, Janzen C, Maginnis S, Walters G (2016) Nature-based Solutions to address global societal challenges. International Union for Conservation of Nature. https://doi.org/10.2305/IUCN.CH.2016.13.en. Accessed 13 April 2020

Cohen-Shacham E, Andrade A, Dalton J, Dudley N, Jones M, Kumar C, Maginnis S, Maynard S, Nelson CR, Renaud FG, Welling R, Walters G (2019) Core principles for successfully implementing and up scaling Nature-based Solutions. Environ Sci Policy 98:20–29. https://doi.org/10.1016/j.envsci.2019.04.014

Deloitte Touche Tohmatsu India LLP / DRA Consultants Ltd (2018) Transaction Advisor for Bulk, Distribution System, and Sewerage System for the Shimla Water Supply and Sewerage Project Revised Situation Analysis Report July, 2018. Greater Shimla Water Supply and Sewerage Circle, Municipal Corporation Shimla. Shimla, Himachal Pradesh

Disaster Management Cell (2012) City disaster management plan- Shimla, City Disaster Management Cell. Municipal Corporation Shimla, http://hp.gov.in/hpsdma/DisasterManagement/CDMP_MCShimla.pdf. Accessed 8 Aug 2020

Earthwatch Institute (2020) Planet earth is seemingly awash with water—we call it 'the Blue Planet'. Freshwater watch. https://freshwaterwatch.thewaterhub.org/content/water-limited-resource-archived#:~:text=But%20how%20much%20accessible%20fresh,stored%20below%20ground%20as%20groundwater. Accessed 8 Aug 2020

Ellis E (2007) Land use and land cover change and climate change, The Encyclopaedia of Earth. http://ecotope.org/people/ellis/papers/ellis_eoe_lulcc_2007.pdf. Accessed 27 April 2020

European Commission (2017) Nature-based solution, European Commission. https://ec.europa.eu/research/environment/index.cfm?pg=nbs. Accessed 13 April 2020

GIZ ASEM (2011) City sanitation plan for Shimla, CDD Society, Alchemy, CRDU. http://www.shimlamc.org/file.axd?file=2011%2F10%2FR-SSRFinal-110517.pdf. Accessed 08 August 2020

Hofste RW, Reig P, Schleifer L (2019) 17 Countries home to one-quarter of the world's population face extremely high water stress. World Resources Institute. https://www.wri.org/blog/2019/08/17-countries-home-one-quarter-world-population-face-extremely-high-water-stress. Accessed 13 April 2020

Jayaram N (ed) (2017) Social dynamics of the urban: studies from India. Springer, New Delhi

Jayaswal SNP (1979) Problems of land use and urban development in Simla. Dissertation, Chhatrapati Sahuji Maharaj University. http://hdl.handle.net/10603/277312

Kaistha KC, Sharma SK (eds) (1998) Population, spatial mobility and environment: issues and challenges for sustainability development. Anamika Publishers & Distributors (P) Ltd., Delhi

Kumar A, Pushplata, (2015) City profile: Shimla. Cities 49:149–158. https://doi.org/10.1016/j.cities.2015.08.006

Kumar P, Thakur CL, Rai P, Attri K (2018) Identification of existing agroforestry systems and socio-economic assessment in Kandaghat block of Solan district Himachal Pradesh India. Int J Curr Microbiol Appl Sci 7(4):3815–3826. https://doi.org/10.20546/ijcmas.2018.704.429

McGranahan G, Satterthwaite D (2014) Urbanisation concepts and trends, IIED Working Paper. http://pubs.iied.org/10709IIED. Accessed 16 March 2020

Municipal Corporation Shimla (2009) Water Supply & Sewerage Department. http://www.shimlamc.org/page/Water-Supply-Sewerage.aspx. Accessed 4 April 2020

NABCONS (2015) District Irrigation Plan 2015–2020 Shimla Himachal Pradesh, Department of Agriculture Himachal Pradesh

NIH and I&PH (2013) Impact of Sewage Effluent on Drinking Water Sources of Shimla City and Suggesting Ameliorative Measures, National Institute of Hydrology Roorkee, I & PH DepartmentShimla. http://nhp.mowr.gov.in/docs/HP2/PDS/Surface%20Water/3043/NIH_SW_Impact%20of%20sewage%20effluent%20of%20drinking%20water%20sources.pdf. Accessed 8 Aug 2020

NRSC (2006) Land Use / Land Cover database on 1:50,000 scale, Natural Resources Census Project, LUCMD, LRUMG, RS & GIS AA, National Remote Sensing Centre, ISRO Hyderabad

NRSC (2019) Land Use / Land Cover database on 1:50,000 scale, Natural Resources Census Project, LUCMD, LRUMG, RSAA, National Remote Sensing Centre, ISRO Hyderabad

Oki T, Quiocho RE (2020) economically challenged and water scarce: identification of global populations most vulnerable to water crises. Int J Water Resour Dev 36:416–428. https://doi.org/10.1080/07900627.2019.1698413

Planning Department (2005) District human development report Shimla, UNDP, Planning Commission GoI, Planning Department Himachal Pradesh. https://www.im4change.org/docs/417shimla_district_human_development_report.pdf. Accessed 4 April 2020

Prakasam C, Raja A, Kanwar VS (2018) Land use land cover change detection & urban sprawl study: a case study on Shimla tehsil Himachal Pradesh India. J Geogr Environ Earth Sci Int 16(1):1–13. https://doi.org/10.9734/JGEESI/2018/42185

Roy PS, Roy A (2010) Land use and land cover change in India: A remote sensing & GIS perspective. J Indian Inst Sci 90(4):489–502

Sharma DK (2003) Urban development administration in Himachal Pradesh: A study of role and performance. Dissertation, Himachal Pradesh University http://hdl.handle.net/10603/121005

Sharma SK, Kansal ML, Tyagi A (2015) Resource assessment and strategic planning for improvement of water supply to Shimla city in India using geo-spatial techniques. Egypt J Remote Sens Space Sci, 1–12.https://doi.org/10.1016/j.ejrs.2015.04.001

Shekhar S (2011) Urban Sprawl and other spatial planning issues in Shimla, Himachal Pradesh. Institute of Town Planners India, pp 53–66

Shiao T, Maddocks A, Carson C, Loizeaux E (2015) 3 Maps explain India's growing water risks, World Resource Institute. https://www.wri.org/blog/2015/02/3-maps-explain-india-s-growing-water-risks. Accessed 9 Aug 2020

Snyder A (2014) Shortage in the mountains of plenty: water supply in mountain and hill cities throughout the Hindu-Kush Himalayan region, The World Food Prize. https://www.worldfoodprize.org/documents/filelibrary/images/youth_programs/2014_interns/2014_br_research_papers/SnyderAbigail_LONGReport_56ED38F157B76.pdf. Accessed 4 April 2020

TERI (2015) Green growth and water in Himachal Pradesh, The Energy and Resources Institute. https://www.teriin.org/projects/green/pdf/HP-Water.pdf. Accessed 8 Aug 2020

Thakur HK (2018) Shimla water crisis symptomizes a greater malaise. Int J Sci Res 7:1357–1360

UNDP (2006) Human development report beyond scarcity: power, poverty and the global water crisis. United Nations Development Programme. https://www.undp.org/content/dam/undp/library/corporate/HDR/2006%20Global%20HDR/HDR-2006-Beyond%20scarcity-Power-poverty-and-the-global-water-crisis.pdf. Accessed 13 April 2020

UN Environment-DHI, UN Environment, IUCN (2018) Nature-based solutions for water management: a primer. United Nation Environment Programme. http://wedocs.unep.org/handle/20.500.11822/32058. Accessed 2 March 2020

UNESCO (2015) The United Nations world water development report 2015: water for a sustainable world, United Nations Educational Scientific and Cultural Organization. https://sustainabledevelopment.un.org/content/documents/1711Water%20for%20a%20Sustainable%20World.pdf. Accessed 13 April 2020

U.S. Department of the Interior (2020) Water facts—worldwide water supply. Bureau of Reclamation. https://www.usbr.gov/mp/arwec/water-facts-ww-water-sup.html. Accessed 11 Aug 2020

Vasudeva V (2018) When Shimla queued up for water. The Hindu. https://www.thehindu.com/news/national/when-shimla-queued-up-forwater/article24175811.ece. Accessed 13 April 2020

Veolia (2016) Water Stress. https://www.veolia.in/about-us/meeting-indias-challenges/water-stress. Accessed 8 Aug 2020

Vihervaara P, Ronka M, Walls M (2010) Trends in ecosystem service research: early steps and current drivers. Royal Swedish Acad Sci 39:314–324. https://doi.org/10.1007/s13280-010-0048-x

Wu P, Tan M (2012) Challenges for sustainable urbanization: a case study of water shortage and water environment changes in Shandong China. Procedia Environ Sci 13:919–927. https://doi.org/10.1016/j.proenv.2012.01.085

WWAP (2018) The United Nations world water development report 2018: nature-based solutions for water, United Nations Educational Scientific and Cultural Organization. https://unesdoc.unesco.org/ark:/48223/pf0000261424. Accessed 16 March 2020

WWDR (2020) The United Nations world water development report 2020: water and climate change, United Nations Educational Scientific and Cultural Organization. https://unesdoc.unesco.org/ark:/48223/pf0000372985.locale=en. Accessed 13 April 2020

Chapter 16
Application of Remote Sensing Image in ECO-DRR for Dehradun City

Atul Kumar, Jeevan Madapala, Mahua Mukherjee, Shirish Ravana, and Sandeep Sharma

Abstract The case study of Dehradun city is an example to show the applications of remote sensing in Eco-DRR. Geographically the town is located in the Doon Valley, which is prone to multi hazards, rapid and unplanned urbanization in the city is adding the concern of high risk for the communities and people. Combined with inadequate land-use planning and the inability of local authorities to control building standards, the unplanned growth of cities raises insecurity in all dimensions. This study tries to map Blue-Green infrastructure in the municipal boundary of the town using the space-based application and suggest an Eco-DRR map that can help identify location and hazard existence and contribute to sustainable development. The imagery of freely available or low-cost satellite is used to develop indices for Vegetation, water bodies and build-up and analyzed Blue-Green infrastructure and physical component of the urban ecosystem for the municipal boundary of the city. The geospatial information generated using remote sensing imagery is georeferenced and calibrated for further actions that led to the integration with the existing LULC map and city master plan. Finally, integrated maps use to suggest a holistic approach of Eco-DRR in the urban area for sustainable development.

Keywords Eco-DRR · Blue-green infrastructure · GIS · Remote sensing

A. Kumar (✉) · J. Madapala · M. Mukherjee · S. Sharma
Indian Institute of Technology Roorkee, Roorkee, India
e-mail: akumar5@ar.iitr.ac.in

M. Mukherjee
e-mail: mahuafap@iitr.ac.in

S. Sharma
e-mail: ssharma@ar.iitr.ac.in

S. Ravana
UN-SPIDER, Vienna, Austria
e-mail: shirish.ravan@un.org

© The Author(s), under exclusive license to Springer Nature Singapore Pte Ltd. 2021
M. Mukherjee and R. Shaw (eds.), *Ecosystem-Based Disaster and Climate Resilience*, Disaster and Risk Research: GADRI Book Series,
https://doi.org/10.1007/978-981-16-4815-1_16

16.1 Introduction

The belief in nature-based solutions in the growing age of urbanization to reduce the impact of risk has been well discussed and accepted in major international agreements such as Paris Agreement, several SDGs (Sustainable development goals), etc. All such organizations have prioritized the ecosystem based approach as the first list for urbanization and development. Well-managed ecosystems such as natural infrastructure in a built premise of urban areas reduce risk from many hazards and increase the socio-economic resilience of the communities. Numerous scientific studies have also confirmed the role of the ecosystem in the process of adaptation to adverse effects of environment, protection from various extreme events, to develop safe, resilient and sustainable cities. Seven global targets by the Sendai framework focused on the "ecosystem based approaches to build resilience and reduce risk." The 5th assessment report of IPCC (2014) states that "successful adaptation will depend on our ability to allow and facilitate natural systems to adjust to a changing climate, thus maintaining the ecosystem services on which all life depends" (IPCC 2014). The world expert's communities also have consent on how the natural environment has evolved as a tool to buffer against natural hazards in real, sustainable and cost-effective ways through several case studies globally.

Ecosystem based disaster risk reduction (Eco-DRR) is defined as "sustainable management, conservation and restoration of ecosystems to reduce disaster risk, with the aim of achieving sustainable and resilient development." (Estrella 2013). An urban ecosystem is the same as other ecosystems having biological as well as physical components that interact with each other. It gained attraction because of its win–win characteristics as it provides multiple benefits to all urban scales from landscape scale to regional level. It is cost-effective, promote community participation and the use of traditional and local knowledge systems comparing to conventional infrastructure. Eco-DRR participatory nature and cross-sectoral approaches fill the gap of policy objectives among national and local strategies. Its impact on risk reduction often has low regret to no regret options, in this way Eco-DRR enhances resistance to fight against risk and its components like hazard, vulnerability and exposure as well as in the post disaster reconstruction period offers an opportunity to "Build Back Better" only need to plan the natural infrastructure strategically.

16.2 Disaster Risk Management (DRM) and Its Relation with Ecosystem Based and Remote Sensing

Over the last 30 years, Asia as a continent suffered the most due to disasters. Asia alone comprised more than 88% of the global disaster victims and 39% of the global fatalities. With its high density of population and necessity for economic development, the story with India is also either very much similar or little bit worse, due to issues such as poverty and high-density population. Therefore, it has become of utmost importance to incorporate the ideas of Disaster Risk Management into the development activities and everyday life. Where will disaster strike next? Are we ready? In view of these questions and circumstances, it is important to exploit the best technologies available. Remote sensing offers such capabilities, with high temporal resolution, spatial resolution and spectral resolution. Remote Sensing data combined with applications such as Geographical Information Systems (GIS), offers the best support to disasters planners. With climate change, we witness an increasing human and financial cost from disasters; the geospatial capabilities offer an advantage to Disaster Risk Management (DRM). Responding to, recovering from and preparing for disasters, has always been a State responsibility. However, the active stakeholders such as R&D institutions, educational institutions and private players can play major role in actions towards DRM. Therefore, the remote sensing and geo-intelligence capabilities can play a major role, that help predict, warn and measure the disasters. In DRM, there is an undoubted correlation between the remote sensing data, geo-intelligence capabilities and the informed decision-making. In this regard, the literature review tries to focus on the opportunities to develop an integrated and holistic system incorporating the advantages to remote sensing and geo-intelligence in the endeavours towards the DRM. The focus of the study is also to tackle the DRM, through a sustainable mechanism such as the Ecosystem Based Solutions, as long-term strategy to tackle the DRM.

Remote Sensing (RS) in DRM: Remote Sensing technologies use the electromagnetic radiations to remotely study the properties of the objects. Among all RS technologies, satellite based RS technologies are the best. They can be used to study large areas, operating in all weathers. With very high-resolution capabilities, satellite based technologies can be used to study the environmental problems, monitor disasters, etc. In the year 2000, International Charter Space and Major Disasters (hereinafter referred to as "the International Charter") and the United Nations (UN) came into effect, encouraging the applications of satellite based RS technologies in the field of DRM came into action.

Since 2000, the Indian Space Research Organization (ISRO) also begin exploring the idea of remote sensing based DRM. Over the past two decades, remote sensing capabilities have been extensively used to forecast the hazards such as cyclones and floods (especially the natural hazards), develop and analyze hypothetical scenarios, plan rescue and recovery operations, estimate the extent of damage and develop several other tools to aid disaster managers. India's greatest progress of remote sensing based DRM can be witnessed in the cyclone prediction operations and in planning the relief operations in the flood affected areas. For example, when the 1999 Odisha Super cyclone occurred, India lost more than 10,000 people officially. However, during cyclone Phailin (2013) and cyclone Fani (2019) which were of almost the same intensity, the fatalities were about 100. In the view of the recent cyclone Amphan (may, 2020) that hit the eastern coast of India, the Indian metrological department was praised by the world organization for its maximum accuracy of tracking. This kind of achievement in DRM operations was only possible due to extensive developments in DRM operations. With its advanced capabilities, ISRO has been an active contributor to the DRM activities in the country.

An earth observation system usually has 5 segments for studying cause, impact and solution for any disasters. Five segments of earth observation units are the source of radiation (e.g. Sun, Earth and artificial radiation source), atmospheric window and interaction, surface interaction, space and ground. For disaster risk reduction using remote sensing systems, the ground segments of the system play an important role as major human factors are directly related to the ground segment. Other segments are responsible for better technical analysis and data providing for a better action plan for disaster management support (Kaku 2019). The remote sensing based tools have their application in all the stages of DRM operations namely, rescue and relief operations, quantifying post disaster damages and planning capacity building measures and most recently in developing resilient cities. One main reason for this rapid progress has been the introduction of high-resolution, commercially available satellite imagery. The development of high-resolution capabilities and active sensors (e.g. synthetic aperture radar, or more commonly known as SAR and light detection and ranging or LIDAR), further demonstrated the strength of remote sensing applications in DRM activities. The system can promise to provide regular and updated data on the status of risk and hazards on global, regional and national levels as well as with high-resolution satellite data neighbourhood to micro level condition can easily be monitored (Petiteville 2015).

Remote sensing study facilitates the process of studying these processes and making amends in order to act towards DRM. Most remote sensing studies concerned with natural hazards have been about an area's vulnerability to a disaster, the monitoring of events that could precipitate a disaster and the magnitude, extent and duration of a disaster. This chapter attempts to discuss various types of remote sensing information and the suitability for identifying and assessing particular natural hazards and where to look for them. With urbanization, high economic cost and high-density population, climate change and anthropogenic influence over the natural ecosystems, the vulnerability of any region is simple increasing (Barbosa et al. 2015). Therefore, the idea of integrated disaster management, which focuses on reducing social, economic and physical vulnerabilities is the necessity of the day. This integrated planning requires identifying both the hazard and vulnerability at an early stage which is accurately possible due to remote sensing applications. This integrated planning also requires real time data, which is also by remote sensing data with high temporal resolution.

The most important aspect of DRM is also identifying the key hotspots, regions with high vulnerability, regions with possibility of interventions. High spatial resolution data offered by remote sensing applications, facilitates identifying and designing solutions to the key hotspots. Risk reduction measures in identified vulnerable regions can mitigate the effects of disasters.

The efficiency of the remote sensing also depends on temporary factors such as clouds and heavy haze, adequate textural and tonal or colour contrast and technological factors such as resolution of the image, the acquisition scale of the sensor data, the working scale, etc., The interpretation of the data can be further enhanced using technologies such as stereo-models, etc., While there are several remote sensing technologies categorized based on the platforms such as aerial remote sensing, satellite based remote sensing, etc., the latter one is much more preferred due to economic budgets, large datasets and flexibility of operations.

While the existing remote sensing data with the best resolutions, is quite adequate for the DRM activities, tools and procedures to extract the information are still being developed. In this regard, the chapter attempts to develop methodologies, incorporating the holistic design of DRM actions. Besides identifying and analyzing the vulnerability, the methodology also attempts to design the ecosystem based solutions as solutions to DRM.

Main causes of increased disaster risk lack of holistic data sets. As elements of the ecosystem can not only be beneficial for the specific place but can impact a large area, major concerns of risk especially in the urban areas are

1. Spatial planning and expansion of the urban limits into river beds and natural forests, illegal Encroachments, Poor Master Planning,
2. Absence of effective urban policies, lack of effective DRM guidelines
3. Minimum participation and responsibility of local bodies in DRM activities. Also absence of individual responsibilities, absence of contingency plans, etc.,
4. Lack of sufficient basic infrastructure such as drainage facilities and adequate green cover within city limits, etc.

Conventional Engineering Measures Versus Ecosystem Based Measures

Hydraulic engineering system	Risk mapping
Detention basin	Permanent protection areas
Retention basin	Reforestation projects
Polders and dikes	River restoration projects
Canalization and bank stabilization	Alternative land-use systems
Slope stabilization	
Terracing and drainage of embankments	
Drain pipes in settlements and along roads	

So far, there has been a high reliance on conventional engineering solutions to offer disaster risk reduction services. In many cases, these solutions only provided temporary relief and in some cases, these grey engineering measures lead to further exploitation of the natural systems. Besides the economic cost of these engineering measures is also very high. However, lately, the ecosystem based solutions are gaining importance in planning and policy making process. Ecosystem solutions offer long-term solutions, permanent benefits at low cost and secondary benefits such as aesthetics and clean environment. In this background, the paper attempts to suggest ecosystem based solutions, wherever possible. Also, in this section, we tried to list various conventional engineering measures and ecosystem based measures for a clear understanding.

16.3 Necessity and Implications of Ecosystem Based DRR in Hill Cities

Hill cities are very different from the rest of the urban centres. While the hill cities have dense population, high economic activity like any other cities, they offer incredible challenges to disaster managers. The biggest challenges are the slope and instability issues. Landslides are very frequent in hill cities. Hill cities are mostly situated within green zones such as protected forests. Therefore, protecting wildlife, green cover are also the challenges. While most hill cities have fast flowing rivers at their young stage, some hill cities may also have unique features like springs. These water bodies are very sensitive to anthropogenic influence at this stage of flow. Even providing basic facilities such as providing adequate drainage, waste management solutions, transportation connectivity is also a challenge in hill cities. Incorporating these challenges and the importance of the ecosystem dependence of the system in hill cities, the paper proposes the ecosystem based DRM solutions as the best. In risk perception survey it was highlighted that, while people are aware of vulnerabilities and risks of settlements in high-risk zones, they lack knowledge about capacity building measures. Also, they lack resources to build resilience measures (Lange et al. 2013). The degradation of ecosystem by human activities has increased the risk of landslides, mudslides and floods during extreme events (Nehren 2014). Intense summer rainfall is the main landslide-triggering factor (Fernandes et al. 2004; Smyth and Royle 2000). The likelihood of landslides increases with the degradation of natural ecosystems due to uncontrolled land use, fires and land clearance. Efforts for ecosystems recovery cannot cope with the high level of degradation.

Landscapes and ecosystem systems as a whole aim at long-term effects to achieve resilience and support sustainable development. Wide range of actions from planning urban forestation and river planning need more time to show positive effects and often require continuous maintenance. This ecosystem based systematic approach to addressing disaster risks and participation of local authorities and communities in the planning and implementation process, often result in the long-term results which are more effective than pure technical solutions. Costs are usually lower and there are co-benefits, such as climate, biodiversity, watershed and soil protection. Multi hazard scenarios incorporating hazards such as earthquakes, landslides, forest fires, fire accidents, stampedes and floods are considered for the purpose of disaster mitigation studies in this report. In an effort to act towards disaster risk reduction (DRR) through the use of Remote Sensing (RS) technologies, the study attempts to develop methodologies to exploit the potential of remote sensed data, especially in the most challenging environments, such as densely populated hilly regions.

Considering the necessity of Sustainable development (in principle with the Sustainable Development Goals), the paper also attempts to develop methodologies to utilize the remote sensing data to plan and develop ecosystem based solutions. Applications of remote sensing data can effectively and efficiently help understand the scenario of Disaster Risk and the Ecosystem Services of any particular area. Therefore, developing a remote sensing based tool to increase the collaborations among the efforts of Disaster Risk Reduction and Ecosystem Based Measures is the current necessity and hence the focus of the paper. An effective methodology can also be used to provide guidance to the stakeholders on how to reduce Disaster Risk through ecosystem measures considering critical zone and interventions zone.

16.4 Geospatial Tools for Risk Assessment and Ecosystem Based Solutions

The process of remote sensing can play an important role in mapping and maintaining the existing green infrastructure of the city as well as it also helps in the process of planning; it helps in risk hotspots identification as well as provide base data for strategic networking and planning of natural infrastructure as shown in the Fig. 16.1. Remote sensing can be a good tool for risk profiling and thus led stakeholders to make wise decision spatially and temporally both. Earth observation data are the paradigm shift for the analysis of either natural or human caused landscape or the components of ecosystem. It can provide a holistic approach from Data-Information-Knowledge-Action. The basic principle behind earth observation data is capturing spectral reflectance, since each object on the surface show a specific spectral signature when observed in narrow spectral bands, for example, vegetation show high reflectance in near infrared spectrum compare with other bands. The study of these features can be done easily using remote sensing. As the world is urbanizing at a very high rate, urban area is facing the various micro to macro scale issues, to combat

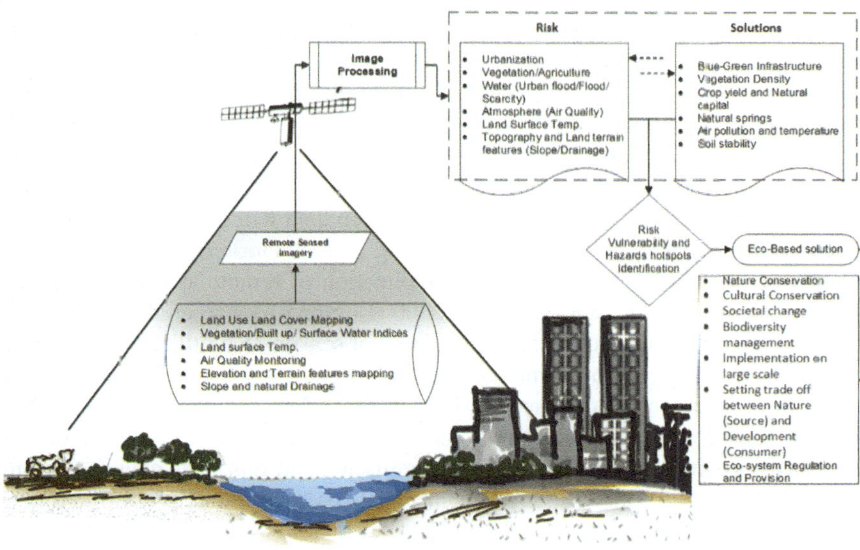

Fig. 16.1 Remote Sensing application in ecosystem based solutions

this, the stakeholders and actors required actionable risk information in an accessible, credible and useful format. The remote sensing can help from the supply side as raw data can access via cloud platforms (e.g. Copernicus Data and Information Access Services [DIAS]), open data source (e.g. ESA-Sentinel mission), from demand side it provides information from the raw data after processing, that can be used with knowledge to action. Remote sensing application allow all the stakeholders to identify and characterize risk and its components, assets livelihood and ecosystems and support mitigation measures at each scale, as small as location specific (Barbosa et al. 2015). It is required to understand urban ecosystem for supportive DRR actionable information for urban area, this information can be extracted by the application of remote sensing and earth observation data. The application helps in analyzing dynamic characteristics of the land features even that are inaccessible. The remotely sensed data can be used to study various urban ecosystem parameters such as height, extent, pattern, density, impact on its neighbourhood etc. that can be further used by stakeholders for decision-making. Remote sensing applications make Eco-DRR approaches easier for complex urban environment interactions between natural and built systems; remote sensing can be used to map characteristics of ecosystem and its functions. Remote sensing technique used to analyze reflected radiation to estimate properties of earth's surface such as land cover type, biomass and leaf area index for each pixel (spatial resolution) which provide spatially dynamic maps and other ecosystem parameters can infer and map through contextual relations across regions. Because of technical constraints, different spatial, temporal and spectral resolutions are the limiting factor for the utilization of the satellite image data for different applications; and the sensors go for a resolution trade-off. Primary ecosystem components have direct relations with the reflected signals like chlorophyll contents, urban vegetation demarcation etc. but there are some parameters that required some extra data sets to delineate like ground water table, air quality, urban heat island etc. Urban ecosystem can understand better by delineating urban blue-green infrastructure, for this through remote sensing the first step is to identify the nature of the relationship between spectral information and ecosystem components. Land-use land cover classification and various vegetation indices developed using remote sensing and GIS are one of the simple and important approach for Eco-DRR. The emergence of new and more sophisticated products, Earth observation data will continue to contribute extensively to research on modelling, mapping and valuation of ecosystem goods and services (Barbosa et al. 2015). The application of remote sensing to disaster management support has started around 2000 (Lorenzo-Alonso et al. 2019). Earth observation satellite data collected by hundreds of sensors that utilize the narrowest band of electromagnetic spectrum and provide different approaches to measure the change from micro to regional scale. Much of these data are available freely from the open source of various organizations these data are ready for operation. The Copernicus programme of the European commission and European agency, various platforms operated by the US Government, BHUVAN an Indian platform etc. are the multiple sources from where satellite data can acquire. UN-SPIDER knowledge portal is a leading platform for earth observation data sets, it provides dataset of various resolutions, baseline data, thematic data. Satellite data can help in hazards mapping from global to national or regional basis both temporally and spatially. In the year 2008, the US government has opened the archive of Landsat data series, since then about 20 million imagery has been downloaded all around the world (Gaetani et al. 2015). Further the European Space Agency also began to share Sentinel data. This was a major boost by the agency in space application analysis as the Sentinel has high spatial resolution that Landsat and other various factors. Bhuvan is an Indian platform hosted by India Space Research Organization that provides 2D and 3D data sets for India only using Cartosat sensors. Some commercial platform provides very high-resolution data sets ranges from 0.4 m to 4 m (QUICKBIRD, IKONONS, WORLDVIEW1 etc.) can be used.

16.5 Urban Risk Assessment in Dehradun City

Study Area: Dehradun is the district headquarters, administrative centre and the capital city of the newly formed state of Uttarakhand. The city is located in the southern part of the Dehradun district between 78 00′E to 78 15′E latitude and 30 15′N to 30 25′N longitude as shown in Fig. 16.2. The hill city Dehradun is well known for its salubrious climate and natural beauty has large historical documentation. The district is associated with Ramayana and Mahabharata; one can find a number of religious places associated with lord Rama and Lakshmana. In the era of Aurangzeb,

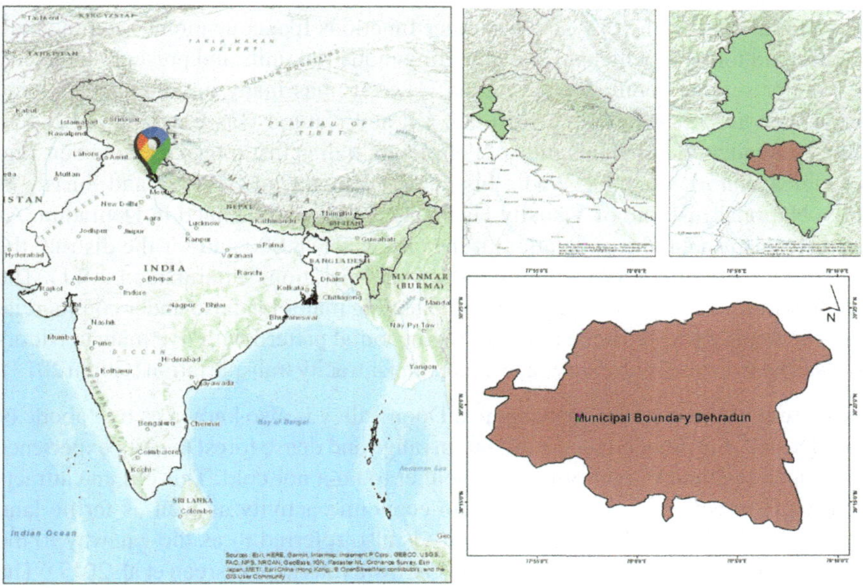

Fig. 16.2 Location map

the city developed as a place for temporary staying, from where the word 'Dehra' came since the city is at the foothill of the Himalayan valley the word 'Doon' which means valley is added and the city is named as Dehradun. In the past, the whole district was covered with dense forest and biodiversity. After the annexation of the British Empire, several administrative changes took place and the city starts growing; after 1951, the district was merged with the Meerut division until 1968 when it was transferred to the Garhwal division. In the year 2000 November, the Indian Govt. has announced Uttarakhand as the 27th state of India and Dehradun as a capital city for the states (Directorate of Census Operations Uttrakhand 2011). Geographically the town is in the watershed of Song River. Situated at the height of 2200 feet above sea level, the valley has Shivalik range of Himalayas to its south, Ganga river to its east and the river Yamuna to its west. The city is surrounded by dense forest all around and have a network of natural springs and canals flowing from north to south direction. Majorly the city region falls in the watershed of the Song river, it is in the east and the river Tons in the west Fig. 16.8. The Himalayan range in the north and Sal forest in the west such geographical features give a picturesque scenario to the city and temperate climatic condition (DMMC 2012). Two erratic river streams named Ripsana and Bindal define the municipal boundary of the city. The area of the administrative control of Dehradun municipal Board is approx. 34 km^2 as per 2001 Census (Gupta 2013). Total area of masterplan boundary as proposed for 2030 of the city is 487 km^2. Dehradun is located 263 km north of the national capital region falls under the close and accessible proximity as one of the counter magnets. The 70 km long Dehradun valley once was a fertile area abundant in agriculture and horticulture products, dense forest and enriched biodiversity experience the urbanization. Other than administrative, the city also has other functions like educational, commercial, defence and tourism, which are the attracting industrial units and presenting the city as a major service centre for the region. The city has many national institutional assets such as Forest Research Institute; Oil and Natural Gas Corporation; Indian Military Academy; Indian Institute of Petroleum, Indian Institute of Remote Sensing, Zoological Survey of India, Wadia Institute of Himalayan Geology and Survey of India, National Institute of Visually Handicapped, etc. are located in Dehradun. As per the Census data, the population of the city was 578,000 and for the district, the figure was 1.69 million (Directorate of Census Operations Uttarakhand 2011). The central part of the city is densely urbanized having major land use under commercial use. The roads network of the city follows the radial pattern with five main corridors originating from the centre serving the inter and intracity transportation (Gupta 2013).

Ecosystem of the city: The bow-shaped Doon valley is also known as the 'abode of Guru Drona'. Surrounded by the mountain range and dense forest the city experience the temperate climate condition and is neither too hot nor cold. This climate attracts more tourism activity and helps in socio-economic activity as well as fertile land led to more agricultural practice. The city is also referred to as the gateway to the queen of the mountain the Mussoorie and Garhwal interior (Nasreen et al. 2017). The steep northern Himalayan slopes and deep and dense Shivalik slopes led to abundant rainfall. The network of the perennial and ephemeral streams can able to irrigate the land and recharge the groundwater. Dehradun city is enriched in biodiversity with four major types of forest occurring in different localities, Moist Shivalik Sal Forests, Moist Bhabar Doon Sal forests, West Gangatic Moist Deciduous forest can be found in higher Shivalik slopes and south-east of the municipal boundary. The district administration identifies three protected areas named Rajaji National Park about 800 square kilometres, Benog Wildlife Sanctuary of around 11 square kilometres and One Conserve Reserve (Nasreen et al. 2017).

16.5.1 Risk Profile of the City

Dehradun city is located in a valley at the Himalayan foothill. Under JNNURM mission, the city is designated as disaster-prone area and the city was lacking behind in disaster Management mechanism. The entire city is under high-risk zone and facing several hazardous events year by year (Urban Development Department—Government of Uttarakhand 2007). The city is lacking in economic base, high tide of urbanization, poor infrastructure facilities (Fig. 16.3a), housing shortage and apart from all these physical components, the city is also losing its environment and facing climate change problems due to pollution, depletion of natural assets and loss of biodiversity and ecosystem (Bansal 2015). Since the city lies in a valley, it is more prone to floods and unplanned development is exacerbating the risk of urban flood. The city falls under the Seismic Zone IV as per the seismic zoning map of India (IS-1893-2002) and highly prone to landslides hazards. The city experience more than one hazard and that makes the city at risk to multi hazards (Bansal 2015). Urban flood is the major concern for the region and the city is majorly affected by water logging of roads. The city's two main rivers Bindal and Rispana take care of the all-small and major drainage, these rivers are usually dry except during monsoon (Fig. 16.3b, d). Surface runoff generated due to reduction of permeable area and choked by encroached urban setting creates the problem of urban flood and water logging scenario. Flooding in the Dehradun city is associated with high intensity rainfall for short durations (Bansal 2015). The rising of urbanization creating imbalance in the city ecosystem and degrading its natural capital. It has been reported by the authorities that the city's population growth rate is high; to contain this population the flood of the high-rise apartment and commercial centres are eroding the ecosystem and biodiversity of the land. Reduction in the availability of fresh waters that affects the agriculture practises, a report state that 270 acres of area with rich water bodies including water streams and natural canals have been encroached upon for construction activities (Joshi and Joshi 2020). With the situation of flood, the city is also facing water scarcity. Due to the expansion of the city limits and increase in population, the burden on the city's water supply system has increased. Considering threshold set by Environmental Hygiene Committee of India, the average consumption in litres per capita per day should be 220, it has been forecasted that by the year 2021 an additional supply of 120 MLD need to provide (Urban Development Department—Government of Uttarakhand 2014). The density map indicates that there are

Fig. 16.3 **a** Ghanta-Ghar urban centre of the city. **b** Ripsana river course in the city. **c** Pathways with Vegetation cover in Dehradun University. **d** Ripsana river course in the city

few high-density areas and these are concentrated in the centre of the city (Durga Rao 2005). Flooding and waterlogging of roads are major problems and are getting worse by the year due to high rate of urban green infrastructure degradation. Traffic congestion and rapid construction practices are adding heat and pollution to the atmosphere and summers are getting unbearable. The entire ecology has changed combined with inadequate land-use planning and the inability of local authorities to control building standards, the unplanned growth of cities raises insecurity in all dimensions. The consequence is the low quality of life and urban infrastructure is declining. The city still has several patches of green cover due to the presence of institutional campuses (Fig. 16.3c) such as Forest Research Institute of India, Indian Military Academy, Wildlife Institute of India, Forest Survey of India and the Indian Institute of Petroleum Dehradun, along with many other training and educational campuses. However, these green covers are not sufficient to cater the urban ecosystem of the city. After the declaration of the city as a state capital, urbanization rate increase and most of the conversion into built up took place from agricultural land, forest areas and open spaces (Singh et al. 2013) and this led to increase in temperature of the urban region or Dehradun city creating the problem of urban heat island. The temperature is crossing the highest value of 35 °C in urban centres touching some places as high as 46 °C. Industries, construction practice and other anthropogenic activity adding pollutant into the atmosphere this led to degradation of air quality. A study states that the city exceeds the National Ambient Air Quality standards the level of SPM has increased from 250 to 400 μg/m3 (Singh et al. 2013).

Climate Change and Futuristic Scenario: Dehradun city is facing several environmental challenges embedded in its unique geographical features. The rising urbanization tide creating imbalance in the city ecosystem and degrading its natural capital. The valley is experiencing rise in temperature in each season however it also reported that there will be decrease in the temperature in winter by 2.6 °C in the year 2030 comparing to the year 1970 (Habeeb and Javaid 2019). All the seasons other than Monsoon in India are experiencing warming trend because of that the frequency of hot days are more than cold days and it is predicted that the temperature will be more 2 °C warmer by 2030. The Climate Change Assessment Report by Ministry of Environment and Forest, Government of India states that in the valley the net increase in temperature will range to increase from 0.9 ± 0.6 °C to 2.6 ± 0.7 °C in 2030, The temperature will have increased by 1.7 °C to 2.2 °C comparing to the temperature of the year 1970. As per the climate change assessment report temperature-humidity index got a significant hike in the Himalayan regions and can be experienced easily in the month of March to September. Thermal discomfort or stress is likely to increase by 2030. Rainfall trend has increased in north-west, west coast and peninsular of India. The assessment report, has highlighted the increase in 3–7% overall increase in the all summer monsoon rainfall with respect to 1970. The region is also facing the problem of change in precipitation pattern; it has increased in numbers and the intensity of rainfall increased by 1–2 mm/day. The natural resources of the region are the life supporting elements and regulate the other climatic events. Some of the noticed impacts of climate changes are receding glaciers, depletion of natural resources, extreme rainfall events, shifting of cultivation zone, fluctuation of crop yield, reduction of snow in winter, increase in the frequency of flash floods, drying up of the natural streams. Impact of climate change in the valley has pushed the stakeholders to go for separate action plan preparation as impacts can differentiate between plains and hills.

16.6 Applications of Tools and Techniques

Considering build-up, Vegetation and water bodies as the elements of blue-green infrastructure and holistically a major component of urban ecosystem, the case study tries to map Blue-Green infrastructure in the municipal boundary of the town using the space-based application. Along with these remote sensed map data Land Surface

Temperature (LST), Ground Water Table (GWT), Master plan boundary, Topographical features map is developed as a GIS layer. Finally, an overlay matrix of map is generated and by visual interpretation of the map with basic concepts and relation among various map components, the hotspots zones of risk and intervention is identified. For risks, Identification a conditional approach is followed and for intervention zone, demarcation an approach is used to create a network of blue-green infrastructure in the city considering proposed master plan of the city and identification of most probable zone where BGI elements can propose. Specification of BGI elements can be proposed as per area based development.

The study used Sentinel-2 and Landsat-5, 8 data series for Mapping of Ecosystem components primarily BGI as discussed above and Cartosat-1 DEM data for terrain and stream pattern extraction, along with various other baseline data like demography (by govt. Census), Proposed Master Plan (by Development Authority), GWT interpolation map and LST for the analysis. Application of remote sensing techniques allow us to develop different types of indices by classifying spectral signatures of different landscape features such as Vegetation (good reflectance in Near Infrared band (NIR) so subtracting the nearest red band spectrum from NIR band will provide vegetation reflection) and other indices are discussed in further sections. The study uses NDVI and NDBI for Vegetation (Green infrastructure) and Built up (Grey infrastructure), for water bodies (Blue infrastructure) of urban regions an interpolation map of GWT map is used as listed in Table 16.1. Degradation of the ecosystem has various adverse effects in the urban region of Dehradun city like an increase in temperature. To map the temperature gradient in the city LST map is developed using the thermal band of the Landsat sensor. An overlay matrix of the GIS layer of map data is developed and by visual interpretation hotspots is identify and for intervention zone, a conditional statistic is applied. The relationship among all the considered ecosystems and other phenomena like ascent in temperature trend and depletion in the GWT is studied referring to various experts' study as explained in the Table 16.1 to identify hotspots zones. Finally, an Ecosystem based solution like the one inclusion of blue-green infrastructure is suggested for urban risk mitigation or in the whole Eco-DRR approach for area based development (ABD). Further, for future study, one can validate and proposed area based BGI elements.

Table 16.1 Applied remote sensing layers to assess ecosystem mapping

GIS layer	Urban ecosystem components
NDVI	Vegetation
NDBI	Built up
GWT map	Ground water
LST	Surface temperature (Skin)
Streams & terrain	Natural drainage
Other base data	Municipal boundary, population
Overlay matrix	

16.6.1 Green Infrastructure (Vegetation)

The Normalized Difference Vegetation Index (NDVI) is most commonly used to monitor the health condition of Vegetation. Numerically it is defined as the ratio of NIR band and Red band as in Eq. (16.1). NDVI quantifies Vegetation by measuring the difference between Near Infrared (NIR) and Red wavelength range as NIR is strongly reflected by healthy Vegetation while Red is absorbed. Dried Vegetation tends to reflect more Red and absorb less NIR as compared to healthy green vegetative cover. The results of NDVI can range from −1 to +1. Vegetated surfaces tend to have positive values while values corresponding to bare soil tend to zero and water surfaces tend to have negative values.

$$NDVI = (NIR - Red)/(NIR + Red) \tag{16.1}$$

Two land cover types can broadly characterize urban ecosystems: impervious surface and Vegetation. In this study, NDVI is used to quantify the density of Vegetation in the area of interest and to check the trend of vegetation spatial extent. Vegetation (Green infrastructure) is an integral part of the urban ecosystem. Vegetation provides enriched biodiversity and is a significant driver of biophysical components. The composition of the urban landscape is a trade-off between built up and Vegetation while developing the land mass, their weightage is strongly negatively correlated. It is easier to map vegetation cover as the bare soil and built up has similar spectral behaviour. Maps in Fig. 16.4 is showing the vegetation cover of the Dehradun municipal boundary for the year 1991 and 2019.

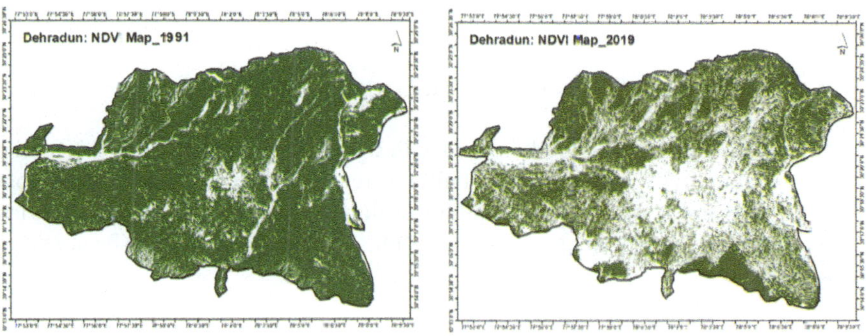

Fig. 16.4 Comparing NDVI map of year 1991 and 2019

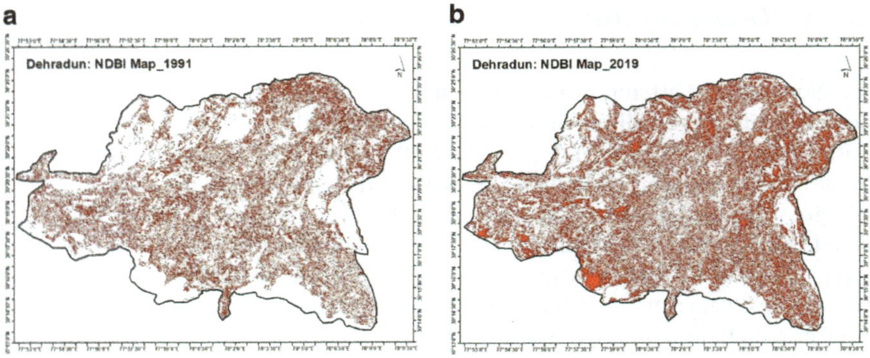

Fig. 16.5 Comparing NDBI map of the year 1991 and 2019

16.6.2 Urbanization (Built up Region)

The Normalized Difference Built up Index (NDBI) is used to identify the built features and areas. Numerically NDBI values correspond to the reflection in the short-wave infrared spectrum. It uses the SWIR and near infrared spectral bands to define built up features in any region. In this study, NDBI is used to check the changing trend of built up spatial extent. The NDBI brings out the built up area data with high accuracy in almost all cases. This index works on the principle that most built up areas and barren land experience a drastic increment in their reflectance of SWIR (Short-wave infrared / re-radiation) when compared to NIR (Near Infrared) while Vegetation has a slightly larger or smaller reflectance of SWIR when compared to NIR.

$$\text{NDBI} = (\text{SWIR} - \text{NIR})/(\text{SWIR} + \text{NIR}) \tag{16.2}$$

Although NDBI provides built up area with significant accuracy but in this particular study, limitations, while mapping built up area are differencing between expose rocky surface and actual built up region is validated and corrected by comparing NDBI with google earth imagery. Urbanization growth in the city has been shown in the Fig. 16.5.

16.6.3 Blue Infrastructure (Water Bodies, Natural Springs, Natural Drainage)

Digital Elevation Map (DEM) from CARTOSAT-2 is used to delineate slope gradient, stream features and catchment area for the city area, sub basin has also been delineated. Ground water data for the year 2019 from CGWB Govt. of India (Table 16.2) is

Table 16.2 Ground water level data by CGWB Govt. of India

Frequency table of depth to water level (m bgl), May 2019-Uttaranchal

No of wells in depth range (May 2014) and its percentage from total number of wells monitored

S. No.	State	District	No of wells monitored	Minimum	Maximum	0 to 2	in %	2 to 5	in %	5 to 10	in %	10 to 20	in %	20 to 40	in %	> 40	in %
1	Uttaranchal	Almora	1	10.21	10.21		0		0		0	1	100		0		0
2	Uttaranchal	Champawat	3	9.73	45.04		0		0	1	33	1	33		0	1	33
3	Uttaranchal	Dehradun	47	5.10	89.75		0		0	9	19	20	43	10	21	8	17
4	Uttaranchal	Haridwar	39	3.19	67.70		0	9	23	13	33	13	33	3	8	1	3
5	Uttaranchal	Nainital	11	3.34	73.58		0	1	9	2	18	3	27	2	18	3	27
6	Uttaranchal	Pauri Garhwal	3	25.85	91.10		0		0		0		0	1	33	2	67
7	Uttaranchal	Udham Singh Nagar	45	1.94	18.21	2	4	20	44	15	33	8	18		0		0
s	Uttaranchal	Ultarkashi	7	10.75	45.50		0		0		0	4	57	2	29	1	14
	Uttaranchal Total		1S6	1.94	91.10	2		30		40		50		18		16	

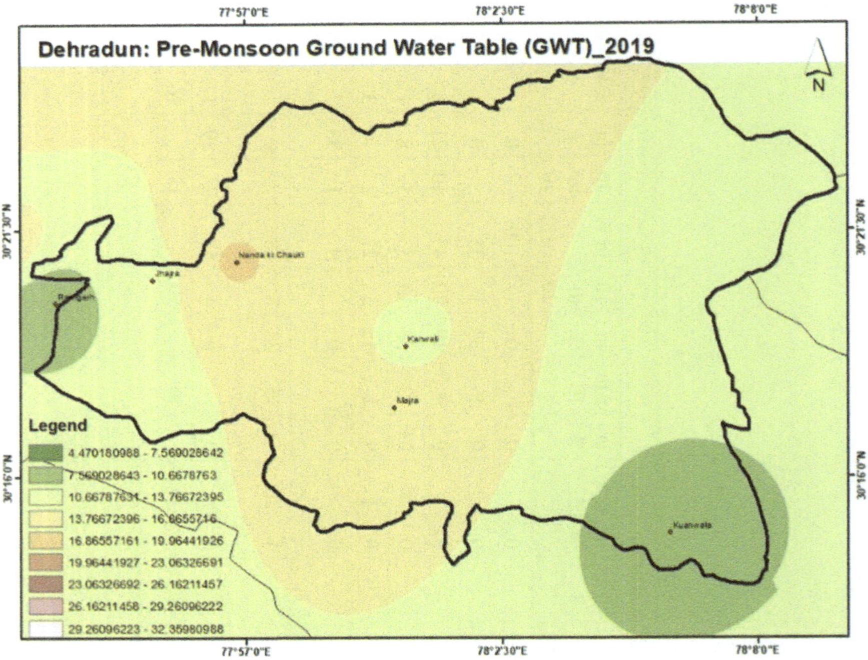

Fig. 16.6 Ground water table map

used to generate interpolation map as shown in Fig. 16.6. Hydrological components mapping is important for water resource management especially if the urban region is located on a high slope terrain or nearby a channel of natural springs.

Topographical map of urban area provides various components of blue infrastructure such as water streams, lakes, ponds and wetlands. Urban ecosystem changes with movement of water as it has strong positive correlation with the living system of organism. The shape size and connectivity of these water bodies have important effects on ecological communities and ecosystem processes (Steele et al. 2014). Intensive alteration of water bodies due to rapid urbanization led to water scarcity and depletion of other ecosystem components like Vegetation and further has an impact on ground water table. Since the urban region is located on a high, slope terrain. Streams and other terrain features are delineated using DEM as shown in Fig. 16.7. Mapping of such topographical features can be used in geomorphology science to model surface runoff, water flow, soil humidity etc. Table 16.3, is the data of Ground Water Table (GWT) for the Dehradun district for the month of April-June that is pre-monsoon months (source: CGWB govt. of India). The data used to develop an interpolated map of ground water table for the year 2019 pre-monsoon months. Since the pre-monsoon months fall under the summer seasons and this the critical time period when the city faces the problem of water scarcity.

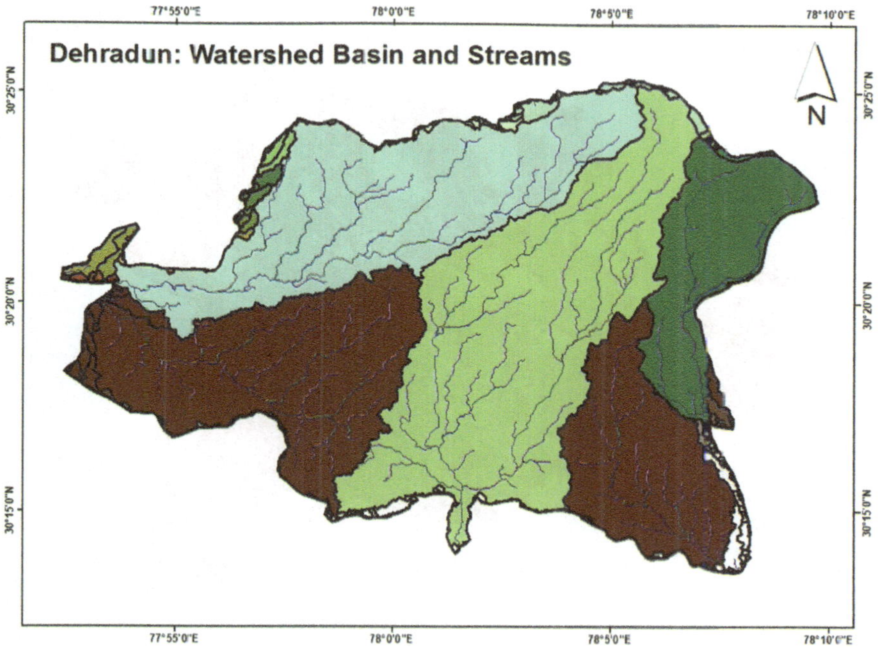

Fig. 16.7 Watershed delineation and streams features

Table 16.3 GWT_Pre-Monsoon

WELL station name	GWT (in m)
Chhorba	32.36
Jhajra	13.4
Kanwali	13.14
Kuanwala	8.9
Lal tapar	17.48
Majra	16.59
Motichur	11.03
Nanda ki Chauki	17.3
Ramgarh	7.8
Rampura	9.8
Redapur	6.74
Rishikesh	4.47
Sabhawala	9.18
Selakui	16.62
Singhniwala	9.37

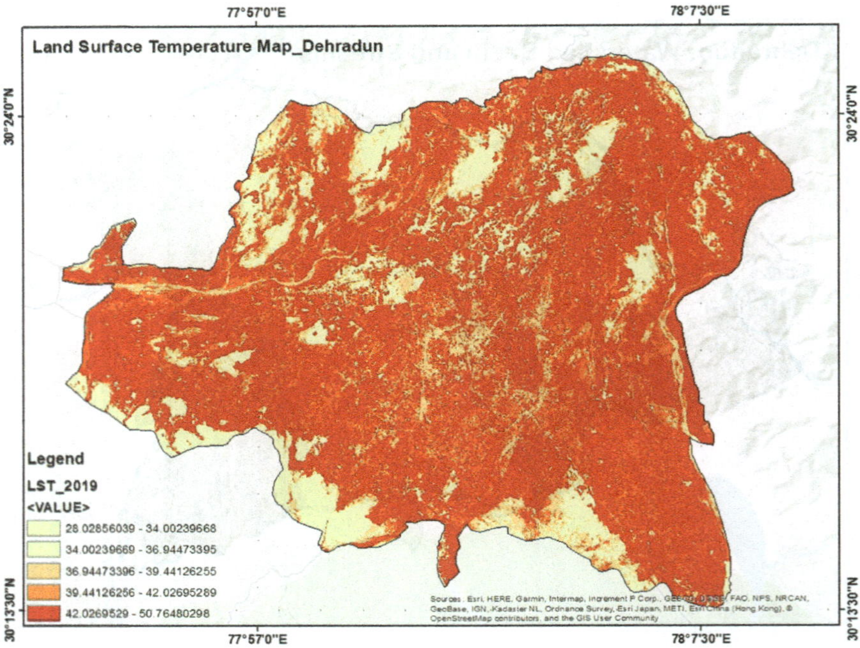

Fig. 16.8 Land surface temperature map of Dehradun

16.6.4 Land Surface Temperature (LST)

Land surface temperature (LST) is an important parameter to analyze skin temperature of urban landscapes. The measurement of surface energy of physical components of urban ecosystems will help in the study of thermal behaviour of various components. The relative cooling and warming characteristics of various surfaces help in identification of heat sink zone. The LST can process using thermal band of satellite imagery on GIS platform using various models, the study used the radiative transfer equations model to delineate land surface temperature. To determine the effective radiometric temperature of the Earth's surface "skin" in the instrument field of view ("skin" refers to the top surface in bare soil conditions and top of the canopies in vegetative cover and built up). This method assumes that there is liner relation between surface temperature and Brightness.

$$\text{LST} = A_0 + B_0 * \text{Thermal band} \tag{16.3}$$

where a_0, b_0 and c_0 are classes of coefficients that depend on atmospheric water vapour, satellite-viewing angle and land surface emissivity.

Figure 16.8 is showing the LST map of the Dehradun municipal boundary for the year 2019.

16.6.5 Overlay Matrix

Overlay matrix analysis follow some basic assumption as per suggested by experts throughout the world in their research, such as high the built up higher will the surface temperature built up or constructed land cover possess high temperature than natural land cover. The conditional correlation among all GIS layers to identify hotspots is stated as: con ("NDVI < 0", critical, con ("NDBI > 0", critical, con ("NDWI < 0", critical, con ("GW $\geq$ maximum", con ("LST $\geq$ Maximum temp", critical))))) (Mukherjee and Ravan 2021). The conditional correlation to identify hotspot state that a region is under critical zone when the value of NDVI is minimum or zero; the value of NDBI is maximum; the value of GWT and LST is maximum. Figure 16.9 is the schematic flow chart of the methodology adopted for the study. The data sets used for the map development are of the month of April to June, as these months are the summer season in India and consider to be the critical months for mapping BGI. One of the challenges that appeared in the analysis was the fact that the 1991 data was sourced from Landsat 5 and therefore some regions appeared in both the NDBI as exposed rock surfaces and NDVI as moss and lichen covered areas. This had to be manually controlled for the calculations of the areas by comparing map data with google earth historical imagery. Since the region cannot be considered as built up that occurred due to human activity and is in fact a natural rock formation, it was counted as part of the NDVI. Sentinel data is used for developing indices map for year 2019 as it provides data with high resolution of 10 m. Landsat data is used to check the trend of built up and Vegetation. Land Surface temperature map is developed using thermal band that is band 10 of 30 m resolution. Most of the data available for the Dehradun is having cloud cover so the data is downloaded for the month of May considering May as the hottest month of the year.

16.7 Analysis

Urbanization and depletion of Vegetation: After the announcement of Dehradun as a capital city, the city is experiencing high rate of urbanization and commercial development. The proposed master plan development boundary of area 487 km^2 has been selected for the case study and delineating various land surface features. The NDVI and NDBI map shows the built up and vegetative landscape region of the city, the trend analysis of built up, Vegetation and Water bodies delineated and calculated by counting the pixel number of the polygon of the specific features and findings are represented in graphs in Fig. 16.10. Urban built up area has increased from 89 km^2 in 1991 to 150 km^2 in the year 2019 and for the green region that is vegetative part in the region was 289 km^2 in year 1991 which gets reduced to 209 km^2 in year 2019 as shown in Figs. 16.11 and 16.12 and Table 16.4. The Vegetation is replaced by concrete structure and the loss of permeable surfaces are generating high surface runoff, with high rate of urbanization city is losing its water bodies too along with

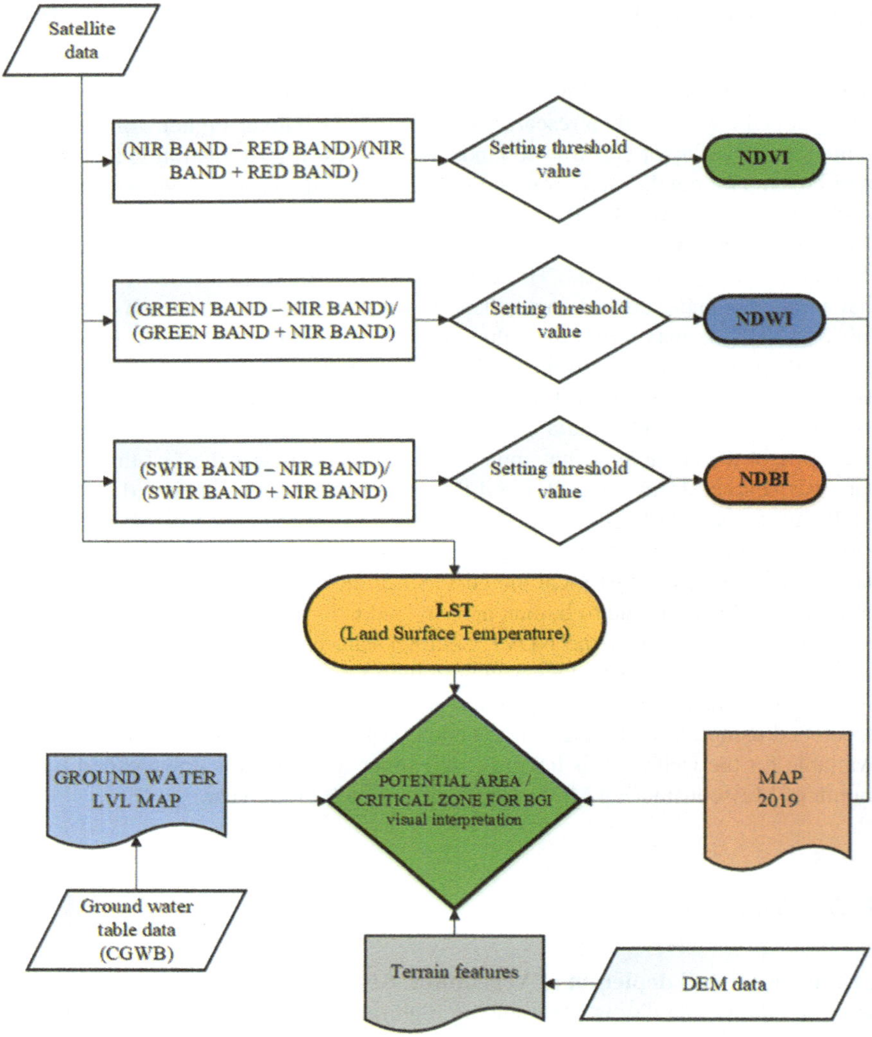

Fig. 16.9 Schematic diagram of overlay matrix for identification of hotspots

various hydrological concern in the city such as water logging, water scarcity. The green flux and water bodies were responsible for temperature control and ground water recharging and many cultural and social services of the city due to degradation of natural infrastructure city is losing all its ecosystem services. Dense urban regions developed are creating barriers in networking of Blue-green infrastructure. As proposed in mater plan 2025 the city will have green cover in two part, one on the north of the municipal region and on the south. Central part of the city has no significant size of green patches.

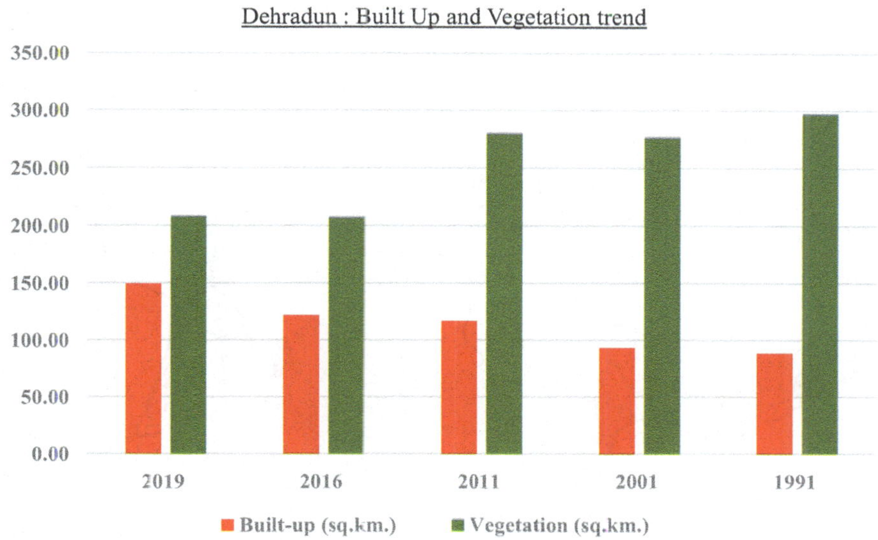

Fig. 16.10 Graphical representation of built up and vegetative area status in the City Master Plan Boundary

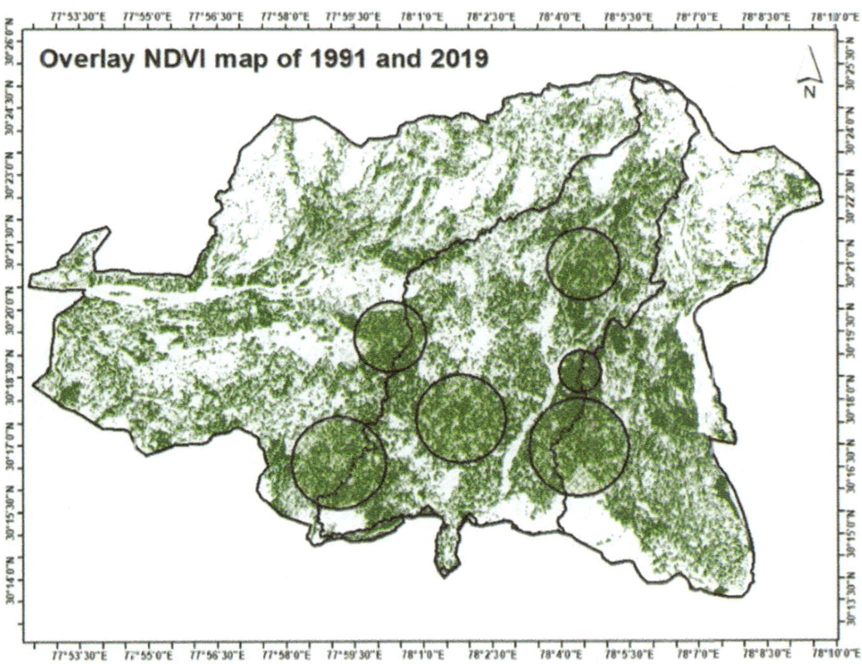

Fig. 16.11 Overlay map showing (in green) depleted vegetation area from 1991 to 2019 and demarcated area with high risk because of major loss in green component of ecosystem

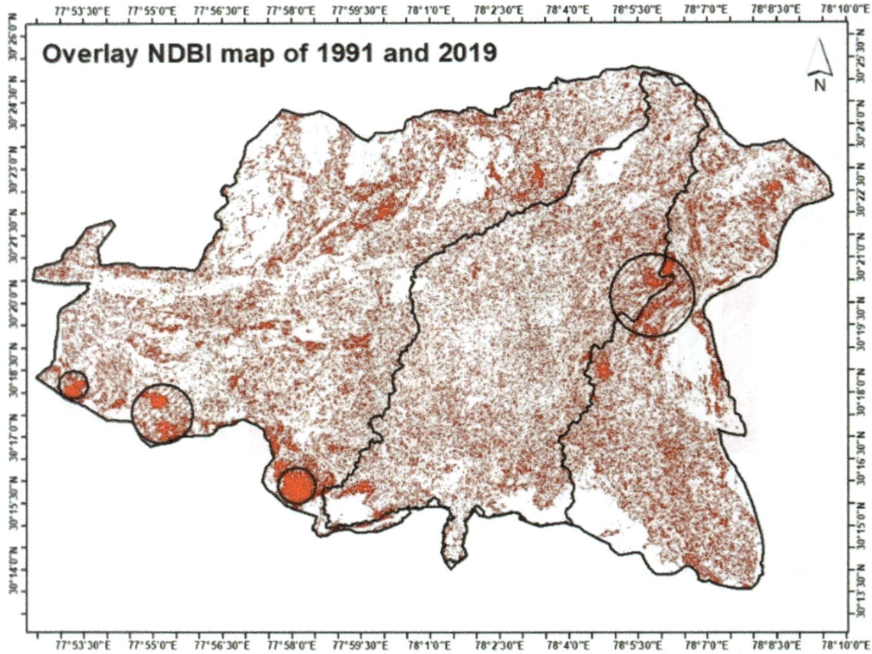

Fig. 16.12 Overlay map showing (in red) increase in built up from 1991 to 2019 and demarcated zone of risk with respect to change of land surface features and ecosystem

Table 16.4 Area calculation using RS INDICES

Area calculation	2019	2016	2011	2001	1991
Built up (km^2)	150.13	122.59	117.68	94.24	89.42
Vegetation (km^2)	208.82	207.99	280.68	277.64	297.92

Surface Water and Ground water: As per the map used for Ground water mapping of the city from CGWB govt. of India, it can be easily said that the range of ground water level in the whole district is 20–40 m as shown in Table 16.2. Again the municipal boundary region is showing the range of ground water table as 16 to 19 m as shown in the Fig. 16.6, map developed on GIS platform using the data of each bore well of the city from CGWB data archive section. Depletion of ground water, as well as surface water bodies such as ponds and a numbers of natural springs and increment of built up region shows that BGI has been encroached by the urban built up over the period of time, thus the city is losing its blue infrastructure. That is a potential tool for controlling various ecosystem services like provision of water for livelihood and agriculture practices and other social cultural activity. Although some major streams can be found at boundary demarcation of, the city but they are also drying up due to encroachment and climate change impact.

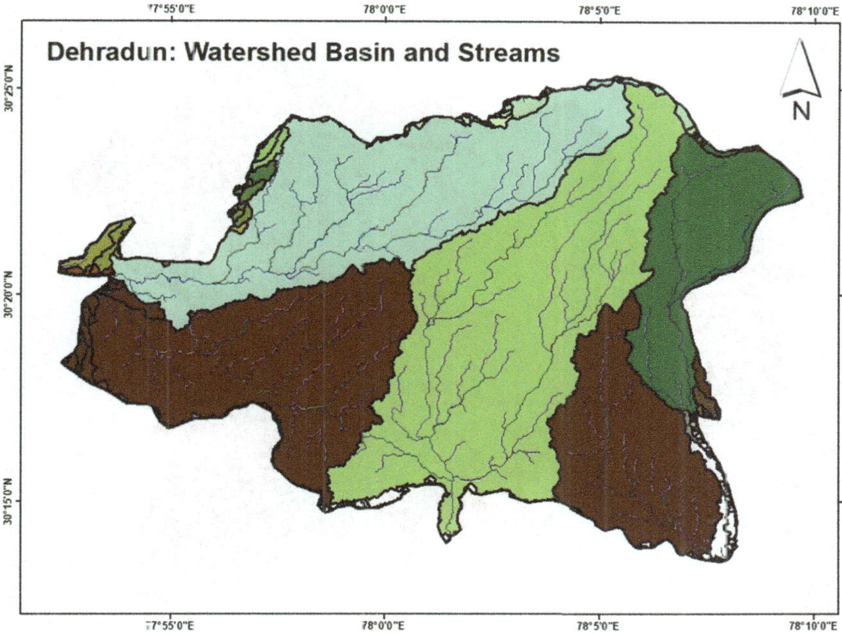

Fig. 16.13 Watershed basin and streams

Watershed and slope: The region has five major sub-basins as shown in Fig. 16.13, four out of five basins are comparatively less affected by urbanization, population density and vegetation degradation. The central basin shows dense urbanization and degradation of Vegetation especially due to the encroachment on the natural spring path (Fig. 16.14). This has led to an increase in the severity of urban floods. In addition, the diversion of the stream path has led to a flood situation in other basin areas downstream. Stream order up to 4 can be identified when considering a pixel value of 500 while defining stream features. Considering a pour point at the bottom of the central basin, it appears that the runoff generated by about 80 Sq. Km of the surface passes through this point and stream order 3 can be delineated at this point. Higher the stream order more will be the volume of water and higher will be the chance of urban flood.

Land surface temperature: This tool can be used to identify hotspots region based on thermal environment, although this is not directly related to the phenomenon of urban heat islands but identification of zone with higher surface temperature will lead to selection of proper land surface features. Mapping of land surface temperature shows an interesting feature, some part of the city at higher altitude is also showing high signature of land surface temperature as it is in densely packed urban regions and loss of show line. The temperature is ranging from 28 °C to a maximum of 50 °C as shown in Fig. 16.15. While validating with temperature data reported by

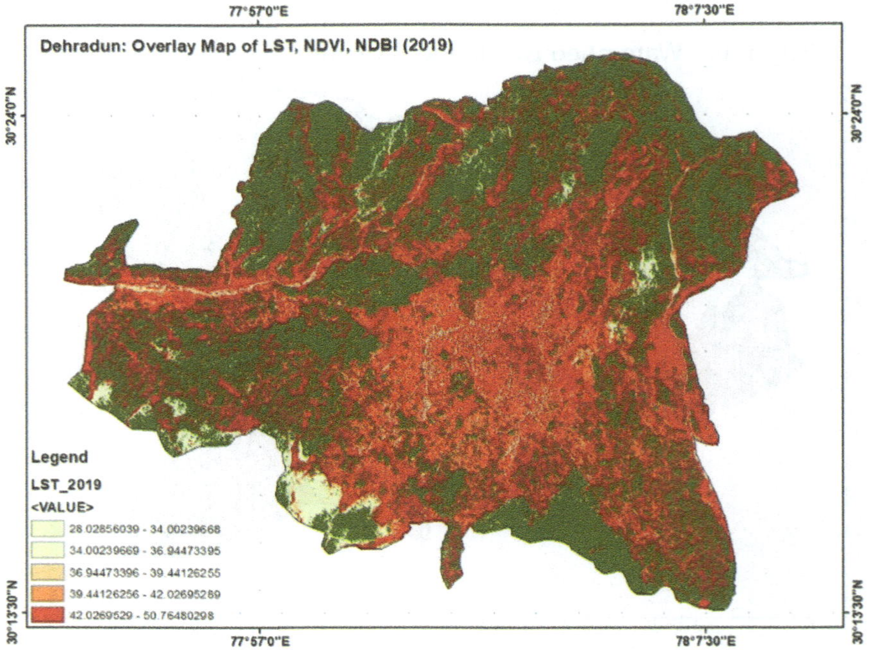

Fig. 16.14 Central basin overlay with built up (NDBI)

the weather department of the city its seems very close relation. Higher surface temperature is responsible for loss of biodiversity of the city fast melting of snow and lower snow lines. Surface of rivers can be identifying in the mid-range, for snow regions, temperature is lower and urban region has higher range of temperature.

Overlay map and identification of hotspots: Finally, an overlay matrix map of the entire GIS map (NDVI, NDBI, GWT, Streams, Basins, LST, Municipal Boundary and Proposed Master Plan) is developed to identify hotspots with respect to high-risk zone, intervention zone as shown in Fig. 16.16. Further ground validation and Area Base Development approach will help in findings specific elements of BGI that is ecosystem components to reduce the impact of risk.

Risk zone identification: Risk zone identified by following the condition matrix as discussed above in the section overlay matrix. The risk zone identification follows the process of demarcating zone showing depletion of Vegetation or green zone, depletion of water channels, then comparing this zone with increment of urban region, high land surface temperature as shown in Figs. 16.11, 16.2, 16.3 and 16.4. In Fig. 16.16, Risk zone has been mapped while overlaying all maps with the proposed master plan 2025 of Dehradun. Overlaying map brings the focus to the central basin of the municipal boundary region as shown in Fig. 16.14, the central basin part is showing maximum depletion of Vegetation and high rate of urbanization. The basin has 4th

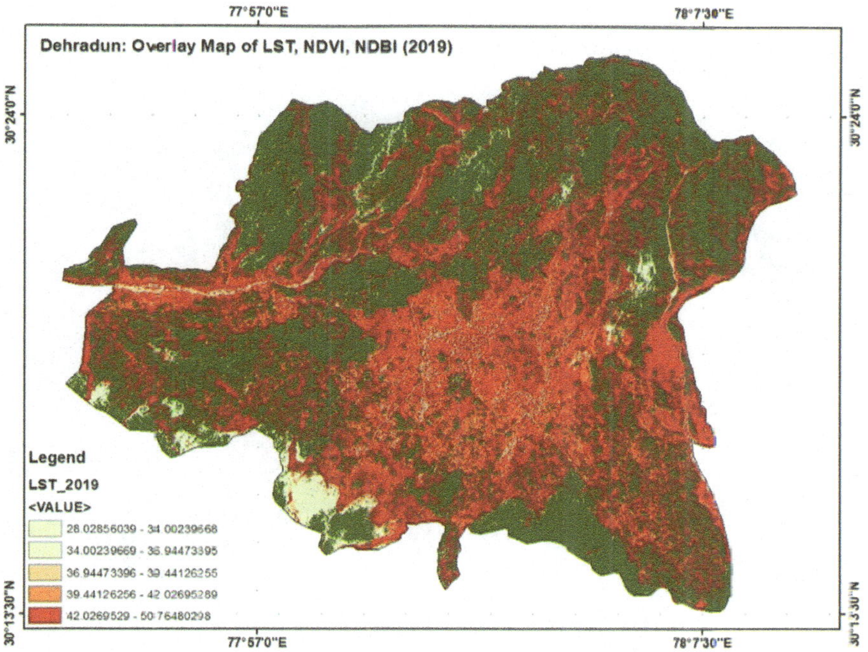

Fig. 16.15 Overlay maps showing vegetation (in green), built up (in black) and higher temperature zone (in red), high temperature zone can be identified in built up region

order streams at the southern part and they are responsible for caring runoff generated by about 80 km^2 urban region. The natural springs channels have been either deviated or encroached by built up zone (commercial, residential). GWT map is showing the water table for the entire region is in the range of 16–19 m whereas as per CGWB the entire Dehradun district has average ground water table in the range of 20–40 m. The LST map shows the region with exposed concrete or rock surface shows high temperature, the central basin of the municipal boundary is highly vulnerable with respect to temperature, surface runoff and depletion of Vegetation and loss of water bodies.

Intervention zones: The demarcation of the intervention zones is done by adding other baseline data maps such as the master plan, GWT, in relation to the NDVI, NDBI and LST and Streams mapping. The selection was focused on linking zones that showed a healthy vegetative cover (in NDVI) and open ground availability (based on NDBI). Some areas were also selected along rivers especially where some vegetative cover was seen nearby. It is evident from the NDBI trends that areas to the centre of the city used to be green and have eventually been converted into contiguous built up. This led to creation of areas with UHI and low ground permeability causing both urban floods and a significant drop in the ground water table. To develop a network of BGI in the city, intervention zone has been identifying nearby commercial zone.

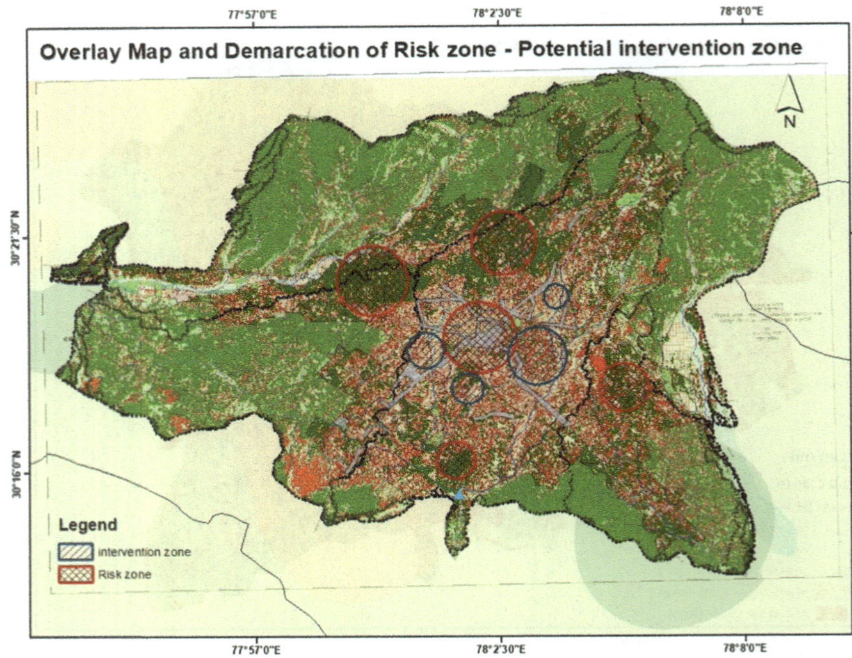

Fig. 16.16 Overlay map demarcating intervention zone

Since the commercial zone shows linear zoning going from north to south and east to west and creating a junction at the centre.

16.8 Ecosystem Based Solutions for Dehradun

Dehradun is also called the 'Adobe of Drona', the city embodied with scenic grandeur and beautiful surroundings. Once the flourishing ecosystem and biodiversity was an integral part of the city and known for its natural capital. The city witnessed a dramatic transformation from a quiet sub-Himalayan town to a bustling commercial centre (Ukpp et al. 2015). Identifying hotspots region and adopting ecosystem based approaches including conservation, sustainable management and restoration can help communities in hill cities like Dehradun to reduce vulnerability. The two major rivers Rispana River in the east and Bindal River in the west demarked the municipal boundary of the city. The ecosystem based solutions for the urban risk of Dehradun city include; preservation, management and restoration aspects. Preservation provide the conservation and protection of natural assets like forests and other Vegetation, wetlands, agricultural land and natural springs etc. Preservation will lead to development of city in a holistic manner and increase the social and cultural values of the communities, it will also help in enhancement of socio-economic condition of the city. Rise in temperature is also a major concern for the city, urbanization and commercial activity causing in the rise of the temperature trend and urban heat island phenomenon, once the city was known for its pleasant weather and one of the favourite tourist destinations in summer is now facing the problem of high temperature and uncomfortable environment for dwellers. Management of ecosystem means using ecosystem service in an optimum way without causing any degradation; it will help in urban flood situation of the city and help in recharging ground water. Network of ecosystem-based solutions will provide the city a holistic approach of development and reduce the impact of risk. Management of ecosystem majorly focuses on creating a network of physical components of ecosystem that can take care of a number of urban risks such as temperature, flood and water logging, water scarcity.

Application of remote sensing can be identifying in all aspects of ecosystem based DRR either in a direct way as well as in an indirect way. Direct application is mapping of all physical components of ecosystem in the urban region by the image processing process, it is basically the transforming of spectral reflectance of the remotely sensed raw data to identify specific land surface features along with specific components of ecosystem such as Vegetation, water, urban region. Other components like crops yield, agriculture practice, types of forest and urban park can map in an indirect way by adding or subtracting various spectral reflectance. In the present case study remotely sensed data is used to map some important components of the ecosystem, its trend and present scenario as discussed in the process section.

16.9 Discussion and Conclusion

Through this case study, application of remote sensing in Eco-DRR is emphasized. Remote sensing can help in either way of risk identification or suggesting measures to mitigate the impact. It can be a holistic approach from knowledge to action plan. The data used in the studies are mostly freely available and low resolution; high-resolution data will increase the accuracy of assessment. For identification of hotspots zone overlay matrix, follow the conditional assumptions referring to various research and setting correlation analysis among various GIS maps.

The city of Dehradun is prone to multi hazards like earthquakes, landslides etc., apart from this urban centre is facing various risks like urban floods, increase in temperature, degradation of urban Vegetation, depletion of ground water and encroachment of natural streams. The issues are emerging primarily because of unplanned urbanization and depletion of the natural landscape and this is leading to socio-economic vulnerability of the people. The methodology of using remote sensing application in Eco-DRR consider several significant steps. The first step is to map ecosystem components such as Vegetation, built up, water bodies followed by mapping of temperature, ground water and topographical features map. The mapping of these components shows the increasing trend in built up and temperature, degradation of Vegetation and ground water. These maps also help to set the relationship among the various features, like vegetation and built up are related negatively, whereas with the increment of built up region more and more ground water are been extracted for various purposes and this lead to depletion of ground water table, here built up is positively related to ground water table. The identified hotspots zone map is overlaid with a master plan to suggest the intervention zone.

- Compact cities planning and approach, like the cities in Japan, South Korea may be the best way forward to resilient Indian cities.
- Ecosystem based solutions may the answer to order reduce the gap between the pristine conditions and the urban areas with huge anthropogenic influence.
- It is important that we address this gap, while also working towards the goal of Disaster Risk Reduction (DRR).
- Applications of Remote Sensing can be extensively utilized for the purposes of Ecosystem Based Disaster Risk Reduction. Remote sensing can help in either way of risk identification or suggesting measures to mitigate the impact.
- Through this case study of Dehradun, we have been able to highlight the issues with Indian cities and how remote sensing-based ecosystem based-solutions can be developed to tackle the issues. Through this case study, the application of remote sensing in Eco-DRR is emphasized.
- Remote Sensing applications for great for analyzing and developing solutions at macro scale. Dehradun was selected for case study purposes accordingly. It can be a holistic approach from knowledge to action plan. The data used in the studies are mostly freely available and low resolution; high-resolution data will increase the accuracy of assessment.
- The city of Dehradun is prone to multi hazards like earthquakes, landslides etc., apart from this urban centre is facing various risks like urban flood, increase in temperature, degradation of urban vegetation, depletion of ground water and encroachment of natural streams. The issues are emerging primarily because of unplanned urbanization and this is leading to socio-economic vulnerability of the people.
- The methodology of using remote sensing application in Eco-DRR consider several significant steps. The first step is to map ecosystem components such as vegetation, built up, water bodies followed by mapping of temperature, ground water and topographical features map.
- For identification of hotspots zone overlay matrix, follow the conditional assumptions referring to various research and setting correlation analysis among various GIS maps.
- The mapping of these components shows the increasing trend in built up and temperature, degradation of vegetation and ground water. The identified hotspots zone map is overlaid with a master plan to suggest the intervention zone.

References

Bansal N (2015) Multi hazard Urban risk assessment. PhD Thesis, IIT Roorkee

Barbosa A, De CC, Atkinson PM, Dearing JA (2015) Remote sensing of ecosystem services: a systematic review. Ecol Ind 52:430–443. https://doi.org/10.1016/j.ecolind.2015.01.007

Directorate of Census Operations Uttrakhand (2011) District census handbook Dehradun. Census of India 2011

DMMC (2012) State disaster management action plan for the State of Uttarakhand. http://dmmc.uk.gov.in/files/pdf/complete_sdmap.pdf

Durga Rao KHV (2005) Multi-criteria spatial decision analysis for forecasting urban water requirements: a case Study of Dehradun City, India. Landsc Urban Plan 71(2–4):163–174. https://doi.org/10.1016/j.landurbplan.2004.03.001

Estrella M (2013) The ecosystem-based disaster risk reduction

Fernandes NF, Guimarães RF, Gomes RAT, Vieira BC, Montgomery DR, Greenberg H (2004) Topographic controls of landslides in Rio de Janeiro: field evidence and modeling. Catena 55(2):163–81. https://doi.org/10.1016/S0341-8162(03)00115-2

Gaetani F, Petiteville I, Pisano F, Rudari R, St Pierre L (2015) A synergy framework for the integration of earth observation technologies into disaster risk reduction, 17: 15780.

Gupta K (2013) Unprecedented growth of Dehradun urban area: a spatio-temporal analysis. Int J Adv Remote Sens GIS Geogr 1(2):47–56

Habeeb R, Javaid S (2019) “Social Inclusion of Marginal in the Great Climate Change Debate: Case of Slums in Dehradun, India.” *SAGE Open* 9 (1). https://doi.org/10.1177/2158244019835924.

IPCC (2014) “Climate change 2014 synthesis report IPCC.” Managing the risks of extreme events and disasters to advance climate change adaptation: special report of the Intergovernmental Panel on Climate Change, vol 9781107025. https://doi.org/10.1017/CBO9781139177245.003

Joshi A, Joshi N (2020) Sanitation Park Nathuwawala , Dehradun : a study of solid waste management in Uttarakhand, vol 15

Kaku K (2019) Satellite remote sensing for disaster management support: a holistic and staged approach based on case studies in Sentinel Asia. Int J Disaster Risk Reduction 33(September 2018):417–32. https://doi.org/10.1016/j.ijdrr.2018.09.015

Lange P, Driessen PPJ, Sauer A, Bornemann B, Burger P (2013) Governing towards sustainability-conceptualizing modes of Governance. J Environ Policy Plan 15(3):403–425. https://doi.org/10.1080/1523908X.2013.769414

Lorenzo-Alonso A, Utanda Á, Aulló-Maestro ME, Palacios M (2019) Earth observation actionable information supporting disaster risk reduction efforts in a sustainable development framework. Remote Sensing 11(1):1–8. https://doi.org/10.3390/rs11010049

Mukherjee M, Ravan S (2021) GeoSM-NatE for urban risk resilience, Working paper series of the United Nations Office for Outer Space Affairs, 2021 (under review)

Nehren (2014) The ecosystem-based disaster risk reduction case study and exercise source book, Geneva and Cologne: partnership for environment and disaster risk reduction and center for natural resources and development cover image: © Philippa Terblanche Design and Lay

Petiteville I (2015) CEOS earth observation handbook: satellite earth observations in support of disaster risk reduction. ESA, Www. Eohand-Book. Com, 84. https://doi.org/10.13140/RG.2.2.29441.45926.

Rizvi AN (2017) Faunal Diversity of Dehradun District : an overview, no January

Singh O, Arya P, Chaudhary BS (2013) On Rising temperature trends at Dehradun in Doon Valley of Uttarakhand, India. J Earth Syst Sci 122(3):613–622. https://doi.org/10.1007/s12040-013-0304-0

Smyth CG, Royle SA (2000) Urban landslide hazards: incidence and causative factors in Niterói, Rio de Janeiro State, Brazil. Appl Geogr 20(2):95–118

Steele MK, Heffernan JB, Bettez N, Cavender-Bares J, Groffman PM, Grove JM, Hall S (2014) Convergent surface water distributions in U.S. cities. Ecosystems 17(4):685–697. https://doi.org/10.1007/s10021-014-9751-y

Ukpp, The, Uttarakhand Parivartan Party, Climate Change and Climate Change (2015) UKPP's Dehradun resolution on climate change in the Himalayan region September 2015, no September, pp 20–22

Urban Development Departmen—Government of Uttarakhand (2007) City Development Plan : Dehradun Revised, no May, pp 1–244. http://udd.uk.gov.in/files/CDP_DDUN.PDF

Chapter 17
Ecosystem-Based Approaches for Water Stress Management—Lessons from Nagpur Metropolitan Area, India

Vibhas Sukhwani, Kamakshi Thapa, Rajib Shaw, Sameer Deshkar, Bijon Kumer Mitra, and Wanglin Yan

Abstract Urban areas around the world are today witnessing remarkable development transformations, paralleled by the growing influx of populations. The expanding city boundaries and their associated development activities are, however, altering the surrounding ecosystems. Against the growing water demands in urban areas and its declining availability, water stress is becoming a global concern. While the drinking water needs in urban areas are often met on priority, the co-dependent rural areas are disproportionately affected, and the anticipated change in climate is bound to further exacerbate the urban–rural water conflicts. To address these concerns, the importance of ecosystem-based approaches is increasingly being realized as they generate additional environmental, economic, and social benefits. To further understand their significance, this chapter discusses the case of Nagpur Metropolitan Area in Central India, which has recently experienced severe water stress. The chapter underlines the rising concerns in Nagpur due to rapid urbanization, changing climate, and transboundary developments. It mainly puts forward the key lessons derived through the India–Japan Bilateral Research Project (2018–2020), which worked toward developing new paradigms in urban–rural linkages and fostering collective resilience. Acknowledging the need for integrated management of shared water resources, the chapter emphasizes on implementing ecosystem-based approaches through transboundary cooperation at regional level.

Keywords Ecosystem-based approaches · Water stress · Urban–rural conflict · Multi-stakeholder engagement · Transboundary cooperation

---

V. Sukhwani (✉) · R. Shaw · W. Yan
Graduate School of Media and Governance, Keio University, Fujisawa, Japan
e-mail: vibhas@sfc.keio.ac.jp

K. Thapa · S. Deshkar
Department of Architecture and Planning, Visvesvaraya National Institute of Technology, Nagpur, India

B. K. Mitra
Institute for Global Environmental Strategies (IGES), 2108-11 Kamiyamaguchi, Hayama, Japan

© The Author(s), under exclusive license to Springer Nature Singapore Pte Ltd. 2021
M. Mukherjee and R. Shaw (eds.), *Ecosystem-Based Disaster and Climate Resilience*, Disaster and Risk Research: GADRI Book Series,
https://doi.org/10.1007/978-981-16-4815-1_17

17.1 Introduction

By 2050, the world population is projected to increase from 7.6 billion (in 2018) to around 9.8 billion, with approximately 68% population (up from 55% in 2018) residing in urban areas. The future population growth is likely to be centered in the fast-developing countries of Africa and Asia (UN DESA 2019). Against the rapidly growing population, the global resource consumption in terms of water, food, and energy are also projected to correspondingly increase by 55, 60, and 80% (IRENA 2015). Particularly for urban areas, the freshwater demands are projected to increase by around 80% (Florke et al. 2018). The growing demands of food and energy resources also lead to indirect water demands, as both the sectors are water-intensive (Djehdian et al. 2019). In that manner, the projected direct and indirect increase in water demands have raised serious concerns for sustainable urban development, as one in four large cities globally is already facing direct water stress (ARUP 2018). Water stress, as defined by the European Environment Agency (2020), occurs when the demand for freshwater exceeds the available amount during a specific period of time or when poor quality restricts its use.

Urban areas have for long been dependent on ecological systems outside their physical boundaries for a wide array of ecosystem services like potable water (Heard et al. 2017). Over the years, the rapid expansion in urban populations and haphazard growth in terms of gray infrastructure development, reduced green spaces, wetland reduction, etc., have led to a decline in availability of freshwater resources within and around the urban areas (World Bank 2018). The growing concerns of urban drought and their significant implications to 2030 Sustainable Development Goals (SDGs) have also been underlined by Zhang et al. (2019). It has been pointed that more and more water is now been fetched from distant water sources, mainly from the reservoirs constructed for irrigation purposes, to sustain the progressing socioeconomic developments in urban areas (McDonald et al. 2014; Garrick et al. 2019). The increasing water transfer to urban areas and inadequate wastewater treatment are also raising water stress concerns for upstream and downstream rural areas within the wider metropolitan regions (Hoekstra et al. 2018). While the rapid urbanization trends, growing resource demands, and the fast-changing climate are outpacing the coping capacities of human societies, it is important to realize that the degradation of natural environment will lead to further consequences, as they provide a range of essential services and help people to adapt to climate-related risks (Sudmeier-Rieux et al. 2006; Renaud et al. 2013). Earlier, Millennium Ecosystem Assessment (2005) produced clear evidence that healthy ecosystems such as mangroves and wetlands help in mitigating the impact of natural hazards, while supporting people's day-to-day livelihoods.

In recent decades, ecosystem-based approaches to climate change adaptation (EbA) and disaster risk reduction (Eco-DRR) have received increasing traction at the global platform as they are proven to be cost-effective, socially equitable, and environmentally sustainable, in contrast to 'hard' infrastructure-based approaches

(Triyanti and Chu 2018). Ecosystem-based approaches, as defined by the Convention on Biological Diversity (CBD), refer to the integrated management of land, water, and living resources that promotes conservation and sustainable use in an equitable way. The 2018 edition of the United Nations World Water Development Report (WWAP and UN-Water 2018) strongly emphasized on Nature-based Solutions (NbS) to tackle the challenges of water stress management. NbS is basically recognized as an umbrella concept that covers a range of ecosystem-based approaches (Cohen-Shacham et al. 2016). These approaches address various risks to ecosystems and dependent human systems in an integrated manner, present more options to address climate change adaptation and enable broader considerations for potential trade-offs.

Over the years, several international organizations like International Union for Conservation of Nature (IUCN), Partnership for Environment and Disaster Risk Reduction (PEDRR), the World Bank, United Nations Environment Programme, World Wide Fund for Nature have stressed on the need for integrating ecosystem-based approaches with development planning, which has brought renewed attention to this issue. Today, most of the global policy agreements on climate change and biodiversity such as the Paris Agreement on Climate Change, Sendai Framework for Disaster Risk Reduction, The Ramsar Convention, The New Urban Agenda, SDGs, etc., have recognized the role of ecosystems in building resilient communities (also explained in CBD 2019). Also, many countries are now integrating ecosystem-based approaches into their national strategies, disaster management plans, drought relief policies, etc. Several examples have been synthesized in CBD Technical Series No. 85 (Lo 2016). However, despite their growing recognition, a number of studies (Renaud et al. 2013; Voulvoulis et al. 2017; PEDRR 2019) have pointed that there is very limited progress in their application at local level and most of the ongoing efforts are generally limited to project or pilot demonstration levels. The differential understanding of ecosystem-based approaches and the insignificant evidence base are recognized as few of the major constraints that hinder the large-scale implementation of these approaches.

This chapter deliberates on the need and significance of ecosystem-based approaches for water stress management at regional level. At the outset, the chapter provides a brief understanding of different ecosystem-based approaches and their progressive research trends. The key terminologies are explained, before discussing their overlapping focus areas. Thereafter, these approaches are discussed specifically in context of water stress management. The key challenges in implementing the ecosystem-based approaches at local level have been underlined alongside the potential pathways. To further highlight the existing challenges and the key opportunities for implementing ecosystem-based approaches at local level, the chapter discusses a specific case of Nagpur Metropolitan Area 'NMA' in Central India, which has recently experienced severe water stress concerns. The varied challenges of water scarcity, water quality, declining groundwater levels, etc., have been discussed to set the context. Thereafter, based on the findings from India–Japan Bilateral Research

Project (2018–20), the chapter provides suitable directions for water stress management in NMA by implementing ecosystem-based approaches. Toward the end of chapter, the authors summarize the key conclusions and underline the way forward.

17.2 Overview of Ecosystem-Based Approaches

'Ecosystems' are basically defined as the dynamic complexes of living organisms, their physical environment, and their interrelationships within a particular space. With enormous variations in size and scale, there are several different types of ecosystems ranging from terrestrial to aquatic settings like forests, grasslands, wetlands, mangroves, deserts, agriculture, urban ecosystems, mountains, and coastal (comprehensively discussed in CBD 2019). Serving as the basis of all life and livelihoods (be it in urban or rural areas), the varying ecosystems provide vital services that include provisioning services (like food, water, and fuel), regulating services (like waste treatment, water filtration, and flood mitigation), supporting services (like nutrient cycling and soil formation), and cultural services (like aesthetics, recreational, and spiritual services). The multifaceted characteristics and benefits of ecosystems are also explained by Martin-Ortega et al. (2015). Particularly, the regulating and supporting services help people adapt to the adverse effects of climate change and reduce disaster risk. A number of other studies (e.g., Sudmeier-Rieux and Ash 2009; GWP 2016) have also discussed the wide-ranging benefits of healthy ecosystems that include livelihood benefits and cost-effective natural buffers to hazard events.

While ecosystems produce the bulk of renewable resources and ecosystem services utilized by the human societies, they are also water-dependent. Falkenmark (2003) precisely described the water linkages between human activities and ecosystems from a drainage basin perspective. As shown in Fig. 17.1, the precipitation inside the water divide (or catchment area) forms the key water source for co-dependent human societies and ecosystems. In the land surface, a part of rainwater is consumed in crop production and evapotranspiration (as green water flow) by forests, croplands, grasslands, etc. The surplus water moves from upstream to downstream areas recharging the aquifers and rivers (blue water), that support the aquatic ecosystems and human societies. In context of the expanding urban populations, freshwater aquatic ecosystems (lakes, ponds, rivers, groundwater, etc.) serve for varying purposes including drinking water needs, industries, food production, fisheries, etc. From food production to freshwater storage to energy generation, the progressing urban development is highly reliant on the continued supply of these varying ecosystem services. However, the growing emphasis on hard infrastructure development often impacts the water ecosystems within and outside the city boundaries. The profit-driven land management decisions being made by various stakeholders (industries, farmers, consumers, etc.) at different scales (like urban and rural) often overlook the crucial role of natural environment.

Due to the growing concerns of water stress, a genuine need for mainstreaming ecosystem-based approaches has been recognized to lower the hydrological impacts

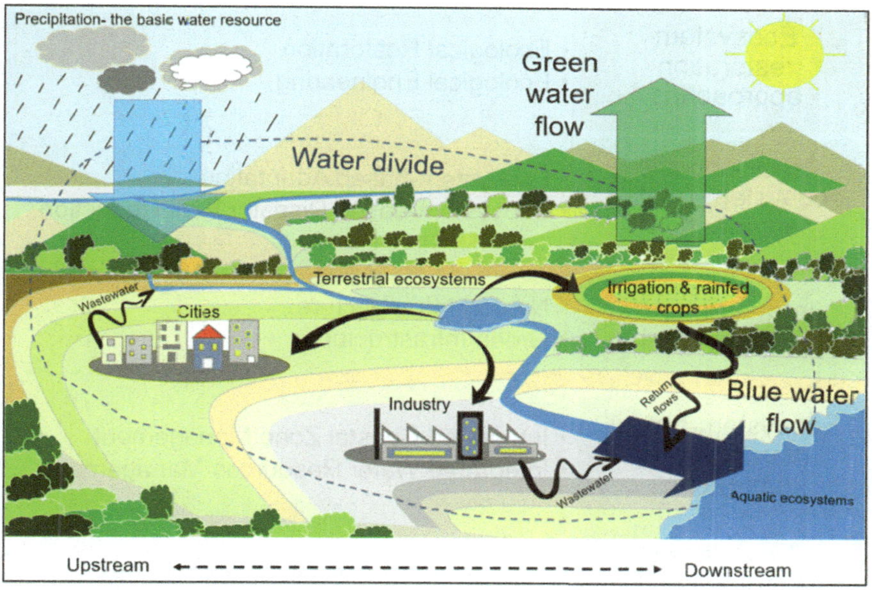

Fig. 17.1 Water cycle from a drainage basin perspective (Prepared by the Authors, in reference to Falkenmark 2003)

of current urbanization trends. However, their implementation at local level is often hampered due to differential understanding of the variety of concepts. Also, the definitions of numerous ecosystem-based approaches share similar terms, underlying principles, and rationale (like restoration, conservation, and management), which often leads to confusion (GWP 2016). Over the years, several terminologies and frameworks have emerged with a common focus on addressing the societal issues linked to ecosystems. These terms are often used interchangeably by practitioners, academic, and research organizations. Under the recently introduced umbrella concept of NbS, Cohen-Shacham et al. (2016) explained that a variety of ecosystem-based approaches (highlighted in Fig. 17.2) have emerged from different spheres ranging from scientific research (like ecological engineering) to practice (like ecological restoration) and policy contexts (like EbA). Also, some of the approaches are developed based on the combination of preexisting approaches and tools (like Eco-DRR).

As such, a methodical search (query-based) was conducted on Scopus, the world's largest database of peer-reviewed research literature, to understand the progressive research trends related to the variety of NbS approaches. The stacked area chart shown in Fig. 17.3 highlights the year-wise research trends associated with various NbS approaches. It can be seen that selected approaches like ecological restoration, ecological engineering, and climate adaptation services have been in practice for a long time (since 1960s). However, in the last two decades, there has been a

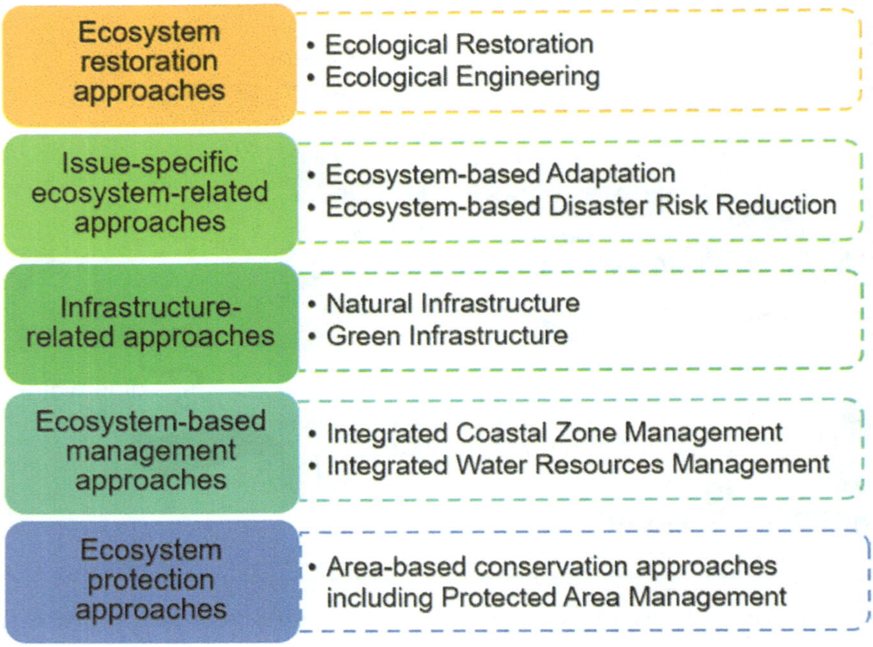

Fig. 17.2 Various categories and examples of NbS approaches (Prepared by the Authors, in reference to Cohen-Shacham et al. 2016)

notable increase in the scientific research related to these varying ecosystem-based approaches. Other NbS approaches like Green Infrastructure, EbA, and ecosystem-based management have also evolved over the years and are increasingly been adopted for policy implementation.

17.3 Ecosystem-Based Approaches for Water Stress Management

Freshwater is one of the most fundamental resources, that is essential for sustaining ecosystems, biodiversity, economic growth, and human societies as a whole. However, the gradually changing climate is increasing variability in the water cycle, inducing extreme weather events, and threatening sustainable development (UN-Water 2019). An estimated 3.6 billion people (around half the global population) worldwide are presently living in areas that experience water scarcity every year, and the affected population is expected to increase to 4.8–5.7 billion people by 2050 (WWAP and UN-Water 2018).

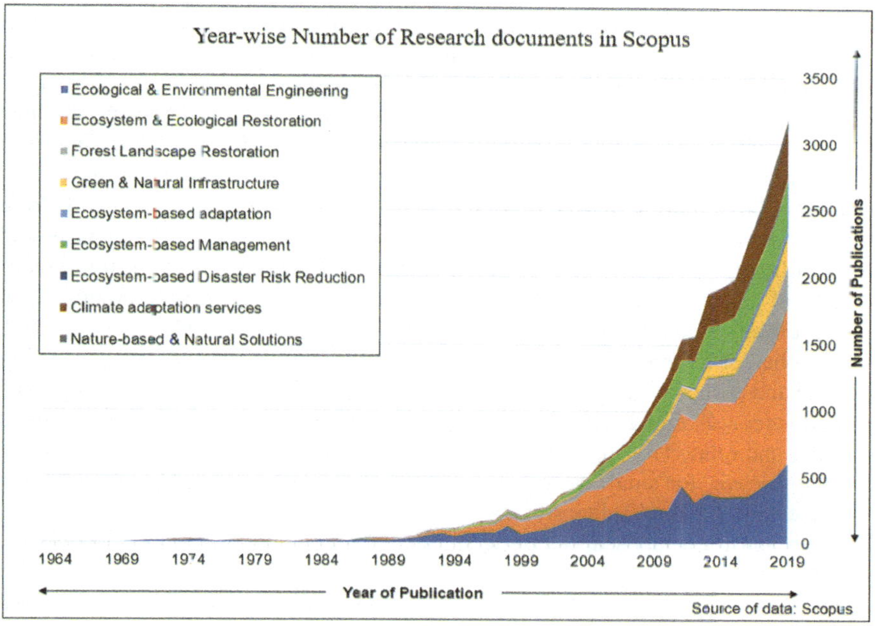

Fig. 17.3 Historical research trends for various ecosystem-based approaches (Image *source* Authors)

In the backdrop of growing water demands and increasing pressure on the finite water resources, the sustainability of freshwater ecosystems is being threatened globally (Odume 2017; Boelee et al. 2017). The wide range of water users across different sectors (like agriculture, industry, energy, and transport) and at different administrative scales (like urban and rural areas) are experiencing unprecedented competition for the limited stock of water resources, especially in the water-stressed areas. The shifting precipitation patterns may further intensify the situation, as more rivers, lakes, wetlands, and aquifers are expected to dry up (OECD 2013; US AID 2017). Since the variety of ecological processes taking place in different landscapes influence the quality of water and the way it moves through the drainage basin (including soil formation, sediment transport, etc.), the continuing ecological degradation has become a matter of huge concern for water resource management. Over the last century, around 64–71% of natural wetlands around the world have reportedly been lost to human activities (Davidson 2014).

In face of the multifaceted challenges being experienced by the human societies, the 2030 Agenda for Sustainable Development, through its 17 SDGs, has laid strong emphasis on water resource management. SDG 6 is specifically focused on ensuring sustainable access to water and sanitation for all. Under the same goal, Target 6.6 specifically emphasizes on protecting and restoring water-related ecosystems. The continued access to water resources is also one of the key determinants for several other SDGs focused on food, energy, industry, sustainable communities, etc. In that regard, the wide range of ecosystem-based approaches (explained in Sect. 17.2) offers promising solutions for enhancing the water sustainability while striving to preserve the intrinsic value of ecosystems (several case study examples highlighted in Sonneveld et al. 2018). Investing in freshwater ecosystems is becoming even more important, as they are inextricably linked to water cycle and hydrology. From managing precipitation to water storage, ecosystem-based approaches can help to address water stress concerns in several different ways, which are sustainable as well as cost-effective (also explained by Oral et al. 2020). Like for example, the importance of natural and constructed wetlands within urban areas has been recognized for a long time to mitigate the impact of polluted stormwater runoff and wastewater. Approaches of Green Infrastructure have also recently gained high prominence to manage and reduce water pollution from urban areas, while facilitating for improved groundwater recharge. Furthermore, the IWRM approach promotes the management of water and related resources (land, biodiversity, etc.) on a watershed basis, which is more relevant for small catchments and transboundary basins. Moreover, all the ecosystem-based approaches serve as win–win solutions to enhance water security in urban and rural landscapes, while simultaneously generating social, economic, and environmental benefits.

The successful implementation of ecosystem-based approaches requires cooperation among multiple institutions and stakeholders, which is often challenging as ecosystems function at different scales, and very often their boundaries do not correspond with the administrative boundaries like those of village, city, region, district (Roy et al. 2011; ELAN 2012). Since the conventional approaches have partly failed to ensure sustainable management of water resources, a new paradigm of broader multi-stakeholder involvement with wider recognition of cultural, ecological, and economic values is evolving in research and policy practice (Schoeman et al. 2014). Transboundary water cooperation has also been strongly emphasized in the SDGs (Target 6.5) for enabling integrated management of water resources that cross administrative boundaries. With growing consideration of upstream–downstream water users and nexus-based implications (like for food and energy sectors), the notion of 'payment for ecosystem services' is increasingly been adopted to facilitate collective urban–rural action at regional level (Bleeker and Vos 2019). There is also a growing emphasis on 'collaborative water management' to bring together stakeholders from different sectors across wider geographic areas to determine feasible solutions for addressing the growing water-related concerns (Ostovar 2019). Martin-Ortega et al. (2015) emphasized on mainstreaming monetary incentives, watershed policies and regulations, to realize collective action for water stress management.

17.4 Case of Nagpur Metropolitan Area, India

Nagpur is one of the fastest-growing urban agglomerations in central India. It is the third-largest city (after Mumbai and Pune) and the winter capital of Maharashtra

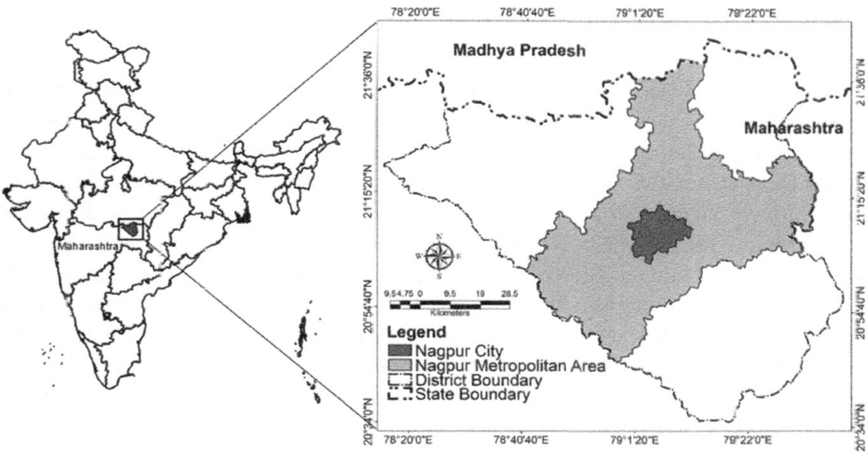

Fig. 17.4 Location map of Nagpur City and Nagpur Metropolitan Area in India (Image *source* Authors)

state. Figure 17.4 highlights the location of Nagpur City and Nagpur Metropolitan Area 'NMA' in India. Spread over an area of 3567 km^2, NMA comprises of 721 villages and 24 Census Towns. Apart from its commercial and political significance, Nagpur (also referred to as Orange City) is widely recognized as a major trade center of oranges (NIT 2015).

Due to the ongoing rapid development transformations including Multimodal International Cargo Hub and Airport at Nagpur (MIHAN), Smart city development, and Metro project, Nagpur is projected to be one of the fastest-growing cities in the world for the period of 2019–35 (Holt 2018). However, at the same time, it has recently experienced serious water stress situation. During the year 2019, the Nagpur Municipal Corporation 'NMC' was forced to restrict the water supply in Nagpur City to alternate days (Nation Next 2019; India Today 2019), as the water levels in the key reservoirs came down due to less rainfall and upstream developments. Due to the declining water availability in shared water resources and growing water demands, regional level concerns pertaining to socioeconomic development, agriculture, industrial demands, etc., at different administrative scales have come into limelight. It is therefore important to establish a regional level understanding of water resource flow and transboundary interlinkages. The following three subsections provide an overview of the environment profile of NMA, the transboundary water-related concerns, and water governance issues.

17.4.1 Environmental Profile of NMA

The topography of NMA is typically of the Deccan traps, having flattop and terraced features. With an average elevation between 350 and 260 m, the greater part of NMA is an undulating plateau. More than ninety percent of NMA ranges within zero to three percent slope, and the northern and southwestern parts of NMA have comparatively steeper slopes in the range of 10–35%. Typical to central Deccan Plateau region, the climate of NMA is dry and tropical. It receives an average rainfall of 1200 mm, but has recently experienced high climate variations. As pointed out by MPCB (2019a), Nagpur City recorded its highest temperature as 48 °C on 19th May 2015, and the lowest temperature was recorded on 16th January 2016 as 3.9 °C. Further, based on the geomorphological features, the western region of Nagpur City is occupied by Deccan trap formation and eastern side is occupied by metamorphic and crystalline series. In geological formations ranging from Archean to recent ages, various minerals (including manganese and coal) are also mined within NMA (NIT 2015).

Within NMA, the forest areas (particularly southern tropical dry deciduous types) cover about 14.5% of the lands, mainly in the northern and southwestern areas. The widespread forest cover in the region is also a boon for tourism sector, as Nagpur is referred to as 'Tiger Capital of the World.' Further, lands within NMA are very fertile as they are dominated by black cotton soils. As seen from Fig. 17.5, agriculture is practiced in about 66.10% of the NMA earth's surface. Although a large proportion of farmlands in the northeastern part fall under irrigation command areas, most of the cultivated land in NMA is dependent on monsoon rainfall. Apart from that, NMA is also endowed with abundant water resources including lakes, dams, rivers, etc., that cover around 6.4% of the land surface. In the recent years, there have been considerable developments around the centrally located Nagpur City. Due to increase in urban sprawl and hardscapes, water percolation and green cover around the city boundaries have reduced greatly. Previous studies, including Dhyani et al. (2018), have also highlighted the rapid change in built environment around Nagpur and underlined the growing urbanization pressures on the rich biodiversity of the area.

17.4.2 Freshwater Resources in NMA and Transboundary Concerns

With natural slopes in two directions, north to northeast and southwest to southeast, the topography of NMA provides an excellent natural drainage pattern. NMA is drained by several rivers including Kanhan, Pench, Wardha, and Wainganga rivers. Approximately 70% of NMA drains into river Wainganga through Kanhan Rivers and its tributaries, including Pench River. Other tributaries to these rivers, viz. Pili and Nag, originate in the western fringe of Nagpur City and carry its sewage as

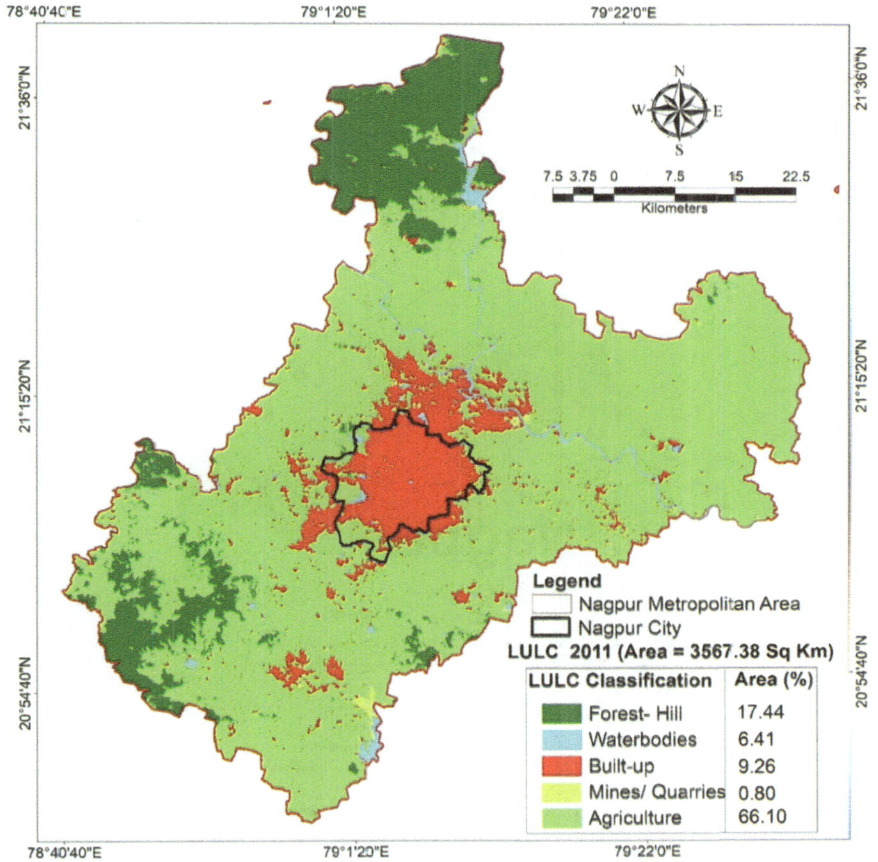

Fig. 17.5 Landuse/landcover (LULC) map of Nagpur Metropolitan Area (Processed by the Authors; Satellite Image courtesy of the US Geological Survey)

they traverse through the city. Further, groundwater in NMA is available from both confined and unconfined aquifers, with yields ranging between 1 and 5 L/s. As observed from the hydrogeological map of NMA (Fig. 17.6), the underground aquifers are mostly due to secondary porosity and fractures in the rocks. Although the water depth in the region varies during different seasons, all tehsils (blocks) within NMA presently fall in the 'safe' category in terms of the overall stage of groundwater development (Bhatnagar 2018).

Water supply in NMA is reliant upon both surface and groundwater sources. For urban areas, the water demands are primarily met through surface water sources like lakes, rivers, and reservoirs, and to a lesser extent through groundwater. On the other hand, the water supply in rural areas is mostly based on the groundwater resources, wherein several rural water supply schemes are implemented in terms of developing

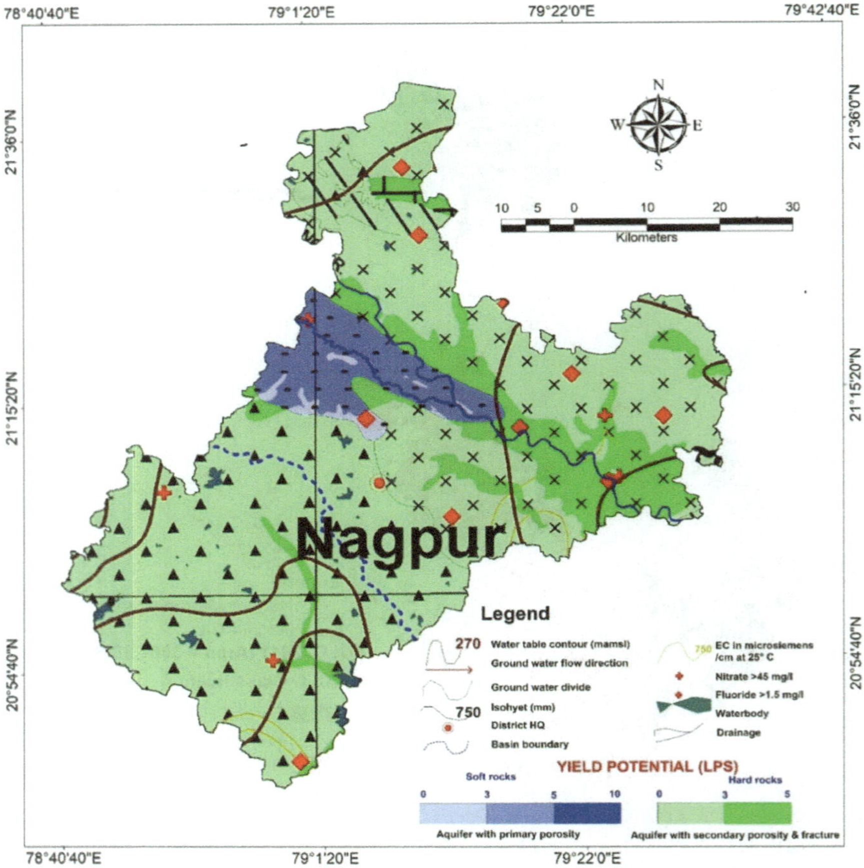

Fig. 17.6 Hydrogeological map of Nagpur Metropolitan Area. Modified from the Hydrogeological map of Maharashtra State, which the authors received from the office of Central Ground Water Board, Nagpur in September 2019

tube wells, hand pumps, and bore wells. Notably, the irrigation of agricultural lands in NMA is done through a combination of ground and surface water sources such as canals, tube wells, and bore wells (NIT 2015).

With changing climate, Nagpur has recently experienced high fluctuations in rainfall patterns and surface water availability (Behl 2019). The situations of water stress in summer and flooding in rainy season have recently become the new normal. Deshkar (2019) also underlined the varying water-related concerns in Nagpur from a resilience perspective. While Pench reservoir in northern part of NMA presently serves as the main source (caters to more than 70% water demands) of water for Nagpur City, it also provides water to the Koradi and Khaperkheda Thermal Power Stations and caters for irrigation in more than 400 villages in Pench command areas.

The recent decline in Pench reservoir has therefore raised cross-sectoral concerns. Since Pench reservoir receives water from Totladoah reservoir situated in the neighboring state of Madhya Pradesh (as shown in Fig. 17.7), the ongoing water supply concerns in Nagpur are also affected by transboundary water conflicts (MWRRA 2018).

Fig. 17.7 Key water sources in Nagpur Metropolitan Area and transboundary linkages. Prepared by the Authors; the background image has been sourced from Esri, HERE, Gramin, USGS, Intermap, INCREMENT P, NR Can, Esri Japan, METI, Esri China (Hong Kong), Esri Korea (Thailand), NGCC, OpenStreetMap contributors, and the Community

17.4.3 Water Governance

Water is a state subject in India, and the water-related issues within Maharashtra are mainly handled by the State government. Within NMA, the water supply for Nagpur City is managed by the NMC, under a Public–Private Partnership agreement with a private company 'Orange City Water Pvt. Ltd.' For the rural areas at large, several agencies including Zilla Parishad, The Maharashtra Water Supply and Sewerage Board (MWSSB), Maharashtra Jeevan Pradhikaran (MJP), the Groundwater Survey and Development Agency (GSDA), Environmental Engineering Department, etc., are responsible for the planning, design, and implementation of water supply schemes. The State Irrigation Department does not have any direct role in water supply, but it supplies water in bulk quantity to agencies like MWSSB, and NMC from the dams constructed for irrigation purpose. In case of Nagpur, the Vidarbha Irrigation Development Corporation (river basin agency) determines the annual allocation for NMC, irrigation in command areas, and other water users like Thermal Power stations, as per the norms of Maharashtra Water Resources Regulatory Authority (MWRRA). In addition to that, other committees and associations are also functioning at watershed level. Overall, there are several agencies responsible for water resource management at various governance levels within NMA. The presence of multiple institutions and their overlapping responsibilities often hinders the progress of transboundary development projects and integrated water resource management.

17.5 Discussion

Under the India–Japan Bilateral Research Project (2018–20), intensive evidence-based research was conducted within NMA to study the water linkages between urban and rural areas, across different administrative scales. With a thorough understanding of existing governance structures, specific policy level recommendations were also put forward for consideration by the local governments (TLN Team 2019; Ahmed 2020). To enable transboundary cooperation for integrated management of water resources at regional level, the research findings were also widely disseminated through various platforms like stakeholder consultation workshops and policy forums. The following subsections highlight the key project-related findings as well as the identified key areas of policy interventions.

17.5.1 Varied Water-Related Concerns and the Need for Water Conservation

Figure 17.8 highlights the contrasting water-related challenges from different urban and rural areas within NMA. Human settlements in different parts of NMA, be it

Fig. 17.8 **a** Picture from Pipriya Village in NMA **b** Picture from Parsheoni Town in NMA **c** Nag River carrying the sewage generated from Nagpur City (*Image source* Authors)

urban, rural, or forest areas, are found to be experiencing unique water security-related concerns ranging from declining water availability, degrading water quality, to growing water stress. The key determinants that drive these concerns include the varying geomorphological characteristics and upstream–downstream settings. While Nagpur City is increasingly sourcing more and more water from outside their physical boundaries, the priority allocation to city water demands has in contrary led to water stress near the water source areas. The growing pressure on groundwater resources is also apparent through declining water levels in all parts of NMA. Further, the growing water demands in the city are also leading to a corresponding increase in wastewater generation.

To address the wide-ranging water-related challenges, there is a genuine need for enhancing water conservation and management through strategic planning at local level. Climate action planning needs to be mainstreamed in various developmental initiatives at regional level to counter the emerging climatic variations. While urban areas need to work toward reducing their external water dependence (and water losses), more emphasis needs to be put on watershed-level approach for effectively managing the available water resources. Particularly in rural areas, there is a need to foster active community engagement to tap the rainwater flows and rejuvenate the indigenous water conservation practices for water demands during summer season. Over the years, several types of ecosystem-based approaches have been recognized for water conservation and management (as discussed in Sect. 17.3). However, a thorough understanding of local context is imperative to determine the feasibility of any specific approach and for enabling their policy implementation. In reference to the existing scientific studies (e.g., Reid 2016; IUCN 2017), the recognition and uptake of local traditional knowledge play a key role in successful implementation of ecosystem-based approaches. The importance of stimulating community engagement and knowledge sharing has also been recognized for their large-scale implementation.

17.5.2 Need for Addressing Urban–Rural Linkages and Nexus Considerations

Urban and rural areas in NMA are geographically dispersed, but they are dependent on shared surface and groundwater resources, as also explained in Sect. 17.4. Apart from the domestic water needs, the freshwater resources in NMA serve for various other purposes including irrigation, industries, fisheries, energy production, etc. In consideration of the multifaceted water linkages across different sectors and administrative scales in NMA, there is a core need for nexus considerations to water stress management. Sukhwani et al. (2019) underlined several good practices of urban–rural partnerships, based on a variety of nexus-based approaches like payment for ecosystem service, incentivization, water source conservation, public–private partnerships. As water resources in NMA are already stressed, there is a need for enhancing water efficiency, as well as water conservation. In view of the growing water demands in Nagpur, the city government needs to initiate benefit-sharing schemes to compensate for the rising water stress concerns in upstream water source areas. Emphasis should also be put on reducing the wastewater generation and water quality degradation to prevent any negative implications for the downstream rural areas. In context of the ongoing development trends (mining, smart city, etc.) in NMA, a nexus-based approach to water stress management will not only reduce the pressure on water resources but will also open up new avenues for urban–rural collaboration and partnerships at regional level.

17.5.3 Enhancing Wastewater Reuse and Resource Recovery from Nag River

Previously considered to be a liability, treated municipal wastewater is now recognized as a valuable resource for addressing the growing concerns of water scarcity around the world. The 2017 edition of the United Nations World Water Development Report (WWAP 2017) specifically underlined the importance of wastewater management for generating social, environmental, and economic benefits to achieve the 2030 Agenda for Sustainable Development. Rich in carbon and nutrients, domestic wastewater can be reused after adequate treatment for a variety of purposes including industry, irrigation, ecosystem health, generating energy, etc.

As of 2015, the city of Nagpur was generating over 450 Million Liters per Day 'MLD' sewage of which NMC was treating only 80 MLD, through its only sewage treatment plant at Bhandewadi. The residual untreated wastewater was directly released in the natural drains which pollute the streams and rivers (mainly Nag River) flowing toward downstream areas. To enhance wastewater reuse, NMC has already been supplying (from 2015 onwards) around 130 MLD of wastewater to Maharashtra State Power Generation Company (at the price of 150 Million Indian rupees 'INR'/year) for cooling purposes in thermal power stations (The World Bank

2019; MPCB 2019b). With the growing urban population, the wastewater generation from Nagpur City is constantly on the rise, but the treatment capacities are still limited. As such, wastewater reuse is imperative not only to enhance water use efficiency but also to reduce its negative implications on downstream areas.

17.5.4 Enabling Transboundary Cooperation Through Multi-stakeholder Engagement

As climate change is altering the natural environment at various scales, finding better ways to collectively govern and manage the limited water resources has become indispensable. Effective implementation of approaches like EbA and Eco-DRR at regional level requires effective engagement and coordination between a range of stakeholders including governments, academic experts, private sector, community members, etc. (as also emphasized in Sect. 17.3). Throughout the project duration, a range of initiatives were taken to bring together stakeholders (as highlighted in Fig. 17.9) from different sectors and enable collective action. Several key areas of policy interventions were also identified through these multi-stakeholder consultations including the need for mobilizing urban–rural partnerships, facilitating context-specific environmental conservation measures, water audits, and community engagement. A genuine need for a multi-level governance approach among Nagpur City, District Council (Zilla Parishad) and other agencies at regional level was realized to collectively address the transboundary water-related concerns emerging due to the rapid urbanization. The key policy recommendations that were put forward for consideration by the local governments in Nagpur included the need for the establishment of an urban–rural coordinating entity at governance level for dealing with water conflicts in NMA. While numerous agencies are involved in water management in Nagpur region, a coordinating governance entity can potentially enhance policy

Fig. 17.9 **a** Decision theater workshop conducted in Visvesvaraya National Institute of Technology, Nagpur, in June 2019; **b** Range of participating stakeholders (*Image source* Authors)

cohesion at various levels in NMA and pave the way for coordinated urban–rural development at regional level.

17.6 Conclusion

Upscaling ecosystem-based approaches for addressing the range of water-related challenges like water shortage, water pollution, floods, and drought has become the need of the hour. Their growing significance has also been apparent through the recent scientific research, and global policy agreements. However, their large-scale implementation is faced with many cross-sectoral and transboundary challenges, including the discrete governance structures, varied stakeholders, and poor evidence base, as also evident from the case of NMA. The chapter provides a simplistic understanding of the various ecosystem-based approaches, and their significance in context of addressing water stress management. Based on the findings from India–Japan Bilateral Project, the chapter also provides a precise understanding of regional level transboundary water linkages in NMA, alongside the potential measures that could facilitate transboundary cooperation and collective action at regional level.

Overall, the effective implementation of ecosystem-based approaches at regional level is found to be reliant upon the improved understanding of water resource flow at drainage basin level. While urban areas around the world are witnessing rapid transformations, it is increasingly important to recognize their transboundary implications on upstream and downstream areas. Addressing the emerging water stress concerns at regional level is also imperative to ensure long-term sustainable development. Learning from the real-world experiences, there is a need to mainstream ecosystem-based approaches through innovative practices like watershed policies and regulations, monetary incentives, payment for ecosystem services, and urban–rural coordinated governance approach.

Acknowledgements All the authors sincerely acknowledge the support received from the Japan Society for the Promotion of Science (JSPS) and the Indian Council of Social Science Research (ICSSR), under the India–Japan Bilateral Research Project. The first author (V.S.) is also thankful to the Ministry of Education, Culture, Sports, Science, and Technology (MEXT), Japan, for the provided scholarship.

Conflicts of Interest The authors declare that they have no conflict of interest.

References

Ahmed S (2020) Nagpur: VNIT study suggests new body, measures for metro region water woes. The Times of India. https://timesofindia.indiatimes.com/city/nagpur/vnit-study-suggests-new-body-measures-for-metro-region-water-woes/articleshow/73179086.cms. Accessed 15 June 2020

ARUP (2018) Cities alive: water for people. ARUP/International Water Association (IWA). https://www.arup.com/perspectives/publications/research/section/cities-alive-water-for-people. Accessed 12 June 2020

Behl M (2019) Climate change impacting city weather too: expert. Available online: https://timesofindia.indiatimes.com/city/nagpur/climate-change-impacting-city-weather-too-expert/articleshow/68209049.cms. Accessed 12 June 2020

Bhatnagar SK (2018) Ground Water Brochure, Nagpur District, Maharashtra, Ministry of Water Resources, River Development and Ganga Rejuvenation, Government of India, Central Ground Water Board (CGWB): Central Region, Nagpur

Bleeker S, Vos J (2019) Payment for ecosystem services in Lima's watersheds: power and imaginaries in an urban-rural hydrosocial territory. Water Int 44(2):224–242. https://doi.org/10.1080/02508060.2019.1558809

Boelee E, Janse J, Le Gal A, Kok M, Alkemade R, Ligtvoet W (2017) Overcoming water challenges through nature-based solutions. Water Policy 19(5):820–836. https://doi.org/10.2166/wp.2017.105

CBD (2019) Voluntary guidelines for the design and effective implementation of ecosystem-based approaches to climate change adaptation and disaster risk reduction and supplementary information. Secretariat of the Convention on Biological Diversity (CBD) Technical Series No. 93. Montreal, Canada, 156 pp. https://www.cbd.int/doc/publications/cbd-ts-93-en.pdf

Cohen-Shacham E, Walters G, Janzen C, Maginnis S (eds) (2016) Nature-based solutions to address global societal challenges. The International Union for Conservation of Nature (IUCN), Gland, Switzerland, 97 pp. https://www.iucn.org/sites/dev/files/content/documents/nature-based_solutions_to_address_global_societal_challenges.pdf

Davidson NC (2014) How much wetland has the world lost? Long-term and recent trends in global wetland area. Mar Freshwater Res 65(10):934–941. https://doi.org/10.1071/MF14173

Deshkar S (2019) Resilience perspective for planning urban water infrastructures: a case of Nagpur city. In: Ray B, Shaw R (eds) Urban drought: emerging water challenges in Asia. Springer Nature Singapore Pte Ltd., Singapore, pp 131–154. https://doi.org/10.1007/978-981-10-8947-3

Dhyani S, Lahoti S, Khare S, Pujari P, Verma P (2018) Ecosystem based Disaster Risk Reduction approaches (EbDRR) as a prerequisite for inclusive urban transformation of Nagpur City, India. Int J Disast Risk Re 32:95–105. https://doi.org/10.1016/j.ijdrr.2018.01.018

Djehdian LA, Chini CM, Marston L, Konar M, Stillwell AS (2019) Exposure of urban food–energy–water (FEW) systems to water scarcity. Sustain Cities Soc 50:101621. https://doi.org/10.1016/j.scs.2019.101621

ELAN (2012) Integrating community and ecosystem-based approaches in climate change adaptation responses. Ecosystem & Livelihoods Adaptation Networks (ELAN). https://d2ouvy59p0dg6k.cloudfront.net/downloads/integrating_community_and_ecosystem_based_approaches_in_climate_change_adaptation_res.pdf

European Environment Agency (2020) Water stress. HelpCenter definition. https://www.eea.europa.eu/archived/archived-content-water-topic/wise-help-centre/glossary-definitions/water-stress. Accessed 12 June 2020

Falkenmark M (2003) Water management and ecosystems: living with change. Global Water Partnership, TEC Background Papers No. 9. https://www.gwp.org/globalassets/global/toolbox/publications/background-papers/09-water-management-and-eco-systems.-living-with-change-2003-english.pdf

Flörke M, Schneider C, McDonald RI (2018) Water competition between cities and agriculture driven by climate change and urban growth. Nat Sustain 1:51–58. https://doi.org/10.1038/s41893-017-0006-8

Garrick D, Stefano LD, Yu W, Jorgensen I, O'Donnell E, Turley L, Aguilar-Barajas I, Dai X, Leao RdS, Punjabi B, Schreiner B, Svensson J, Wight C (2019) Rural water for thirsty cities: a systematic review of water reallocation from rural to urban regions. Environ Res Lett 14:043003. https://doi.org/10.1088/1748-9326/ab0db7

GWP (2016) Linking ecosystem services and water security—SDGs offer a new opportunity for integration. Perspectives Paper. Global Water Partnership (GWP). https://www.gwp.org/globalassets/global/toolbox/publications/perspective-papers/gwp_pp_-ecosystemservices.pdf
Heard BR, Miller SA, Liang S, Xu M (2017) Emerging challenges and opportunities for the food-energy-water nexus in urban systems. Curr Opin Chem Eng 17:48–53. https://doi.org/10.1016/j.coche.2017.06.006
Hoekstra AY, Buurman J, van Ginkel KCH (2018) Urban water security: a review. Environ Res Lett 13, 053002. https://doi.org/10.1088/1748-9326/aaba52
Holt R (2018) Global cities, which cities will be leading the global economy in 2035? Oxford Economics. https://workplaceinsight.net/wp-content/uploads/2018/12/Global-Cities-Dec-2018.pdf
IRENA (2015) Renewable energy in the water, energy and food nexus. International Renewable Energy Agency (IRENA), Abu Dhabi, United Arab Emirates. https://www.irena.org/documentdownloads/publications/irena_water_energy_food_nexus_2015.pdf
India Today (2019) Nagpur: water cuts imposed for 3 days due to less rainfall. https://www.indiatoday.in/india/story/nagpur-water-cuts-imposed-for-3-days-due-to-less-rainfall-1571125-2019-07-19. Accessed 12 June 2020
IUCN (2017) Ecosystem-based adaptation: issues brief. International Union for Conservation of Nature (IUCN). https://www.iucn.org/sites/dev/files/import/downloads/ecosystem-based_adaptation_issues_brief_final.pdf
Lo V (2016) Synthesis report on experiences with ecosystem-based approaches to climate change adaptation and disaster risk reduction. Secretariat of the Convention on Biological Diversity (CBD) Technical Series No. 85. Montreal, Canada, 106 pp. https://www.cbd.int/doc/publications/cbd-ts-85-en.pdf
Martin-Ortega J, Ferrier RC, Gordon IJ, Khan S (eds) (2015) Water ecosystem services—a global perspective. Cambridge University Press, United Kingdom, International Hydrology Series
McDonald RI, Weber K, Padowski J, Flörke M, Schneider C, Green PA, Gleeson T, Eckman S, Lehner B, Balk D, Boucher T, Grill G, Montgomery M (2014) Water on an urban planet: urbanization and the reach of urban water infrastructure. Global Environ Chang 27:96–105. https://doi.org/10.1016/j.gloenvcha.2014.04.022
Millennium Ecosystem Assessment (2005) Ecosystems and human well-being: synthesis. Island Press, Washington, DC. https://www.millenniumassessment.org/documents/document.356.aspx.pdf
MPCB (2019a) Action plan to control air pollution in Nagpur City. Maharashtra Pollution Control Board (MPCB). http://www.mpcb.gov.in/sites/default/files/pollution-index/severly-report/Nagpur_Action_Plan07112019.pdf
MPCB (2019b) Note on domestic wastewater reuse project at Nagpur. Nagpur Municipal Corporation to Supply Water For (3 × 660 Mw) Koradi Thermal Power Plant. Maharashtra Pollution Control Board (MPCB), Nagpur. http://www.mpcb.gov.in/sites/default/files/focus-area-reports-documents/NMC_%26_KTPS_success_story_28052019.pdf
MWRRA (2018) Petition filed by Adv. Ashish Jaiswal at Nagpur bench of Honorable High Court of Bombay as regards Making Provision for Irrigation Purpose by Curtailing the Use of Water from Pench Project Complex by Nagpur Municipal Corporation- Case No. 8 of 2017. Maharashtra Water Resources Regulatory Authority (MWRRA). https://mwrra.org/wp-content/uploads/2018/11/Case-No-8-of-2017-Final-Order.pdf
Nation Next (2019) Water crisis in Nagpur: citizens may get water on alternate days for four months. Nation Next Newsroom. https://nationnext.in/water-crisis-in-nagpur-citizens-may-get-water-on-alternate-days-for-four-months/. Accessed 12 June 2020
NIT (2015) Nagpur metropolitan area development plan: 2012–32. Draft Development Plan Report. Nagpur Improvement Trust (NIT). http://www.nitnagpur.org/pdf/Metro_Region_DP.pdf
Odume ON (2017) Ecosystem approach to managing water quality. In: Tutu H (ed) Water quality. Intech, Croatia, pp 3–22. https://doi.org/10.5772/65707

OECD (2013) Water and climate change adaptation: an OECD perspective. The Organization for Economic Co-operation and Development (OECD). https://www.oecd.org/env/resources/Water%20and%20Climate%20Change%20Adaptation-%20brochure.pdf

Oral HV, Carvalho P, Gajewska M, Ursino N, Masi F, van Hullebusch ED, Kazak JK, Exposito A, Cipolletta G, Andersen TR, Finger DC, Simperler L, Regelsberger M, Rous V, Radinja M, Buttiglieri G, Krzeminski P, Rizzo A, Dehghanian K, Nikolova M, Zimmermann M (2020) A review of nature-based solutions for urban water management in European circular cities: a critical assessment based on case studies and literature. Blue-Green Syst 2(1):112–136. https://doi.org/10.2166/bgs.2020.932

Ostovar AL (2019) Investing upstream: watershed protection in Piura, Peru. Environ Sci Policy 96:9–17. https://doi.org/10.1016/j.envsci.2019.02.005

PEDRR (2019) Issue brief- integrating risk management ecosystems and water-related risks. Partnership for Environment and Disaster Risk Reduction (PEDRR), Global Platform for Disaster Risk Reduction (GP2019). Geneva, Switzerland, 13–17 May 2019. https://www.preventionweb.net/files/globalplatform/5cca9d868ecb8WS14-Issues_Brief_-_PEDRR.pdf

Reid H (2016) Ecosystem-and community-based adaptation: learning from community-based natural resource management. Clim Dev 8(1):4–9. https://doi.org/10.1080/17565529.2015.1034233

Renaud FG, Sudmeier-Rieux K, Estrella M (eds) (2013) The role of ecosystems in disaster risk reduction. United Nations University (UNU) Press, Tokyo, Japan. https://collections.unu.edu/view/UNU:1995. Accessed 12 June 2020

Roy D, Barr J, Venema HD (2011) Ecosystem approaches in integrated water resources management (IWRM), a review of transboundary River Basins. International Institute for Sustainable Development (IISD) and the UNEP-DHI Centre for Water and Environment. https://www.iisd.org/pdf/2011/iwrm_transboundary_river_basins.pdf

Schoeman J, Allan C, Finlayson CM (2014) A new paradigm for water? A comparative review of integrated, adaptive and ecosystem-based water management in the Anthropocene. Int J Water Resour D 30(3):377–390. https://doi.org/10.1080/07900627.2014.907087

Sonneveld BGJS, Merbis MD, Alfarra A, Ünver O, Arnal MA (2018) Nature-based solutions for agricultural water management and food security. The Food and Agriculture Organization of the United Nations (FAO) Land and Water Discussion Paper no. 12. Rome, Italy, 66 pp. http://www.fao.org/3/CA2525EN/ca2525en.pdf

Sudmeier-Rieux K, Ash N (2009) Environmental guidance note for disaster risk reduction: healthy ecosystems for human security, Revised Edition. The International Union for Conservation of Nature (IUCN), Gland, Switzerland, 34 pp

Sudmeier-Rieux K, Masundire H, Rizvi A, Rietbergen S (eds) (2006) Ecosystems, livelihoods and disasters: an integrated approach to disaster risk management. The International Union for Conservation of Nature (IUCN), Gland, Switzerland and Cambridge, UK, 58 pp. https://www.iucn.org/content/ecosystems-livelihoods-and-disasters-integrated-approach-disaster-risk-management. Accessed 12 June 2020

Sukhwani V, Mitra BK, Takasawa H, Ishibashi A, Shaw R, Yan W (2019) Urban-rural partnerships: a win-win approach to realize Regional CES (Regional Circular & Ecological Sphere), Compendium of Good Practices from Japan. IGES, Yokohama, Japan, p 27. https://www.iges.or.jp/en/pub/urban-rural-partnerships-win-win-approach/en

The World Bank (2019) Wastewater: from waste to resource: the case of Nagpur, India. http://documents.worldbank.org/curated/en/847531576610020104/Wastewater-From-Waste-to-Resource-The-Case-of-Nagpur-India

TLN Team (2019) Urban-rural partnership for resilient future of Nagpur city. Metropolitan Region mooted at VNIT. The Live Nagpur (TLN). https://thelivenagpur.com/2019/06/20/urban-rural-partnership-for-resilient-future-of-nagpur-city-metropolitan-region-mooted-at-vnit/. Accessed 15 June 2020

Triyanti A, Chu E (2018) A survey of governance approaches to ecosystem-based disaster risk reduction: current gaps and future directions. Int J Disast Risk Re 32:11–21. https://doi.org/10.1016/j.ijdrr.2017.11.005

UN DESA (2019) World urbanization prospects: the 2018 revision. United Nations, Department of Economic and Social Affairs, Population Division, New York, United States of America. Available online: https://population.un.org/wup/Publications/Files/WUP2018-Report.pdf

UN Water (2019) UN-water policy brief on climate change and water. https://www.unwater.org/publications/un-water-policy-brief-on-climate-change-and-water/

US AID (2017) Ecosystem-based adaptation and water security, evidence summary. https://pdf.usaid.gov/pdf_docs/PA00MXNJ.pdf

Voulvoulis N, Arpon KD, Giakoumis T (2017) The EU water framework directive: from great expectations to problems with implementation. Sci Total Environ 575:358–366. https://doi.org/10.1016/j.scitotenv.2016.09.228

World Bank (2018) Water scarce cities: thriving in a finite world. World Bank Group, Washington DC. http://documents.worldbank.org/curated/en/281071523547385102/. Accessed 12 June 2020

WWAP (2017) The United Nations World water development report 2017. Wastewater: the untapped resource. United Nations World Water Assessment Programme (WWAP), Paris, UNESCO. https://unesdoc.unesco.org/ark:/48223/pf0000247153. Accessed 12 June 2020

WWAP, UN-Water (2018) The United Nations World water development report 2018: nature-based solutions for water. United Nations World Water Assessment Programme (WWAP), Paris, UNESCO. https://unesdoc.unesco.org/ark:/48223/pf0000261424. Accessed 12 June 2020

Zhang X, Chen N, Sheng H, Ip C, Yang L, Chen Y, Sang Z, Tadesse T, Lim TPY, Rajabifard A, Bueti C, Zeng L, Wardlow B, Wang S, Tang S, Xiong Z, Li D, Niyogi D (2019) Urban drought challenge to 2030 sustainable development goals. Sci Total Environ 693:133536. https://doi.org/10.1016/j.scitotenv.2019.07.342

Chapter 18
Challenges in Decision-Making for Building Resilience to Climate Risks

Anamitra Anurag Danda, Nilanjan Ghosh, Jayanta Bandyopadhyay, and Sugata Hazra

Abstract Climate risk to something of value is not uniform even in the same geography despite exposure to the same climate-related events or trends or their impacts. Therefore, decision-making on climate resilience is challenged by multiple variables. In the wake of a devastating event or a creeping process, decision on whether to rebuild and persist in situ or relocate people may be necessary. This paper explores mechanisms to enhance resilience to climate risks through 'accommodate', 'protect' and 'strategic and managed retreat' approaches and analyses the outcomes using a three criteria framework. We argue that where non-diminishing socio-economic wellbeing can be assured, in-situ adaptation is the option, provided the cost of in-situ adaptation is lower than the value of the business-as-usual economy; the occurrence of an event or a creeping process is not expected to become more likely in future; and political risk of the climate resilience option is lower than the climate risk. The framework was tested using the case of the Indian Sundarbans. It revealed that decision-making on climate resilience options involves challenges beyond the immediate climate risk. Effective policy development requires due consideration of political risks in the absence of which retreat options are unlikely to be implemented

Keywords Climate resilience · Strategic and managed retreat · Political risk · Sundarbans · Adaptation

18.1 Introduction

There is significant overlap between disaster risk reduction (DRR) and climate change adaptation (adaptation). The IPCC Working Group II adopted the concept of risk of climate change impacts akin to the approach and practices of risk assessment in the disaster risk reduction community. While some of the terms in the DRR and

---

A. A. Danda · N. Ghosh · J. Bandyopadhyay (✉)
ORF Kolkata, Kolkata, India
e-mail: jayanta@iimcal.ac.in

S. Hazra
School of Oceanographic Studies, Jadavpur University, Kolkata, India

© The Author(s), under exclusive license to Springer Nature Singapore Pte Ltd. 2021
M. Mukherjee and R. Shaw (eds.), *Ecosystem-Based Disaster and Climate Resilience*, Disaster and Risk Research: GADRI Book Series,
https://doi.org/10.1007/978-981-16-4815-1_18

adaptation realms are the same, the definitions differ in breadth and focus. Since this chapter is written from the perspective of climate change, it follows the IPCC AR5 definitions.

Climate risk is understood as the potential for consequences of climate change impacts where something of value is at stake. Risk is often represented as probability of occurrence of hazardous events or *trends multiplied by the impacts if these events or trends occur*. Risk results from the interaction of vulnerability, exposure and hazard. Here hazard implies the potential occurrence of a natural or human-induced physical event or *trend* or physical impact that may cause loss of life, injury or other health impacts, as well as damage and loss to property, infrastructure, livelihoods, service provision, *ecosystems and environmental resources*. In adaptation literature, the term hazard usually refers to climate-related *physical events or trends or their physical impacts*. The presence of people, livelihoods, *species or ecosystems, environmental functions, services, and resources*, infrastructure, or economic, social or cultural assets in places and settings that could be adversely affected is defined as exposure (Field et al. 2014).

Climate-related hazards in the Sundarbans considered in this paper are tropical cyclones (physical event) and sea-level rise (physical impact of temperature rise). Vulnerability of the human systems emanates from development deficit in the region, high incidence of poverty and choice of livelihoods (predominantly rain-fed paddy agriculture followed by fishing and fishery). We take the iconic species and the apex predator—the tiger—to represent the natural systems of the Sundarbans. Though a sturdy swimmer, the vulnerability of the tiger emanates from loss of land due to sea-level rise (effect of physical impact of temperature rise) and other factors. Being densely populated, a population of over five million are exposed to tropical cyclones.[1] Within this population, about 0.1 million are currently exposed to sea-level rise and loss of land.

Tigers of the Sundarbans are the only ones found in a mangrove habitat. On the Indian side, it is estimated that there are only about ninety of them (Jhala et al. 2020), all exposed to sea-level rise and loss of land (see Fig. 18.1). Mukul et al. (2019), using maximum entropy (MaxEnt) modelling, current distribution data, land use/land cover and bioclimatic variables, modelled the likely future distribution of tiger in the Sundarbans in Bangladesh under RCP6.0 and RCP8.5 scenarios, projected that there will be a dramatic decline in suitable Bengal tiger habitat in the Bangladesh Sundarbans by 2070. The fate of tigers in Indian Sundarbans will possibly be no different, if not worse due to the fact that the Indian side receives less freshwater and sediment from upstream sources due to an easterly tilt in the Bengal Basin induced by neotectonic movements during 12–15 AD (Morgan and McIntire 1959). Personal experience of one of the authors during cyclone Aila of 2009 shows that natural areas and wildlife including tigers generally remain unaffected by cyclones. Tiger deaths have also not been reported after landfall of cyclone Amphan in May 2020. Therefore, in case of tigers, loss of land poses higher risk while for humans it is tropical cyclones[2] although negative consequences of loss of land for the affected in the absence of any state mechanism to ameliorate the situation are of the highest

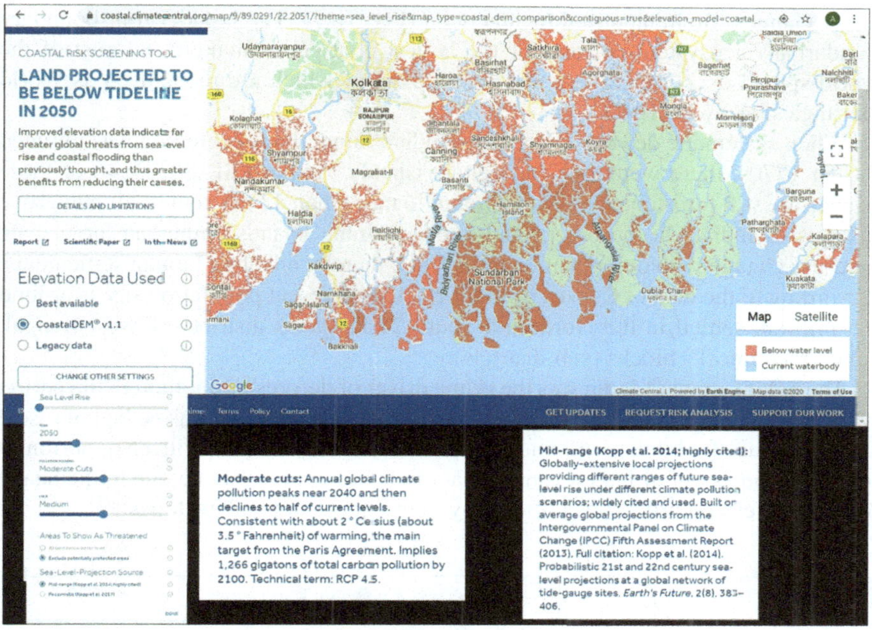

Fig. 18.1 Land projected to be below tide line in 2050. *Source* Climate Central (Open access)

order (Mortreux et al. 2018); about a quarter of a million could be faced with such a situation around the middle of the century (see Fig. 18.1).

The questions that arise are: (i) how climate risks can be reduced for humans and tigers, and (ii) how to decide on the reduction pathway to climate risks.

The rest of the article is organised as follows. Section 18.2, following our earlier publication (Danda et al. 2020), provides a brief description of the Sundarbans, initiation of human settlement by the East India Company and impacts of cyclones on the largest island during that period. Section 18.3 discusses the various options available to state actors and the affected humans for reduction of climate risks in the Indian Sundarbans. Section 18.4 presents a decision framework for policymakers and administrators to help rationalise between choices of in-situ risk reduction through 'manage or accommodate', 'resist or protect' and 'strategic retreat'. Section 18.5 embeds the illustrative case of the Indian Sundarbans in the context of the global practices and literature and provides the concluding remarks.

18.2 The Sundarbans

As previously described by Danda et al. (2020), the Indian Sundarbans ecoregion is the Indian component of the larger Sundarbans ecosystem and is an integral part of the lower Ganges–Brahmaputra–Meghna delta (Fig. 18.2). The broader transboundary Sundarbans ecosystem consists of the largest contiguous mangrove forest of the world spread over Bangladesh and India, parts of which are recognised as World Heritage sites by UNESCO.

When systematic deforestation of the Indian Sundarbans ecoregion was initiated during 1770s, there were 100 odd deltaic islands. As per FSI (2019), 46 of these islands are still forested, that is tantamount to more than 42 per cent of all mangrove habitats of India. On the other hand, the remaining islands witnessed large-scale land-use change with deforestation until 1943, thereby enabling human settlements. According to the data of Census of India 2011, the present population size is around 4.5 million, settled in the North and South 24 Parganas districts of West Bengal, subdivided into 19 blocks (sub-districts).

The colonisation of India was the prime driver of deforestation: though the period witnessed large-scale forest clearance, the process of settlement was no bed of roses! Creating space for human settlement provided the colonial rulers with an opportunity to earn incomes out of land revenues that often had to be foregone due to the heavy losses to lives and properties inflicted by the frequent tropical cyclones.

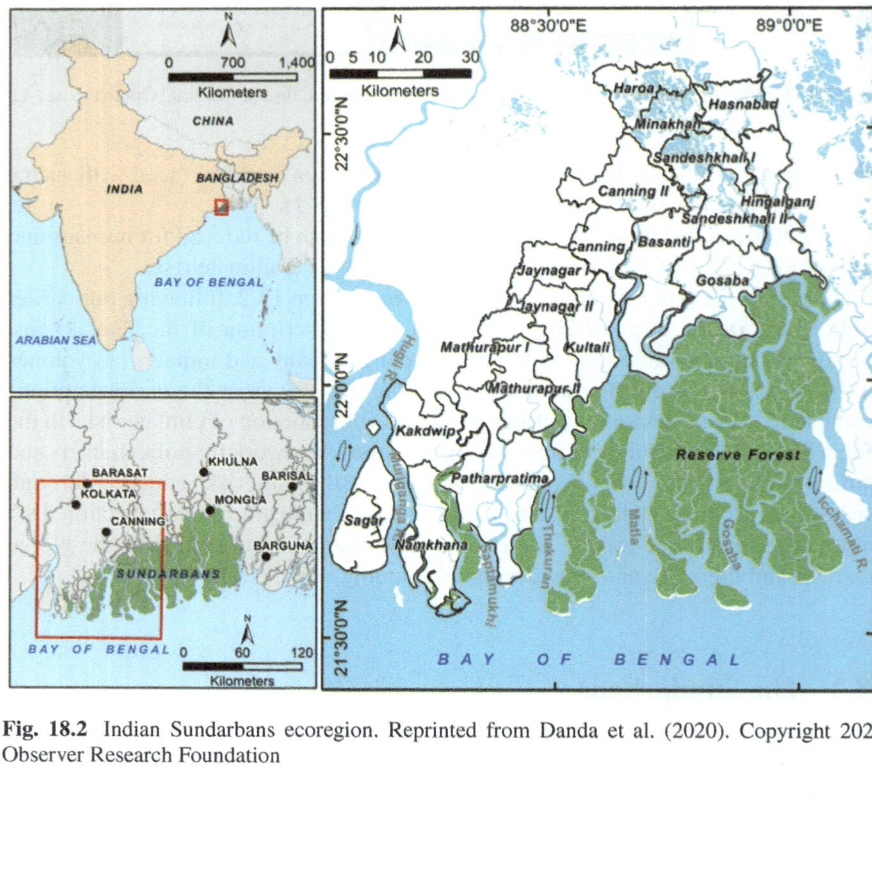

Fig. 18.2 Indian Sundarbans ecoregion. Reprinted from Danda et al. (2020). Copyright 2020 Observer Research Foundation

A historical tracking of such developments will clarify the above contention. The initial plan of creating settlements in the westernmost and the largest island, Sagar Island, in 1811, was faced with obstacles on various grounds of feasibility. Reclamation was re-initiated in 1819 with embankment construction, forest clearance, and freshwater tanks and settling the farm community. There was significant progress in at least five settlements when in 1833, a cyclone in May destroyed everything. Subsequent storms in June 1842, October 1848 and in June 1852 forced the settlers to retreat from the island. Yet, the colonial rulers created incentives for settlers to stay back, by removing all taxes until 1863. After 1863, just as fresh surveys were conducted for tax assessment, another cyclone on 5 October 1864 created havoc and left the system in doldrums. Another cyclone in November 1867 led to a manifold decline in the agricultural acreage to about 440 ha from ~5000 ha (Pargiter 1934). The series of natural disasters made the dispensation contemplate of making the island free of land taxes. Simultaneously, new land covering an area of ~8500 ha was granted during 1875–77 for six settlements. These were also made free of rent or cesses. Yet, by 1880, as noted by Ascoli (1921), forest regenerated organically in much of the unutilised land (Ascoli 1921).

The primary occupation of the settlers ever since the colonial times has been rain-fed paddy agriculture (Danda 2007) followed by fishing and fishery. The various indicators of human development are dismal in the region that is afflicted with a poverty level as high as 34%. (CSE 2012). The stresses go beyond being socio-economic and are also enforced by natural factors. Erosion and land-losses are prominent especially in the south-western parts of the archipelago. Hazra et al (2010) estimate net land-loss at an average rate of 550 ha per year between 2001 and 2008, leading to an overall loss of 4400 ha. Further, large-scale breaching of embankments results in salinity ingression and brackish water inundation, thereby making land unproductive (Ghosh et al. 2014).

These are all evidences of the hazards posed by the forces of global warming and climate change. As such, the south-western blocks of Sagar, Namkhana and Patharpratima are more vulnerable than the rest, as may be evidenced from the existence of villages where agriculture is rendered infeasible due to recurrent ingression of the brackish waters or due to land-losses because of submergence. With the land being physically non-existent due to submergence, the opportunity to sell such plots, liquidate them, and earn property incomes is lost. Without adequate provisions of financial and social capital, such households are incapacitated to migrate elsewhere, despite being occupationally and physically displaced.

18.3 Climate Resilience Options for the Indian Sundarbans

Developing climate resilience options involves two sets of tasks. The first set involves studies to determine which parts of the target area is likely to be at risk. The second set involves the development of options. The first set of tasks is relatively well-understood and tractable through the use of probabilistic numerical models. In West

Bengal, this task has been initiated from the perspective of disaster risk rather than climate risk but yet to be completed (Pal and Ghosh 2018). The second set of tasks involves agreeing on approaches and options based on a decision-making analysis or procedure. In this section, the discussion is on the climate resilience approaches that are typically categorised as 'manage/accommodate', 'resist/protect' and 'retreat' (Nicholls et al. 2007). Using the information from a recent study by us (Danda et al. 2020), we will examine the possible options under each of the categories.

18.4 Manage/Accommodate

The vulnerability of the human community of the region is evident from the fact that the cyclonic storms have generally resulted in losses in property (homes, cattle, crops, etc.) and lives. The high incidence of poverty leads to low coping capacity: most houses are temporary structures (*kutcha* houses) with mud walls and thatched (or tiled or sheet metal) roofs that can easily get destroyed by a strong wind or a storm.[3]

The solution therefore is palpable: constructing more permanent structures can increase resilience and *reduce* vulnerability. One leeway from this impasse can be sought in the Government of India launched housing programme for the rural poor, The Pradhan Mantri Awas Yojna-Gramin (PMAY-G). This is beneficial for many including the landless, provided the concerned state government provides land for the avowed purpose. Under the scheme, with each rural household being eligible for an assistance of Rs. 0.12 million,[4] a conservative estimate suggests that the cost of converting temporary structures to permanent ones in the Sundarbans is approximately Rs. 63 billion. This figure is obtained on the basis of the 2011 Census figure of 884,073 households across 19 sub-districts of the region considering 60% temporary structures.

Flooding and inundation due to storm surges increase the salinity of the agricultural fields and freshwater ponds. This not only leads to crop losses, but renders the land unsuitable for agriculture, thereby making the community bereft of its fundamental livelihood means of freshwater paddy agriculture. Hazra et al (2019) report that this has made prospects of farming poor in the region due to low irrigation intensity between 2.36 and 19.05%, and cropping intensity between 101.31 and 171.05%. Given the shrinking arable acreage, the person-cropland ratio stands at 14 persons per hectare of gross cropped area that is clear reflection of the *surplus labour* in agriculture (the point from where marginal returns to labour are zero) (ibid.). The failure of the in-situ economy to provide means of livelihoods has compelled around 18% of working adults to migrate away from their village to seek a living.[5]

No doubt that building resilience to climate risks is the biggest challenge. While salt-tolerant paddy varieties (*reduce sensitivity*) or grain seeds treated for salinity tolerance could have helped in agricultural seasons following a storm surge event, the yields of salt-tolerant paddy varieties are low and seed treatment for salinity tolerance is proprietary (SALICROP 2019). Climate resilience can be enhanced through imparting requisite knowledge of farming (*enhanced capacity*), and providing with salt-tolerant paddy seeds or treated seeds for a range of soil salinity conditions that compensates for the production loss due to land-loss with a higher yield than the traditionally used seeds.[6] Unfortunately, such a 'disruptive intervention' through technical innovation and breakthrough is yet to happen in the region.

The low elevation and gradient of the islands with human habitat make the problem further complex. The impacts of inundation due to storm surge and sea-level rise are even more prominent in these parts. If coexisting with the inundated land is to be seen as an adaptation option, the floor levels of the residential and other constructions need to be elevated to avoid flooding (*reduce sensitivity*). This has to be complemented with enhanced drainage and sanitation and drainage network linking the system with the already existing sink in the form of the inland wetlands (*improve capacity*).

18.5 Resist/Protect

Irrigation and Waterways Department of the Government of West Bengal maintains about 3221 km of embankment in the Indian Sundarbans to protect inhabited islands spread across ten of the 19 sub-districts. While these embankments make it possible for human habitations to be maintained on these low-lying islands, Ghosh and Boyd (2019) are of the opinion that these in their current form jeopardise long-term sustainability of both humans and non-humans in the socioecological system. Under the 'resist/protect' approach this entire length of embankment would have to be upgraded (*reduce sensitivity*) with dredged riverbed materials and/or by local earth from countryside, along with block pitching on riverside slope. About 183 km length of embankment has been upgraded or is underway at a cost of Rs. 13.89 billion according to 2010 prices.[7] If the entire length of embankment were to be upgraded, the total cost would be Rs. 245 billion as per 2010 prices.

18.6 Retreat

Under the 'retreat' option, certain areas in the Sundarbans would have to be relinquished pre-emptively (*reduce exposure*). These places are experiencing rapid erosion due to which maintaining embankments at these locations is increasingly becoming more challenging and expensive. These locations are prone to inundation (*vulnerability*) by surge height of less than two metres.[8] Villages across 28 Gram Panchayats (primary self-governance units) with a total population of about 0.6 million as per Census 2011 from eight of the sub-districts are faced with this situation (*exposure*). However, the retreat approach is not preferred either by the residents or the elected executives despite the fact that resilience of the communities is heavily jeopardised and gains made over years are often lost during the next high-intensity weather event (ibid.). There is also opposition to retreat among a section of academicians. Given the number of households at risk, the cost to governments is also high and potentially beyond the means of public finances. The challenge is daunting if retreat is perceived as movement of population akin to development-induced displacement and relocation, rather it should be viewed as organic movement of people in pursuit of individual and societal development objectives, facilitated by state and non-state actors.

In 2009, Bandyopadhyay had articulated such organic movement from extremely vulnerable areas of the Indian Sundarbans as managed retreat, a long-term approach to relocation (Danda et al. 2011). The Municipal Government of Tubigon, Philippines, views education as an integral part of voluntary relocation and offers scholarships to qualified school children from vulnerable islands to equip its beneficiaries with university degrees or vocational skills that would enable them to find jobs in the mainland and settle down (Jamero et al. 2019). A similar program, called migration with dignity, is being implemented in Kiribati. Providing scholarships for studying within the country or in New Zealand and Australia, the program aims to improve the skills of I-Kiribati and increase their job opportunities in foreign countries (Yamamoto and Estenban 2017). Recognising that 'retreat' is already occurring as an individual goal to reduce risk, Bangladesh is working towards increasing the capacity of cities to accommodate migrants (GIZ n.d). The project offers climate migrants (*reduced exposure*) income-generating opportunities (*increase capacity*) via training measures (vocational and trainings on entrepreneurship and basic final education), assistance to improve their access to public social services and finances (*increase capacity*), and it aims at increasing the cities' capacity to accommodate migrants through strengthening of administrative structures.

Implemented in this manner, retreat gets integrated into long-term development goals and addresses social, cultural, psychological and long-term economic consequences of the process for those who retreat (source population) as well as those who receive people who retreat (host population) (Siders et al. 2019).

Initiatives as in Bangladesh, complemented by retreat legislation to prevent sale, transfer, reconstruction and bequeathing of existing properties of residents who retreat with state facilitation will help gradually disestablish these properties (*reduce exposure*) when present owners are deceased or structures are damaged. At such time, government could progressively buy back the properties as they become available (Gibbs 2016). This way, retreat would not only deliver economic development but also deliver nature conservation if the vacated land is repurposed for mangrove regeneration with the possibility of deriving private benefits if buy-back option is not exercised.

18.7 Climate Resilience Decision-Making Framework

Here we present a decision framework to enhance resilience to climate risk for policymakers and administrators to help rationalise between choices of in-situ adaptation (manage/accommodate and resist/protect), and strategic retreat. Our argument is that in locations where non-diminishing socio-economic well-being can be ensured, in-situ adaptation is the option. This, however, needs to satisfy the conditions: (i) that the cost of in-situ adaptation is lower than the value of the business-as-usual economy at that location, (ii) that the occurrence or severity of an event or a creeping process is not expected to become more likely in future and (iii) that political risk of the climate resilience option(s) is lower than the climate risk.

In the previous section, we have seen that a minimum expenditure of Rs. 500 billion is required in the Indian Sundarbans for combining manage/accommodate and resist/protect approaches to make homes and inland livelihoods resilient to climate risks. Both of the approaches applied singularly have residual risks and non-diminishing socio-economic well-being cannot be ensured in-situ. Moreover, the Sundarbans' GDP is less than half that amount.[9]

Northern and eastern coasts of the Bay of Bengal are witnessing the largest sea-level changes in the North Indian Ocean, and over the past two decades (2000–2018), a significant rise (+ 0.86 per decade) in the frequency of very severe cyclonic storms (VSCS) has been observed during the post-monsoon season (October–December) in the Bay of Bengal (Krishnan et al. 2020). With continued global warming, the activity of VSCS over the NIO is projected to further increase during the twenty-first century (ibid.).

From the foregoing, it is apparent that i) the cost of in-situ adaptation is higher than the value of the business-as-usual economy, and (ii) the occurrence or severity of an event or a creeping process is expected to become more likely in future that may cause loss of life, injury, or other health impacts, as well as damage and loss to property, infrastructure, livelihoods, service provision, ecosystems and environmental resources. It might appear that retreat from at-risk locations is the obvious choice since retreat reduces exposure to coastal hazards by moving people, assets and human activities out of the exposed coastal area and eliminate residual risks. However, neither the population at risk nor the political actors are in favour of the option. This is primarily because retreat is perceived as relocation where the target population has no agency and little control over the choice of timing and destination. Obviously, political risk associated with such retreat is high and fraught with multiple governance challenges due to the fact that a large section of the population does not have clear title to property that raises equity concerns with regard to the distribution of and access to benefits of relocation. While governance challenges can be addressed through public participation, conflict resolution practices and enabling conditions, political risks are not altogether eliminated.

18.8 Concluding Remarks

Gibbs (2016) has noted that it is common in democracies for major policy decisions and initiatives to be approved by elected executives. It is therefore logical that elected executives will consider the political risk of policy initiatives and reject initiatives that carry unacceptable political risk. Political risk is understood as erosion of political capital. In the context of coastal inundation risk, he has observed that manage/accommodate and resist/protect are relatively less expensive for governments but also less effective. While retreat has the highest effectiveness, it is also the most expensive option for governments and carries the highest risk to political actors. Gibbs has argued that effective policy development, therefore, requires due consideration of political risk, in the absence of which retreat options are unlikely to be implemented.

Alternatively, governments could consider assisting migrants through policy reforms to enable relocation to the fast-growing economic regions in the country while ensuring that competition for resources and within labour markets, and pressure on frontline services do not overwhelm destination locations. Economically fast-growing destination areas may actually gain as populations and capital relocate and provide a new source of labour, capital and innovation. It would, therefore, be appropriate for governments to adopt an approach that integrates retreat into long-term development goals (Danda et al. 2011; Siders et al. 2019).

Experience in the Philippines and elsewhere (Jamero et al. 2019) shows that even if retreat is the most appropriate approach, it is time-consuming. Therefore, while the negotiated process is on, addressing some of the underlying socio-economic problems of the affected communities aids retreat since a community's ability to move is directly related to their level of social, economic and political capital. Encouraging organic movement of younger individuals yields positive outcomes not only for the beneficiary individual but for the retreat process itself since remittances from those who have relocated provide flexibility to their families and communities in terms of livelihood options and capital for investment, besides encouraging participation in the negotiated process. In eastern societies, it is usual for migrating children to bring over their parents to live with them once they are settled in their new homes. It could, therefore, one day lead to the natural depopulation of at-risk locations provided others are restricted from occupying the vacated vulnerable spaces. On the other hand, it is also possible that instead of relocating together with their families; individual migrants simply provide financial support through remittances, enhancing the capacity of such families to continue to reside at their locations. Nevertheless, strategies like improving accessibility to higher education and vocational training increase the possibility and accessibility of jobs beyond traditional livelihood activities, as well as the willingness of the communities to participate in relocation. This also ensures that political risk is not immediate; on the contrary, such strategies yield immediate gains in political capital.

Decision-making on building resilience to climate risk should be guided by the lowest indicator of improved livelihoods of the at-risk population. By being better off, it is understood that the population at risk have access to appropriate livelihood options and are gainfully employed/engaged in economic activities, have adequate and robust housing, have access to community assets and services either in situ, or at safer locations where they have undergone a calibrated process of social inclusion and have access to health care that is significantly better than what was available in at-risk locations.

Notes

1. Human population density in the Indian Sundarbans at over 1100 persons per sq. km is higher than the West Bengal state average which itself is the second-most densely populated state in India. https://censusindia.gov.in/2011-prov-results/data_files/india/Final_PPT_2011chapter7.pdf.
2. "The intensification rate during November, which accounts for highest number of intense cyclones in the north Indian Ocean, has registered a steep rise of 26% per hundred years, implying that a tropical depression forming in the Bay of Bengal during November has a high probability to reach to severe cyclone stage." Singh (2007).
3. According to Socio Economic and Caste Census 2011, in West Bengal, about 60% of the houses are of impermanent nature. This figure in the Sundarbans region cannot be lower than the state average. https://secc.gov.in/statewiseHousingDwellingWallTypeReport?reportType=Housing%20Dwelling%20Wall%20Type.
4. http://nadia.gov.in/DISHA/DISHASchemeGuidelines/PMAY/PMAY(Gramin)%20Draft%20Guideline.pdf.
5. DECCMA 2018, New Insights: Climate Change, Migration and Adaptation in the Indian Bengal Delta.
6. A study by Jadavpur University School of Oceanographic Studies observed conversion of agricultural land to settlements during 2001–2009. Agricultural land shrank from 2149.615 to 1691.246 km^2. https://ideas.repec.org/p/ess/wpaper/id10526.html.
7. Irrigation & Waterways Department, Government of West Bengal. Flood Management—Aila. https://wbiwd.gov.in/index.php/applications/aila.
8. Storm Surge of about 4–6 m above Astronomical Tide was predicted by the India Meteorological Department in case of cyclone Amphan that made landfall in the Sundarbans region on 20 May 2020.
9. In 2009, Sundarbans' GDP was estimated to be Rs. 134 billion. 'INR 6.7 billion … is equivalent to about 5% of the Sundarbans' gross domestic product (GDP) in 2009' p. 15. *Building Resilience for Sustainable Development of the Sundarbans.* 2014. Washington, DC: The International Bank for Reconstruction and Development/The World Bank.

References

Ascoli FD (1921) A revenue history of the sundarbans from 1870 to 1920. The Bengal Secretariat Book Depot, Calcutta

Centre for Science and Environment (2012) Living with changing climate: impact, vulnerability and adaptation challenges in Indian Sundarbans. Center for Science and Environment, New Delhi

Danda AA (2007) Surviving in the Sundarbans: threats and responses (Doctoral dissertation). Retrieved from https://ris.utwente.nl/ws/portalfiles/portal/14252552/surviving.pdf

Danda AA, Sriskanthan G, Ghosh A, Bandyopadhyay J, Hazra S (2011) Indian Sundarbans Delta: a vision. WWF-India, New Delhi

Danda AA, Ghosh N, Bandyopadhyay J, Hazra S (2020) Strategic and managed retreat as adaptation: addressing climate vulnerability in the Sundarbans. ORF Issue Brief No. 387. https://www.orfonline.org/research/strategic-and-managed-retreat-as-adaptation-addressing-climate-vulnerability-in-the-sundarbans/

Field CB et al (2014) Technical summary. In: Field CB et al (eds) Climate change 2014: impacts, adaptation, and vulnerability. Part A: global and sectoral aspects. Contribution of working group II to the fifth assessment report of the intergovernmental panel on climate change. Cambridge University Press, Cambridge, United Kingdom and New York, NY, USA, pp 35–94

Forest Survey of India (2019) Mangrove cover. In India State of Forest Report 2019. Accessed 20 July 20 2020. https://fsi.nic.in/isfr19/vol1/chapter3.pdf

Ghosh A, Boyd E (2019) Unlocking knowledge-policy action gaps in disaster-recovery-risk governance cycle: a governmentality approach. Int J Disaster Risk Reduction 39:101236. https://doi.org/10.1016/j.ijdrr.2019.101236

Ghosh T, Hajra R, Mukhopadhyay A (2014) Island erosion and afflicted population: crisis and policies to handle climate change. In: Filho WL (ed) International perspectives on climate change. Springer, Berlin, pp 217–225

Gibbs MT, Thebaud O, Lorenz D (2013) A risk model to describe the behaviours of actors in the houses falling into the sea problem. Ocean Coast Manag 80:73–79

Gibbs MT (2016) Why is coastal retreat so hard to implement? Understanding the political risk of coastal adaptation pathways. Ocean Coast Manag 130:107–114

GIZ (nd) Urban management of internal migration due to climate change (UMIMCC)/Urban Management of Migration and Livelihoods (UMML). Accessed 10 June 2020. https://www.giz.de/en/worldwide/31936.html

Hazra S, Samanta K, Mukhopadhyay A, Akhand A (2010) Temporal change detection (2001–2008) study of Sundarban. Jadavpur University, Kolkata, School of Oceanographic Studies

Hazra S, Bhadra T, Ray SPS (2019) Sustainable water resource management in the Sundarban biosphere reserve, India. In: Ray SPS (ed) Ground water development—issues and sustainable solutions. Springer, Singapore

Jamero ML et al (2019) In-situ adaptation against climate change can enable relocation of impoverished small islands. Mar Policy 108:103614. https://doi.org/10.1016/j.marpol.2019.103614

Jhala YV, Qureshi Q, Nayak AK (eds) (2020) Status of tigers, copredators and prey in India, 2018. National Tiger Conservation Authority, Government of India, New Delhi, and Wildlife Institute of India, Dehradun

Krishnan R et al (eds) (2020) Assessment of climate change over the Indian region. Springer, Singapore

Mills M, Mutafogula K, Adams VM, Archibald C, Bell J, Leon JX (2016) Perceived and projected flood risk and adaptation in coastal Southeast Queensland, Australia. Clim Change 136:523–537

Morgan JP, McIntyre WG (1959) Quaternary geology of the Bengal Basin, East Pakistan and India. Bull Geol Soc 70:319–342

Mortreux C, de Campos RS, Adger NW, Ghosh T, Das S, Adams H, Hazra S (2018) Political economy of planned relocation: a model of action and inaction in government responses. Glob Environ Chang 50:123–132

Mukul SA et al (2019) Combined effects of climate change and sea-level rise project dramatic habitat loss of the globally endangered Bengal tiger in the Bangladesh Sundarbans. Sci Total Environ 663:830–840

Nicholls RJ et al (2007) Coastal systems and low-lying areas. In: Parry ML et al (eds) Climate change 2007: impacts, adaptation and vulnerability, contribution of working group II to the fourth

assessment report of the intergovernmental panel on climate change. Cambridge University Press, Cambridge, UK. https://www.ipcc.ch/site/assets/uploads/2018/02/ar4-wg2-chapter6-1.pdf
Pal I, Ghosh T (2018) Risk governance measures and actions in Sundarbans Delta (India): a holistic analysis of post-disaster situations of cyclone Aila. In: Pal I, Shaw R (eds) Disaster risk governance in India and cross cutting issues. Disaster risk reduction (Methods, approaches and practices). Springer, Singapore. https://doi.org/10.1007/978-981-10-3310-0_12
Pargiter FE (1934) Revenue history of the Sundarbans from 1765 to 1870. Bengal Government Press, Alipore
Siders AR, Hino M, Mach KJ (2019) The case for strategic and managed climate retreat. Science 365(6455):761–763
SALICROP (2019) White paper: SALICROP—Maharashtra India rice field trials 2018–2019. https://www.salicrop.com/
Singh OP (2007) Long-term trends in the frequency of severe cyclones of Bay of Bengal: Observations and simulations. Mausam 58(1):59–66
Yamamoto L, Esteban M (2017) Migration as an adaptation strategy for atoll island states. Int Migr 55(2):144–158

Anamitra Anurag Danda a social anthropologist, is a Visiting Senior Fellow at ORF Kolkata, India. He has particular interest in the practice of sustainable development. His experience of over 20+ years spans collective action and institution dynamics, adaptation to climate change, and nature conservation.

Nilanjan Ghosh is an economist, and Director of ORF Kolkata. He is among the leading ecological economists and development analysts of South Asia. He is a pioneer in the application of neoclassical and heterodox economics in water governance.

Jayanta Bandyopadhyay is Visiting Distinguished Fellow at ORF Kolkata. He is a researcher and author on science and the natural environment, and has worked in ICIMOD (Kathmandu), IAE (Geneva), and IIMs at Bangalore and Kolkata previously.

Sugata Hazra is a geologist and Professor at the School of Oceanographic Studies, Jadavpur University, Kolkata, India. Over last two decades, he has done pioneering research on Sundarbans, Climate Change and Biophysical impact. He has lead international and national research projects on Sundarbans and coastal oceans.

Chapter 19
A "Greener" Alternative: The Sri Lankan Experience of Eco-DRR

Deepthi Wickramasinghe

Abstract With increase in number, frequency and intensity, and its link to climate change, natural hazards are becoming complicated to manage. As a solution, ecosystem-based approach has gained much attention in many parts of the world. Sri Lanka is no exception. Being a tropical island nation Sri Lanka is vulnerable to a variety of disasters with many socioeconomic and environmental impacts. This chapter provides context for ecosystem-based disaster risk reduction (Eco-DRR) approaches and experiences in Sri Lanka. It also highlights how fully functioning natural habitats increase disaster resilience especially to cope up with extreme weather events. Examples from good practices and ancient wisdom are discussed with a special reference to traditional home gardens and protected area management. Recent case studies on wetland management and coastal disaster resilience through mangrove restoration are stressed. How community resilience, especially in rural areas, is enhanced through ecosystem services is explored with examples. Finally, this chapter addresses the issues and challenges in mainstreaming Eco-DRR into national policy and action agendas to offer a "greener alternative" for better disaster management.

Keywords Ecosystem-based approach · Disasters · Sri Lanka

19.1 Introduction

No place on earth is free from natural hazards. Yet, due to increased exposure and vulnerability as well as reduced reliance on communities, hazards can be turned into disasters (UNISDR 2009). Impacts of disasters are growing worldwide (UNISDR 2013). The global annual average death toll of natural disasters is close to 90,000 with various impacts on 160 million people (IUCN French Committee 2019). During the period of 2005–2014, 20% of the world's population was affected by disasters

D. Wickramasinghe (✉)
Department of Zoology and Environmental Science, University of Colombo, Colombo, Sri Lanka
e-mail: deepthi@zoology.cmb.ac.lk

© The Author(s), under exclusive license to Springer Nature Singapore Pte Ltd. 2021

M. Mukherjee and R. Shaw (eds.), *Ecosystem-Based Disaster and Climate Resilience*, Disaster and Risk Research: GADRI Book Series,
https://doi.org/10.1007/978-981-16-4815-1_19

(UNISDR 2015). From the early ages, communities have lived amidst natural disasters with significant impacts on their lives. To protect people and property from exposure to hazards, use of structural measures was the common solution. With time, negative impacts of engineering solutions were experienced that included high expenses in designing, building, and maintaining (Lacambra et al. 2008) which were not viable solutions, especially for the developing nations. World leaders who gathered in Paris at UNCOP21 in 2015 highlighted the importance and benefits of considering ecosystem-based approach to disaster risk reduction, which encourages using natural ecosystems to reduce disaster risk (SCBD 2015). In this context, the use of "free" services provided by ecosystems gained attention worldwide as a viable option to reduce disaster impacts (Uy and Shaw 2012).

An ecosystem is any natural entity with communities of animals, plants, and microbes (living components) distributed in large areas in the land or in the sea. The living component interacts with each other as well as with the non-living components in the area including water, air, and soil. Ecosystems are known to perform significant services to humankind (Costanza and Daly 1992). Among such services, ecosystems and their ability to help reduce disaster impacts have gained much attention in recent times. Eco-DRR refers to managing ecosystems to ensure their healthy and proper functioning, which will help reduce disaster risk and enhance community resilience (Nelson et al. 2009; Estrella and Saalismaa 2013). On the other hand, ecosystem services are not infinite and taken for granted. Due to various natural and anthropogenic impacts, ecosystems' ability to perform services could be weakened and deteriorate. Hence, any degradation in ecosystems can lead to a loss of effectiveness in their ability to mitigate the impacts of natural disasters.

This chapter provides detailed information on ecosystem-based disaster risk reduction in the Sri Lankan context which is organized as follows: Firstly, an overview of the natural hazards of Sri Lanka is presented. In the next part, different traditional knowledge and good practices that support healthy ecosystems and environmental conservation are discussed. The third part is dedicated to recent development related to Eco-DRR practices. Challenges related to the successful implementation of Eco-DRR approach are included as the fourth part. Finally, concluding remarks are included.

19.2 An Overview of Natural Disasters in Sri Lanka

Sri Lanka is an island country situated in the Indian Ocean between 5′ 55° and 9′ 51° N and 79′ 41° and 81′ 54° E. A land area of 65,000 km^2 is demarcated by its coastline which is of 1585 km.

Sri Lanka has been identified as a vulnerable country to various types of natural disasters. Given the fact that the government's strategy to elevate the socioeconomic status of the country is of top priority, (with many activities that could contribute to disaster vulnerability) disaster management has become even more important. On top of this, the impacts of climate change make the island more susceptible to natural disasters.

The Indian Ocean Tsunami wave which hit the island on December 26th, 2004, was a unique event: It was the most devastating natural hazard that affected Sri Lanka in recent history. Furthermore, it was a turning point that redefined "disasters and their management" in the country. The wave with an average height of 5 m along the southwest coast of Sri Lanka, among many other parts (Tomita et al. 2006) was the most frightening experience to coastal residents. With more than 30,000 deaths and nearly 4000 gone missing, tsunami disaster left a dreadful memory in many people (Wijetunge 2006). The socioeconomic impacts and disturbances to daily lives were enormous (Rossetto et al. 2007). The government nor the society were equipped with adequate resources and technical ability to manage such sudden and large-scale disasters. It called for overseas assistance, in monetary terms and otherwise, thereby adding its first-ever international influence/association to disaster management in the country in a significant way. However, as every cloud could have a silver lining, there was a positive that came out in the aftermath of this terrible tragedy. The Disaster Management Centre (DMC) was established as a new mechanism to tackle and deal with disasters. The main aim of the DMC is to minimize disaster risk with a set of policies to mitigate the impacts of disasters.

Apart from the Tsunami, Sri Lanka is situated in the Intertropical Convergence Zone (ITCZ) frequently experiences hydrometeorological disasters, floods in particular (DMC 2005).

Floods are one of the major events in the island which pose threats on the society with many deaths and thousands of people affected annually (Karunaratne 2001; Releifweb 2018). Droughts of serious nature occur in about 3–4 years and severe droughts of national significance once in about 10 years (De Silva and Kawasaki 2018). Yet, the impacts of droughts are felt more frequently in the community due to their impacts on the hydro-electricity generation on which a major portion of country's electricity supply depends on. Nevertheless, during each drought season, many farmers in the dry zone suffer from lack of water supply to sustain and maintain their paddy fields which is the main agricultural crop in the country.

Recently, landslides have become a recurring devastation with the increase in rainfall in some sensitive areas, especially in the central highlands. Landslides are mostly connected to vulnerable landscape characteristics such as excessive rainfall, loss of forest cover, and unsustainable human activities (Jayawardane 2006). The frequency of such disasters together with short- and long-term socioeconomic and environmental impacts raise the need for strengthened disaster management planning and implementation at the local level (Jayasuriya et al. 2005).

19.3 Sri Lankan Experience in Indigenous Knowledge, Ancient Wisdom, and Good Practices Related to Disaster Risk Reduction

This part is dedicated to present how Sri Lanka as a country having doing well with using and managing natural infrastructures, the ecosystems to reduce disasters and increase community resilience.

Sri Lanka is a country which is geared with natural resources as well as good practices to get benefits from nature. From ancient times, the natural as well as social capitals in Sri Lanka have been interconnected and interdependent, as such, the human culture thrived in harmony with nature (Gunatilleke 2015). In this context, indigenous knowledge and traditional practices, among others, helped to build a disaster-resilient environment.

19.3.1 Indigenous Knowledge, Wisdom, and Good Practices

Sri Lanka has a long history of environmental conservation and sustainable use of natural resources which is reflected in the livelihood of people. One good example present in agricultural practices. Paddy cultivation has been the single most large-scale agricultural activity since the ancient times. To support year-long cultivation, from ancient times a wide array of man-made water tanks or village-level reservoirs have been existing. A tank cascade system, which is a unique characteristic of dry zone irrigation system, comprises of water tanks of different sizes which are interconnected for the water to be recycled and reused. The practices aimed at not allowing a single drop of water goes to the sea without being used. Construction of such cascade systems by ancient Sri Lankan kings goes back to more than 2000 years, and some of them are still in operation (Abeywardena et al. 2018). Even though the tanks were man-made, with many centuries of existence, they have now turned into natural networks of wetlands hosting a wide array of biodiversity. For many months of the year, these tanks and surroundings host a plethora of biodiversity including plants, fish, and birds.

The cascade system serves in two main ways to reduce disaster risk: They act as a network of wetlands that are interconnected to store rainwater (Fig. 19.1) when the precipitation is abundant. Conversely, during the dry season, they function as a water storage facility to feed the irrigation canals and supply water to the paddy fields (Dharmasena 2004). Figure 19.2 shows the Tekkam Barrage in the Malwathu River, Mannar, in the dry zone, which is believed to be built by King Dhatusena (459–477 A.D.). Apart from sustaining crops, the tank system provides water for daily needs of communities such as bathing and washing. Another salient feature of the cascade system is the support to retain soil moisture during the dry periods, enabling the crops to grow and feeding domestic wells of the area. In addition, tanks, due to their

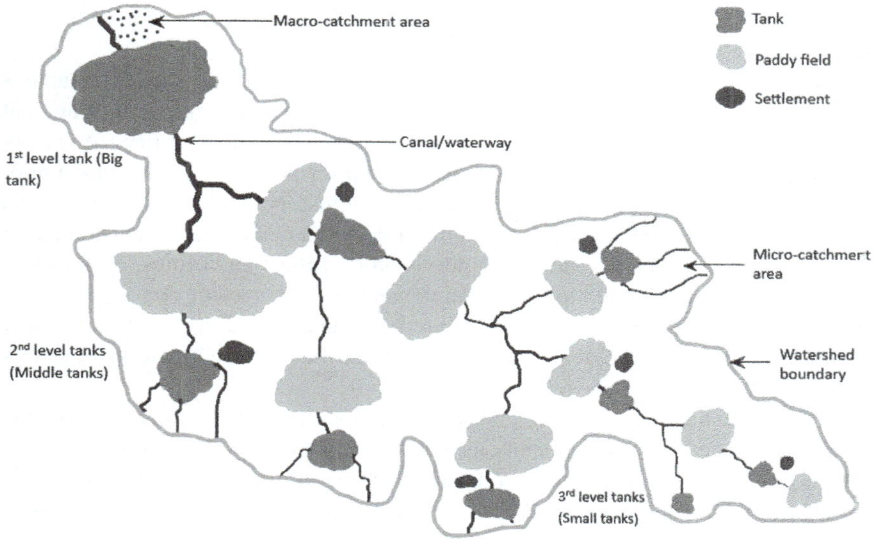

Fig. 19.1 Schematic diagram (not to scale) of tank cascade system depicting different levels of water tanks connected to each other and with paddy fields

Fig. 19.2 Tekkam Barrage at Malwathu River, Mannar (*Photo credit* Eng. Badra Kamaladasa)

open waters, nourish the atmosphere with moisture which helps to reduce elevated temperature in the area.

One of the key factors that made the cascade system successful and sustainable is the attention paid to its management which was a priority from the time of ancient kings (Bandara 1995). Records reveal the involvement of different tiers of society in management actions. According to a few stone inscriptions from ancient Anuradhapura Kingdom and key chronicles of Sri Lankan history, Maha Wansa and Chula Wansa, a well-planned hierarchical system for decision making and management of activities, were apparent. Under the supervision of village authorities, each farmer had a role to play in operation and maintenance of the system due to which its maintenance and sustainability were ensured (Vidanage et al 2005). Thus, a fully fledged network of tank cascade systems has contributed to reducing disaster risk while enhancing community resilience.

On the other hand, conservation ethics, even though not recognized by its name, has been practiced in ancient society as a norm, which enhanced ecosystem health. Village-level communities paid strong attention to minimize pollution, maintain the health of habitats and ecosystems, and were always ready to take actions to restore degraded habitat using indigenous knowledge. However, it is also evident that increased negligence of the ancient eco-friendly agricultural system may have been due to the modern practices of commercial agriculture.

19.3.2 Traditional Home Gardens as Good Practices

A wide array of home gardens acting as a united natural habitat is common in rural Sri Lanka. A home garden can be described as a combination of multi-species and multilayered collection of vegetation including usually food, medicinal, timber, and ornamental plants (Peyre et al 2006; Kumar 2020). Sri Lanka has a long history of community involvement in conservation of forests and other natural vegetation both in small and large scales, the later begins at domestic level as traditional home gardens (Ariyadasa 2002). In Sri Lanka, home gardens offer food (fruits, nuts, leaves), medicines, flowers, firewood as well as shelter, protection, natural beauty and helps develop tranquility in mind (Pushpakumara 2000). Some trees are planted which are selected by residents, whereas other vegetation would naturally colonize according to the characteristics of the eco-climatic zone (Mahendrarajah 2003). In the rural settings in Sri Lanka, most of the home gardens are continuous other than in the areas covered by infrastructures and croplands. Although maintenance is not routinely practiced in many home gardens, the very nature of being "wild" is beneficial in relation to ecosystem services. Areas where home gardens are healthy and intact, help maintaining the water cycle, prevention of droughts and landslides. Especially in the rural areas, traditional home gardens provide ecosystem services that include protection by floods (Calvet-Mir et al. 2012). Conversely, in the Sri Lankan settings, a wide array of traditional home gardens has been serving to minimize disaster risk while enhancing community resilience (Fig. 19.3). In addition, home garden habitats

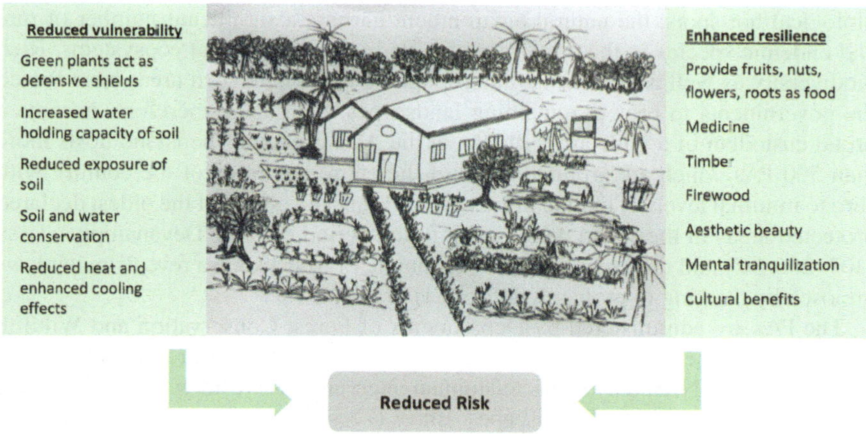

Fig. 19.3 Traditional home garden contribution to reduce disaster risk (*Drawing credit* Merrica Fernando)

act as effective adaptive measure in combating impacts of climate change, which to is linked to increased disaster impacts (Weerahewa et al 2012).

19.3.3 The Role of Protected Areas as a Good Practice in Eco-DRR

Conservation of ecosystems relies principally on the approach of protected areas. With clearly defined boundaries and set of laws and regulation for effective management, it is not unfair to argue that probably the highest level of ecosystem services is provided by protected areas (PAs). PAs are declared, owned, and governed by the central or local government of many countries. The concept of PAs could have been evolved purely to conserve nature and biodiversity. Yet, while protecting nature, these landscapes or seascapes inevitably provide many ecosystem services including the potential to mitigate disaster impacts.

A protected area (PA) can be defined as a "geographical location, identified, dedicated, and managed, to accomplish long-term conservation of nature together with associated ecosystem services and cultural values which could be achieved through regulations and legal measures" (IUCN 2008). Many sustainable development goals (SDGs) of the UN emphasize the protection of nature and natural resources. For instance, SDG target 15 highlights the need to "protect, restore, and promote sustainable use of terrestrial ecosystems, sustainably manage forests, combat desertification, and halt and reverse land degradation and halt biodiversity loss".

Sri Lanka has a long history of protecting forests and natural vegetation, which is tightly linked to the livelihood of communities. Being identified as the world's 34th biological hot spots, the natural environment harbors a significant number of rare and endemic species in the country with a wide array of natural ecosystems. Rich biodiversity as well as increasing anthropogenic impacts in such areas have forced the governments to protect remaining landscapes. Sri Lanka, therefore, became a proud custodian of a wide array of PAs in the Asian region. The island hosts more than 500 PAs which are around 26.5% of the entire land area of the country with zero to minimal levels of human influence. Sri Lanka has some of the oldest declared protected areas in the world dating back to the period of King Devanampiya Tissa (307–266 BC) and the great Sinhalese chronicle of Mahawamsa reveals many laws imposed by the king to protect wildlife (CHM 2017).

The PAs are administered by Departments of Forest Conservation and Wildlife Conservation where categories of parks range from strict nature reserves (SNR) to Sanctuaries. In SNRs, only restricted human interventions are allowed, and areas are maintained as pure natural landscapes. Entry is restricted only to researchers who are allowed to carry out studies under the supervision of the Department of Wildlife Conservation staff with prior permission. Another category, the national parks are open for visitors for restricted tourist activities, which are often for viewing purposes and under the supervision of authorities. Sanctuaries are the least regulated PAs which allow protection of wildlife in privately owned lands. In sanctuaries, protection of habitats and human activities occurs simultaneously.

The network of PAs provides multiple benefits: significant protection to biodiversity and ecosystems as they are regulated by strict laws. Therefore, human activities such as misuse of land resources, landuse changes, habitat destruction, and fragmentation are minimized.

Generally, all ecosystem services-provisioning services, regulating services, cultural services, and supporting services are provided by PAs (Millennium Ecosystem Assessment 2005). Protected areas are geared with suitable natural infrastructure to reduce disaster vulnerability (Dolidon et al. 2009). Firstly, healthy vegetation provides support to soil stability and integrity. The root systems of trees are beneficial in maintaining soil integrity and stability and help reinforcing soil layers and to anchor the bedrock. Deep-rooted trees possess the ability to help stabilizing and firming shallow soil layers, as well as improving drainage, which can decrease the likelihood of shallow, rapidly moving landslides (Sidle et al. 2006). Vegetation, especially closely occurring tree lines can act as a natural defense structure that facilitates attenuating and blocking debris flows and rockfalls, which reduces the chance of sliding earth segments. In general, thick vegetation in forests helps retain soil moisture by interception, evaporation and transpiration, the primary mechanisms in reducing drought risks.

In this context, many PAs in Sri Lanka serve as areas of "disaster-resilient heavens." No moist tropical rainforests in Sri Lanka have ever experienced droughts. In fact, some PAs such as Flood plains National Park, situated in the lower reaches of Sri Lanka's longest river, Mahaweli, serves as an area to store floodwaters, thereby reducing flood risks. Similarly, due to the ability to store water and maintain soil moisture, the flood plain helps its surroundings to reduce the impacts of droughts. Depending on the area, many PAs accommodate a wide array of water resources that include rivers, streams, and wetlands and thus serve to mitigate impacts of drought, and also reduce flood risks.

19.4 Ecosystem Services and Disaster Risk Reduction: Recent Case Studies

The Eco-DRR concept has been widely accepted as an alternative strategy to support disaster management (Millennium Ecosystem Assessment 2005; Yanagisawa et al 2009). Similarly, in Sri Lanka, there have been several interventions to make natural ecosystems healthier, to reap its benefits. In the recent past, increasing evidence has emerged, to support the notion of nature and ecosystem-based disaster risk reduction and increased resilience. As in many other countries, government agencies have taken steps to conserve and manage ecosystems, to get maximum support to act as defensive barriers in combating disaster.

19.4.1 Improving Wetland Network and Infrastructure to Reduce Urban Flood Risk

Colombo, the commercial capital of Sri Lanka underwent a significant change in landuse, landcover patterns in the past couple of decades and is suffering by losing natural habitats and their services (Saparamadu et al. 2018). Specifically, the loss of low-lying areas including wetlands and their connectivity has contributed to the increase of floods. A study (Wickramasinghe et al. 2018) revealed that the lower Kelani River basin in Colombo has experienced floods more frequently in the recent four decades due to loss of wetlands. Similarly, Dammalage and Jayasinghe (2019) have reported that loss of natural habitats and rise in built-up areas have contributed to increased flood events in Colombo. Furthermore, an increase in the damages of floods has been shown to inversely proportionate to the decrease in community resilience (Hettiarachchi et al. 2014).

Colombo is increasingly experiencing flash floods and particularly the ones in 2010 and 2017 have been extremely devastating. The Government of Sri Lanka, following many investigations, has identified the role played by urban wetlands in reducing disaster risk. One major issue with the increased intensity and frequency of urban floods is the poor infrastructure to handle flood water and direct water flows to waterways and rivers. Inundation of the cities leads to many damages including socioeconomic disruptions. The government of Sri Lanka identified the need to develop low-lying areas and wetlands to store floodwater and reduce impacts. The regulator, Sri Lanka Land Reclamation and Development Authority, under its program on conservation management of urban wetlands took steps to restore degraded wetlands and establish connectivity among these habitats in selected parts of Colombo (Releifweb 2018). Under this program, many natural wetlands were rehabilitated with an artificial canal system, to connect water bodies while retention capacities of wetlands were enhanced and managed. As a result of programs to develop and conserve wetlands, the city of Colombo has been named one of the worlds' eighteen "wetland cities" in 2018, by the Ramsar Convention Secretariat (WWF 2020).

19.4.2 Mangroves to Protect the Coast from Hazards

Sri Lanka is surrounded by a 1620 km long coastline (Samaranayake 2005). The coastal zone provides economic resources contributing to 40% of national GDP. It harbors a range of ecosystems such as coral reefs, mangrove forests, seagrass beds, salt marshes, sand dunes, estuaries, lagoons, and coastal wetlands (Samaranayake 2005). However, the coastal zone is prone to many natural hazards including cyclones, storm surges, coastal erosion, and sea-level rise. The 2004 tsunami event had the deadliest impacts. Some hazards are partly influenced by anthropogenic activities such as sand mining, coral mining, urbanization, overexploitation, and destruction of habitats (Shanmugaratnam 2005; Garcin et al. 2008; Ranasinghe 2012).

Perhaps the most common and effective natural barrier to protect the shoreline from any coastal disasters is the mangroves (Quartel et al 2007). Accordingly, one study indicates the impact of coastal vegetation patches along the estuary of the Madu River in the southern coast of Sri Lanka which was severely hit by the Indian Ocean Tsunami in 2004. According to Sudmeier-Rieux and Ash (2009), coastal plants acted as a shield to lessen the tsunami wave energy, thereby protecting houses in the inland coastal area. The effect was significant as houses and other buildings that were located in the area where there was no such coastal vegetation were heavily damaged. With their unique roots system which appears as a "live shield" to reduce the impacts of wave energy, mangroves have gained attention worldwide to protect the coast and the inland landscape.

Sri Lanka could be the first nation in the world to protect island-wide mangrove belt with a project of long-term multiple impacts (Guardian 2015). As mangroves are increasingly being destroyed in the country, the "Seacology", a nonprofit environmental conservation organization, organized an island-wide mangrove protection, replantation, and conservation programme.

The salient feature of this project is the involvement of women: more than 7500 underprivileged females, who used to remove mangrove trees or parts for different domestic uses, were given microloans to start small-scale businesses as income-generating activities. The recipients were expected to stop cutting mangroves and conserve these habitats near their homes. Altogether, nearly 14,500 ha of mangrove patches have been demarcated in more than 50 coastal lagoons and river basins for protection.

Women and youth groups were trained to conserve mangroves with hands-on training in mangrove propagation, reforestation, and conservation. While proving protection against coastal natural hazards, the co-benefits that the project result includes sustainable livelihood training, enhanced economic stability, and better nutrition and health. Due to the multiple benefits generated, the mangrove conservation project "Seacology" won the prestigious UNFCCC Climate action award (UNFCC 2018).

19.5 Enhanced Community Resilience Through Ecosystem Services

A woman carrying firewood, a boy climbing trees seeking fruits and nuts, and a man directing water from a lake to his paddy fields are common sights in rural areas in the country. This depicts how nature and ecosystems are tightly linked to communities and their livelihoods. Especially, in Sri Lanka, living in a village enjoys many services that are provided free of charge by nature. One study carried out in the Knuckles region in the central mountains of the island revealed that more than 60% of the income of communities are generated by the nearby forests (Gunatilake et al. 1993). Many communities who live in the periphery of Ritigala dry zone forests, collect indigenous fruits, medicinal plants and bee honey from the forest, which accounts for a significant fraction of their income (Wickramasinghe et al. 1996). Collection of non-timber forest products by the villagers in the periphery to Sinharaja MAB reserve has been one of their main ways of supplying for their day-to-day requirements. Many wetlands such as rivers help in livelihood for residents. For instance, a study carried out in Madu River Ramsar site revealed that nearly one-third of the principal income of these communities depends on the ecosystem, where tourism is the key sector (Manusinghe 2009).

Even though the link between rural communities and ecosystems is often highlighted in the literature, some urban habitats are not different when delivering ecosystem services. As mentioned elsewhere, Colombo is a city with a wide array of interconnected wetlands. Under the wetland management strategy implemented by the Sri Lanka Land Reclamation and Development Authority, a Rapid Assessment of Wetland Ecosystem Services (RAWES) was carried out at 60 different wetland sites. Accordingly, 30 different wetland ecosystem services operating from local to a larger scale were revealed. The largest number of services fall into the regulatory category (thirteen) followed by seven cultural and six supporting services (McInnes and Everard 2017).

Similarly, Muthurajawela, the largest wetland in the Western province, has gained much attention by the scientists and policymakers due to the wide array of services the ecosystem provided to the community. A study on valuing ecosystem services was carried out by Wattage and Mardle (2005) reports the most valued service was its contribution to flood attenuation. However, according to them, the total value of ecosystem services, which includes industrial and domestic wastewater treatment, inland fisheries, and freshwater supply, could be as large as 726.49 million rupees per year, indicating how ecosystems contribute to community wellbeing.

19.6 Toward an Effective Eco-DRR Approach: Proper Care and Management for the "Greener Alternative"

Ecosystem approach to disaster risk reduction is still in its infant stages in Sri Lanka due to many reasons. The island is yet to achieve a significant positive outcome with ecosystem-based disaster risk reduction. Many challenges to be included in policy and action agendas which are common to developing countries need to be overcome in order to get there (Table 19.1). In addition, some solutions to circumvent difficulties in the long run and in appropriate special and geographic scales are also included.

Table 19.1 Challenges in achieving Eco-DRR for developing countries and possible solutions

	Challenges	Solutions
Socioeconomic	Poor awareness about benefits of ecosystems and the services they offer as many are invisible Non-existence of incentives for the communities to conserve nature	Programs to enhance community-based conservation of ecosystems Raising awareness on multiple benefits that ecosystems offer and their link to livelihoods
	Inadequate community participation	Launch programs to motivate communities
Political	Inadequate political support and willpower to conserve ecosystems	Awareness raising with evidences
	Lack of attention paid to conservation and restoration of ecosystems in developmental activities	Including environmental conservation and development in the same domain Strengthening EIA processes
	Developmental decision making without a clear scientific basis	Encourage more interdisciplinary scientific studies
Institutional	Insufficient institutional capacity and technical knowhow to connect and include DRR and natural resources management	Training and capacity building
	Responsibility of ecosystem management lie with different institutes	Integrated planning and implementation
	Poor monitoring of impacts of development on ecosystem health	Regular monitoring, follow-up activities, documentation
	Poor enforcement of law in managing ecosystems	Strengthening policy implementation and appropriate legal action

19.7 Concluding Remarks

The essential element in achieving efficient Eco-DRR strategies is improved care for "ecosystems" which act as a "greener" alternative for engineered defense solutions. It becomes "greener" as the ecosystems provide more than a "green" environment by providing many benefits. The concept is grounded in the understanding that ecosystems, when functioning fully, could help the communities in a variety of ways: They range from protecting communities and their properties from exposure to hazards, lessening disaster impacts and strengthening coping up capacities of the society. Sri Lanka, as a nation has centuries-old traditions and good practices of caring for natural ecosystems and managing environment sustainably which have indirectly help reducing disaster risk. The case studies reveal that by conservation and sustainable management alone could help to achieve many services that the ecosystem offers to reduce disaster risk. Even though the island is facing developmental and other pressures resulting ecosystem degradation, it is now evident that the country is making a case for the implementation of Eco-DRR approaches.

Acknowledgements The author is grateful for the support extended by many individuals in the preparation of the manuscript. University of Colombo has acknowledged for various support received.

References

Abeywardana N, Bebermeier W, Schütt B (2018) Ancient water management and governance in the dry zone of Sri Lanka until abandonment, and the influence of colonial politics during reclamation. Water 10(12):1746

Ariyadasa KP (2002) Assessment of tree resources in the homegardens of Sri Lanka. ECFAO Partnership Program on Information and Analysis for Sustainable Forest Management, Bangkok

Bandara M (1995) Tank cascade systems in Sri Lanka: Some thoughts on their development implications (No. H016795). International Water Management Institute

Calvet-Mir L, Gómez-Baggethun E, Reyes-García V (2012) Beyond food production: ecosystem services provided by home gardens. A case study in Vall Fosca, Catalan Pyrenees Northeastern Spain. Ecol Econ 74:153–160

CHM (2017) http://lk.chm-cbd.net/?page_id=302. Accessed on 22nd May 2020

Costanza R, Daly HE (1992) Natural capital and sustainable development. Conserv Biol 6(1):37–46

Dammalage T, Jayasinghe N (2019) Land-use change and its impact on urban flooding: a case study …. Eng Technol Appl Sci Res

De Silva MMGT, Kawasaki A (2018) Socioeconomic vulnerability to disaster risk: a case study of flood and drought impact in a rural Sri Lankan community. Ecol Econ 152:131–140

Dharmasena PB (2004) Small tank heritage and current problems. In: Small tank settlements in Sri Lanka. Kobbakaduwa Agrarian Research and Training Institute, Colombo, Sri Lanka, pp 31–39

DMC (2005) Towards a safer Sri Lanka—Road map for disaster risk management. Disaster Management Centre, Ministry of Disaster Management, Government of Sri Lanka, December

Dolidon N, Hofer T, Jansky L, Sidle R (2009) Watershed and forest management for landslide risk reduction. In: Landslides–disaster risk reduction. Springer, Berlin, Heidelberg, pp 633–649

Estrella M, Saalismaa N (2013) Reduction (Eco-DRR): an overview. The role of ecosystems in disaster risk reduction, 26
Garcin M, Desprats JF, Fontaine M, Pedreros R, Attanayake N, Fernando S, Siriwardana CHER, De Silva U, Poisson B (2008) Integrated approach for coastal hazards and risks in Sri Lanka
Gunatilake HM, Senaratne DMAH, Abeygunawardena P (1993) Role of non-timber forest products in the economy of peripheral communities of knuckles national wilderness area of Sri Lanka: a farming systems approach. Econ Bot 47(3):275–281
Gunatilleke N (2015) Forest sector in a green economy: a paradigm shift in global trends and national planning in Sri Lanka. J Nat Sci Foundation Sri Lanka 43(2):101–109
Hettiarachchi M, Morrison TH, Wickramsinghe D, Mapa R, De Alwis A, McAlpine CA (2014) The eco-social transformation of urban wetlands: a case study of Colombo, Sri Lanka. Landsc Urban Plan 132:55–68
IUCN (2008) International Union for conservation of Nature. Geneva
IUCN French Committee (2019) Nature-based solutions for climate change adaptation and disaster risk reduction. France, Paris
Jayasuriya S, Steele P, Weerakoon D (2005) Post-tsunami recovery: issues and challenges in Sri Lanka. Report Presented to the Prime Minister of Sri Lanka. November. Institute of Policy Studies, Colombo
Jayawardane AKW (2006) Disaster mitigation initiatives in Sri Lanka. In: Proceedings of international symposium on management systems for disaster prevention, pp 09–11.
Karunaratne G (2001) Sri Lanka experience on natural disaster mitigation, seminar on Sri Lanka & India: cooperation in technology for development, 1st and 2nd September in Chennai, India, a collaboration of Institution of Engineers, India and Institution of Engineers, Sri Lanka
Kumar V (2020) Biodiversity, conservation and sustainable development (issues and approaches) vol II. New Academic Publishers, New Delhi ISBN: 978-8186772751 Importance of Homegardens Agroforestry System in Tropics Region
Lacambra C, Spencer T, Moeller I (2008) Tropical coastal ecosystems as coastal defences. ProAct network. The role of environmental management and eco-engineering in disaster risk reduction and climate change adaptation
Mahendrarajah ES (2003) Agroforestry as a means of alleviating poverty in Sri Lanka; proceedings The XII World Forestry Congress, held from 21 to 28 September 2003 in Québec, Canada (www.fao.org/docrep/ARTICLE/WFC/XII/0964-A1.HTM)
Manusinghe P (2009) Changing socio economic pattern and conservation of Maduganga wetland. In: Proceedings of international forestry and environment symposium, vol 14.
McInnes RJ, Everard M (2017) Rapid assessment of wetland ecosystem services (RAWES): an example from Colombo, Sri Lanka. Ecosyst Serv 25:89–105
Millennium Ecosystem Assessment (2005) Ecosystems and human well-being: synthesis. Island Press, Washington, DC
Nelson E, Mendoza G, Regetz J, Polasky S, Tallis H, Cameron D et al (2009) Modeling multiple ecosystem services, biodiversity conservation, commodity production, and tradeoffs at landscape scales. Front Ecolo Environ 7(1):4–11
Peyre A, Guidal A, Wiersum KF, Bongers F (2006) Dynamics of homegarden structure and function in Kerala, India. Agrofor Syst 66:101–115
Pushpakumara, DKNG (2000, November) Kandyan homegardens: promising land management system for food security, biodiversity and environmental conservation. In: Advances in Land Resources Management for 21st Century. Proceedings of the International Conference on Land Resources Management for Food, Employment, and Environmental Security held from, pp 9–13
Quartel S, Kroon A, Augustinus P, Van Santen P, Tri NH (2007) Wave attenuation in coastal mangroves in the Red River Delta Vietnam. J Asian Earth Sci 29(4):576–584
Ranasinghe TS (2012) Mangrove restoration efforts in Sri Lanka. Sharing Lessons on Mangrove Restoration, p 75
Reliefweb (2018) https://reliefweb.int/disaster/fl-2018-000060-lka. Accessed on 1st May 2020

Rossetto T, Peiris N, Pomonis A, Wilkinson SM, Del Re D, Koo R, Gallocher S (2007) The Indian Ocean tsunami of December 26, 2004: observations in Sri Lanka and Thailand. Nat Hazards 42(1):105–124

Samaranayake RADB (2005) Pre-and post-tsunami coastal planning and land-use policies and issues in Sri Lanka

Saparamadu S, Yi Z, Zongping Z (2018) Temporal changes of land use land cover and environmental impacts: a case study in Colombo, Sri Lanka. Int J Earth Environ Sci 2018:2018

SCBD: Secretariat of the Convention on Biological Diversity (2015) Governments encouraged to use biodiversity and ecosystem services as strategy for climate change adaptation and disaster risk reduction. Press release, Montreal, Canada. https://www.cbd.int/doc/press/2015/pr-2015-12-09-biodiv-cop21unfccc-en.pdf

Shanmugaratnam N (2005 Nov) Challenges of post-disaster development of coastal areas in Sri Lanka. In: Consultative workshop on post-tsunami reconstruction experiences of local NGOs, vol. 23

Sidle RC, Tani M, Ziegler AD (2006) Catchment processes in Southeast Asia: atmospheric, hydrologic, erosion, nutrient cycling, and management effects. For Ecol Manage 224(1–2):1–4

Sudmeier-Rieux K, Ash N (2009) Environmental guidance note for disaster risk reduction: healthy ecosystems for human security, Revisedth. IUCN, Gland

The Guardian 2015 https://www.theguardian.com/environment/2015/may/12/sri-lanka-to-become-the-first-nation-in-the-world-to-protect-all-its-mangroves

Tomita T, Imamura F, Arikawa T, Yasuda T, Kawata Y (2006) Damage caused by the 2004 Indian Ocean tsunami on the southwestern coast of Sri Lanka. Coast Eng J 48(02):99–116

UNFCCC (2018) https://unfccc.int/climate-action/momentum-for-change/planetary-health/sri-lanka-mangrove-conservation-project

UNISDR (2013) United Nations international strategy for disaster reduction post-2015 framework for disaster risk reduction (HFA2): report from 2013 global platform consultations. UNISDR, Geneva

UNISDR 2015: United Nations International Strategy for Disaster Reduction (2015) The economic and human impact of disasters in the Last 10 Years. UNISDR, Geneva, Switzerland. http://www.unisdr.org/files/42862_economichumanimpact20052014unisdr.pdf

UNISDR (2009) 2009 UNISDR Terminology on disaster risk reduction. Geneva. http://www.unisdr.org/we/inform/terminology

Uy N, Shaw R (2012) The role of ecosystems in climate change adaptation and disaster risk reduction. In: Uy N, Shaw R (eds) Ecosystem based adaptation. Emerald Group Publishing Limited, Bingley, pp 41–59

Vidanage S, Perera S, Kallesoe M (2005) The value of traditional water schemes: small tanks in the Kala Oya Basin, Sri Lanka. IUCN Water, Nature and Economics Technical Paper No. 6, IUCN-The International Union for Conservation of Nature. Ecosystems and Livelihoods Group Asia, pp 1–76

Wattage P, Mardle S (2005) Stakeholder preferences towards conservation versus development for a wetland in Sri Lanka. J Environ Manage 77(2):122–132

Weerahewa J, Pushpakumura G, Silva P, Daulagala C, Punyawardena R, Premalal S, Miah G, Roy J, Jana S, Marambe B (2012) Are homegarden ecosystems resilient to climate change? An analysis of the adaptation strategies of homegardeners in Sri Lanka. APN Sci Bull 2:22–27

Wickramasinghe A, Pérez MR, Blockhus JM (1996) Nontimber forest product gathering in Ritigala Forest (Sri Lanka): household strategies and community differentiation. Hum Ecol 24(4):493–519

Wickramasinghe DD, Chathumani D, Manawadu L (2018) Floods find their own path: A case study from Kelani. In: Proceedings of the world water week, Stockholm, Sweden

Wijetunge JJ (2006) Tsunami on 26 December 2004: spatial distribution of Tsunami height and the extent of inundation in Sri Lanka. Sci Tsunami Haz 24(3):225–239

WWF (2020) https://wwf.panda.org/wwf_news/?337222/Ramsar-announces-first-18-Wetland-Cities. Accessed on 26 May 2020

Yanagisawa H, Koshimura S, Goto K, Miyagi T, Imamura F, Ruangrassamee A, Tanavud C (2009) The reduction effects of mangrove forest on a tsunami based on field surveys at Pakarang Cape, Thailand and numerical analysis. Estuar Coast Shelf Sci 81:27–37

Chapter 20
The Watarase Retarding Basin—A Historical Example of Ecosystem-Based Disaster Risk Reduction in Japan

Tomohiro Ichinose, Jun Ishii, and Ikuko Imoto

Abstract The Watarase Retarding Basin, located 60 km north of Tokyo, is the largest retarding basin in Japan. The retarding basin was constructed at the beginning of the twentieth century to store toxic pollution from the nearby Ashio Copper Mine. The copper mine is located in the upper stream of the Watarase River and is the best known example of environmental pollution in Japan. We analyzed land use changes over the past 100 years in the area. At the time construction began in 1907, ponds, grassland, and marshes were evident. By 1979, the Watarase Retarding Basin was a wetland or water area. A large part of the retarding basin had become arid after that. In 2001, the land use in the retarding basin was a mosaic of wetlands with common reed (*Phragmites australis*) and amur silver grass (*Miscanthus sacchariflorus*). Most of the area became a Ramsar site in 2012 because the area is representative of a reed-dominated low moor wetland in Japan and has a high level of diversity of wetland flora and fauna. Typhoon Hagibis struck eastern Japan on 12–13 October 2019. The Watarase Retarding Basin stored 250 million m^3 of water and prevented flooding downstream in the Edo River.

Keywords Ashio Mine · Biodiversity · Flood control · Ramsar site · Land use change · Pollution · Tone River

T. Ichinose (✉)
Faculty of Environment and Information Studies, Keio University, 5322 Endo, Fujisawa 252-0882, Kanagawa Prefecture, Japan
e-mail: tomohiro@sfc.keio.ac.jp

J. Ishii
Fukui Prefectural Satoyama-Satoumi Research Institute, 122-12-1 Torihama, Wakasa-cho, Mikatakaminaka-gun 919-1331, Fukui Prefecture, Japan
e-mail: j-ishii-81@pref.fukui.lg.jp

I. Imoto
Keio Research Institute at SFC, Keio University, 5322 Endo, Fujisawa 252-0882, Kanagawa Prefecture, Japan
e-mail: imotoi@sfc.keio.ac.jp

© The Author(s), under exclusive license to Springer Nature Singapore Pte Ltd. 2021
M. Mukherjee and R. Shaw (eds.), *Ecosystem-Based Disaster and Climate Resilience*, Disaster and Risk Research: GADRI Book Series,
https://doi.org/10.1007/978-981-16-4815-1_20

20.1 Introduction

The Watarase Retarding Basin, located 60 km north of Tokyo, is the largest such basin in Japan. It covers an area of 33 km^2 and has a total water storage capacity of 264 million m^3, covering four prefectures: Tochigi, Gunma, Saitama and Ibaraki. The basin is located at the confluence of Watarase River, Uzuma River, and Omoi River (Figs. 20.1 and 20.2). The Watarase River flows south and joins the Tone River, one of Japan's three major rivers. In 2012, the retarding basin was listed as a Ramsar wetland, a valuable habitat for a variety of species, including animals and plants. Typhoon Hagibis hit Japan in October 2019, causing approximately 250 million m^3 of flooding in this area, or about 95% of the basin's total water storage capacity. Although the Tone River flows partly downstream into the Edo River, which empties into Tokyo Bay, the Edo River did not reach flood level. The water storage function of the Watarase Retarding Basin was one of the factors that prevented flooding in the lowland areas east of Tokyo during the typhoon. Thus, it has recently been attracting attention as a historical example of ecosystem-based disaster risk reduction (Eco-DRR) in Japan.

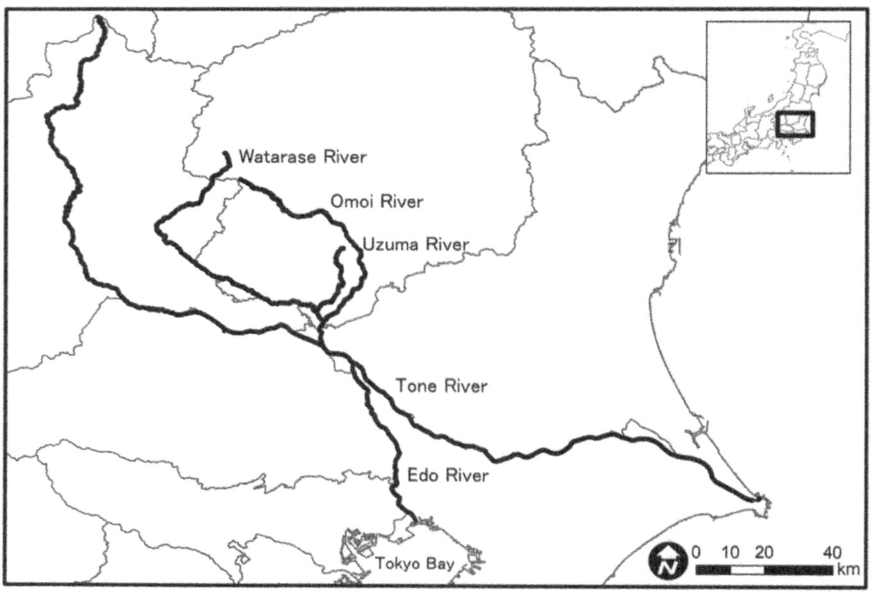

Fig. 20.1 The location of Watarase, Uzuma, Omoi, Tone and Edo Rivers

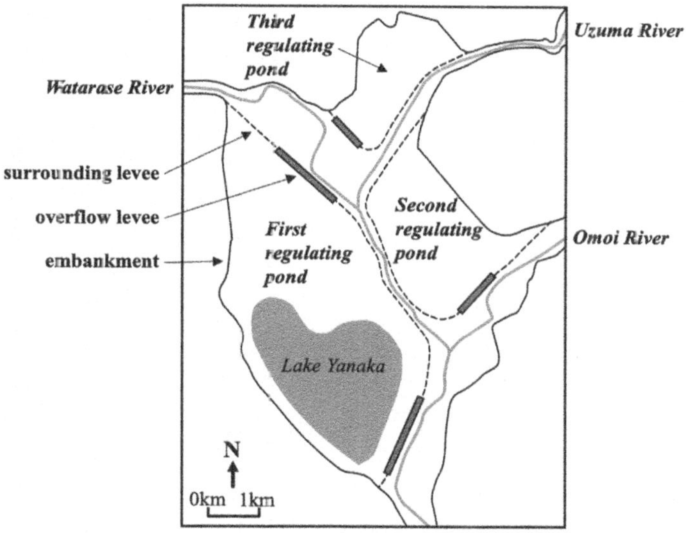

Fig. 20.2 The present map of the Watarase Retarding Basin. Lake Yanaka is an artificial lake: the three rivers (delineated by solid gray lines) flow into each other

20.2 Ashio Copper Mine and Watarase River

The area where the retarding basin is now located was once a flood-prone lowland. A topographical map drawn by the Meiji government in 1884 (Fig. 20.3) shows that it was a wetland with a large number of swamps and ponds of various sizes, the largest of which was Akama Pond. Farmers in the area had a long history of suffering from flooding. An embankment was built in 1595, but it broke or overflowed 49 times in the 207 years from 1704 to 1910, so flooding occurred about once every four years (Matsuura 2003). In 1871, the fourth year after the Meiji government was inaugurated, villages in the area submitted a petition to their administrative districts for a plan to improve the Watarase River.

In the late 1880s, the Ashio Mine Poisoning Incident, said to be Japan's first modern environmental disaster, came to the fore. The Ashio area is located in the upper reaches of the Watarase River and has been mined for copper since ancient times. Mining in the area reached its peak in the first half of the Edo period (1603–1867), but there was little activity by the end of the Edo period. In 1884, a large vein of ore was discovered, making the Ashio Mine the largest copper producer in East Asia. At the time, copper became one of Japan's major exports and supported the country's modernization (Matsuura 2003). The effects of mineral poisoning began to be seen in the Watarase River in the late 1870s. By the 1880s, it was reported that fishing in the Watarase River had become nearly impossible and forests in the Ashio area were dying (Takaishi et al. 2015). The large-scale floods of 1890 and 1896 contaminated farmland in the lower reaches of the mine, and in 1897, the First Mineral Poisoning

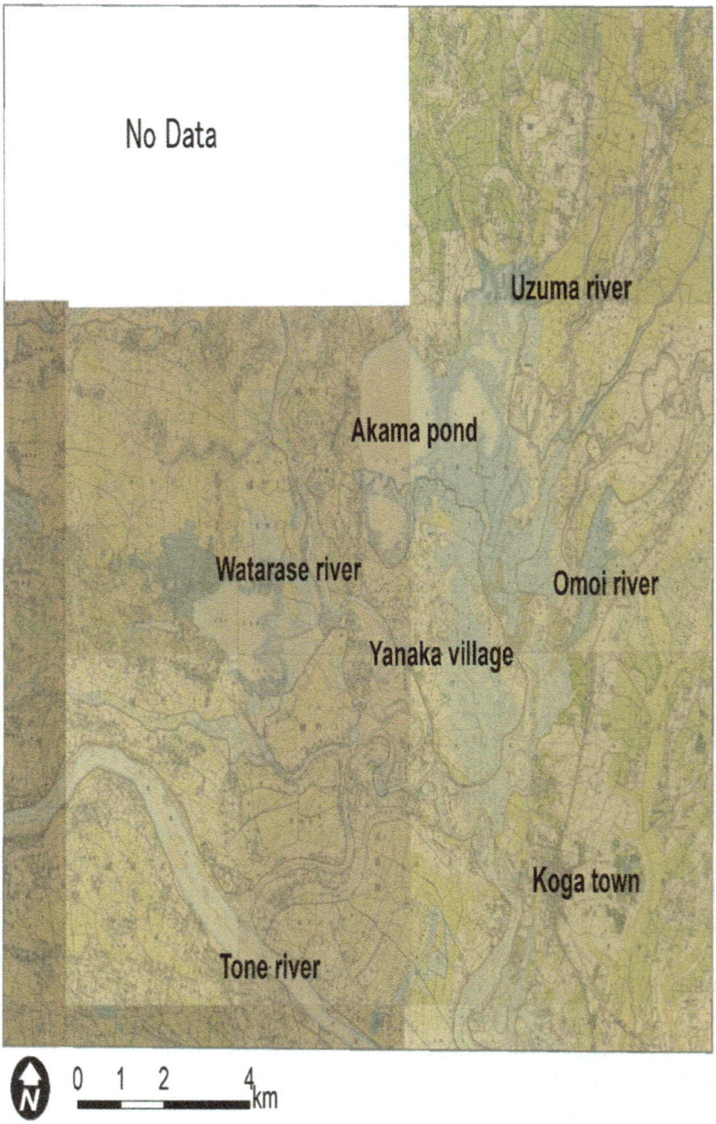

Fig. 20.3 Yanaka Village, rivers, ponds, and wetlands on the provisional survey map of Kanto Plain (1884) published by the army land survey (Japan Map Center 2013). Created by editing the Rapid Survey Map of Kanto Plain[1]

[1] Institute for Agro-Environmental Sciences, NARO. "Kanto Rapid". Accessed Dec 10, 2020. https://aginfo.cgk.affrc.go.jp/mapprv/wms.html.en.

Incident Investigation Committee was established under the direct control of the Cabinet. Based on the report of this committee, works were carried out, including the construction of ponds in the Koga area for depositing toxic waste from the mine. However, a flood in 1898 damaged the siltation pond and the surrounding area. A residents' movement called for the suspension of mining at the Ashio Mine, and residents clashed with police in 1900, in what is known as the Kawamata Incident. In order to deal with these worsening social problems, the government established the Second Mineral Poisoning Investigation Committee in 1902, which examined a river improvement plan for the Watarase River, including the construction of the Watarase Retarding Basin. The following year, the committee submitted a report on the Ashio Copper Mine to the prime minister. The report proposed that Yanaka Village, which was centrally located, be abolished and a huge retarding basin be built. Due to fierce opposition from the residents of Yanaka Village, land expropriation was granted by Tochigi Prefecture in 1907, and all remaining houses were forcibly demolished. Finally, in 1910, construction on the Watarase Retarding Basin began as a project directly under the Meiji government's supervision (Matsuura 2003).

20.3 Flood Control on the Tone and Edo Rivers, and Watarase Retarding Basin

The Watarase Retarding Basin was developed not only to deposit waste from the Ashio Mine but also to prevent flooding on the Tone and Edo Rivers. The Tone River, which joins with the Watarase River, once flowed into Tokyo Bay, but between 1621 and 1654 (in the early Edo period), a channel was constructed to connect the two rivers with the Tokiwa River, which flows into the Pacific Ocean. It is generally said that the purpose of this major civil engineering project was flooding control in the capital city, but it has been pointed out that the two rivers were also connected to ensure water transport (Okuma 1991). At the time, the canal was only about 18 m wide, and it was difficult to drain the floodwaters during a flood. The former Tone River was renamed the Edo River, and a portion of it still flows into Tokyo Bay. Mt. Asama erupted in 1783, and the ashfall caused the rivers in the Tone River basin to rise, causing frequent flooding in the basin. In 1868, the Meiji Restoration occurred, and the Meiji government was inaugurated. Flooding control in the Tone River, as well as the Watarase River, was a major issue for this government (Okuma 1991).

The 1885 flood of the Tone and Edo Rivers was the first time that modern flow measurements were made. The flows were 3700 m^3/s at Nakata, where the Watarase River and Tone River meet, and 2500 m^3/s in the Tone River, and the backflows from the Tone River to the Watarase River were 83 m^3/s (Matsuura 2015). In the great flood of September 1896, as mentioned earlier, a significant amount of mineral pollution flowed into the Watarase River basin, but there was also flooding damage in the lower reaches of the Tone and Edo Rivers. Although the mineral pollution did not reach the lower reaches of the Tone and Edo Rivers, it is believed that the

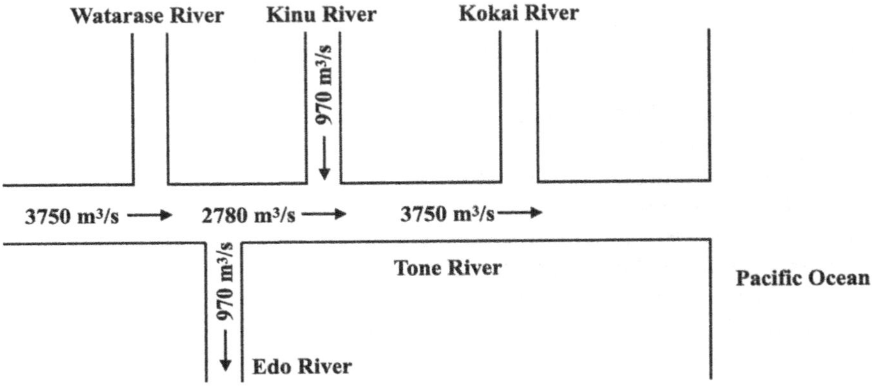

Fig. 20.4 Flood discharge design of the Tone River system in 1900

Meiji government was afraid that the pollution could spread to the capital city of Tokyo (Matsuura 2015). To prevent flooding of the Tone River system, including the Watarase River, it was desirable to lower the bed of the Tone River and increase the flow of the Edo River, but such an approach had the potential to extend the flow of mineral pollution into the lower reaches of the Edo River (Okuma 1991).

The first river improvement work was started on the Tone River in 1900. The flood control plan is shown in Fig. 20.4. The designed flood discharge was very low, and the scale of the floods was such that they occurred once every two to three years (Okuma 1991). As mentioned earlier, in 1902, the Second Mineral Poisoning Investigation Committee proposed the establishment of a 3000-hectare retarding basin at the confluence of the Watarase River and Tone River so that flooding in the Watarase River would not affect flooding in the Tone River. In August 1910, the Tone and Watarase Rivers experienced one of their largest floods, with a peak flood discharge estimated to be from 11,000 to 14,000 m^3/s at the confluence of the Tone River and Karasu River, which joins the Tone River upstream from the Watarase River. About 2300 km^2 of the Kanto Plain, including Tokyo, were inundated. The damage was estimated to be about 42 million yen, equivalent to one-sixth to one-seventh of the national budget at the time (Okuma 1991).

Immediately after the flood of 1910, the Meiji government revised the flood control plan for the Tone River. The revised flood discharge design is shown in Fig. 20.5. To reduce the construction budget, flood discharge at the confluence of the Tone and Karasu Rivers was set at 5570 m^3/s, which was much lower than the 1910 flood flow. The Watarase River relocation project, which began in 1910, was completed in 1918, and the entire area of the former Yanaka Village was converted to a retarding basin. The construction of an embankment was completed in 1922. As a result, the amount of water flowing down from the retarding basin into the Tone River at the time of flooding was estimated to be 0 m^3/s (Fig. 20.5). The Tone River improvement work was completed in 1930; the total cost over 30 years beginning in 1900 was about 74,620,000 yen (Okuma 1991).

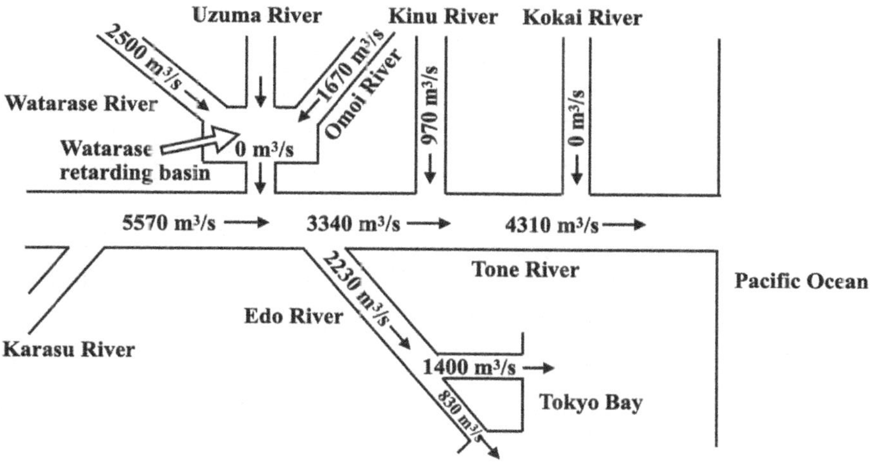

Fig. 20.5 Revised flood discharge design of the Tone River system in 1911

Floods occurred in September 1935, June to July 1938, and August to September 1938 after the Tone River improvement work. There were no bank breaches or inundations in the upper Tone River or the Edo River. In the lower reaches of the Tone River, a series of breaches due to backflow from the Tone River occurred in the lower reaches of the Kokai River. According to the flood control plan of the Tone River system, the diversions to the downstream Tone River and Edo River were roughly 3:2, but in reality, most of the Tone River floodwater went downstream to the lower Tone River. Therefore, a new Tone River system flood control plan focusing on the lower reaches of the river was launched in 1939 (Fig. 20.6). In this plan, the planned flood discharge to the Tone River was increased to 10,000 m^3/s, and flood discharge from the Tone River to the Watarase Retarding Basin was estimated to be 500 m^3/s.

The flood control plan was revised in 1939, but little work was done as the country entered World War II. The Tone River system was severely flooded by Typhoon Kathleen in September 1947. In this flood, the estimated maximum flood discharge at the confluence of the Tone and Karasu Rivers was 15,000–17,000 m^3/s (Okuma 1991). The flood control plan was revised again in 1949 as a result of this major flood (Fig. 20.7). In this plan, the designed flood discharge at the confluence of Tone and Karasu Rivers was raised to 14,000 m^3/s and cut by 3000 m^3/s by dams upstream of Tone River to allow for Typhoon Kathleen-scale flooding. The planned flood discharge of the Watarase River, the Uzuma River, and the Omoi River, which flow into the Watarase Retarding Basin, was also increased, while the designed flood discharge from the retarding basin to the Tone River was set at 0 m^3/s (Okuma 1991). Since then, the flood control plan for the Tone River system has been revised often, but the flood discharge design of the Watarase Retarding Basin, the Watarase River, the Uzuma River, and the Omoi River has remained essentially unchanged. In the

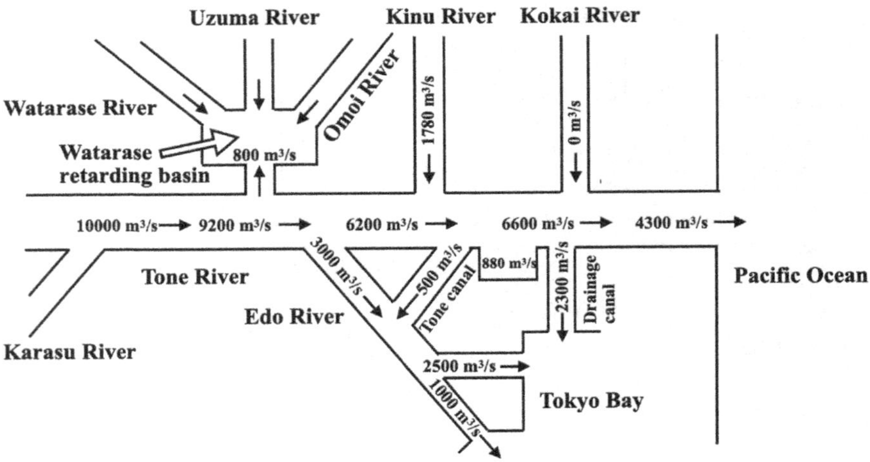

Fig. 20.6 Revised flood discharge design of the Tone River system in 1939

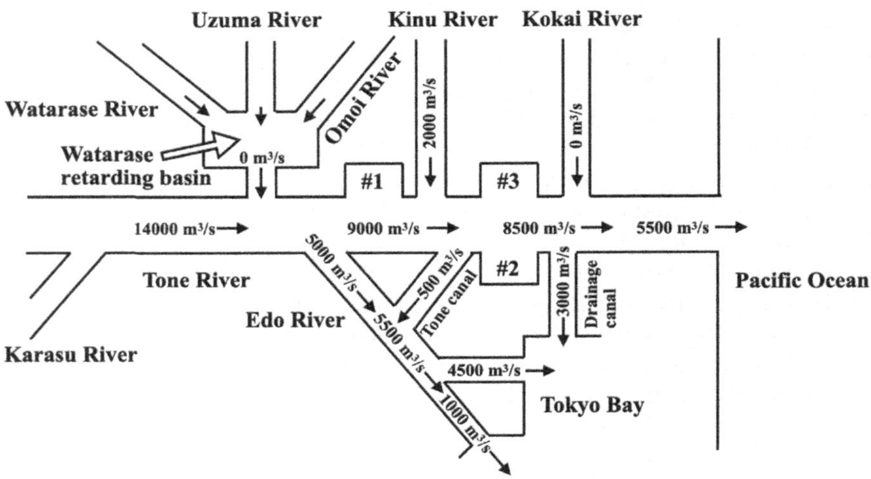

Fig. 20.7 Revised flood discharge design of the Tone River system in 1949. The three retarding basins (1, 2, and 3) control 2000 m^3/s

event of floods in the Tone River system, the retarding basin catches the flood flows of the three rivers and discharges them into the Tone River.

20.4 Development of the Watarase Retarding Basin After World War II

Mining at the Ashio Copper Mine began to decline after 1955, and the mine was closed in 1973 A study that investigated water quality in the Watarase River and the retarding basin from 1992 to 1993 found that the total arsenic concentration in the water was higher than the environmental standard of 0.01 g/l in some places (Morishita and Tsukigi 1996). On the other hand, in 1953, it was reported that the irrigation water for paddy fields in the village of Morita, Yamada-gun, Gunma Prefecture (present-day Ota City, Gunma Prefecture), upstream of the retarding basin, contained copper when it rained, but that water was not usually used for agricultural purposes (Deguchi 1955). The Watarase Retarding Basin has not experienced large-scale flood damage since the 1958 flood caused by Typhoon Ida, so it can be said that the retarding basin has, to a certain extent, achieved its goal of preventing the spread of toxic pollution from the mine to the downstream area.

After the completion of the perimeter embankment in 1922, the plan was to build a regulating pond to increase the water storage capacity of the retarding basin, but this was interrupted by World War II. When the flood control plan for the Tone River system was revised in 1949, the plan to expand storage at the retarding basin was also revised. Construction of the first regulating pond began in 1963 and was completed in 1970. Since then, the plan has been revised several times, and construction of a third regulating pond was finally completed in 1997 (Editorial Committee of 100 Year History of the Tone River and Japan Institute of Construction Engineering, 1987). The location of each of the regulating ponds is shown in Fig. 20.2, and their areas and water storage capacities are shown in Table 20.1. A flood flow analysis showed that, when the peak flood discharge under various discharge hydrographs is about 14,000 m^3/s at the Kurihashi Observatory in Kuki City, Saitama Prefecture (where the Watarase River joins the Tone River after leaving the Watarase Retarding Basin), these ponds contribute to reducing peak flood discharge and delaying the peak occurrence time at Kurihashi by storing flood discharge that flows back to the Watarase River from the Tone River (Matsumoto et al. 2014).

Typhoon Hagibis arrived at Izu Peninsula in Shizuoka Prefecture on 12 October 2019. Just before arriving, the central atmospheric pressure was 955 hPa, and the maximum wind speed was 40 m/s. It crossed the Kanto region and turned into a tropical depression off Sanriku at 12:00 p.m. on 13 October. The typhoon brought record-breaking heavy rainfall. Active rain clouds intermittently developed before

Table 20.1 Area and water storage capacity of three regulating ponds in the Watarase Retarding Basin

	Area (km^2)	Storage capacity (million m^3)
First regulating pond	15.0	117.1
Second regulating pond	5.0	35.6
Third regulating pond	2.8	19.1

the typhoon arrived, and rain continued to fall over a wide area. In particular, the town of Hakone in Kanagawa Prefecture, located on the north side of the Izu Peninsula, recorded 942.5 mm of precipitation in 24 h, the largest amount in Japanese history. The number of dead and missing was 89 as a result of numerous floods and landslides across eastern Japan. About 1 million buildings were damaged nationwide, and 142 levee failures occurred in 71 rivers across Japan. In the Tone River system, 18 levees broke, but there was no levee failure or flooding in the watershed downstream from the Watarase Retarding Basin. The basin stored approximately 250 million m^3 of floodwater, or about 95% of its planned capacity.[2] At the Kurihashi Observatory, the water level was above the flood risk level of 8.9 m between 1:00 a.m. and 10:00 a.m. on the 13th, but the Edo River did not reach its flood risk level at about that same time. The Yanba Dam upstream of the Tone River was to be completed in 2019, and trial water-logging had begun on 1 October. About 75 million m^3 of water from Typhoon Hagibis was stored in this facility, and it is said that this also contributed to the prevention of flooding. The water storage capacity of the Watarase Retarding Basin area, however, is much larger and will continue to play a major role in flood prevention in the future.

20.5 Land Use Changes Since Before the Construction of the Watarase Retarding Basin to the Present

To clarify how the retarding basin and the surrounding area have changed and how the wetland environment has been conserved, we analyzed land use from before the construction of the Watarase Retarding Basin to the present.

We selected the 1907 topographic map drawn just before the construction of the retarding basin, another topographic map from 1952 (just after the end of World War II), and two vegetation maps made in the years 1979 and 2001 (published by the Ministry of Environment). We interpreted the vegetation classifications to correspond with the land use types, and classified eight land use types based on the old topographic maps: forest, grassland, wetland, bare land, open water, rice paddy, dry field, and urban land use.

The land use maps are shown in Fig. 20.8. By the time construction began in 1907, ponds, grassland, and marshes, including Akama Pond, were evident in the middle of the map. The area is located at a low elevation within the Kanto Plain, where the Omoi and Watarase Rivers meet, and it has long had widespread lowland wetlands. Between two meandering rivers in the lower center of the map was the Yanaka Village, which was officially abolished in 1906. The village is said to have been established in the fifteenth century; it was a farming community with an abundant

[2] Tonegawa-jyoryu Office of Rivers of Kanto Regional Development Bureau, Ministry of Land, Infrastructure, Transport and Tourism, Japan. 2019. "Flash Report: The largest ever flood volume of 250 million m^3 was stored in the Watarase retarding basin during Hagibis in 2019". Accessed May 13, 2020. https://www.ktr.mlit.go.jp/kisha/kyoku_s_00000358.html (in Japanese).

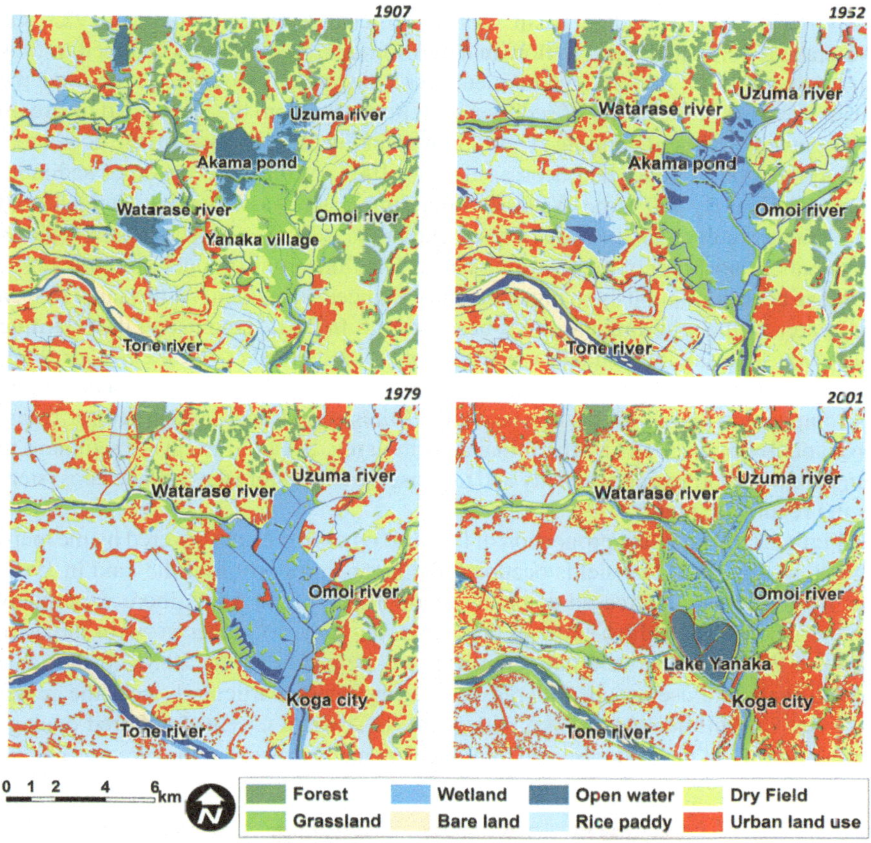

Fig. 20.8 Land use in 1907, 1952, 1979, and 2001 at the Watarase Retarding Basin and surrounding area. Based on the Topographic Map 50,000, 1907 and 1952, by the Geospatial Information Authority of Japan, and the Vegetation Map of Japan, 1979 and 2001, by the Ministry of the Environment

source of fish from the nearby swamps and marshes. Sedges from the swamp were also used to make sedge hats (Matsui et al. 2004). As noted previously, in the 1880s, the Ashio Copper Mine upstream of the Watarase River suffered water damage, and sediment from the mine caused soil contamination, and the fields in this area were severely damaged. Responding to the pollution and social unrest, the government planned to turn the area in and around Yanaka Village into a retarding basin, as noted previously. The 1907 map gives no information on the location of the reservoir and indicates that the area was then a grassland (wasteland) and a dry field. A detailed map (a provisional survey map) prepared by the army in 1884 (Fig. 20.3) shows that the area corresponding to the grassland is marked as having thatch plants, along with the color and symbol used for wetland, suggesting that it was a semi-humid grassland with growths of sedge (Matsui et al. 2004).

When the first retarding basin was constructed between 1914 and 1922, the Watarase River, which used to flow in a north–south direction, was changed to flow into Akama Pond, and an embankment was constructed. However, even after the basin was built, a great deal of sediment still flowed in from the devastated upstream area, causing frequent flooding. The 1952 map in Fig. 20.8 shows that the pond area, including Akama Pond, had decreased since 1907, primarily because of sediment accumulation. By 1979, the Watarase Retarding Basin was already a wetland with some open water areas. In 1989, a reservoir called Lake Yanaka was constructed in the southernmost part of the retarding basin. The flood control function of the reservoir has been continuously maintained and enhanced since then. At the same time, a large area of the retarding basin has become arid. In 2001, land use in the retarding basin was a mosaic of wetlands consisting of common reeds (*Phragmites australis*) and amur silver grasses (*Miscanthus sacchariflorus*), and the wetland area decreased as compared to that in 1979. As a result, the Ministry of Land, Infrastructure, Transport and Tourism (MLIT), which manages the area, began to excavate sediment to improve reservoir storage capacity and to conserve and restore the wetland ecosystem in the regulating ponds.

Figure 20.9 shows the proportion of land use types in the area covered by the maps shown in Fig. 20.8. Dry fields occupied the largest proportion of the land in 1907, but they gradually decreased as paddy fields increased. In 1907, there were many mulberry orchards, which grew in dry fields, but they no longer grow in the area. Most of the dry fields were later converted to rice paddies (Fig. 20.8). At the same time, urban land use has also increased over the years, particularly near the center of Koga City in the southeastern part of the area. Koga City had suffered damage from flooding of the Watarase and Omoi Rivers many times, but the city was able to

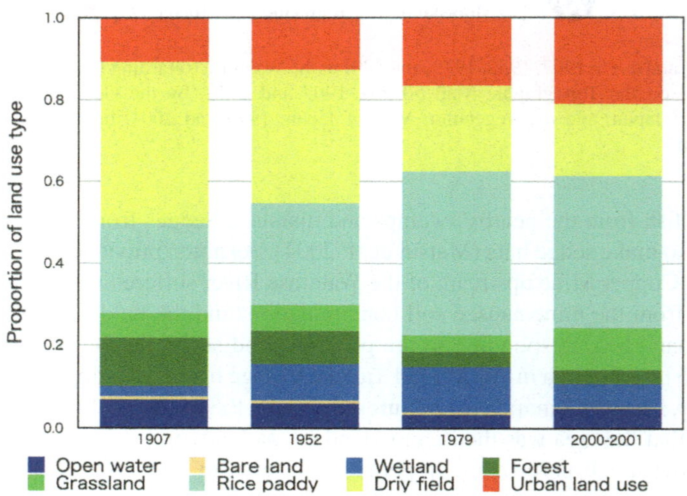

Fig. 20.9 Proportion of land use type in 1907, 1952, 1979, and 2000–2001

Fig. 20.10 A moist tall grassland dominated by common reeds and amur silver grasses in the Watarase Retarding Basin

make great strides in its development thanks to flood control measures, such as the development of regulating ponds and the construction and repair of embankments.

20.6 Biodiversity in the Watarase Retarding Basin

Because flood control was the priority for the Watarase Retarding Basin, the basin's use has been strictly limited for a long time. Therefore, despite being in the capital region, the wetland within the retarding basin has been maintained, which is in stark contrast to development in the surrounding areas (Figs. 20.8 and 20.9). During the twentieth century, 61.1% of the total area of natural wetlands in Japan was lost, mainly through land reclamation.[3] The total area of the Japanese wetlands was approximately 821 km^2 in 1999. Most of the remaining wetlands have lost their ecological integrity and/or been subjected to environmental deterioration. In particular, lowland floodplains are considered to be threatened ecosystems because of intensive urbanization and agricultural reforms (Washitani 2001, 2007; Yoshimura et al. 2005). The Watarase Retarding Basin, the largest lowland floodplain wetland in Honshu (Japan's mainland), has been conserved, and the wetland is currently mainly composed of moist tall grassland dominated by common reeds and amur silver grasses (Fig. 20.10).

[3] Geospatial Information Authority of Japan (Ministry of Land, Infrastructure, Transport and Tourism). 2000. "Lake and wetland survey in Japan". Accessed April 1, 2020. https://www.gsi.go.jp/kankyochiri/shicchimenseki2.html (in Japanese).

Fig. 20.11 The overflow levee in the third regulating pond. Each picture is a view from the north, and the Watarase River flows on the right side of the levee. The photo on the right shows overflow from the river on September 7, 2007

As described above, when the embankment around this area was built to create the retarding basin, the course of the Watarase River was also changed so that the river water could flow into the upstream side of the retarding basin (Figs. 20.3 and 20.8). Consequently, a large amount of sediment was stored in the retarding basin, and most of the open water areas such as ponds, including Akama Pond, and swamps disappeared (Fig. 20.8). Subsequently, levees were built within the retarding basin to divide it into three parts, the first, second and third regulating ponds, and the three rivers flowed among these regulating ponds (Fig. 20.2). Flood control was achieved in that the river water overflowed from the rivers into each pond during flooding through overflow levees, which are in the lower parts of the surrounding levees (Fig. 20.11).[4] Each of three ponds is rarely flooded under ordinary conditions, but the system works as a detention basin in the case of flooding.[5] Even so, overall flooding frequency is low in the wetland. Furthermore, drainage canals were also built within the ponds. Such riparian works and the resulting environmental changes probably caused the shift to drier conditions (Fig. 20.8), but the area never completely dries out.

The environmental change to moister conditions with a decreasing intensity of flooding disturbance is likely to provide favorable habitats for common reeds and amur sliver grasses, resulting in the development of grasslands typical to the area (Fig. 20.10). Local people have harvested the dead straw of both species for commercial use in reed blinds, roof thatching, and other products, and have also traditionally conducted winter burning to maintain and conserve the vegetation (Fig. 20.12). The traditional vegetation management of cutting and prescribed burning in autumn or winter can retard vegetation succession and maintain common reed and amur silver grass grasslands, including the native herbaceous plant species growing in the

---

[4] Ministry of Land, Infrastructure, Transport and Tourism, Japan. "Overview of Watarase-yusuichi". Accessed May 9, 2020. https://www.ktr.mlit.go.jp/tonejo/tonejo00081.html (in Japanese).

[5] Ministry of Land, Infrastructure, Transport and Tourism, Japan. 2019. "Airphotos of Watarase-yusuichi". Accessed April 9, 2020. https://www.ktr.mlit.go.jp/tonejo/tonejo_index051.html (in Japanese).

Fig. 20.12 Cutting of the dead straw of common reeds and amur silver grasses for commercial use (left) and burning for vegetation management (right)

Fig. 20.13 Threatened plant species *Ophioglossum namegatae* (left), *Galium tokyoense* (middle), and *Viola raddeana* (right) inhabiting the Watarase Retarding Basin

understory (Gryseels 1989; Cowie et al. 1992; Ostendorp 1999; Ishida et al. 2008). Cutting and prescribed burning remove herbaceous litter and constrain the occurrence and performance of woody species (Güsewell and Nédec 2004; Ishida et al. 2008; Yoshikawa et al. 2013), thereby creating well-illuminated conditions, a high cumulative ground-surface temperature, and a wide range of daily temperature fluctuations (Wang et al. 2015). Those environmental conditions promote germination and establishment of herbaceous understory species (Cowie et al. 1992; Ishida et al. 2008; Yoshida and Nishiyama 2008; Yoshikawa et al. 2013; Wang et al. 2015). Such processes would have enhanced maintaining the tall, moist grassland dominated by common reeds and amur sliver grasses along with a rich and diverse group of understory plants in the Watarase Retarding Basin. More than 700 plant species (Ohwada and Ogura 1996), including more than 50 threatened species listed in the national Red List[6] (e.g., *Ophioglossum namegatae*, *Galium tokyoense*, and *Viola raddeana*; Fig. 20.13), have been recorded in this area. Many seedlings of such species can be seen every spring in the grassland.

At first glance, a tall moist grassland (Fig. 20.10) may seem to be homogenous, but actually, it is a heterogenous and spatially complex mosaic of vegetation types,

[6] Watarase-yusuichi Acclimation Promoting Foundation. "Plants of Watarase-yusuichi". Accessed April 9, 2020. https://watarase.or.jp/plant/ (in Japanese).

contributing to a great diversity of understory plant species (Ishii and Washitani 2007; Ishii et al. 2009). Within the grassland, shoot densities of common reed and amur silver grass vary spatially due to the heterogeneity in the environment, such as groundwater level and natural and anthropogenic disturbances. Relative dominances of these species can be associated with the habitats of understory species. For example, *O. namegatae* and *G. tokyoense* occur more frequently in mixed stands of common reed and amur silver grass. Both *Carex cinerascens* and *Amsonia elliptica* are nationally threatened species, and the former occurs very often in nearly pure common reed stands, whereas the latter rarely occurs in such stands. Well-illuminated sites of common reed and amur silver grass stands appear to be a favorable habitat for *V. raddeana* (Ohwada 2011).

More than 250 bird species have been observed in the retarding basin. These include more than 40 threatened species listed in the national Red List,[7] such as marsh grassbirds (*Locustella pryeri pryeri*) and eastern marsh harriers (*Circus spilonotus spilonotus*). Some grassland species such as great reed warblers (*Acrocephalus arundinaceus*), zitting cisticolas (*Cisticola juncidis*), and marsh grass birds breed during spring and summer. Barn swallows (*Hirundo rustica*) and sand martins (*Riparia riparia*) roost communally in aggregations, which often contain more than tens of thousands of individuals, in the grassland from late summer to early autumn. Some waders use the retarding basin as a stopover site in spring and autumn. Thousands of ducks, buntings, and eastern marsh harriers spend winter in the retarding basin. In addition to the eastern marsh harriers, other birds of prey, such as the species of Accipitridae, Pandionidae, Falconidae, and Strigidae, are also observed in the area; some use the area for breeding, while others overwinter. Furthermore, in Japan the native population of oriental white storks (*Ciconia biyciana*) was extirpated in 1971, and an experimental release for reintroduction began in 2005 (Naito and Ikeda 2007). Interestingly, some oriental white storks have been observed visited the retarding basin since 2016.[8] Habitats of this species are paddy fields, ditches, and rivers in rural landscapes, and their conservation and restoration have become of increasing concern (Naito and Ikeda 2007; Yamada et al. 2019). The retarding basin is expected to provide favorable habitat for this species.

Many insect species have also been recorded in the retarding basin. This includes about 50 threatened species listed in the national Red List,[9] as well as some species that rarely are observed in Japan. For example, *Anadastus pulchelloides*, a nationally threatened beetle, has recently been recorded at only two localities (Ohmomo et al. 2011; Ohkawa 2013). Other beetle species, *Platamartus jakowlewi*, was also reported to have occurred in a few localities including the retarding basin. This species was

[7] Watarase-yusuichi Acclimation Promoting Foundation. "Wild birds of Watarase-yusuichi". Accessed April 10, 2020. https://watarase.or.jp/bird/ (in Japanese).

[8] Oyama City Web Site. "Watarase-yusuichi". Accessed April 10, 2018. https://www.city.oyama.tochigi.jp/site/wataraseyusuichi/382.html (in Japanese).

[9] Watarase-yusuichi Acclimation Promoting Foundation. "Insects of Watarase-yusuichi". Accessed April 9, 2020. https://watarase.or.jp/insect/ (in Japanese).

recently newly recorded in Japan in a revision of the family Kateretidae of Japan (Hisamatsu 2011).

Several species that have been recorded in the retarding basin include "watarase" as a place name in part of their names. For example, *Impatiens ohwadae* is named "watarase-tsurifunesou" in Japanese; it is a wetland plant species growing in a moist tall grassland and was reported as a new species of Balsaminaceae in 2009, and its type locality is the Watarase Retarding Basin (Watanabe and Serizawa 2009). *Elaphrus sugai* is a ground beetle named "watarase-hanmyoumodoki"; it inhabits marsh and riparian environments, and its type locality also is the retarding basin (Sasakawa 2016). This species is listed in the national Red List, and it has currently been found only in the Watarase Retarding Basin, so conservation measures are urgently needed for this species (Sasakawa 2016). *Anthicus watarasensis* is an ant-like flower beetle inhabiting marsh land (Sakai and Ohbayashi 1994; Ohkawa 2013) and is named "watatarase-mizugiwaarimodoki" in Japanese (Sakai and Ohbayashi 1994). *Laccophilus dikinohaseus* is a water beetle described as a new species in 2005 (Kamite et al. 2005), and its Japanese name is "watarase-tsubugengorou".

20.7 Designation as a Ramsar Site and "Wise Use" of the Watarase Retarding Basin

Most parts of the Watarase Retarding Basin (28.6 km^2) became a Ramsar site called "Watarase-yusuichi" in 2012.[10,11] The site qualifies under Criterion 1 of the Ramsar Convention. The following reasoning was given:

> Watarase-yusuichi is representative of a reed (*Phragmites australis*) dominated low moor wetland in the Japanese Forest biogeographic ecoregion. The extensive reedbed is one of the largest in the biogeographic region and supports a diversity of wetland flora and fauna. The site also has an important flood control function by retarding the flood water from the Watarase, Uzuma, and Omoi Rivers that flow into the site, and then slowly releasing the water into the Tone River that flows downstream.

The purpose of the Ramsar Convention is the conservation and wise use of wetlands as a contribution toward achieving sustainable social development. Actually, the Watarase Retarding Basin provides various ecosystem services, and it also plays a social role. As noted previously, flood control is the priority, and this function has led to increasing paddy and urban areas in the surrounding region. The retarding basin has also been valued highly as habitat for diverse animals and plants and a harvesting site of *P. australis* and *M. sacchariflorus* for commercial use.

During periods when there is no flooding, many local residents and tourists from urban areas visit the retarding basin, which provides a vast open space within the

---

[10] Ramsar Sites Information Service. 2012. "Watarase-yusuichi". Accessed April 9, 2020. https://rsis.ramsar.org/ris/2061.

[11] Ministry of the Environment, Japan. "RAMSAR SITES in JAPAN". Accessed April 9, 2020. http://www.env.go.jp/en/nature/npr/ramsar_wetland/pamph/index.html.

capital region, for diverse purposes (Matsui et al. 2004). Some people are engaged in small-scale fisheries, while most enjoy outdoor recreation. Recreational activities include nature observation (birdwatching is very popular), environmental education activities, and recreational fishing, all of which involve direct contact with nature. Indirect activities, in which people enjoy doing something in nature, include walking, jogging, and cycling within the retarding basin. In addition, other activities are allowed in designated sites, including picnics and sports such as skydiving and hot air balloon rides. Visitors can also learn about the history of Yanaka Village in the historic site conservation zone. Lake Yanaka is a human-made lake designed for flood control and water utilization in the downstream sector and offers a wide variety of recreational activities, including nature observation, recreational fishing, and water sports.[12]

At present, there are primarily two issues to be resolved for the conservation and wise use of the Watarase Retarding Basin.[13] The first issue is to improve the flood control function by increasing reservoir storage capacity. The second issue is to restore a diversified wetland community that has been damaged because of habitat loss for aquatic and short-lived plant species, aridification of the wetland, and reduced flood disturbance (Ishii et al. 2011). An additional issue is that the invasive plant species *Solidago altissima* has invaded the tall moist grassland (Ishii and Washitani 2013). Topsoil removal is one wetland restoration measure (Klimkowska et al. 2007); it creates wet conditions, facilitates recolonization of aquatic and short-lived plants on bare soils, and removes alien plants. Removal of deposits of earth and sand can also increase reservoir storage capacity. After confirming the effectiveness of topsoil removal by excavation for wetland restoration in a pilot experiment conducted in the second pond (Fig. 20.14; Ishii et al. 2011), the "Basic Plan for Wetland Conservation and Restoration of Watarase-yusuichi" was formulated to cope with the deterioration of the wetland environment and the decline of the flood control function in 2010, and restoration by excavation in the second pond was initiated.[14] This plan also aims to restore habitats of diverse species of birds, fishes, and insects.

Since 2010, the Committee for Monitoring of Wetland Conservation and Restoration Projects in Watarase-yusuichi has monitored and evaluated the results of investigations on wetland conservation and restoration projects. Conservation measures including biodiversity monitoring, habitat restoration of paddy fields for floodplain species (Washitani 2007), and the control of invasive species have been implemented

---

[12] Tonegawa-jyoryu Office of Rivers of Kanto Regional Development Bureau, Ministry of Land, Infrastructure, Transport and Tourism, Japan. "Let's go to Lake Yanaka!". Accessed April 19, 2020. https://www.ktr.mlit.go.jp/tonejo/tonejo_index029.html (in Japanese).

[13] Ramsar Sites Information Service. 2012. "Watarase-yusuichi". Accessed April 9, 2020. https://rsis.ramsar.org/ris/2061.

[14] Tonegawa-jyoryu Office of Rivers of Kanto Regional Development Bureau, Ministry of Land, Infrastructure, Transport and Tourism, Japan. "A Committee for Wetland Conservation and Restoration of Watarase-yusuichi". Accessed April 21, 2020. https://www.ktr.mlit.go.jp/tonejo/tonejo_index027.html (in Japanese).

Fig. 20.14 A pilot experimental site to examine the effectiveness of topsoil removal by excavation on the recolonization of aquatic and short-lived wetland plants in the Watarase Retarding Basin. Photo by L. Hashimoto

in collaboration among local residents, non-profit organizations, educational institutions, academic experts, and local governments. The Council for Wetland Conservation and Utilization of Watarase-yusuichi was established in 2013, after the designation of the Watarase Retarding Basin as a Ramsar site, to support collaboration and to enhance conservation efforts and utilization of wetland resources.[15] The Council for the Promotion of Ecological Networks in the Watarase-yusuichi Area was established in 2015 to restore ecological networks in the Kanto region, where the Watarase Retarding Basin is located, using oriental white storks and Japanese crested ibises as indicator species. Projects reintroducing these species are ongoing at the national and regional level in Japan.[16] The council aims to stimulate the economy and to revitalize the community through conservation activities. In these ways, many stakeholders are participating in conservation and restoration activities in the Watarase Retarding Basin with the goal of enhancing sustainable development in the future.

[15] Tonegawa-jyoryu Office of Rivers of Kanto Regional Development Bureau, Ministry of Land, Infrastructure, Transport and Tourism, Japan. "A Council for Wetland Conservation and Utilization of Watarase-yusuichi". Accessed April 21, 2020. https://www.ktr.mlit.go.jp/tonejo/tonejo_index028.html (in Japanese).

[16] Tonegawa-jyoryu Office of Rivers of Kanto Regional Development Bureau, Ministry of Land, Infrastructure, Transport and Tourism, Japan. "A Council for the Promotion of Ecological Networks in the Watarase-yusuichi Area". Accessed April 21, 2020. https://www.ktr.mlit.go.jp/tonejo/tonejo_index042.html (in Japanese).

20.8 Conclusion

While we were writing this article, news reports confirmed the first breeding of storks at the Watarase Retarding Basin. As mentioned previously, there had been records of storks flying into the retarding basin, but none of them were breeding. A male bird released in Noda City, Chiba Prefecture, and a female born in the field in Naruto City, Tokushima Prefecture, became the first known breeding pair. In March 2020, they began breeding in an artificial nest tower at the second regulating pond. As of the end of June 2020, two juveniles have been reported to be growing well. Since the reintroduction of the stork, many field breeding events have been recorded in western Japan, but this is the first time that the storks have bred in the field in eastern Japan. It is significant that the Watarase Retarding Basin was selected as the site for this breeding. During Typhoon Haggis in October 2019, it played a major role in preventing flooding of the Tone River system, including the Edo River, and it prevented flood damage in the Edo River basin in eastern Tokyo. The 100-year-old Watarase Retarding Basin is one of Japan's leading historical examples of Eco-DRR.

However, it should also be remembered that the Watarase Retarding Basin was originally built as a response to the first large pollution event in Japan. Yanaka Village, which was located in this area, suffered from floods on many occasions throughout its history, but the floods left the rich farmland in a state of flux. Fishing and the production of hats made of sedge were also popular livelihoods. Pollution from the Ashio Copper Mine destroyed the livelihoods of the residents of Yanaka Village. Although there is no documentation of the health hazards to the residents of Yanaka Village, in the village of Ueno in Tochigi Prefecture, the number of deaths was higher than the number of births after the flood of 1896. It was also reported that the number of stillbirths and deaths of children under the age of two increased markedly (Takaishi et al. 2015). To prevent the mine's toxic waste from spreading to the capital city and to implement flood control in the Tone River system, the government decided to abandon Yanaka Village and to establish a vast retarding basin. Residents of the former Yanaka Village were forced to relocate, some into the wilderness of Hokkaido, where the environment was so completely different and so severe that many of the inhabitants later abandoned the settlement (Matsuura 2015). Building the Watarase Retarding Basin stopped the mine's toxic waste from spreading farther downstream and prevented flooding throughout the Tone River basin. The retarding basin alone did not achieve all of these results, but the residents of the former Yanaka Village were the ones who were forced to bear the heaviest burden. So even though we now regard the Watarase Retarding Basin as a proud example of Japan's Eco-DRR, we must remember that it was built on the hardship of local people.

In 2019, Typhoon Haggis did not cause significant damage in the Tone River basin, but flooding and landslides in other parts of eastern Japan harmed as many residents and a large number of buildings. Most of these areas have long been identified on flood hazard maps as being at high risk of flooding. Japan's population is now rapidly declining, and there is no longer a need to continue to use areas at high risk of disaster. The Location Normalization Plans drawn up by municipalities to achieve a "compact city" have been designed in part to reduce exposure to disaster risk. The Ministry of Land, Infrastructure, Transport and Tourism (MLIT) has urged that high disaster risk areas be avoided as residential areas. Restoring natural areas in high-risk areas should act as a buffer against disaster, thereby reducing disaster risk. A major challenge for the future is how to apply Eco-DRR to land use planning. The decision-making process is important, particularly incorporating the wishes of residents living on high-risk land. A bottom-up planning approach is necessary. In addition, nature restoration will bring benefits not only to the residents in the vicinity of the restored areas, but also to the wider region. Securing the funds necessary for Eco-DRR is an important issue for consideration.

Acknowledgements We are grateful Y. Sakai, Japan Institute of Construction Engineering (JICE) for providing data on historical flood discharge designs of the Tone River system. We thank K. Nakamura of the Public Works Research Institute (PWRI) for providing information about the Watarase Retarding Basin. We appreciate diverse advice and helpful suggestions on field study from I.Washitani. We thank R Hashimoto and M. Ohwada for field surveys. We also thank the Upper Tone River Office of the Japanese Ministry of Land Infrastructure and Transport for providing the study opportunity and the data for analyses. This research was performed with aid from the Environment Research and Technology Development Fund (JPMEERF20154005, JPMEERF19S20530) of the Environmental Restoration and Conservation Agency of Japan.

Appendix 1. Chronological Order of Events Related to the Watarase Retarding Basin

Year	Date and month	Event
1879		A ban on the sale of fish in the Watarase River because of toxic water pollution
1884	November	A huge copper vein was discovered at the Asio Mine, and it became the largest copper mine in Japan
1885		Ashio mineral toxicity spread to the Watarase River, causing massive fish death
1890	22-Aug	Flood at the Watarase and Tone Rivers; 16.4 km^2 of rice paddies affected by mineral toxicity
1896	9-Sep	Great flood of the Watarase, Edo and Tone Rivers; 456 km^2 of rice paddies affected by mineral toxicity
1897	24-Mar	First Mineral Poisoning Incident Investigation Committee was established
1898	September	Siltation pond in Koga area was broken by flooding
1900	13-Feb	Kawamata Incident (resident's movement in Tokyo calling for the suspension of mining)
		First flood control plan of the Tone River system
	October	First Tone River improvement work started

(continued)

(continued)

Year	Date and month	Event
1902	17-Mar	Second Mineral Poisoning Incident Investigation Committee established
1903	2-Mar	Second Committee submitted a report to the Prime Minister of Japan
1906	1-Jul	Yanaka Village was abolished
1907	29 June to 5 July	All remaining houses in the former Yanaka Village demolished
1910	1-Apr	Construction work on the Watarase River started
	10-Aug	Largest flood in the Tone River system during the Meiji Era Revised flood control plan of the Tone River system
1911	April	137 residents (16 families) from the former Yanaka Village moved to Hokkaido
1918	August	Watarase River relocation project completed
1922		Embankment construction of the Watarase Retarding Basin completed
1930		First Tone River improvement work completed 1935
	September	Flood in the Tone River
1938	June to July	Flood in the Tone River
	August to September	Flood in the Tone River
1941		Revised flood control plan of the Tone River system
1947	September	Great flooding in the Tone River system caused by Typhoon Kathleen
1949		Revised flood control plan of the Tone River system
1963		Construction of the first regulating pond in the Watarase Retarding Basin began
1970		Construction of the first regulating pond completed
1972		Construction of the second regulating pond completed 1973
	February	Asio Mine closed
1989		Lake Yanaka constructed in the first regulating pond
1997		Construction of the third regulating pond completed
2019	12–13 October	Typhoon Haggis hit eastern Japan; Watarase Retarding Basin stored 95% of designed storage

References

Cowie NR, Sutherland WJ, Ditlhogo MKM, James R (1992) The effects of conservation management of reed beds. II. The flora and litter disappearance. J Appl Ecol 29:277–284. https://doi.org/10.2307/2404496

Deguchi M (1955) Studies on mineral pollution of the river Watarase (part 1): on properties of irrigation water and injurious status of paddy rice plant. Jpn J Soil Sci Plant Nutr 26:81–87. https://doi.org/10.20710/dojo.26.3_81

Editorial Committee of 100 Year History of the Tone River and Japan Institute of Construction Engineering (eds) (1987) 100 years history of the Tone River: flood control and irrigation. Kanto Regional Development Bureau, Ministry of Construction (in Japanese)

Gryseels M (1989) Nature management experiments in a derelict reedmarsh. I: Effects of winter cutting. Biol Conserv 47:171–193

Güsewell S, Le Nédic C (2004) Effects of winter mowing on vegetation succession in a lakeshore fen. Appl Veg Sci 7:41–48. https://doi.org/10.1111/j.1654-109X.2004.tb00594.x

Hisamatsu S (2011) A review of the Japanese Kateretidae fauna (Coleoptera: Cucujoidea). Acta Entomologica Musei Nationalis Pragae 51:551–585

Ishida S, Nakashizuka T, Gonda Y, Kamitani T (2008) Effects of flooding and artificial burning disturbances on plant species composition in a downstream riverside floodplain. Ecol Res 23:745–755. https://doi.org/10.1007/s11284-007-0434-4

Ishii J, Washitani I (2007) Habitat analyses of six threatened plant species in a moist tall grassland based on hyperspectral-remotely sensed data. Glob Environ Res 11:161–169

Ishii J, Lu S, Funakoshi S, Shimizu Y, Omasa K, Washitani I (2009) Mapping potential habitats of threatened plant species in a moist tall grassland using hyperspectral imagery. Biodivers Conserv 18:2521–2535

Ishii J, Hashimoto L, Washitani I (2011) Early vegetation growth in an experimental restoration site in the Watarase wetland. Jpn J Conserv Ecol 16:69–84 (in Japanese with English abstract)

Ishii J, Washitani I (2013) Early detection of the invasive alien plant *Solidago altissima* in moist tall grassland using hyperspectral imagery. Int J Remote Sens 34:5926–5936

Japan Map Center (2013) Q&A of survey and map. Japan Map Center, Tokyo (in Japanese)

Kamite Y, Hikida N, Satô M (2005) Notes on the Laccophilus kobensis species-group (Coleoptera, Dytiscidae) in Japan. Elytra 33:617–628

Klimkowska A, Van Diggelen R, Bakker JP, Grootjans AP (2007) Wet meadow restoration in western Europe: a quantitative assessment of the effectiveness of several techniques. Biol Conserv 140:318–328

Matsui K, Tanji T, Kato H (2004) Diverficication of public use for open space in the Watarse retarding basin. Ann Hum Reg Geogr 26:151–182 (in Japanese)

Matsumoto T, Nakai T, Fukuoka S, Sumi T (2014) Considerations of the flood control functions of the Watarase Retarding Basin. J Jpn Soc Civ Eng Ser B1 (Hydraul Eng) 70: I_1477-I_1482 (in Japanse with English abstract)

Matsuura S (2003) Ashio poisoning incident and Watarase River relocation. Water Sci 47:31–59. https://doi.org/10.20820/suirikagaku.47.3_31

Matsuura S (2015) Asio Mine poisoning incident and the Watarase River. Shinkoronsha, Tokyo (in Japanese)

Morishita T, Tsukigi H (1996) Behavior of contaminant elements in sediment, suspended solid and running water of the Watarase River. Environ Sci 9:357–368

Naito K, Ikeda H (2007) Habitat restoration for the reintroduction of Oriental White Storks. Glob Environ Res 11:217–221

Ohkawa H (2013) The visual dictionary of insects of Watarase-yusuichi. Watarase-yusuichi Acclimation Promoting Foundation, Tochigi (in Japanese)

Ohmomo S, Takahashi K, Nishiyama A (2011) A list of Coleoptera recorded in Ukishima area of Lake Kasumigaura, Inashiki, Ibaraki Prefecture, central Japan. Bull Ibaraki Nat Museum 14:75–92 (in Japanese with English abstract)

Ohwada M, Ogura H (1996) A floristic study of the Watarase retarding basin. Bull Tochigi Prefect Museum 13:31–108 (in Japanese with English abstract)

Ohwada M (2011) The visual dictionary of plants of Watarase-yusuichi. Watarase-yusuichi Acclimation Promoting Foundation, Tochigi (in Japanese)

Okuma T (1991) Evolution of flood control system in the Tone River. Jpn J Water Poll Res 14:146–150 (in Japanese)

Ostendorp W (1999) Management impacts on stand structure of lakeshore *Phragmites* reeds. Int Rev Hydrobiol 84:33–47

Sakai M, Ohbayashi N (1994) A new Anthicid beetle (Coleoptera, Anthicidae) associated with ditch reeds in marshy places of Japan. Jpn J Entomol 62:555–561

Sasakawa K (2016) Notes on the reproductive ecology and description of the preimaginal morphology of *Elaphrus sugai* Nakane, the most endangered species of *Elaphrus* Fabricius (Coleoptera: Carabidae) ground beetle worldwide. PLoS ONE 11: e0159164

Takaishi M, Oshima H, Asano S (2015) Pollution caused by Ashio Copper Mine and its effects on environment and human health. J Int Univ Health Welfare 20:59–69 (in Japanese with English abstract)

Wang Z, Nishihiro J, Washitani I (2015) Effects of traditional vegetation usage and management on the growth of facilitator keystone species in a moist tall grassland. Landsc Ecol Eng 11:101–109. https://doi.org/10.1007/s11355-013-0240-9

Washitani I (2001) Plant conservation ecology for management and restoration of riparian habitats of lowland Japan. Popul Ecol 43:189–195. https://doi.org/10.1007/s10144-001-8182-8

Washitani I (2007) Restoration of biologically-diverse floodplain wetlands including paddy fields. Glob Environ Res 11:135–140

Watanabe M, Serizawa S (2009) *Impatiens ohwadae*, a new species of Balsaminaceae from Japan. Shidekobushi 1:61–70 (in Japanese)

Yamada Y, Itagawa S, Yoshida T, Fukushima M, Ishii J, Nishigaki M, Ichinose T (2019) Predicting the distribution of released Oriental White Stork (*Ciconia boyciana*) in central Japan. Ecol Res 34:277–285. https://doi.org/10.1111/1440-1703.1063

Yoshida K, Nishiyama K (2008) Effects of periodical cutting management of *Phragmites communis* on species composition of plant communities. Izunuma-Uchinuma Wetl Res 2:89–96

Yoshikawa M, Izumi D, Hoshino Y (2013) Floristic characteristics of a floodplain tall-grass vegetation managed by winter burning in central Japan. Veg Sci 30:1–15. https://doi.org/10.15031/vegsci.30.1

Yoshimura C, Omura T, Furumai H, Tockner K (2005) Present state of rivers and streams in Japan. River Res Appl 21:93–112. https://doi.org/10.1002/rra.835

Chapter 21
Self-efficacy for EbA and Human Health in a Post-disaster Recovery Phase

Ai Tashiro

Abstract This part discusses the relationships between residents' health conditions and their self-efficacy for greenspace management in a post-disaster recovery phase in rural communities, which are environmentally vulnerable to natural disasters. Self-efficacy of residents is crucial to enhance local communities' inherent capabilities to build such a resilient living environment by promoting residents' participation in green activities. The prioritized issue is to restore post-disaster environments. Residents' health would be the key to enhance their self-efficacy and manage their neighboring living environment. Thus, this study aimed to examine residents' health conditions and self-efficacy for managing a post-disaster environment in Japan. Results showed no significant association of self-efficacy with physical activity or self-reported health (SRH). In contrast, good SRH was significantly associated with a higher rate of self-efficacy. Additionally, weeding experience and higher awareness of Eco-DRR had a positive association with self-efficacy. This study highlights how residents' good SRH influences self-efficacy for green environmental management at the individual level. If the residents who conduct community-based management are unhealthy, they would not feel confident about managing their green environments. This study implies the importance of integrating public health approach into post-disaster environmental management based on EbA strategies in a post-disaster rural context.

Keywords Disaster risk reduction · Ecosystem-based adaptation · Health statuses · Self-efficacy · Green environmental management

21.1 Introduction

After natural disasters occur, post-disaster environmental conditions directly or indirectly affect communities, human health, and quality of human life (Institute of Medicine 2015). Although a local neighborhood or the living environment is usually

---

A. Tashiro (✉)
Graduate School of Environmental Studies, Tohoku University, Sendai, Japan
e-mail: ai.tashiro.q7@dc.tohoku.ac.jp

© The Author(s), under exclusive license to Springer Nature Singapore Pte Ltd. 2021

M. Mukherjee and R. Shaw (eds.), *Ecosystem-Based Disaster and Climate Resilience*, Disaster and Risk Research: GADRI Book Series,
https://doi.org/10.1007/978-981-16-4815-1_21

discussed within sustainability, it is equally important for residents' quality of life. Indeed, degraded living environmental conditions after natural disasters may have adverse effects on ecosystems, biodiversity, and, in some cases, even more, extreme consequences (St. Cyr 2005).

Altogether, the interactions between healthy natural environments and healthy people are multiple and complex (Maller et al. 2006) and require interdisciplinary attention and action for full understanding and resilient development of both nature and human beings (Caillon et al. 2017). The post-disaster environmental management tries to protect ecosystems, sustain development, and adapt to or recover the post-disaster environment (Davidsson 2020). Nature conservation continues to decline, building a commitment to ecological restoration through volunteer travel, which is the key to addressing a wide range of environmental concerns (Strzelecka et al. 2017).

Ecosystem-based approaches are needed to sustainably manage coastal and marine resources (Berkes 2015; Foley et al. 2010). The sustainability of small-scale social-ecological systems (SESs) after natural disasters or climate change is challenged by external socio-economic and environmental drivers' changes that interact with endogenous drivers unexpectedly and in complex ways. The meanings ascribed to coastal green environments (Burley et al. 2007) and cultural factors (Martin et al. 2016) can impact the natural resources management in coastal SESs. Thus, to sustainably manage the coastal SESs and their ecosystem services, capturing and considering the range of cultural interrelationships, cultural meanings, values, and identities are required. Knowledge and practice in local communities are also related to coastal SESs (Poe et al. 2014).

Understanding how traditional ecological knowledge (TEK) is deployed is also useful for restoring coastal SESs. The cultural ecosystem services derived from coastal SESs can provide important insights into the underlying values and local community's connection with seascapes (Chakraborty et al. 2020). However, these underpinning cultural factors and values are often poorly understood (Rivero and Villasante 2016).

More recently, the success of ecological restoration projects globally depends on an individuals' willingness to contribute to the well-being of the endangered ecosystems through various activities like conservation careers, political involvements in nature conservation, or environmental volunteering behaviors (Ganzevoort and van den Born 2020). Indeed, given the negative ecological consequences of anthropogenic pressures on the natural environment (Evans et al. 2014), environmental volunteers are present at the core of organizations tasked to manage environmental programs (Bramston et al. 2010). Thus, a community's current volunteering movement has become an essential resource for ecological restoration (Ryan et al. 2001). Therefore, attracting and retaining an adequate supply of environmental volunteering behavior is a crucial task (Jerome et al. 2017).

After a natural disaster, environmental rehabilitation at a small-scale community level generates significantly higher wealth benefits like reduction and optimum natural resource utilization. For post-disaster environmental recovery, considering the utilization of a self-efficacy mechanism is useful because self-efficacy is related to people's ability to intervene for improving the community environment (Wu and

Mweemba 2010). Self-efficacy is defined as individuals' actions based on their judgment about what they can do and their beliefs about their actions' expected consequences (Sawitri et al. 2015). Driven by one's belief that participation in restoring a community environment is a worthwhile activity, hedonic values fostered by self-efficacy depend on the promise of environmental management experiences for ecological restoration (Wu and Mweemba 2010). However, only a few studies focused on community management and residents' environmental self-efficacy as drivers in the post-disaster phase (Monday and Sunday 2019).

Self-efficacy can improve the degraded environment and mitigate natural disasters. Environmental health management by self-efficacy is the intentional modification of the natural resources and built environment for providing opportunities for health promotion (Monday and Sunday 2019; Sawitri et al. 2015). However, few studies attempt an integrated approach of human health, self-efficacy for environmental management, and awareness of ecosystem-based disaster risk reduction (Eco-DRR). To maintain good self-efficacy, residents need to be healthy and be capable of undertaking some environmental management tasks and being aware of Eco-DRR.

Thus, this study aims to examine the association between residents' health status, capacity related to green management, and awareness of Eco-DRR, and self-efficacy in a post-disaster recovery phase.

Population decline provides ample opportunities for implementing Eco-DRR. Eco-DRR takes advantage of ecosystems' multi-functionality, including the capacity to mitigate disaster risks while providing multiple ecosystem services (Nehren et al. 2016). The adaptation to disaster risks is vital for local communities to coexist with nature and community development. As one of the strategies, it is a key for local communities to reconcile the conservation and utilization of the local natural resources with their livelihoods and economic concerns (Okello et al. 2009). Lessons from previous disasters, such as the recent earthquakes in Pakistan and China, the Indian Ocean tsunami, and hurricanes Katrina, provide evidence that integrates the environment's role in the immediate days after the disaster. The restoration of livelihoods is essential for a successful long-term recovery (Mainka and McNeely 2011). Empirical studies suggest that economic compensation for the loss of access to natural resources and interventions in income-generating activities are associated with ecotourism (Masterson et al. 2019). The interventions are used as "win–win" opportunities with ecological and socio-economic benefits. However, this approach overlooks the cultural, historical, and personal values of nature for local people (Masterson et al. 2019). The natural infrastructure provided by well-functioning ecosystems has several advantages over physical infrastructures, such as dams and levees (Sutton-Grier et al. 2015).

21.2 Post-disaster EbA Solution in the Japanese Context

Natural disasters have been increasing over the years, partly due to the current climate change. The adaptation to disaster risks is vital for rural communities that are environmentally vulnerable (Paton and Johnston 2001). Natural disasters affect not only people but also the ecosystems in which they live. On March 11, 2011, devastating earthquakes and tsunami hit Japan, which reduced 250 ha of sand beach and dune vegetation to 100 ha compared to prior the disaster in the most affected three prefectures in the Tohoku region (Ministry of the Environment 2015). The impact of the 2011 tsunami disaster in Japan and several coastal communities in the Tohoku region was described as fragile (Koshimura et al. 2014). With a high dependency on a severely depleted and overfished natural resource base and badly degraded coastal ecosystems, few coastal communities in the disaster-affected areas have found sustainable routes out of the 2011 tragedy. For such communities, the added onslaught brought by a natural disaster can prove to be an almost intolerable burden that leaves the existing livelihood options inadequate (Koshimura et al. 2014).

After the Great East Japan Earthquake and Tsunami in 2011 (GEJET), many of the disaster-affected coastal communities in the Tohoku region, northeast Japan, are experiencing shrinking populations (Murakami et al. 2014). The Tohoku region has been prone to natural disasters for a long time (Tashiro 2020). There is a long-standing issue on how to align the conservation of ecosystems with poverty alleviation and development in communities with the accelerated rate of shrinking populations. However, people have long cultivated a sense of reverence for nature, which can be both nurturing and destructive, and fostered wisdom for adapting to and living in harmony with nature instead of conquering it (Tashiro et al. 2018). Rural communities also have a traditional way of utilizing ecosystems to mitigate disaster: maintaining forests to prevent soil erosion; planting pine trees along the coast to mitigate winds and blown sand; planting bamboo trees along river banks to reduce flooding; and using rice paddies for temporary water retention (Ministry of the Environment 2016; Tashiro 2020). Such TEK effects reduce opportunities for capital investments related to building infrastructures for disaster prevention and reduction. TEK also makes up for the labor shortage through social cohesion.

Since the Sendai Framework for Disaster Risk Reduction (SFDRR) was advocated for ecosystem-based adaptation (EbA) in 2015 from Eco-DRR/Climate Change Adaptation perspective (Kelman 2015). The post-disaster reconstruction in disaster-affected rural communities has been garnering increased attention. By the SFDRR, the role of ecosystems and the environment is underpinned in a post-disaster context. Implementing integrated ecosystem management and risk reduction strategies is gaining importance in the Tohoku region (Renaud and Murti 2013). Eco-DRR takes advantage of ecosystems' multi-functionality, including their capacity to mitigate disaster risks while providing multiple ecosystem services.

So far, in the Tohoku region, local people have coexisted with nature to reconcile the conservation and development of a local community for the loss of satisfying their basic needs and improving their well-being (Tashiro 2020; Tashiro and

Nakaya 2020). However, after the GEJET, building infrastructure projects, such as giant seawalls, raising roads, and banking plans, are some of the main causes of habitat loss in the disaster-affected areas (Tashiro and Sakisaka 2015). As a result, an intricate interaction between humans and their natural environment has caused a change in the ecosystem. Thus, a lack of environmental management brought environmentally vulnerable lands (Zaré and Ghaychi Afrouz 2012). Changes in ecosystem services are almost always caused by multiple interacting drivers (Ioan and Gradinaru 2011). Therefore, this research focuses on post-disaster rehabilitation lessons to draw general principles for rehabilitating poor coastal communities' livelihoods. This research contends that avoiding past mistakes requires a framework for understanding the diversity of coastal people's livelihood strategies and endogenous drivers' behavior for rehabilitating sustainable communities.

After the GEJET, the growing recognition of EbA by trees and forests as well as their vulnerability has moved rural forest definitions toward a more holistic and Eco-DRR concept in rural communities. After the disaster, the disaster-affected coastal communities faced seawall construction problems. Local governments planned to decide the height of concrete seawalls to deal with Level 1 tsunamis (Tashiro and Sakisaka 2015), roughly once every several decades to one hundred years plus several decades. A seawall is a planned gigantic concrete wall with a maximum height of about 15 m and a width of around 100 m, extending several hundreds of meters along the seashore with huge taxes (Nakajima coast in Koizumi District) (Tashiro and Sakisaka 2015). However, such a gigantic seawall is suggested to adversely impact tourism and inshore fishing, reduce disaster awareness by blocking the ocean's view, lead to a decline in seaside lifestyle and culture, and negatively affect the ecosystem (Martini and Minca 2018). Over the seawall issue, the government made changes lowering the height in some places, whereas per their judgment, there are no assets behind the seawall that should need protection. However, opinions on what should be protected from residents differ depending on where they live, how they make a living, and so on (Kimura 2016).

21.3 Case Study of EbA in a Post-disaster Rural Community

21.3.1 Study Site

Maehama community (141.54°15′–141.55°16′ E, 38°80′–38°81′ N) in Motoyoshi-cho in the southern region of Kesennuma City in the Miyagi Prefecture is an ecological community where people coexist with nature (Fig. 21.1). However, when the GEJET occurred, local people faced the danger of being separated from seascapes. The tsunami struck the Maehama community on March 11, 2011. About 30 min before the earthquakes, the high waves had already reached the Maehama Fishing Port. Offshore within the port that had now turned into an inlet, the two break walls

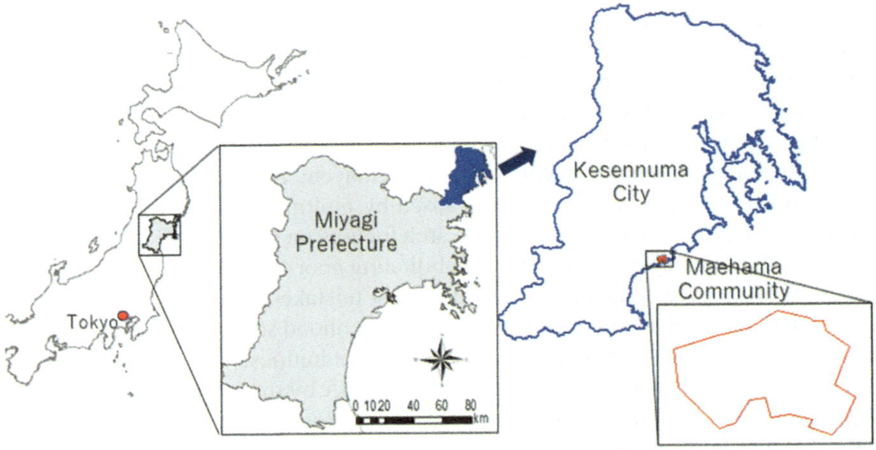

Fig. 21.1 Study site, the Maehama community, Kesennuma City, Miyagi Prefecture. Created by author

sticking out from both east and west sides of the coast were about to be swallowed by tsunami waves. By the disaster, 39 households out of 130 (population of 450), 20 completely or partially destroyed households, four dead, and five missings (Hiroshige 2014). The number of households decreased to 105 in April 2019.

Like other communities, seawall construction was discussed over the seawall issues, such as the height, cost, and seawalls materials in this community to defend the future tsunamis. After several years of discussion, it was decided to use local coastal vegetation as a seawall. Instead of constructing a concrete higher seawall, the local community marked the highest wave of the 2011 tsunami with trees and stakes. These signatures remain visible, keeping the memory of the tsunami's dimensions alive (Tashiro 2020).

Utilizing the characteristics of conventional ecological communities, residents started the Camellia Forest Project. In this project, Japanese camellia trees were planted as ecological seawalls (Fig. 21.2). In this community, camellia seeds drifted from the sea and grew naturally in this area for a long time (Hiroshige 2014). Most Japanese camellia trees' height is around 3–5 m, and some even grow up to 10 m. The ancestors in the Maehama community knew that camellia trees withstand high tides, and this traditional wisdom for Eco-DRR-based seawall was passed down from the ancestors to the current residents (Chiba 2015). Camellia trees not only benefit the Eco-DRR but also work in multiple ways, which contribute to a sustainable landscape/seascapes and provide social-economic benefits. Residents extract camellia oil from the camellia trees and produce and sell it as local products. The "Camellia Forest Project" will enable the following effects: self-supply of oil (local production for local consumption) through collaborative work and sharing among residents, create social participation, and health benefits of ingesting oleic acid of camellia oil (Hiroshige 2014).

Fig. 21.2 **a** Camellia planting; **b** Camellia forests management with local people and volunteers. *Source* Photos by Dr. Takeshi Hiroshige

Camellia trees can also be useful in tidal forests, windbreak forests, forest parks for disaster prevention and mitigation, leisure, and cultural and conservation services. This local project provides the possibility of exiting the regional economy of the anemia structure. Its reflection on administrative dependence is linked to the supply from the vast conventional system utilization of local ecological resources like camellia trees (Chiba 2015). It is important to encourage all residents in a community to participate in the camellia project. However, some residents and outside volunteers participated in the project by planting trees, collecting seeds, raising seedling, and improving the planting area (Hiroshige 2014).

Due to depopulation related to aging, until the disaster, TEK of camellia had been abolished. Still, the GEJET strengthened the community's cohesiveness again. During discussions regarding future recovery plans, residents re-recognized the value of local natural resources, the ancestors' wisdom, and decided to utilize it for community recovery (Hiroshige 2014). Using TEK, residents' various ideas, abilities, and "age tech" are also meaningful for sustainable resource usage in a homogenous rural community.

In disaster-affected areas, residents' mental recovery tends to delay (Harada et al. 2015). Mental distress negatively affects the motivation of ecosystem management (Bratman et al. 2019). Thus, understanding residents' connection and self-sufficient living with a stable life is needed to propose co-management by the local actors and residents. However, few studies attempted to understand the potential possibilities of residents' ownership or motivation for managing local resources with local livelihoods, residents' health, and disaster awareness. For that reason, this study examined the association between residents' health conditions, disaster awareness, and local resource management motivation.

21.3.2 Data Set

The survey was conducted in Maehama district in March 2019. To clarify the diversity of coastal people's livelihood strategies and endogenous drivers' behavior in the post-disaster community rehabilitation, a quantitative survey and face-to-face interviews for local actors were conducted. Interviews, particularly, focused on the project's opinions through a short narration about the interviewee's involvement with the project and his/her understanding of its aims and impact, including perceptions of how the project was perceived and played out in the community.

During the survey period, the author spent a few days walking and driving through the local landscape and seascapes accompanied by local informants, which helped in describing the local endogenous bio-cultural classification of types of places that forms the basis for asking the interviewees about the emotions they associate with the Camellia Forest Project. Based on the interview survey, a mail survey was conducted for all the households in the Maehama community ($N = 105$) with five questionnaire sheets between 6 and April 25, 2019. In the questionnaire, the participants were asked to fill in demographic characteristics like age, sex, occupation, number of years of living in the current place, household number, disaster damage of housing, and household economic condition. The questionnaire also had questions regarding health conditions, such as stress [no (1, 2, 3, and 4) or yes], regular exercise (30 min or more twice a week), and self-rated health answered with 4 Likert scales (poor 1, 2, 3, 4 good).

Additionally, questions related to local environmental green management, mowing experience, self-efficacy (the degree of individual contribution to the improvement of the community environment through various greening activities, such as flower planting along the road, picking up garbage, weeding, with 7-grade evaluation), and feeling solastalgia on the change of neighborhood nature (do you feel anxiety due to environmental degradation in Maehama? With 5-grade evaluation), and Eco-DRR (do you think that greening management raises the residents' awareness of DRR? With 7-grade evaluation) were also included in the questionnaire. The targeted participants of the survey were residents above the age of 7. The households' collecting rate was 45% (47/105), whereas the respondents' obtained number was 110. Of these, invalid data of age and sex were excluded, which left us with valid data of 90 respondents.

21.3.3 Statistical Method

The ordered logit model analysis was used to examine the association between residents' characteristics, health conditions, and attitudes toward coastal green management. The dependent variable was self-efficacy on green management, and the independent variables were personal characteristics, health conditions, green management behaviors and motivation, and awareness of Eco-DRR. After the ordered logit model analysis, predicted probabilities change (dy/dx) of self-efficacy level was calculated to examine the effects of the respondents' health status, green management, and awareness of Eco-DRR ($p < 0.01$).

21.4 Results

The number of respondents with valid responses was 90. Among 90 responses, 79% of the respondents were over the age of 50 years of age. Additionally, 52% of the respondents were female. 37% were unemployed, 16% of respondents were company employees, and 12% of respondents were had a part-time job. Regarding household economic conditions as compared with pre-GEJET, 39% of the respondents answered being on a tight budget, and 56% of respondents answered no change in the household economic conditions. Regarding the housing damage, 60% of the respondents answered completely destroyed (beyond repair), half-destroyed (requires major repairs), or partially damaged (requires minor repairs).

Results from logistic regression models found no significant association of self-efficacy with physical activity or self-reported health. In contrast, good SRH was significantly associated with higher self-efficacy. In addition, weeding experience and higher awareness of Eco-DRR had positive associations with self-efficacy. 73% of the respondents lived with three or four people family members.

Among the respondents, 39% answered feeling stressed regarding health conditions, and 26% answered good or moderate SRH. 50% answered that they have regular exercise habits. Regarding green management, 58% answered four or more for their having self-efficacy. 82% percent of the respondents answered having a weeding experience in their living neighborhood. Furthermore, 22% of the respondents answered feeling solastalgia regarding the change of the neighborhood's nature after the disaster. Lastly, 54% of the respondents answered that greening management raises residents' awareness of Eco-DRR.

Table 21.1 shows the association between self-efficacy and respondents' characteristics, health status based on the ordered logistic regression analysis.

As shown in Table 21.1, regarding in terms of personal characteristics, men were more positively associated with self-efficacy than females (OR 3.40; 95%CI, 0.96–12.01). People aged between 40 and 49 were negatively associated with self-efficacy (OR 0.06; 95% CI, 0.01–0.70). Regarding occupation, when referring to an unemployed group, housekeeping was positively associated with self-efficacy (OR 7.14; 95% CI, 1.27–40.11). Living the number of years living at the current place was negatively associated with self-efficacy (OR 0.21; 95% CI, 0.06–0.82). Regarding health statuses, stress and regular exercise were not statistically not associated with self-efficacy. In contrast, Good SRH was positively associated with self-efficacy (OR 4.19; 95% CI, 1.14–15.45). Regarding green management, solastalgia was not statistically not associated with self-efficacy. In contrast, mowing experience and Eco-DRR awareness were positively associated with self-efficacy (OR 4.55; 95% CI, 1.04, 19.88, OR 5.79; 95% CI, 1.72–19.48, respectively).

Table 21.1 The association between respondents' characteristics, health status, green management, and self-efficacy

	Odds ratio	*p*-Value	95%CI
Personal characteristics			
Sex			
Male	3.40	0.06	(0.96, 12.01)
Female	Ref.		
Age			
7–12	–		
13–15	0.21	0.52	(0.00, 25.56)
16–29	0.08	0.14	(0.00, 2.28)
30–39	Ref.		
40–49	**0.06***	**0.03**	**(0.01, 0.70)**
50–59	0.26	0.29	(0.02, 3.09)
60–69	0.26	0.26	(0.03, 2.68)
70 or more	0.28	0.31	(0.02, 3.27)
Occupation			
Civil servant	0.44	0.49	(0.04, 4.61)
Civil engineering	0.36	0.64	(0.01, 23.9)
Technical staff	0.04	0.11	(0.00, 2.02)
Manufacturer	0.04	0.10	(0.00, 1.85)
Company employees (others)	0.71	0.74	(0.10, 5.19)
Agriculture	0.68	0.84	(0.02, 27.07)
Marine industry	0.57	0.65	(0.05, 6.33)
Transport and communication			
Self-employed	0.17	0.20	(0.01, 2.55)
Housekeeping	**7.14***	**0.03**	**(1.27, 40.11)**
Part-time job	0.94	0.94	(0.19, 4.61)
Student	–		
Unemployed	Ref.		
Others	0.76	0.87	(0.03, 20.01)
Household economic status			
Very poor	Ref.		
Poor	0.66	0.63	(0.12, 3.61)
Normal	1.54	0.63	(0.27, 8.67)
Stable	0.85	0.91	(0.05, 14.94)
Very stable	0.23	0.56	(0.00, 30.83)
Household size			

(continued)

Table 21.1 (continued)

	Odds ratio	*p*-Value	95%CI
Alone	Ref.		
Two	6.91	0.25	(0.26, 181.75)
Three	12.43	0.13	(0.50, 311.4)
Four	8.70	0.20	(0.31, 245.8)
Five or more	4.13	0.51	(0.06, 266.9)
Housing damage			
Completely destroyed	Ref.		
Almost all destroyed	0.65	0.75	(0.05, 9.31)
Half-destroyed	0.36	0.44	(0.03, 4.70)
Partially destroyed	1.08	0.92	(0.24, 4.92)
No damage	0.44	0.32	(0.09, 2.22)
Others	2.89	0.59	(0.06, 139.5)
Living years in the current place	**0.21***	**0.02**	**(0.06, 0.82)**
Health status			
Stress			
Yes	1.04	0.95	(0.35, 3.11)
No	Ref.		
Good self-rated health			
Yes	**4.19***	**0.03**	**(1.14, 15.45)**
No	Ref.		
Regular exercise			
Yes	0.92	0.88	(0.31, 2.73)
No	Ref.		
Green management			
Mowing experience			
Yes	**4.55***	**0.04**	**(1.04, 19.88)**
No	Ref.		
Solastalgia			
Yes	0.53	0.30	(0.16, 1.76)
No	Ref.		
Eco-DRR awareness			
Yes	**5.79****	**0.01**	**(1.72, 19.48)**
No	Ref.		

***$p < 0.001$, **$p < 0.01$, *$p < 0.05$

To examine the effects of the respondents' health status and green management and awareness of Eco-DRR on self-efficacy, predicted probabilities change (dy/dx) of self-efficacy level was calculated by each variable, which is statistically significant in Table 21.1 ($p < 0.01$).

Figure 21.3 shows the results of the marginal effect based on the results of Table 21.1. The vertical axis shows a conditional mean of self-efficacy of change when the binary independent variable of SRH (poor/good), having mowing experience (no/yes), and awareness of Eco-DRR (low/high) changes. As a whole, among the three variables, a larger scale of self-efficacy showed a positive difference within the variables.

Regarding the predicted probability change of self-efficacy of respondents who answered 1–4, dy/dx was negative, but those who answered 5–7, dy/dx, were positive. The result indicates that the predicted values of self-efficacy were greater in good SRH than in poor SRH. The predicted probability change of self-efficacy for having mowing experience was greater than that for not having mowing experience when the respondents answered 4 or more. Eco-DRR showed the same trends regarding

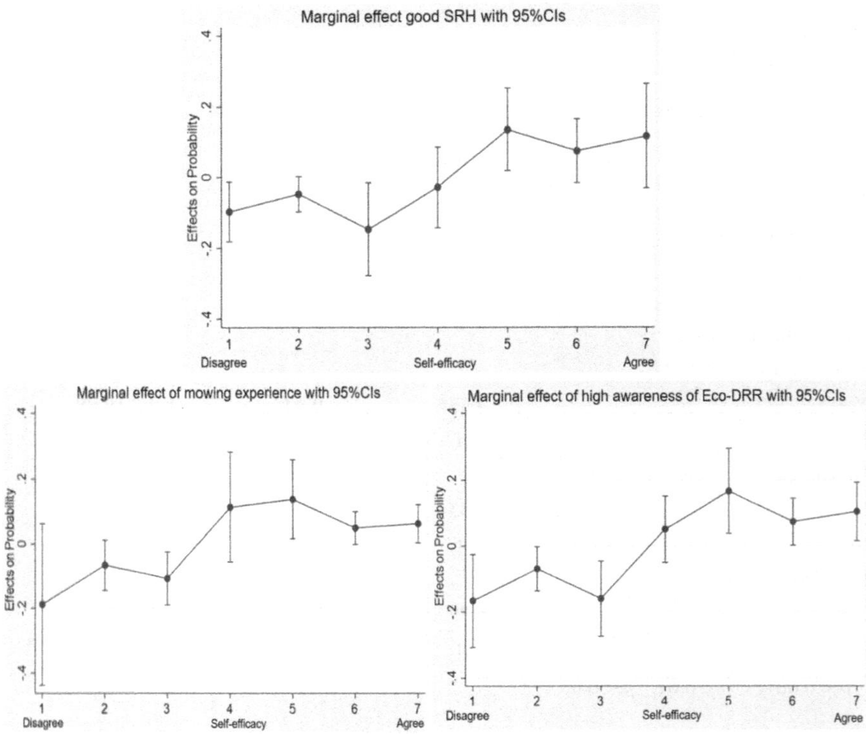

Fig. 21.3 Adjusted prediction of self-efficacy by conditional marginal effects of self-rated health, mowing experience, and awareness of Eco-DRR based on the results in Table 21.1

the mowing experience. The predicted probability change of self-efficacy for having a high awareness of Eco-DRR was greater than that for having a low awareness of Eco-DRR when the respondents answered 4 or more.

21.5 Conclusion

This study highlights the residents' health condition, green management, and awareness of Eco-DRR associated with green environmental management self-efficacy at the individual level. The result of the study shows little association between personal characteristics and self-efficacy. The reason for this weak association is that homogeneity is relatively high in a small community so that demographic variables were not an influential factor in self-efficacy. In contrast, good SRH was associated with self-efficacy. As for green management regarding individual behaviors, mowing experience leads to a more positive self-efficacy. Greater awareness of Eco-DRR would enhance self-efficacy. From these findings, it can be concluded that embedded behaviors and Eco-DRR awareness transcend demographic factors in a rural context. Residents' health status and green management attitude play a significant role in maintaining self-efficacy.

Lessons from the Maehama community, understanding residents' health, potential behavior that enhances green management, and disaster risk awareness are effective strategies to protect local natural resources. Because the residents sustaining community-based management for EbA are unhealthy, they would not feel conficent in managing their green environments. The community's green environments would not be maintained eventually. Thus, this study concludes the importance of rural residents' health and their neighborhood environments for successful EbA in post-disaster communities. Moreover, awareness of Eco-DRR is a significant precursor of environmental self-efficacy and behavior. For successful co-management for the Camellia Forest Project, volunteers, supporters, and other stakeholders must have a clear consciousness of the residents' self-efficacy and related factors' potential capacity.

In the Maehama community, ecological restoration has traditionally involved the local people attempting to refurbish at-risk, damaged, or destroyed ecosystems to some standard of health. However, the success of ecological restoration projects globally depends on the residents' contribution to the disaster's degraded neighborhood environments through activities like the Camellia Forest Project. In this project, a mowing experience is a small individual activity, but mowing or weeding practices have been rooted in rural communities for years (Wu et al. 2019). These practices seem to bother general, but weeding management leads to social cohesion in rural communities and brings a hedonic value. The management and maintenance of camellia trees and ameliorating the neighborhood environment are necessary because camellia trees are united with nature surrounding this community. For this reason, regular environmental maintenance will maintain the local landscape.

This study explored the utilization of self-efficacy theory (Alexander and Ward 2018) in a post-disaster rural context to understand how self-efficacy relates to the EbA solution. Following previous discussions on motivations for green management and voluntary conservation (Cortés-Capano et al. 2020) hedonic experience induced by green management is proposed as a catalyst for fostering environmental self-efficacy. Findings in this study provide a new understanding of how the notions of hedonism (health and well-being), self-efficacy, and disaster risk awareness involvement in environmental stewardship relate to each other in the context of a post-disaster recovery process offers a base for nurturing the residents' commitment to ecological restoration projects. In other words, the results of this study may provide a clearer picture of a social psychological mechanism that prompts residents to devote time and energy to challenge environmental change and promote a more sustainable environment where people coexist with nature in communities. In this sense, the importance of integrating public health approach into post-disaster environment management based on EbA strategies in post-disaster rural communities is implied.

Acknowledgements The author would like to express her sincere gratitude to Koji Hatakeyama, a representative of the general incorporated association, 'Oraho no Totte-Oki' for his assistance in promoting the research and to Shinetsu Kikuchi, a chairman of the Maehama community development association, for his help with acquisition and processing of the necessary data for this research. This study was financially supported by the Organization for Landscape and Urban Green Infrastructure and by the Division for Interdisciplinary Advanced Research and Education, Tohoku University.

References

Alexander AC, Ward KD (2018) Understanding postdisaster substance use and psychological distress using concepts from the self-medication hypothesis and social cognitive theory. J Psychoactive Drugs. https://doi.org/10.1080/02791072.2017.1397304

Berkes F (2015) Coasts for people: interdisciplinary approaches to coastal and marine resource management. Routledge. https://doi.org/10.4324/9781315771038

Bramston P, Pretty G, Zammit C (2010) Assessing environmental stewardship motivation. Environ Behav. https://doi.org/10.1177/0013916510382875

Bratman GN, Anderson CB, Berman MG, Cochran B, de Vries S, Flanders J, Folke C, Frumkin H, Gross JJ, Hartig T, Daily GC (2019) Nature and mental health: an ecosystem service perspective. Sci Adv. https://doi.org/10.1126/sciadv.aax0903

Burley D, Jenkins P, Laska S, Davis T (2007) Place attachment and environmental change in Coastal Louisiana. Organ Environ. https://doi.org/10.1177/1086026607305739

Caillon S, Cullman G, Verschuuren B, Sterling EJ (2017) Moving beyond the human–nature dichotomy through biocultural approaches: including ecological well-being in resilience indicators. Ecol Soc. https://doi.org/10.5751/ES-09746-220427

Chakraborty S, Gasparatos A, Blasiak R (2020) Multiple values for the management and sustainable use of coastal and marine ecosystem services. Ecosyst Serv. https://doi.org/10.1016/j.ecoser.2019.101047

Chiba H (2015) Maehama Camellia Forest Project: traditional ecological knowledge for future. Special issue, Ecological design for disaster risk reduction. Bio City 61:54–61 (in Japanese)

Cortés-Capano G, Toivonen T, Soutullo A, Fernández A, Dimitriadis C, Garibotto-Carton G, Di Minin E (2020) Exploring landowners' perceptions, motivations and needs for voluntary conservation in a cultural landscape. People Nat. https://doi.org/10.1002/pan3.10122

Davidsson Å (2020) Disasters as an opportunity for improved environmental conditions. Int J Disaster Risk Reduct. https://doi.org/10.1016/j.ijdrr.2020.101590

Evans CD, Bonn A, Holden J, Reed MS, Evans MG, Worrall F, Couwenberg J, Parnell M (2014) Relationships between anthropogenic pressures and ecosystem functions in UK blanket bogs: linking process understanding to ecosystem service valuation. Ecosyst Serv 9:5–19. https://doi.org/10.1016/j.ecoser.2014.06.013

Foley MM, Halpern BS, Micheli F, Armsby MH, Caldwell MR, Crain CM, Prahler E, Rohr N, Sivas D, Beck MW, Carr MH, Crowder LB, Emmett Duffy J, Hacker SD, McLeodj KL, Palumbi SR, Peterson CH, Regan HM, Steneck RS (2010) Guiding ecological principles for marine spatial planning. Mar Policy. https://doi.org/10.1016/j.marpol.2010.02.001

Ganzevoort W, van den Born RJG (2020) Understanding citizens' action for nature: the profile, motivations and experiences of Dutch nature volunteers. J Nat Conserv. https://doi.org/10.1016/j.jnc.2020.125824

Harada N, Shigemura J, Tanichi M, Kawaida K, Takahashi S, Yasukata F (2015) Mental health and psychological impacts from the 2011 Great East Japan Earthquake Disaster: a systematic literature review. Disaster Mil Med. https://doi.org/10.1186/s40696-015-0008-x

Hiroshige T (2014) A study on the recovery process from the Great East Japan Earthquake—the case of Maehama area of Motoyoshi town, Kesennuma city and its support activity. Waseda Daigaku Shakai Kagaku Gakkai 15(1): 27–41

Institute of Medicine (2015) Healthy, resilient, and sustainable communities after disasters: strategies, opportunities, and planning for recovery. The National Academies Press, Washington, DC. https://doi.org/10.17226/18996

Ioan I, Gradinaru G (2011) Drivers of ecosystem change. Quality—access to success

Jerome G, Mell I, Shaw D (2017) Re-defining the characteristics of environmental volunteering: creating a typology of community-scale green infrastructure. Environ Res 158:399–408. https://doi.org/10.1016/j.envres.2017.05.037

Kelman I (2015) Climate change and the Sendai framework for disaster risk reduction. Int J Disaster Risk Sci. https://doi.org/10.1007/s13753-015-0046-5

Kimura S (2016) When a seawall is visible: infrastructure and obstruction in post-tsunami reconstruction in Japan. Sci Cult 25(1):23–43. https://doi.org/10.1080/09505431.2015.1081501

Koshimura S, Hayashi S, Gokon H (2014) The impact of the 2011 Tohoku earthquake tsunami disaster and implications to the reconstruction. Soils Found. https://doi.org/10.1016/j.sandf.2014.06.002

Maller C, Townsend M, Pryor A, Brown P, St Leger L (2006) Healthy nature healthy people: "contact with nature" as an upstream health promotion intervention for populations. Health Promot Int. https://doi.org/10.1093/heapro/dai032

Martin CL, Momtaz S, Gaston T, Moltschaniwskyj NA (2016) A systematic quantitative review of coastal and marine cultural ecosystem services: current status and future research. Mar Policy. https://doi.org/10.1016/j.marpol.2016.09.004

Martini A, Minca C (2018) Affective dark tourism encounters: Rikuzentakata after the 2011 Great East Japan Disaster. Soc Cult Geogr. https://doi.org/10.1080/14649365.2018.1550804

Ministry of the Environment (2015) Green reconstruction project—results of the 2014 Tohoku Pacific Coast Natural Environment Survey. Press. https://www.env.go.jp/press/100898.html. Accessed 8 Aug 2020) (in Japanese)

Ministry of the Environment (2016) Ecosytem-vased disaster risk reductoin in Japan. Nature Conservation Bureau, Ministry of the Environment. https://www.env.go.jp/en/nature/biodiv/eco-drr.pdf

Monday IF, Sunday IE (2019) Psychological factors influencing waste disposal behaviour among residents in Yoruba speaking communities of South-Western, Nigeria. J Solid Waste Technol Manag 45(3):315–328. https://doi.org/10.5276/JSWTM/2019.315

Mainka SA, McNeely J (2011) Ecosystem considerations for post-disaster recovery: lessons from China, Pakistan, and elsewhere for recovery planning in Haiti. Ecol Soc. https://doi.org/10.5751/ES-03858-160113

Masterson VA, Spierenburg M, Tengö M (2019) The trade-offs of win–win conservation rhetoric: exploring place meanings in community conservation on the Wild Coast, South Africa. Sustain Sci 14:639–654. https://doi.org/10.1007/s11625-019-00696-7

Murakami K, Wood DM, Tomita H, Miyake S, Shiraki R, Murakami K, Itonaga K, Dimmer C (2014) Planning innovation and post-disaster reconstruction: the case of Tohoku, Japan/reconstruction of tsunami-devastated fishing villages in the Tohoku region of Japan and the challenges for planning/post-disaster reconstruction in Iwate and new planning chal. Plan Theory Pract 237–242. https://doi.org/10.1080/14649357.2014.902909

Nehren U, Dac Thai H, Marfari MA, Raedig C, Alfonso S, Sartohadi J, Castro C (2016) Ecosystem-based disaster risk reduction and adaptation in practice. https://doi.org/10.1007/978-3-319-43633-3

Okello MM, Ole Seno SK, Nthiga RW (2009) Reconciling people's livelihoods and environmental conservation in the rural landscapes in Kenya: opportunities and challenges in the Amboseli landscapes. Nat Res Forum. https://doi.org/10.1111/j.1477-8947.2009.01216.x

Paton D, Johnston D (2001) Disasters and communities: vulnerability, resilience and preparedness. Disaster Prev Manag: Int J. https://doi.org/10.1108/EUM0000000005930

Poe MR, Norman KC, Levin PS (2014) Cultural dimensions of socioecological systems: key connections and guiding principles for conservation in coastal environments. Conserv Lett. https://doi.org/10.1111/conl.12068

Renaud FG, Murti R (2013) Ecosystems and disaster risk reduction in the context of the Great East Japan Earthquake and Tsunami: a scoping study. UNU-EHS Publication Series. https://doi.org/10.1787/9789264065239-en

Rivero S, Villasante S (2016) What are the research priorities for marine ecosystem services? Mar Policy. https://doi.org/10.1016/j.marpol.2016.01.020

Ryan RL, Kaplan R, Grese RE (2001) Predicting volunteer commitment in environmental stewardship programmes. J Environ Plan Manag. https://doi.org/10.1080/09640560120079948

Sawitri DR, Hadiyanto H, Hadi SP (2015) Pro-environmental behavior from a social cognitive theory perspective. Procedia Environ Sci. https://doi.org/10.1016/j.proenv.2015.01.005

Science R (2015) Model analysis of residents' consciousness regarding seawall construction: case study of Kesennuma City in Miyagi Prefecture. Stud Reg Sci 45(4)

St. Cyr JF (2005) At risk: natural hazards, people's vulnerability, and disasters. J Homel Secur Emerg Manag. https://doi.org/10.2202/1547-7355.1131

Strzelecka M, Boley BB, Woosnam KM (2017) Place attachment and empowerment: do residents need to be attached to be empowered? Ann Tour Res 66:61–73. https://doi.org/10.1016/j.annals.2017.06.002

Sutton-Grier AE, Wowk K, Bamford H (2015) Future of our coasts: the potential for natural and hybrid infrastructure to enhance the resilience of our coastal communities, economies and ecosystems. Environ Sci Policy. https://doi.org/10.1016/j.envsci.2015.04.006

Tashiro A (2020) Implementation of green infrastructure in post-disaster recovery. In: Leal Filho W, Azul A, Brandli L, Özuyar P, Wall T (eds) Sustainable cities and communities. Encyclopedia of the UN sustainable development goals. Springer, Cham. https://doi.org/10.1007/978-3-319-71061-7_110-1

Tashiro A, Sakisaka K (2015) Model analysis of residents' consciousness regarding seawall construction: case study of Kesennuma City in Miyagi Prefecture. Stud Reg Sci 45:419–433. https://doi.org/10.2457/srs.45.419

Tashiro A, Nakaya T (2020) Post-disaster health status and coastal infrastructure reconstruction after the great East Japan earthquake and tsunami. ASM Sci J 13(SI 5):198–206

Tashiro A, Uchiyama Y, Kohsaka R (2018) Marine circular economy towards post-disaster reconstruction for sustainability: experiences in a small coastal town of Northeast Japan. Eur J Sustain Dev 7:81–89. https://doi.org/10.14207/ejsd.2018.v7n3p81

Wu B, Wu Y, Aoki Y, Nishimura S, Kashiwagi M (2019) A study on the reduction of mowing work burden for maintaining landscapes in rural areas: experiment design for mowing behaviors analyze. In: Proceedings—IEEE 17th international conference on dependable, autonomic and secure computing, IEEE 17th international conference on pervasive intelligence and computing, IEEE 5th international conference on cloud and big data computing, 4th cyber science and technology congress, DASC-PiCom-CBDCom-CyberSciTech 2019. https://doi.org/10.1109/DASC/PiCom/CBDCom/CyberSciTech.2019.00106
Wu H, Mweemba L (2010) Environmental self-efficacy, attitude and behavior among small scale farmers in Zambia. Environ Dev Sustain. https://doi.org/10.1007/s10668-009-9221-4
Zaré M, Ghaychi Afrouz S (2012) Crisis management of Tohoku; Japan earthquake and tsunami, March 11 2011. Iran J Public Health

Chapter 22
Freshwater Biomonitoring: An Ecosystem-Based Approach (EbA) for Building Climate Resilience Communities in Fiji

Bindiya Rashni

Abstract To establish climate resilience indigenous communities in rural and remote areas of Fiji, a pilot freshwater biomonitoring project was conducted in six villages of the Vanua Levu island using an innovative, user, age, education and gender friendly community-based river health assessment tool—'Traffic Light Bioindicator Guide' (TLBG). This project was aimed at developing local community practitioners for conducting bioassessment and implementing riverine ecosystem-based conservation. To achieve this, a workshop and riverine field training were conducted to educate the villagers on the riverine biophysical structure, river health bioindicator taxa identification, TLBG use and river mapping for identification of catchment threats. Site-specific rapid bioassessment results are discussed. The social empowerment gained through field assessments and workshops resulted in (a) indigenous community-based plans for water source safeguarding (b) establishment of a village specific river monitoring committee (c) establishment of banana circle bio-filter for wastewater treatment (d) local community rubbish pit construction (e) proper toilet construction and (f) the development of a modified TLBG post-pilot field trial. The annual freshwater biomonitoring results complementing the implementation of community-initiated mitigation measures (a–e) were published as a proof of the successful trial of the TLBG in the Fiji RiverCare Toolkit.

Keywords Freshwater biomonitoring · Ecosystem-based adaptation (EbA) · Climate resilience · Indigenous community and river health

22.1 Introduction

While climate change continues to pose threat on freshwater ecosystems globally, Fiji is not an exception whereby human-induced pressures on these tropical dynamic systems could be exacerbated by climate change. Fiji archipelago is part of the

B. Rashni (✉)
South Pacific Regional Herbarium and Biodiversity Centre, Institute of Applied Science,, University of the South Pacific, Laucala Campus, Suva, Fiji

© The Author(s), under exclusive license to Springer Nature Singapore Pte Ltd. 2021
M. Mukherjee and R. Shaw (eds.), *Ecosystem-Based Disaster and Climate Resilience*, Disaster and Risk Research: GADRI Book Series,
https://doi.org/10.1007/978-981-16-4815-1_22

Melanesia group of islands of the Oceania region of the South Pacific. Inhabitants of the volcanic islands are highly dependent on lotic freshwater systems (e.g., Rivers, creeks, streams and brooks) owing to a variety of ecosystem services (regulating, provisioning, cultural and supporting services) essential for livelihood sustenance. However, the ES is continuously threatened by the climatic and non-climatic drivers of freshwater system degradation. Non-climatic drivers include gravel extraction, intensive agriculture, livestock grazing, logging, mining, damming and road and bridge construction. Climatic drivers include frequent floods, tropical cyclones and drought that lead to degradation in water quality, waterborne pollution and increased water demand during drought. While these are the commonly observed climate threats on Fijian lotic systems, the impacts of climate change on freshwater ecosystem, food web and biodiversity loss including species composition structure remain unexplored. Investigation of such aspects holds crucial importance for sustainable use and management of freshwater ecosystem as well as human food security and livelihood sustenance for local riverine user communities. Taking this into consideration, a pilot project titled 'piloting sustainable livelihood opportunities in rural and remote communities of Fiji,' was trialed via the RiverCare program implemented by Live Learn Environmental Education (LLEE) Fiji office across the six Drawa block communities of Wailevu West Coast of Vanua Levu island of Fiji.

The Republic of Fiji is located in the central Pacific Ocean at about 3000 km east of Australia and 2000 km north of New Zealand. The Fijian archipelago includes over 320 Islands with the largest islands being Viti Levu and Vanua Levu where most of Fiji's population is concentrated. As a Pacific island country, Fiji is second only to Papua New Guinea to be the most affected country by natural disasters since 1990. Fiji experiences a range of extreme weather, from cyclones, monsoons, rainfall, flooding and coastal surges. All of which threaten the natural resources, biodiversity and ecosystems that the locals are dependent upon for livelihood sustenance, especially the inland remote riverine communities. Catchment management is seen as an increasingly important step to support local economic activity, to promote and enhance biodiversity, maintain ethno-biodiversity and increase resilience to extreme weather and empower climate resilience communities. The current pilot project was aimed to empower the targeted remote riverine communities on freshwater biomonitoring as a proactive approach toward climate resilience capacity building using nature-based solution in the view of Ecosystem-based Approach (EbA).

22.2 Methodology

22.2.1 Study Site

Drawa block community is located at Wailevu west coast of Vanua Levu island of Fiji. It comprises of six villages illustrated in the diagram given below (Fig. 22.1).

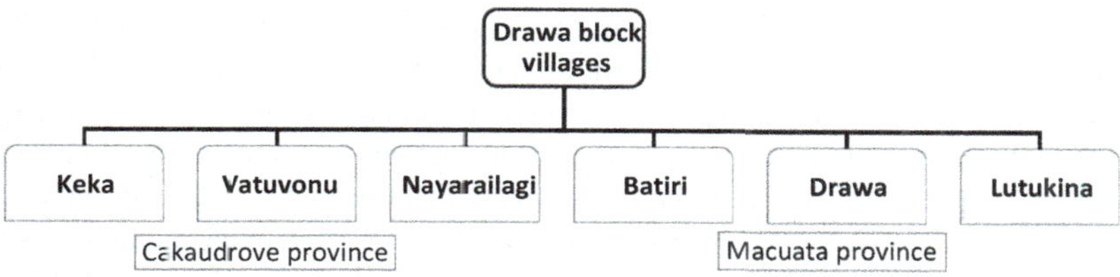

Fig. 22.1 Drawa block community

Vatuvonu, Keka, Nayarailagi and Batiri villages are placed under Cakaudrove province (Fig. 22.1). According to the villagers, the residents of Keka, Nayarailagi and Batiri are originally from Vatuvonu village. Due to the remoteness and in order to access urban services and facilities, the people of Vatuvonu village over the years have migrated to form Keka, Nayarailagi and Batiri village. Vatuvonu village comprises of six households and is the most remote out of the latter. Keka village is made up of over 26 households and is closest to Vatuvonu and within a distance of two hours walk by foot. Nayarailagi and Batiri village share the same riverine system whereby Batiri village is downstream of the main river. Nayarailagi village comprises of over 30 households, and Batiri village comprises of 56 households.

Drawa and Lutukina village are placed under Macuata province (Fig. 22.1) and connected by the Drawa River (Fig. 22.2). Both villages are remote and deprived off urban services and facilities. Drawa village comprises of approximately 15 households while Lutukina village is made up of more than 20 households.

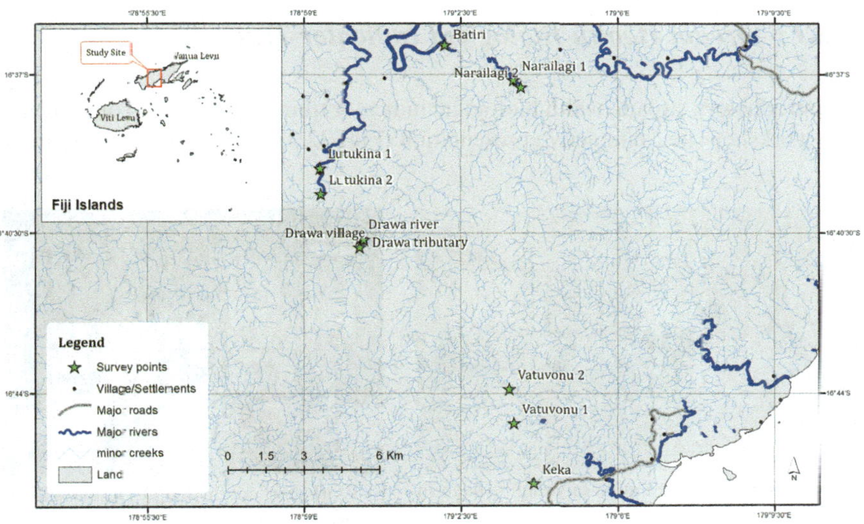

Fig. 22.2 Freshwater biomonitoring training sites across Drawa block community

The six villages (Fig. 22.1) are highly dependent on riverine services. These services include using river water for drinking, bathing, washing clothes and catching riverine fauna such as fishes, prawns and snails for consumption. Based on the high dependency on immediate riverine services, the current community-based biomonitoring training was carried out in the six villages. The training was aimed to increase the understanding on riverine ecosystem functioning and associated catchment activities as well as to explore the rooted traditional knowledge on community-based riverine resource management. The workshop and field work was designed to introduce the participants to basic stream structure and biological communities, stream health monitoring using bioindicators of stream health and water quality and catchment activities that pose threats to their rivers. The presentation and field work was designed for both educated and non-educated participants and could be applicable to remote communities of Fiji.

22.3 Consultation

The Provincial Administration (PA) was consulted by Live Learn Environmental Education Fiji office for implementation of the pilot project on 'sustainable livelihood development in rural and remote communities.' Free Prior Informed Consent (FPIC) and site access permit process was conducted ethically with the PA and the respective village headmen two weeks prior to the workshop.

22.3.1 Stream Health Biological Indicator Training

The introductory stream health biological indicator training comprised of two parts: (1) a workshop, i.e., interactive presentation (Fig. 22.3) to the indigenous commu-

a) Workshop conducted at Naraiyalagi village in the iTaukei language

b) Workshop conducted at Drawa village in the iTaukei language

Fig. 22.3 Workshop conducted in *iTaukei* village halls

a) Rock Wash method

b) Identifying and Matching collected taxa to bioindicator taxa in TLBG

c) Completion of Habitat Assessment Sheet post sampling

Fig. 22.4 Bioassessment conducted by villagers using Rock Wash method and TLBG

nities in their respective village community halls and (2) field work (Fig. 22.4) in the watercourses whereby participant's trial the community-oriented rapid bioassessment method to determine the ecological status of their waterways. To do this, the village representatives sampled freshwater invertebrates via Rock Wash' method and used the color-code field guide to identify and match the bioindicator taxa to assigned ecological status and thus determine the ecological state of their waterways.

22.3.2 Workshop

The workshop involved interactive session with the villagers in their respective community halls. Registration forms were circulated around for participants to fill in. The presentation (Table 22.1) covered basic stream/river physical features, freshwater biotic community ecosystem and functions as well as anthropogenic activities in the catchments that pose threats to freshwater ecosystem (Rashni 2014). The presentation was translated into native language by an *iTaukei* project officer of LLEE (Fig. 22.3) in order to relate the message easily and effectively to the participants.

22.4 Field Work-Sampling and Identification of Biological Indicators of Water Quality

The field work training was conducted upon the completion of theoretical training through PowerPoint presentation (Rashni 2014). Summary of the field work components and associated objectives are presented in Table 22.2. Post Lunch breaks the participants walked to the stream site next to the village where they witnessed the demonstration and trialed the rapid macroinvertebrate sample collection method (Rock Wash method) and the use of the color-coded macroinvertebrate field guide Traffic Light Bioindicator Guide (TLBG) (Fig. 22.4) (Rashni 2014; Rippon et al.

Table 22.1 Summary of components of the presentation and how each component was achieved

Parts of the presentation	Objectives	How were the objectives achieved
Part 1: Introduction to stream structure and biological communities	(a) Introduce the participants to the basic stream features and biotic communities that they directly or indirectly use and impact daily (b) Introduce the participants to relevance of stream health monitoring, how to sample macroinvertebrates and identify the relevant biological indicators plus introduction to color-code guide and its applicability	(a) The participants were shown basic stream features and biotic communities through pictures in PowerPoint slides and asked to identify those that looked familiar to them and if possible point out native names of familiar organisms (b) The participants were given an interactive session to discuss about human health and stream health analogy for clearer understanding of the stream health monitoring concept
Part 2: Local stream threat identification and stream health monitoring plan and development	(a) Introduce the participants to common stream threats in Fiji and assist them to identify any existing stream threats in their catchments (b) Introduce participants to basic stream health assessment using color-code guide and assist them to develop a stream health monitoring plan applicable to their catchment	(a) The participants were familiarized with local stream threats through pictures in PowerPoint slides coupled with appropriate explanation of how each threat affects the stream water quality and biotic community (b) The participants were explained about the process of stream health assessment and importance of monitoring stream health. Participants were also introduced to a general but locally applicable basic color-code macroinvertebrate guide that would be used for stream health monitoring

Table 22.2 Summary of the components of the field work and objectives per component was achieved

Field activities	Objectives	How were the objectives achieved
Rock wash sampling method	To train participants to collect freshwater macroinvertebrates via rock wash sampling method	The author with assistance from LLEE project officer demonstrated the Rock Wash sampling method to the participants. The participants were then supervised as they took turns to collect macroinvertebrates via the demonstrated method
Trialing Traffic Light Bioindicator Guide	Train participants on the usage of color-coded field guide	The author with assistance from LLEE project officer demonstrated the use of the color-coded field guide. The participants were then supervised during the field trial of the guide
Trialing Habitat Assessment Sheet	Train participants on the usage and completion of Habitat Assessment Sheet	The author with assistance from LLEE project officer guided the participants in completing the Habitat Assessment Sheet per site

2015). The Rock Wash sampling method involved placing the sieve downstream but next to the selected rocks and then lifting up the rock to be washed in a bucket of water. The sieve was placed next to the rock in order to catch any loosely attached organism that gets washed while lifting the rock. These were repeated for three rocks each in the upper, mid and lower section of the riffle and run habitat and also on the streamsides. Once the rocks were washed in the bucket of water, they were placed back from where they were removed. The dislodged organisms in the bucket of water were then sieved and washed into the white tray with adequate clean water.

The participants then observed the live organisms in the white tray and matched it with those given in the TLBG. The author then explained the participants about the significance of each bioindicator taxa and the appropriate morphological features for easy identification. Based on the collected bioindicator taxa matched with those in color-codes as per TLBG, the participants ticked the presence or absence of observed bioindicators in the respective columns. The participants were familiarized about the distinguishing features of bioindicators. Together with bioindicators, a basic Habitat Assessment Sheet (Rashni 2014; Rippon et al. 2015) was designed to capture instream habitat, site-specific including riparian and catchment landuse details. This was designed to assist with the understanding of various human or natural causes leading to community composition structure of benthic macroinvertebrates at the sampling sites.

22.4.1 Pilot Stream Health Monitoring Design

Upon discussion with the villagers, a basic stream health monitoring plan was designed referred to as 'three site-based assessment design' (Fig. 22.5). This involved sampling three stations across the stream/waterway draining the village land boundary; (a) station one = below the village and or farm zone (b) Next to the village/farm (i.e., point of disturbance) and (c) above the village or farm areas (close to water source/near intact zone—a control site) (Fig. 22.5). The villagers agreed to sample the stations fortnightly and fill the habitat assessment form accordingly. To do this, two groups per village were established. Group one would be assessing the selected three sites per village in the second week of each month while group two would be assessing in the last week of each month. The data would be recorded in the Habitat Assessment Sheet and deposited with the village headman. A data log book was given to the village headman to record details of stream assessment fortnightly and any village stream or catchment-related significant event.

Placing control sites (near intact zone in a catchment) allows confidence in concluding that changes at trial sites are the result of human impact. The control sites were upstream of the human-impacted zone. Using the proposed 'three site-based assessment' was designed to allow for assessments of impacts within a catchment and be more relevant to the affected community regarding the localized ecological health

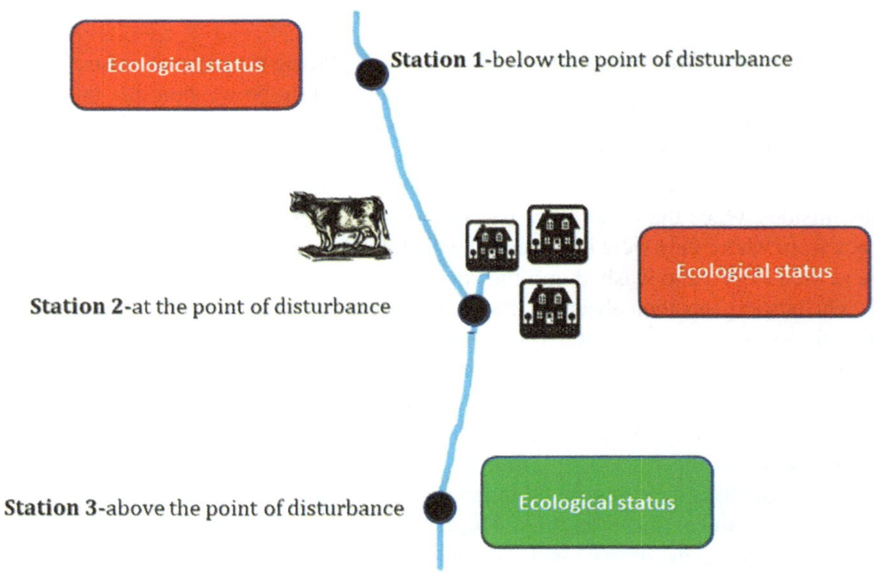

Fig. 22.5 An example of a pilot site-based rapid bioassessment results for upstream and downstream of an indigenous village waterway. Green = ecologically healthy system and good water quality. Red = 'degraded/modified system with poor water quality. Black dots represent the actual sampling stations selected across the waterway draining the village

of sections of waterways with their catchments. This will allow impact assessment of any catchment-related project on the waterways and provide a baseline bioindicator data set to provide early warning signals for changes in catchment health.

The results of the site-based assessment biomonitoring could then be used by the villagers to make wise decisions for any proposed and or ongoing projects in their catchment that could threaten the state of their terrestrial and associated aquatic natural resources. Given the specific assessment periods per month as decided by the villagers, the raw bioindicator and physical data for river/stream can aid in development of a river management plan for (1) maintaining the existing good state of the river and (2) rehabilitation of the already degraded system. The community could be consulted by the RiverCare program team of LLLE which can aid in defining restoration sites based on priority areas for assessment where catchment degrading activities are taking place.

22.4.2 *Rock Wash Sampling and Traffic Light Bioindicator Field Guide*

The sampling instruments (sieve, bucket, twig/forcep, hand lens and white tray) used in the current workshop are locally available and easily accessible. The method used was 'Rock Washing' for specific bioindicator in a known stable aquatic habitat. This simple locally applicable bioassessment method is provided in detail in the RiverCare Toolkit of Fiji (Rippon et al. 2015) and presented as a summary in Fig. 22.6.

The TLBG has been translated into the *iTaukei* language and incorporated with the Habitat Assessment Sheet in the Fiji RiverCare Toolkit (Rippon et al. 2015). This simple one-pager colorful guide comprises of bioindicator organisms placed within the traffic light colors indicative of ecological status of the waterway assessed on the first page and instructions for use in English and *iTaukei* language on the second page. Thus, easily printable and laminated for use as a single page with two-sided print. This color-coded macroinvertebrate guide had specific invertebrate species

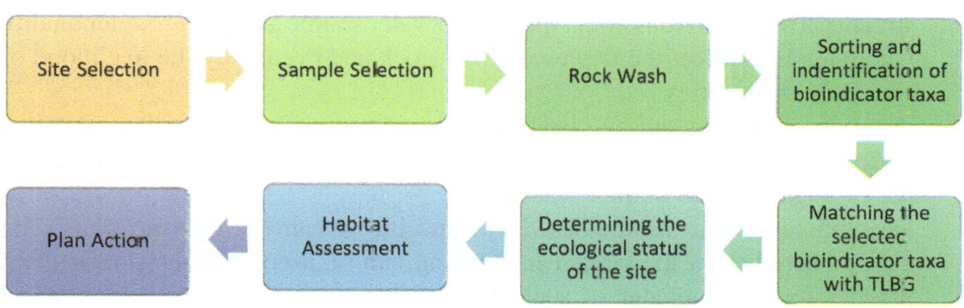

Fig. 22.6 Community-based rapid bioassessment concept for village waterways

(bioindicators) placed in three different color zones depicting the general state of water quality and stream health. The organisms in:

I. Green zone indicated an ecologically healthy waterway;
II. Amber zone indicated a moderately healthy waterway and;
III. Red zone indicated a degraded waterway.

This was designed to help the communities easily connect with the colors and the associated message indicated by each traffic light color which tends to also reflect status of a waterway. The bioindicator-based freshwater eco-status monitoring is based on bioindicator taxa present in the targeted riverine systems with the following criteria:

I. Easily visible and do not require use of complex sampling equipment or preservative measures;
II. Sensitive to environmental change, that is, presence and absence of the bioindicator taxa depicts environmental change;
III. Easily identifiable by village participants, that is, the identification of the bioindicator taxa should not require complex taxonomic work;
IV. Easy to understand, that is, the selected bioindicator taxa and the potential impact on the stream ecosystem it reflects must be comprehensible.

22.5 Results

22.5.1 The Major Workshop Outcomes

1. An improved understanding of basic biotic structure and functioning of streams among the villagers. The interactive workshop introduced the villagers to basic stream structure and biological communities and stream health monitoring using bioindicators.
2. Realization of human-induced pressures such as hillside slope farming, logging, gravel extraction, riparian removal that threaten the ecological state of the village streams. These activities affect the village water source and stream water quality which tends to affect the freshwater fish stock and eventually the livelihood of riverine communities.
3. A group of gender and age balanced trained individuals per village that can collect freshwater macroinvertebrates and identify the bioindicators to assess the ecological status of their waterways.
4. Development of a village-based site-specific stream monitoring plan. The data gathered from the stream monitoring can be used in a decision making process for development of a catchment/river management plan or pre and post-development and rehabilitation process.

22.5.2 Bioassessment Findings Using Traffic Light Bioindicator Guide

A total of nine sites were bio-assessed by the respective villagers according to three station types given in Fig. 22.5. Batiri village creek was not assessed due to heavy showers post-workshop in the afternoon. Table 22.3 presents summary of bioindicator taxa recorded by villagers across three different types of freshwater system with corresponding water quality type as per TLBG including common physical habitat observations (Rippon et al. 2015). Community-based site-specific rapid bioassessment findings indicated a total of 56% of ecologically stable (Good) sites and 22% each of moderately good and degraded sites (Table 22.4).

22.6 Community-Based Adaptations

Post biomonitoring, the respective communities took ownership of mitigating anthropogenic threats on their waterways. The initiatives undertaken within three months included the following;

- Establishment of a community river biomonitoring committee in the six villages.
- Establishment of 'Banana Circles' (Fig. 22.7) as a nature-based solution approach for wastewater/greywater treatment.
- Riverbank livestock relocation.
- Gravel extraction ceases.
- Setting up community rubbish pits.
- Setting up proper toilets especially those closer to creek/stream banks.
- Riparian vegetation rehabilitation planning for phase 2 of the RiverCare Program facilitated by Live Learn Environmental Education

Additionally, an example (Lutukina village downstream site) of an annual community-based biomonitoring results post-training and workshop for 2014 was included on page 17 of the RiverCare Toolkit (Rippon et al. 2015) and is presented as Fig. 22.8. The communities continued to monitor their waterways concurrently with progressive mitigation of obvious human pressures (e.g., shifting livestock further away from river banks, proper rubbish disposal and reduction in in-stream washing spots) and witnessed the disappearance of red zone species and an increase in amber and green zone species by the end of the year.

Furthermore, on an academic level, the initial TLBG was modified to suit area specific bioindicator species for the Drawa block communities and translated into layman's term and the native language to make it user friendly. The improved TLBG and corresponding Habitat Assessment Sheet was later included in the RiverCare Toolkit of Fiji (Rippon et al. 2015).

Table 22.3 Summary of community-based bioassessment results across Drawa block community waterways

Water quality status	River eco-status	Observed physical features
Excellent		-well-vegetated bank (native trees, ferns, shrubs and rootmasses) -No eroded banks -Fairly-highly shaded -instream rocks not covered in silt -Crystal clear water
Fairly good		-Vegetated banks (combination of native trees, grassy vegetation and bamboo trees) -slightly eroded banks or covered with grass -instream rocks somewhat covered with short filamentous algae -clear water -instream rocks and stream edges have silt deposition
Poor		-Poorly vegetated bank -eroded banks -bank undercutting -murky water -evidence of siltation on instream rocks and stream edges -mostly algal/silt covered rocks

Table 22.4 Bioindicator-based ecological status of village waterways across Drawa block community

Sites	Ecological status	Percentage(%) of sites
Vatuvonu village (station 3), Nayarailagi village (station 2), Nayarailagi village (station 3), Drawa village (station 3) & Lutukina village (station 3)	Good	56
Drawa village (station 2) & Lutukina village (station 2)	Moderately good	22
Keka village (station 2) & Nayarailagi village (station 1)	Degraded	22

Fig. 22.7 Establishment of 'Banana Circles'

22.7 Discussion

22.7.1 Community-Based River Monitoring (CBRM) Program Applicability to the Indigenous Community

Piloting Community-Based River Assessment Program (CBRM) across the first six indigenous community in Fiji was designed to directly impact riverine user community at village and catchment connectivity level in order to encourage and empower local RiverCare expertise and use traditional knowledge embedded with taxonomy

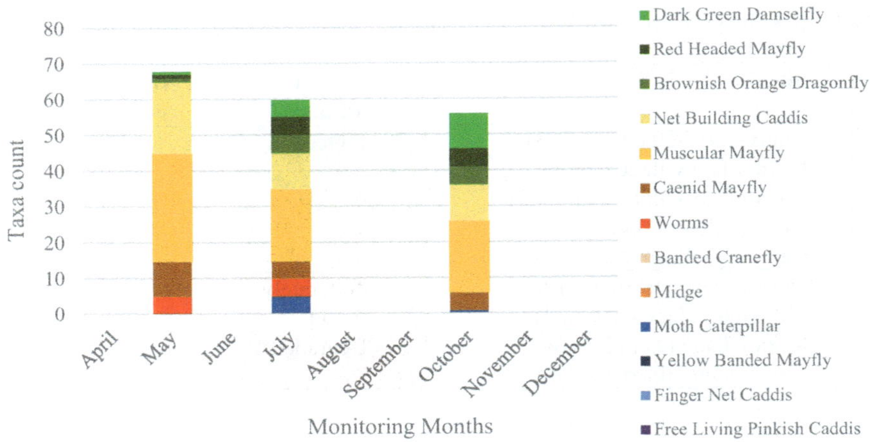

Fig. 22.8 Bioindicator-based monthly biomonitoring results for Lutukina village downstream site. *Source* Excerpt adapted from page 17 of the Fiji RiverCare Toolkit (Rippon et al. 2015)

to generate nature-based solution to provide resiliency against natural hazards, exacerbated by climate change, ensure ecosystem functioning and food security. The rule of thumb in the Pacific culture is that 'inedible biodiversity is rarely known' especially the hidden unexplored freshwater macroinvertebrates and, therefore, the need to introduce the freshwater bioindicator taxa as a 'biological tool' to determine ecological status and develop locally feasible action plans accordingly.

Taxonomy is the basis of management of any ecosystem but the challenge lies in converting the knowledge into grassroots level of understanding as the communities are the custodians of the freshwater ecosystems and in order to make a lasting impact on the conservation of these natural resources which are prone to climate change but more so to anthropic disturbances, the taxonomy and ecology of the freshwater ecosystem bioindicators had to be developed for Fijian freshwater bodies. The development of Traffic Light Bioindicator Guide simplified taxonomy into layman language and added to citizen science approach in Fiji.

22.7.2 *Macroinvertebrate Fauna Assessment, River Health and Human Health*

Macroinvertebrates are widely known to be indicators of ecological health of freshwater bodies (Holt and Miller 2011; Sharma and Rawat 2009). This is because macroinvertebrates are highly dependent on the habitat types and water quality variation that lead to corresponding differences in macroinvertebrate assemblages (Allan 1995). Also, these organisms have limited mobility and are restricted to immediate habitat which makes them vulnerable to degrading conditions in streams/rivers (Allan

2004; Arkle and Pilliod 2010; MacDonald et al. 1991). Additionally, human related activities that cause degrading/improving conditions in freshwater bodies tend to favor certain species over the others and thus place these organisms as bioindicators of such state. For example, in Fiji, siltation due to intensive slope farming in streams of Ovalau led to almost 50% decline in freshwater limpet snails, *Septaria* spp. (Haynes 1989). This is because siltation smothered the algal film on rocks which is the food source of gastropods; gastropods are particularly sensitive to siltation. In contrast, Haynes (1999) found that the endemic caddisfly *Abacaria fijiana* and the clinging mayfly *Pseudocloeon* sp. were persistent during logging period (1989–1990) when thick layer of mud was covering streambed and siltation led to stream discoloration; indicative of resilience to siltation (Haynes 1999).

As bioindicators of water quality which reflects stream health, macroinvertebrates indirectly assist in human health security. Maintaining good water quality through planned catchment activities in a village will help mitigate waterborne diseases or common skin diseases observed in children that bath daily in rivers. At Nayarailagi village, the station 1 (Table 22.4) macroinvertebrate survey indicated a degraded section of the stream. The water in the pool was highly turbid and silt layers were easily visible on stream side rocks and bedrock. Sadly, just before that station was a common bathing and fishing spot. On-site, observation revealed that the children bathing in that spot had skin diseases such as sores and open wounds. Various types of worms/bacteria could possibly enter the circulatory system through these wounds or sores and cause gastrointestinal diseases. Laboratory water quality tests were recommended to confirm the coliform level in the bathing pool to clarify if the water was good enough for secondary contact.

22.7.3 Traffic Light Bioindicator Guide—A Community Friendly Toolkit

The commonly used sampling equipment in Fiji for freshwater macroinvertebrate research purposes includes dip-nets which are bought from foreign countries. These nets are available only at research institutes such as University of the South Pacific and Fiji National University and would be unaffordable by the locals and, therefore, the preference for the simple locally available instruments. Traffic Light Bioindicator Guide is a simple locally applicable age-, education- and cost-friendly tool designed to make freshwater science more accessible to the general public.

In the current study, the participants were directed to sample specific stable aquatic habitats where the specific bioindicator taxa resided by using simple and easily affordable instruments. Although the TLBG was based on general bioindicators of stream health and water quality, it was applicable at all sites except in some cases. This includes sites where targeted species were missing naturally but were represented by ecologically similar species of similar genus or family and or site-specific species were not included in the guide. However, this was not surprising as Vanua

Levu waterways have not been thoroughly studied. Taking this into consideration, the color-code guide was later reviewed in terms of either incorporating additional species found during the survey or removing few species not present at survey sites in order to suite the currently surveyed areas in Vanua Levu. The improved TLBG and corresponding Habitat Assessment Sheet was later included in the RiverCare Toolkit of Fiji (Rippon et al. 2015).

Although the current simplest sampling method (Rock Wash) does not provide quantitative data or species list representative of a stream, it is, however, simplified to suit the education level (mostly primary school graduates) of remote area participants and saves sampling time and taxonomic complications by focusing solely on bioindicators. The data gathered from use of the color-guide could also assist in river management among the villages connected by riverine system; upstream–downstream communities.

22.7.4 Traditional Ecological Knowledge

In the Pacific island countries where indigenous communities are deeply culturally rooted with their environment (i.e., natural resources and biodiversity), climate resilience capacity building is perhaps most effective with the application of the concept of ethno-biodiversity (the cultural link with biodiversity) (McClatchey et al. 2008). In many inhabited islands across Fiji archipelago, customary land ownership accounts for 80–90% of the total land area extending to the nearshore areas known as '*qoliqoli*' boundary (traditional fishing grounds) (Crosetto 2005). Hence, by right the indigenous communities are the primary users and guardians of biological diversity and natural resources.

In the *iTaukei* community, each of the '*mataqali*' (tribe) recognize a plant (*kau*), a terrestrial animal (*manumanu*) which could be a bird or insect and a fish (*ika*) as their cultural totem (Nainoca 2010). To ensure conservation of terrestrial and aquatic systems, these are the keystone species for the individual land-owning tribe. When it comes to ecosystem or habitat conservation, the traditional areas of conservation interest or protected areas would include these species and, therefore, ensures community-generated efforts for conservation or protection. With regards to Drawa block communities, it was important to capture the totemic species within or related to the inland ecosystem, especially the local names given to bioindicator taxa to be able to embed the scientific knowledge with traditional knowledge. For example, a dragonfly naiad was known as '*mama*' across the Drawa block communities while the terrestrial aerial adult was called '*cecewai.*' The villages were familiar with the naiad stage but were not aware that it was the juvenile stage of a dragonfly. Thus, in order to ensure breeding and colonization of a native dragonfly species, the immediate habitat of both forms (aquatic micro-habitats and native riparian vegetation) needed to be conserved or protected which in turn would support ecological integrity (including water quality) of that section of the riverine system. Bioindicator taxa-based ethno-biodiversity discussions such as this then led to the approach of network of riverine systems and riparian systems across upstream and downstream communities and eventually the establishment of a community river biomonitoring committee in the six villages.

22.7.5 *Ecosystem-Based Adaptation Approach*

It is globally understood that taxonomy is the basis of ecosystem management. However, in a country like Fiji which currently lacks a wetland protection legislation, has a selected few freshwater species documented in the Endangered and Protected Species (EPS) ACT BILL and is data deficient for Red Listed freshwater invertebrate species, ethno-biodiversity appears to be more effective (e.g., Drawa case study) for conservation and management of inland wetland ecosystems for ensuring EES provision and maintenance of ecological integrity of riverine systems to be able to offset the impacts of climate change on such aspects of the systems. While the current pilot local project was designed as a proactive EbA approach to empower climate resilience communities for any future climate-related disasters, the concept, however, is complimented by a recent research (Vaughan and Gotelli 2019) across England and Wales which showed that improvement in water quality and reduction of pollution in Rivers can reduce impact of climate warming. '… Using a novel analysis of multiple stressors, an accumulated climatic debt of 0.64 (±0.13 standard error) °C of warming was paid by a water-quality credit equivalent to 0.89 (±0.04) °C of cooling. Although there is finite scope for mitigating additional climate warming in this way, water quality improvements appear to have offset recent temperature increases, and the concept of environmental credit may be a useful tool for communicating climate offsetting…' (Vaughan and Gotelli 2019).

22.8 Conclusion

The current study achieved a successful trial of an EbA approach to empower climate resilience indigenous communities for mitigation of human pressures on tropical inland lotic freshwater systems. This was done through introduction of a simple innovative locally applicable age and education friendly freshwater biomonitoring tool (Traffic Light Bioindicator Guide) to the six rural and remote communities of Fiji for deducing the health of their waterways. The community-based rapid bioassessment results indicated a total of 55% ecologically healthy sites followed by a total of 22% each for moderately healthy and degraded sites. Post rapid bioassessment, the community-based initiatives undertaken for mitigating anthropogenic threats included (a) establishment of a community river biomonitoring committee in the six villages, (b) establishment of 'Banana Circles' as a nature-based solution approach for wastewater treatment, (c) Riverbank livestock relocation, (d) gravel extraction cease, (d) setting up community rubbish pits and (e) setting up proper toilets, especially those closer to creek/stream banks. With the success of community-oriented actions at grassroots level, the current study hopes to establish the similar program across the Oceanic countries in order to encourage an adaptive solution to community climate resiliency and freshwater ecosystem and biodiversity conservation.

Acknowledgements This project was made possible via coordination from Live & Learn Environmental Education (LLEE) Fiji office team and funding from Department of Foreign Affairs & Trade (DFAT) and Fiji Community Development Program (FCDP). Special thanks to the residents of Vatuvonu, Keka, Batiri, Nayarailagi, Drawa and Lutukina for their hospitality and co-operation.

References

Allan JD (1995) Physical factors of importance to the biota stream ecology: structure and function of running waters. Chapman & Hall, London, pp 45–81

Allan JD (2004) Landscapes and riverscapes: the influence of land use on stream ecosystems. Annu Rev Ecol Evol Syst 35:257–284

Arkle SB, Pilliod SD (2010) Prescribed fires as ecological surrogates for wildfires: a stream and riparian perspective. For Ecol Manage 259:893–903

Crosetto J (2005) The heart of fiji's land tenure conflict: the law of tradition and Vakavanua, the customary way of the land. Pac Rim Law Policy J 14:71

Haynes A (1989) Effects of erosion caused by faulty farming practices on the biota of Fijian streams and rivers. In: Haynes RJ, Naidu R (eds) Agricultural development in the Pacific islands in the 90s, Suva, Fiji, 31 March to 1 April 1989. Institute of Agricultural Science, pp 300–304

Haynes A (1999) The long term effect of forest logging on the macroinvertebrates in a Fijian stream. Hydrobiologia 405:79–87

Holt EA, Miller SW (2011) Bioindicators: using organisms to measure environmental impacts. Nat Educ Knowl 3(10):8

MacDonald SB, Mullins WG, Lewis S (1991) Macroinvertebrates as indicators of stream health. Am Biol Teach 53(8):462–466

McClatchey W, Thaman R, Juvik S (2008) Ethnobiodiversity surveys of human/ecosystem relationships. In: Mueller-Dombois D, Bridges K, Daehler C (eds) Biodiversity assessment of tropical island ecosystems. Bishop Museum Press, Honolulu, pp 159–196

Nainoca WU (2010) The influence of the Fijian way of life (bula vakavanua) on community-based marine conservation (CBMC) in Fiji, with a focus on social capital and traditional ecological knowledge (TEK): a thesis presented in fulfilment of the requirements for the degree of Doctor of Philosophy in Resource and Environmental Planning at Massey University, Palmerston North, New Zealand

Rashni B (2014) Piloting stream health assessment in rural & remote communities of Fiji. A case study of Drawa block villages. IAS Environmental Report No. 316

Rippon M, Susau D, Naivalulevu T, Young K, Young W, Rashni B (2015) A water quality monitoring guide and activity manual for teachers and students. Rivers to reef booklet. Live & Learn Environmental Education, Fiji, pp 10–22

Sharma RC, Rawat JS (2009) Monitoring of aquatic macroinvertebrates as bioindicator for assessing the health of wetlands: a case study in the Central Himalayas, India. Ecol Indic 9(1):118–128

Vaughan IP, Gotelli NJ (2019) Water quality improvements offset the climatic debt for stream macroinvertebrates over twenty years. Nat Commun 10(1):1956. https://doi.org/10.1038/s41467-019-09736-3

Chapter 23
Forward-Looking Lens to Mainstream Blue-Green Infrastructure

Mahua Mukherjee and Rajib Shaw

Abstract Healthy natural ecosystem can provide resilience and ecosystem services with socio-economic and environmental benefits. Strategic deployment through systematic connected open spaces, water bodies and vegetated spaces through spatial and fiscal planning can develop blue-green infrastructure (BGI). Performance of any BGI can be designed for targets to achieve. This chapter explores pertinent aspects which influence planning, implementation and maintenance of BGI. The pandemic COVID-19 brought challenges to development; yet, stakeholders agree on preservation and conservation of nature in the form of utilitarian use like BGI to attenuate anthropogenic interventions. Next-generation built environment will look for greater integration of building, services, surrounding open areas, and engineered (grey) and nature-based blue-green infrastructure. An innovative attempt on understanding future climate change scenario with Representative Concentration Pathways (RCPs), Shared Socio-economic Pathways (SSPs), Shared Climate Policy Assumptions (SPAs) and RCP–SSP–SPA framework using lens of Eco-DRR and EbA is discussed for developing future climate policy at country and sub-national levels. To propagate the BGI for Eco-DRR and EbA to arrest surface transformation and fragmentation, adaptive governance through participatory implementation can empower communities. The BGI, no-regret strategy for the leaders, authorities and communities, shall be part of sustainable resilient development schemes globally.

Keywords BGI · Pandemic COVID-19 · RCP–SSP–SPA framework · Adaptive governance

---

M. Mukherjee (✉)
IIT Roorkee, Roorkee, India
e-mail: mahuafap@iitr.ac.in

R. Shaw
Keio University, Fujisawa, Japan

© The Author(s), under exclusive license to Springer Nature Singapore Pte Ltd. 2021
M. Mukherjee and R. Shaw (eds.), *Ecosystem-Based Disaster and Climate Resilience*, Disaster and Risk Research: GADRI Book Series,
https://doi.org/10.1007/978-981-16-4815-1_23

23.1 Introduction

Sustainable development aims to improve quality of urban and rural residential experience; climate protection through pollution and emission reduction, improvement of soil function and water balance and enhancement of plant and animal habitat are facilitating imperatives of sustainability. Anthropogenic development is expected to look at opportunities created by sustainable infrastructure and services which will augment resilience performance even after any disaster or climate change-induced event strikes. Urbanization is globally preferred way of living; cities are adding building, structure and infrastructure made out of non-natural materials. In that particular context, the need to relook at paradigm of environment-friendly place-making and urban design strategies shall be prioritized. Nature-centric place-making, i.e. making places beautiful with distinctive natural characters to contribute towards strong identity of the city, can be primary shared vision for the citizens and urban authorities, and that will help to sustainable and resilient city building.

Growth at the cost of land-use land cover (LULC) changes is one of the convenient strategic pathways adopted by city authorities more often than not. Challenges created by such decisions are surfacing gradually with concrete evidences. Though nature can assist magnificently towards sustainable development of urban world, a perilous trend of reduction/exclusion of natural elements is on rise. Rate of loss or conversion of natural elements in terms of vegetation, water elements and open spaces is too fast; inclusion of blue and green elements at city—or neighbourhood—level helps to imbibe resilience in addition to socio-economic and environmental benefits. Enhanced property values including house prices due to proximity to green space, enticing new businesses and investments by creating attractive settings and saving of energy, water and money for residents and end users are some of the direct benefits of nature-based development. Exemplary global initiatives are successful public transportation, urban environmental infrastructure, livelihood-focused shelter, disaster risk-resilient land-use planning, etc. Bio-mimicry, green built environment development, energy and water conservation, climate-responsive cities, nature-based disaster risk reduction and climate change adaptation are options for implementation which make use of ecosystem-function-based strategies. At building level, both new construction and existing property maintenance shall meet state-of-the-art standards for safety and security. Retrofitting of any building's structure, building skin, heating, ventilation, air conditioning, illumination, fire, security and other building system can be optimally integrated based on target performance (De and Mukherjee 2014). The next-generation built environment shall be guided by (a) climate change and disaster risk (natural and man-made) mitigation and adaptation, (b) energy and water security for both new and retrofitted properties, (c) reuse, recycle and disposal of waste which can be turned to wealth, (d) economy of resources and/or resource efficiency and (e) diverse traditional wisdom for ecosystem-based sustainability and resilience. For next-generation neighbourhood scale retrofitting, the BGI will be a very useful tool.

Open space management is more important than before as disaster and climate change-induced risks are escalating. Maintaining and enhancing neighbourhood-level water and vegetative components, permeable parking and pavement, and efficient street and outdoor lighting, implementation of rating systems and effective regulations like 'Berlin Biotope Area Factor' to take care of smart and sustainable surface transformation (2013) are few remarkable approaches to mention. Strategic systemic management for water, energy and waste will require policy-level decision support systems. Urban forestry may facilitate increase in tree canopy for shade; rejuvenation and reintroduction of water bodies will help in moisture balance in environment, and both of these strategies will impact on ambient temperature and lessening UHI and use of energy.

Dynamic and complex urban ecosystem developed very fast during last century and is accelerating till date. Natural ecosystem's slow process of evolution utilizes resources optimally and creates limited or no unwanted waste; waste from any natural process is valuable for another system and that is the way closed-loop natural process continues to self-rejuvenate water, energy, soil, carbon and nitrogen cycles. Anthropocentric urban world is sitting on self-destructive risk of waste despite the fact that several subcomponents of urban ecosystem are modelled after nature. Through trial and error, replication and experimentation on the natural system, urban system successfully managed to create enormous volume of urban components within shrunk time. Yet, the closed-loop model of nature where waste is wealth and not a trouble-maker is a long way for urban ecosystem to achieve. Expansion of knowledge base has outgrown the perceptible limits of human brain, and only computational powers and artificial intelligence can take cognizance to the same. So, it is expected that solutions will come, but question is when and at what cost. More and more irreparable losses to natural ecosystems will happen by then, and dream for resilient sustainable urban future moves away. Nature-based solutions are getting more attention and seeking prominence in the development programmes for this very reason. Relevant documentation of actions and good practices like BGI for Eco-DRR and EbA is a necessity (Waddington 2020), and the present book is addressing that gap profoundly.

23.2 Nature-Based Solutions—Blue-Green Infrastructure

Blue-green infrastructure (BGI) is strategic networking of water and vegetation-based natural elements which brings multiple benefits of ecosystem and resilience services. Ecosystem functions of the nature like evapotranspiration, absorption, permeation, heat and carbon sink enable the BGI to be one of the most viable nature-based tools for Eco-DRR and EbA, especially in urban areas. The BGI can be initiated and integrated at different scales through sustainable and resilient planning, designing, implementation, operation and maintenance. Use of site features and natural resources will influence built environment design, thermal performance of indoor and outdoor spaces and aesthetics among others and will create far-reaching long-term impacts on security of resources like water, health, food, energy and habitat. Collective policies, technology development and adaptive governance will add values to the BGI development process.

Synergies with international commitments such as IPCC Paris Climate protocol, Sustainable Development Goals (SDGs) and Sendai Framework for Disaster Risk Reduction (SFDRR), BGIs are essential components to facilitate overall global efforts and goals. Interlinkages and interdependencies between BGI and Eco-DRR and EbA are important to ensure and sustain resilience; loss of natural ecosystem and spiralling fragmentation enhances human ecological footprint. Mainstreaming of BGI in the national- and local-level policies through its implementation and maintenance will fulfil Target E of the Sendai Framework for Disaster Risk Reduction (Thomalla and Kumar 2020).

In BGI, both biotic and abiotic factors are seriously considered to ensure uninterrupted flow of nutrition, energy and organisms, provided physical connections between places are improved through blue-green network. The community can generate opportunity for senior citizens' social interactions and visibility, and places for children to play through creation of blue and green spaces at neighbourhood level. The BGI is accepted in the scientific community as countermeasures to urban heat island (UHI), urban flooding, air pollution, etc. (Ely and Pitman 2012). Opportune concepts for BGI integration, like affordable green housing, green walkability, urban agriculture, urban forestry, can trigger noticeable changes into different layers of urban life. Trickling effect of such interventions is culturally appropriate and easy to implement, use and maintain. They can regenerate many other linked services which are both tangible and intangible; e.g. ensuring green walkability within neighbourhood will bring positive impact on health and social security of elderly, women and other vulnerable sections alike; this will also boost local small-scale entrepreneurs and improve environment and livability index. Matrix and wider corridors, i.e. larger blue-green spaces, can promote practice of horticultural skills for growing food locally and influence healthy eating; augmenting community participation and volunteering in urban agriculture will bring necessary community health security. Similarly, exercise and physical activities like walking, cycling, sports and play within approachable distance will boost health and well-being of citizens. Provisioning better opportunities for active physical activity and access to nature and attractive green breathing spaces helps mental well-being. Green roads are another important BGI corridor which will bring connectivity with water security by storm-water drainage and groundwater recharge.

Global initiatives like 'The Economics of Ecosystems and Biodiversity (TEEB)' advocate mainstreaming the value integration of biodiversity and ecosystem services into decision-making; thus helps bringing focus on strategies like BGI for Eco-DRR and EbA. The nature-based development may need some guidelines like GUS-3CC model (Mukherjee and Takara 2018). The model subscribes to *contextual concept* for BGI development plans and initiatives to enable urban dwellers to mitigate negative environmental impact. For example, adopting and adapting to eco-services, local-level social and economic sustainability may get lifeline from walkability-centric planning and its implementation. Same for affordable green housing which can boost livelihood options for economically weaker sections yet can provide opportunity to better living quality and fulfil aspirations, thus increasing feasibility for mainstreaming BGI. While planning and implementing BGI, *core competency* on related aspects will enhance the experience of quality urban living. Important role of mangrove forests along coastlines to reduce the impact of cyclones, waves and storm surge and to prevent flooding and inland salt intrusion, preserving wetlands, forests and floodplains is an acknowledged fact. Application of mangrove restoration as BGI will necessarily demand an integral understanding of ecological, political and social changes of the location under considerations (Muthusankar et al. 2018). Long-term impact of large-scale BGI implementation boosts up the economy and social environment; evidence of the same needs to be documented with *contribution calculation.*

Advancement in anthropocentric world has brought destruction in natural world; this raises the pertinent question whether urbanizing world is progressing or not. Destructive human intervention in food web, both in water and on land, is profound to accommodate human beings' apparent sustainability and food security (Strong and Frank 2010). BGI can assist in restoration of interdependent trophic level: primary producers (like green algae), primary consumers (like mollusc) and secondary consumers to tertiary consumers.

BGI can help in restoring natural water cycle, ground water recharge through soil infiltration and storage, maintain balance of atmospheric and soil moisture content, sustainably manage storm-water drainage and canopy interception, flow control, distribution of excess runoff, and natural treatment to water. Thus, it helps towards abetting water insecurity, through sustainable watershed and spring shed management. Similarly, application of BGI for quality improvement of wastewater through non-mechanized/mechanized treatment system is documented at both household and community levels. Climate change impacts on hydrology and water resources are extremely alarming; pluvial and urban flood, drought and degraded water quality can take cognizance of the benefits BGI offers, and accordingly stand-alone and hybrid system can develop.

Recently organized webinar, on 18 November 2020, on *Nature-based solutions (NBS): Strengthening Synergies for Building Resilient Infrastructure* by the Coalition for Disaster Resilient Infrastructure (CDRI 2020) shows the wider approval of critical role nature-based solutions play in enhancing infrastructure resilience to serve rural communities and major cities. CDRI has called for stronger international cooperation to explore NBS contributions and implementations.

23.3 Pandemic

The pandemic COVID-19 has brought an upheaval in the world. Most recent epidemic has exposed our limited knowledge on various aspects of human–nature relationships (www.conservation.org). At initial stages of the pandemic, many countries followed lockdown of different durations during second quarter of 2020 to control the spread.

Again similar lockdown was followed in parts of Europe in the last quarter of 2020 as they experienced another wave. These steps restricted regular anthropogenic activities temporarily. Cleaner air quality recordings from most polluted cities and other few parametric had done round including carbon footprint reduction (Okyere et al. 2020) during these periods which helped to propagate a perception that nature is being left alone to breath.

On ground, the reality is people walked back to rural areas or smaller towns due to uncertainties over urban livelihoods. Other than spreading the risk of COVID-19 transmission from congested cities to these not-so-prepared rural areas, the in-migration created pressure on the natural repository of rural areas. Many rural areas in Asia, Africa and Latin America are facing increased pressure from land grabbing, deforestation, illegal mining, encroaching on indigenous territories and wildlife poaching. Even after travel ban is lifted, people are avoiding touristic places which again creates pressure on livelihoods of dependent people who are taking resort to nature exploitation as the last means to survive.

The chance of spread and severity of pandemic in future is linked with global warming. Improvements in air and water quality, as reported by media, helped to raise people's awareness on important pro-environmental goals; yet, the pandemic cannot reduce CO_2 emission in long term (Link 2020). Attention to long-term conservational regulations after this health crisis is urgent as actions on climate change being subsided to the background (Ambrose and Harvey 2020).

Interlinkage between pandemic crises and psychological aspects with respect to the concept of connectedness with nature is presented by Hassova et al. (2020). The connectedness concept is explored through survey to understand people's thoughts, activities and behaviours during such critical times and role of nature, if any. Selected people from different states of USA took part in surveys responding towards pandemic waves 1 and 2 (March and May 2020) and shared their individual expressions on levels of nature-connect. In daily activities, nature-related outdoor activities such as gardening, walking, hiking and outdoor sports have increased and a significant changed relation the researchers could establish between nature and human activities due to the pandemic-induced restrictions (Hassova et al. 2020). Stronger human–nature connection may be explained as the possibility of people's belief on nature's restorative power and restricted outdoor activities (Hartig et al. 2014; Collado et al. 2017).

The 'zoonotic' diseases, infectious diseases of animals that affect people, are responsible for many recent outbreaks and pandemics including COVID-19, HIV and Ebola (www.wwf.org.uk). Scientists believe people, livestock and wildlife are coming to unusual close contact as forests and natural habitats are destroyed and threatened and giving rise to zoonotic diseases at an alarming rate. The IPBES Bureau and 'Multidisciplinary Expert Panel (MEP)' workshop on biodiversity and pandemics on 27–31 July 2020 (IPBES-3/3, annex I) has echoed similar thoughts. The collective understanding as presented is given below.

1. Pandemics emerge from the microbial diversity found in nature.

2. Human ecological disruption and unsustainable consumption drive pandemic risk.
3. Reducing anthropogenic global environmental change may reduce pandemic risk.
4. Land-use change, agricultural expansion and urbanization cause more than 30% of emerging disease events.
5. The trade and consumption of wildlife is a globally important risk for future pandemics.
6. Current pandemic preparedness strategies aim to control diseases after they emerge. These strategies often rely on, and can affect, biodiversity.
7. Escape from the pandemic era requires policy options that foster transformative change towards preventing pandemics:
    a. Enabling mechanisms
    b. Policies to reduce the role of land-use change in pandemic emergence
    c. Policies to reduce pandemic emergence related to the wildlife trade
    d. Closing critical knowledge gaps
    e. Foster a role for all sectors of society to engage in reducing risk of pandemics.

Thomalla and Kumar (2020) and Mukherjee et al. (2020) reaffirm the need for Eco-DRR and EbA integration in the recovery strategies and continuity from the COVID-19 crisis and to rebuild the economies by governments, industries, service and business sectors.

23.4 Climate Change Scenario and BGI

Plausible description of future climate change is presently anticipated as integrated product of climate, socio-economic and climate-related policy assumptions (IPCC 2014; Nakicenovic et al. 2014). A global Representative Concentration Pathway (RCP), Shared Socio-economic Pathway (SSP) and Shared Climate Policy Assumption (SPA) (RCP–SSP–SPA) scenario framework has been developed by the Intergovernmental Panel on Climate Change Fifth Assessment Report (IPCC-AR5). A scenario explains how the future might unfold based on relevant assumptions that can alter. Probable routes of development over time in policy, governance, socio-economic conditions, etc., are pathways (IPCC 2014). SSPs and RCPs are complementary scenario development process agreed by IPPC to bring clarity in understanding on Climate Change Scenario by the end of the century. Each SSP inspects the probability of achieving different RCPs within narrated context and from the lens of identified attribute of Shared Climate Policy Assumptions, i.e. SPAs.

RCPs paint the variance in the intensity of radiative forcing due to greenhouse gas concentrations with the assumptions that additional warming up is happening primarily due to anthropogenic activities. Anthropogenic forcing of different pathways is measured in terms of increase in heating of the earth's surface due to changes in the atmosphere or to the earth's surface. Warming based on possible future trajectories of population, economic growth and greenhouse gas emissions are ideated through RCPs; accordingly, mitigation action strategies are expected to be adopted. SSPs throw light on societal choices which can influence climate change scenario; greenhouse gas emissions are dependent on behaviour of society at large which will regulate reaching targets set in Paris Climate Agreement. With no consideration for any new climate change policy other than already adopted ones by countries, the SSPs brought the assortment in 'business as usual' cases based on population, economic growth, education, urbanization and technological development paths chosen by different countries and regions for development (Hausfather 2018).

The regional and global scale studies on RCPs were used in AR5 (IPCC 2014); both RCPs and SSPs are expected to be the base for IPCC's next assessment report, i.e. AR6. RCPs do not detail out the type and structure of climate policy interventions including mitigation and adaptation measures. The SSPs show the way on reductions in emissions; i.e. different RCPs will or will not be achieved without future climate policy under different possible socio-eco-political statuses.

The new ongoing assessment report of IPCC (6th assessment report) has recognized ecosystem-based solutions as one of the key governance issues, which is also linked to the urban–rural linkages. The same concept is also referred as the Regional Circular Ecological Sphere (RCES 2018) and its local implementation at the eco-regions. R-CES has specific governance mechanism, geographical niche, stakeholder involvement as well as community ownership. This makes the R-CES model as one of the core-accepted concepts for BGI.

Shared Climate Policy Assumptions, i.e. SPAs for climate policy, are fundamental in linking socio-economic futures (SSPs) with radiative forcing (RCPs) and climate outcomes. Investments and effectiveness of different adaptation and mitigation policies for different combinations of SSPs and RCPs are understood better in association with SPAs. They describe key climate policy characteristics such as targets, instruments and obstacles of mitigation and adaptation policies at global and country scale to provide globally acceptable yet regionally differentiated realistic scenario. Three attributes of SPAs' climate policy characteristics are (Kriegler et al. 2012):

(a) Collective global climate policy goals, e.g. emission reduction target and ambition in limiting residual climate damages.
(b) Collective global policy regimes and measures introduced to reach the policy goals, e.g. carbon taxes, emission trading scheme policy instruments like emission pricing, low carbon technology subsidies, regulatory policies, transport policies and protection of tropical forest.
(c) Implementation limits and obstacles like exclusion of non-achievable policy options for some regions and sectors.

Nationally applicable and feasible scenarios based on BGI shall be an essential area to develop. BGI and other NbSs for Eco-DRR and EbA can help some of the SSPs to be better-suited for limiting warming to 1.5–2 °C above pre-industrial level.

23.5 Adaptive Governance

The BGI implementation can help in multiple ways for Eco-DRR and EbA, e.g. hazard, vulnerability and exposure reduction (risk-sensitive land-use planning) while increasing adaptive capacity (Thomalla and Kumar 2020). Jones et al. (2014) argued in favour of policy instruments and adaptive governance to convert the economically viable BGI and other NbS into deployment at different scales for ensuring the resilience of nature and community.

Folke et al. (2005) explained adaptive governance as an enabler of ecosystem-based management by creating systematic orders and regulations to encourage social participation in implementation of NbS and to empower community to make such decisions.

In post-COVID-19 pandemic context, role of governments to extend legal protections and financial supports for nature's protection and conservation is crucial (www.conservation.org); in the absence of that, accelerated pace of climate change, loss or elimination of sustainable livelihoods and biodiversity loss and deforestation will hike. In several countries, governments are experiencing a rise in deforestation, illegal mining and poaching; these seek urgent attention to enforce controlling efforts, even during the COVID-19 pandemic. It will be equally important for the countries to start planning for rebuilding their economies in a way that fosters green structural transformation and include long-term commitments to public spending and pricing reforms. After restrictions are lifted, governments and development financing institutions should prioritize stimulus efforts that have high economic multiplier effects and reduce carbon emissions. Such investments would have additional benefits for biodiversity and reduce the risk of zoonotic disease outbreaks, thereby addressing an important root cause of the current pandemic. Scaling up the BGI through partnerships among governments, donors, the private sector and community leaders will be a prudent task to carry forward to create resilience.

Webinars organized by the CDRIs (2020) and IPBIES (2020) reaffirmed the importance of adaptive governance; thus, BGI and other NbSs can bring leverages to resilience planning and will boost defences in disaster-prone areas. This is a no-regret strategy for the leaders, authorities and the community.

23.6 Concluding Remarks

The world is looked at as developed and developing world, despite the fact that every society is microcosm of the global power structure and continuously in search for equality and security. Developing countries aspire for development which are modelled after developed countries; instead of critical introspection of the contextual development need which may require alternative model, same mistakes might be repeated yielding any gainful results. Appropriateness of development of any country and/or region may need to succeed against five verticals:

1. People: Growing population with dwindling reach to education, health care, natural resources and livelihood security needs immediate attention. Strong social networking is the key for survival of several sections of society. Rapid shift to virtual society may improve digital penetration but at the cost of social disconnect.
2. Perception: Expression of urbanization and development is preferring similar international image. Differential context is not reflected in either policies or built forms. This defies laws of nature which has contextual relevance of 'local' and 'variation'. Focused anthropogenic targets for development are ruining the chance for ecosystem-centric growth where biotic and abiotic elements create not only web of symbiotic interactions, but also the variation in expression.
3. Planning: Sustainable development flourishes around natural elements; now natural elements are not even present around anthropocentric development. We need to travel distances to be in company of the natural elements. Spatial and physical planning integrated with natural corridors, matrix and patches shall be prioritized along with fiscal planning. The BGI shall get priority among infrastructure development.
4. Waste as wealth: Nature does not corroborate the concept of waste; any excretory end products and by-products from one subsystem will be useful as resource for other subsystem. Thus, they never become source of emission and pollution; rather, they help in flow of nutrition and energy management. Behavioural adaptation for lesser waste creation and appropriate techno-legal regime to combat waste management are to be prioritized.
5. Innovations: The web of industrialization (Industry 1.0 to Industry 5.0) is evidence of human desire for innovations. Role of innovative tools, approaches and policies is colossal for satisfying basic needs of growing population, alternate appropriate perception for development, technology and implementation of resilience planning and waste minimization. Creating new avenues through innovation will help BGI mainstreaming too.

References

Ambrose J, Harvey F (2020) Cop26 climate talks in Glasgow postponed until 2021. In: Proceedings of the Crucial UN Conference Will be Delayed Until Next Year as a Result of the Coronavirus Crisis, Kings Place

Berlin Biotope Area of Factor-Implementation Guidelines Helping to Control Temperature and Runoff Climate Adapt. Available online: https://climate-adapt.eea.europa.eu/metadata/case-studies/berlinbiotope-area-factor-2013-implementation-of-guidelines-helping-to-control-temperature-andrunoff/#challenges_anchor. Accessed on 5 Jan 2020

Coalition for Disaster Resilient Infrastructure (CDRI) (2020) Nature-based solutions (NBS): strengthening synergies for building resilient infrastructure. Webinar, November 2020

Collado S, Staats H, Corraliza JA, Hartig T (2017) Restorative environments and health. In: Fleury-Bahi G, Pol E, Navarro O (eds) Handbook of environmental psychology and quality of life research. Springer International Publishing, Cham, pp 127–148

De B, Mukherjee M (2014) Strategies and challenges for energy efficient retrofitting: study of empire state building. J Inst Eng (india) Series A 94(4):251–256
Ely M, Pitman S (2012) Green infrastructure life support for human habitats. A review of research and literature: Prepared for the Green Infrastructure Project, Botanic Gardens of Adelaide, Department of Environment, Water and Natural Resources
Folke C, Hahn T, Olsson P, Norberg J (2005) Adaptive governance of social-ecological systems. Annu Rev Environ Resour 15:441–473. https://doi.org/10.1146/annurev.energy.30.050504.144511
Haasova S, Czellar S, Rahmani L, Morgan N (2020) Connectedness with nature and individual responses to a pandemic: an exploratory study. Front Psychol 11:2215. https://doi.org/10.3389/fpsyg.2020.02215
Hartig T, Mitchell R, De Vries S, Frumkin H (2014) Nature and health. Annu Rev Public Health 35:207–228. https://doi.org/10.1146/annurev-publhealth-032013-182443
Hausfather Z (2018) Explainer: how 'Shared Socioeconomic Pathways' explore future climate change. https://www.carbonbrief.org/explainer-how-shared-socioeconomic-pathways-explore-future-climate-change. Accessed on 30 Dec 2020
https://www.conservation.org/stories/impact-of-covid-19-on-nature, Accessed on 9 Jan 2021
https://www.wwf.org.uk/nature-and-pandemics, Accessed on 19 Jan 2021
IPBES Workshop Report on Biodiversity and Pandemics Executive Summary: IPBES (2020) Workshop Report on biodiversity and pandemics of the intergovernmental platform on biodiversity and ecosystem services. Daszak P, das Neves C, Amuasi J, Hayman D, Kuiken T, Roche B, Zambrana-Torrelio C, Buss P, Dundarova H, Feferholtz Y, Foldvari G, Igbinosa E, Junglen S, Liu Q, Suzan G, Uhart M, Wannous C, Woolaston K, Mosig Reidl P, O'Brien K, Pascual U, Stoett P, Li H, Ngo HT, IPBES secretariat, Bonn, Germany. https://doi.org/10.5281/zenodo.4147317
IPCC (2014) The IPCC's fifth assessment report (AR5). Climate Change 2014
Jones R, Patwardhan A, Cohen S, Dessai S, Lammel A, Lempert R, Mirza MMQ, von Storch, H (2014) Foundations for decision making. In: Field CB, Barros V, Dokken DJ, Mach KJ, Mastrandrea MD, Bilir TE, Chatterjee M, Ebi KL, Estrada YO, Genova RC, Girma B, Kissel ES, Levy A, MacCracken S, Mastrandrea PR, White LL (eds) Climate Change 2014: impacts, adaptation, and vulnerability. Part A: Global and sectoral aspects. Working Group II contribution to the Fifth Assessment Report of the Intergovernmental Panel on Climate Change. Cambridge University Press, New York, pp 195–228
Kriegler E, O'Neill BC, Hallegatte S, Kram T, Lempert R, Moss R, Wilbanks T (2012) The need for and use of socio-economic scenarios for climate change analysis: a new approach based on shared socio-economic pathways. Glob Environ Chang 22:807–822
Link D (2020) Fact check: COVID-19 crisis has not created decreased long-term human environmental impact. USA Today. Available at: https://eu.usatoday.com/story/news/factcheck/2020/03/25/fact-check-coronavirus-crisis-benefiting-environment/2908300001/. Accessed Jan 2021
Mukherjee M, Takara K (2018) Urban green space as a countermeasure to increasing urban risk and the UGS-3CC resilience framework. Int J Disaster Risk Reduction 2:854–861
Mukherjee M, Chatterjee R, Khanna BK, Dhillon PPS, Kumar A, Bajwa S, Prakash A, Shaw R (2020) Ecosystem-centric business continuity planning (eco-centric BCP): a post COVID19 new normal. Progr Disaster Sci 7:100–117
Muthusankar G, Proisy C, Balasubramanian D, Bautès N, Bhalla RS, Mathevet R, Ricout A, Senthil Babu D, Vasudevan S (2018) When socio-economic plans exacerbate vulnerability to physical coastal processes on the south east coast of India. In: Shim J-S, Chun I, Lim HS (eds) Proceedings from the international coastal symposium (ICS) 2018 (Busan, Republic of Korea). J Coastal Res (Special Issue No. 85):1446–1450. Coconut Creek (Florida), ISSN 0749-0208
Nakicenovic N, Lempert RJ, Janetos AC (2014) A Framework for the development of new socio-economic scenarios for climate change research: introductory essay. Clim Change 122:351–361
Okyere I, Chuku EO, Ekumah B et al (2010) Physical distancing and risk of COVID-19 in small-scale fisheries: a remote sensing assessment in coastal Ghana. Sci Rep 10, 22407 (2020). https://doi.

org/10.1038/s41598-020-79898-4TEEB. Mainstreaming the Economics of nature: a Synthesis of the Approach, Conclusions and Recommendations of TEEB. Accessed on 5 Jan 2021

RCES (2018) Creation of a Regional circular and ecological sphere (R-CES) to address local challenges, available from: https://www.env.go.jp/en/wpaper/2018/pdf/04.pdf. Accessed on 29 Jan 2021

Strong DR, Frank KT (2010) Human involvement in food webs. Annu Rev Environ Resour 35(1):1–23

Thomalla F, Kumar A (2020) Ecosystem-based disaster risk reduction: implementing nature-based solutions for resilience, United Nations office for disaster risk reduction—Regional Office for Asia and the Pacific. Thailand UNDRR, Bangkok

Waddington R, edited, UNDRR (2020) The Words-into-Action guideline on "Nature-based Solutions for Disaster Risk Reduction" Engaging for resilience in support of the Sendai Framework for Disaster Risk Reduction 2015–2030

The manufacturer's authorised representative in the EU is Springer Nature Customer Service Centre GmbH, Europaplatz 3, 69115 Heidelberg, Germany. If you have any concerns regarding our products, please contact ProductSafety@springernature.com

Printed and bound by CPI Group (UK) Ltd, Croydon, CR0 4YY
30/09/2026
02225293-0019